advances in Hopf algebras

LECTURE NOTES IN PURE AND APPLIED MATHEMATICS

1. *N. Jacobson,* Exceptional Lie Algebras
2. *L.-Å. Lindahl and F. Poulsen,* Thin Sets in Harmonic Analysis
3. *I. Satake,* Classification Theory of Semi-Simple Algebraic Groups
4. *F. Hirzebruch, W. D. Newmann, and S. S. Koh, Differentiable Manifolds and Quadratic Forms*
5. *I. Chavel,* Riemannian Symmetric Spaces of Rank One
6. *R. B. Burckel,* Characterization of C(X) Among Its Subalgebras
7. *B. R. McDonald, A. R. Magid, and K. C. Smith,* Ring Theory: Proceedings of the Oklahoma Conference
8. *Y.-T. Siu,* Techniques of Extension on Analytic Objects
9. *S. R. Caradus, W. E. Pfaffenberger, and B. Yood,* Calkin Algebras and Algebras of Operators on Banach Spaces
10. *E. O. Roxin, P.-T. Liu, and R. L. Sternberg,* Differential Games and Control Theory
11. *M. Orzech and C. Small,* The Brauer Group of Commutative Rings
12. *S. Thomier,* Topology and Its Applications
13. *J. M. Lopez and K. A. Ross,* Sidon Sets
14. *W. W. Comfort and S. Negrepontis,* Continuous Pseudometrics
15. *K. McKennon and J. M. Robertson,* Locally Convex Spaces
16. *M. Carmeli and S. Malin,* Representations of the Rotation and Lorentz Groups: An Introduction
17. *G. B. Seligman,* Rational Methods in Lie Algebras
18. *D. G. de Figueiredo,* Functional Analysis: Proceedings of the Brazilian Mathematical Society Symposium
19. *L. Cesari, R. Kannan, and J. D. Schuur,* Nonlinear Functional Analysis and Differential Equations: Proceedings of the Michigan State University Conference
20. *J. J. Schäffer,* Geometry of Spheres in Normed Spaces
21. *K. Yano and M. Kon,* Anti-Invariant Submanifolds
22. *W. V. Vasconcelos,* The Rings of Dimension Two
23. *R. E. Chandler,* Hausdorff Compactifications
24. *S. P. Franklin and B. V. S. Thomas,* Topology: Proceedings of the Memphis State University Conference
25. *S. K. Jain,* Ring Theory: Proceedings of the Ohio University Conference
26. *B. R. McDonald and R. A. Morris,* Ring Theory II: Proceedings of the Second Oklahoma Conference
27. *R. B. Mura and A. Rhemtulla,* Orderable Groups
28. *J. R. Graef,* Stability of Dynamical Systems: Theory and Applications
29. *H.-C. Wang,* Homogeneous Branch Algebras
30. *E. O. Roxin, P.-T. Liu, and R. L. Sternberg,* Differential Games and Control Theory II
31. *R. D. Porter,* Introduction to Fibre Bundles
32. *M. Altman,* Contractors and Contractor Directions Theory and Applications
33. *J. S. Golan,* Decomposition and Dimension in Module Categories
34. *G. Fairweather,* Finite Element Galerkin Methods for Differential Equations
35. *J. D. Sally,* Numbers of Generators of Ideals in Local Rings
36. *S. S. Miller,* Complex Analysis: Proceedings of the S.U.N.Y. Brockport Conference
37. *R. Gordon,* Representation Theory of Algebras: Proceedings of the Philadelphia Conference
38. *M. Goto and F. D. Grosshans,* Semisimple Lie Algebras
39. *A. I. Arruda, N. C. A. da Costa, and R. Chuaqui,* Mathematical Logic: Proceedings of the First Brazilian Conference
40. *F. Van Oystaeyen,* Ring Theory: Proceedings of the 1977 Antwerp Conference
41. *F. Van Oystaeyen and A. Verschoren,* Reflectors and Localization: Application to Sheaf Theory
42. *M. Satyanarayana,* Positively Ordered Semigroups
43. *D. L Russell,* Mathematics of Finite-Dimensional Control Systems
44. *P.-T. Liu and E. Roxin,* Differential Games and Control Theory III: Proceedings of the Third Kingston Conference, Part A
45. *A. Geramita and J. Seberry,* Orthogonal Designs: Quadratic Forms and Hadamard Matrices
46. *J. Cigler, V. Losert, and P. Michor,* Banach Modules and Functors on Categories of Banach Spaces

47. *P.-T. Liu and J. G. Sutinen,* Control Theory in Mathematical Economics: Proceedings of the Third Kingston Conference, Part B
48. *C. Byrnes,* Partial Differential Equations and Geometry
49. *G. Klambauer,* Problems and Propositions in Analysis
50. *J. Knopfmacher,* Analytic Arithmetic of Algebraic Function Fields
51. *F. Van Oystaeyen,* Ring Theory: Proceedings of the 1978 Antwerp Conference
52. *B. Kadem,* Binary Time Series
53. *J. Barros-Neto and R. A. Artino,* Hypoelliptic Boundary-Value Problems
54. *R. L. Sternberg, A. J. Kalinowski, and J. S. Papadakis,* Nonlinear Partial Differential Equations in Engineering and Applied Science
55. *B. R. McDonald,* Ring Theory and Algebra III: Proceedngs of the Third Oklahoma Conference
56. *J. S. Golan,* Structure Sheaves Over a Noncommutative Ring
57. *T. V. Narayana, J. G. Williams, and R. M. Mathsen,* Combinatorics, Representation Theory and Statistical Methods in Groups: YOUNG DAY Proceedings
58. *T. A. Burton,* Modeling and Differential Equations in Biology
59. *K. H. Kim and F. W. Roush,* Introduction to Mathematical Consensus Theory
60. *J. Banas and K. Goebel,* Measures of Noncompactness in Banach Spaces
61. *O. A. Nielson,* Direct Integral Theory
62. *J. E. Smith, G. O. Kenny, and R. N. Ball,* Ordered Groups: Proceedings of the Boise State Conference
63. *J. Cronin,* Mathematics of Cell Electrophysiology
64. *J. W. Brewer,* Power Series Over Commutative Rings
65. *P. K. Kamthan and M. Gupta,* Sequence Spaces and Series
66. *T. G. McLaughlin,* Regressive Sets and the Theory of Isols
67. *T. L. Herdman, S. M. Rankin III, and H. W. Stech,* Integral and Functional Differential Equations
68. *R. Draper,* Commutative Algebra: Analytic Methods
69. *W. G. McKay and J. Patera,* Tables of Dimensions, Indices, and Branching Rules for Representations of Simple Lie Algebras
70. *R. L. Devaney and Z. H. Nitecki,* Classical Mechanics and Dynamical Systems
71. *J. Van Geel,* Places and Valuations in Noncommutative Ring Theory
72. *C. Faith,* Injective Modules and Injective Quotient Rings
73. *A. Fiacco,* Mathematical Programming with Data Perturbations I
74. *P. Schultz, C. Praeger, and R. Sullivan,* Algebraic Structures and Applications: Proceedings of the First Western Australian Conference on Algebra
75. *L Bican, T. Kepka, and P. Nemec,* Rings, Modules, and Preradicals
76. *D. C. Kay and M. Breen,* Convexity and Related Combinatorial Geometry: Proceedings of the Second University of Oklahoma Conference
77. *P. Fletcher and W. F. Lindgren,* Quasi-Uniform Spaces
78. *C.-C. Yang,* Factorization Theory of Meromorphic Functions
79. *O. Taussky,* Ternary Quadratic Forms and Norms
80. *S. P. Singh and J. H. Burry,* Nonlinear Analysis and Applications
81. *K. B. Hannsgen, T. L. Herdman, H. W. Stech, and R. L. Wheeler,* Volterra and Functional Differential Equations
82. *N. L. Johnson, M. J. Kallaher, and C. T. Long,* Finite Geometries: Proceedings of a Conference in Honor of T. G. Ostrom
83. *G. I. Zapata,* Functional Analysis, Holomorphy, and Approximation Theory
84. *S. Greco and G. Valla,* Commutative Algebra: Proceedings of the Trento Conference
85. *A. V. Fiacco,* Mathematical Programming with Data Perturbations II
86. *J.-B. Hiriart-Urruty, W. Oettli, and J. Stoer,* Optimization: Theory and Algorithms
87. *A. Figa Talamanca and M. A. Picardello,* Harmonic Analysis on Free Groups
88. *M. Harada,* Factor Categories with Applications to Direct Decomposition of Modules
89. *V. I. Istrăţescu,* Strict Convexity and Complex Strict Convexity
90. *V. Lakshmikantham,* Trends in Theory and Practice of Nonlinear Differential Equations
91. *H. L. Manocha and J. B. Srivastava,* Algebra and Its Applications
92. *D. V. Chudnovsky and G. V. Chudnovsky,* Classical and Quantum Models and Arithmetic Problems
93. *J. W. Longley,* Least Squares Computations Using Orthogonalization Methods
94. *L. P. de Alcantara,* Mathematical Logic and Formal Systems
95. *C. E. Aull,* Rings of Continuous Functions
96. *R. Chuaqui,* Analysis, Geometry, and Probability
97. *L. Fuchs and L. Salce,* Modules Over Valuation Domains

98. *P. Fischer and W. R. Smith,* Chaos, Fractals, and Dynamics
99. *W. B. Powell and C. Tsinakis,* Ordered Algebraic Structures
100. *G. M. Rassias and T. M. Rassias,* Differential Geometry, Calculus of Variations, and Their Applications
101. *R.-E. Hoffmann and K. H. Hofmann,* Continuous Lattices and Their Applications
102. *J. H. Lightbourne III and S. M. Rankin III,* Physical Mathematics and Nonlinear Partial Differential Equations
103. *C. A. Baker and L, M. Batten,* Finite Geometrics
104. *J. W. Brewer, J. W. Bunce, and F. S. Van Vleck,* Linear Systems Over Commutative Rings
105. *C. McCrory and T. Shifrin,* Geometry and Topology: Manifolds, Varieties, and Knots
106. *D. W. Kueker, E. G. K. Lopez-Escobar, and C. H. Smith,* Mathematical Logic and Theoretical Computer Science
107. *B.-L. Lin and S. Simons,* Nonlinear and Convex Analysis: Proceedings in Honor of Ky Fan
108. *S. J. Lee,* Operator Methods for Optimal Control Problems
109. *V. Lakshmikantham,* Nonlinear Analysis and Applications
110. *S. F. McCormick,* Multigrid Methods: Theory, Applications, and Supercomputing
111. *M. C. Tangora,* Computers in Algebra
112. *D. V. Chudnovsky and G. V. Chudnovsky,* Search Theory: Some Recent Developments
113. *D. V. Chudnovsky and R. D. Jenks,* Computer Algebra
114. *M. C. Tangora,* Computers in Geometry and Topology
115. *P. Nelson, V. Faber, T. A. Manteuffel, D. L. Seth, and A. B. White, Jr.,* Transport Theory, Invariant Imbedding, and Integral Equations: Proceedings in Honor of G. M. Wing's 65th Birthday
116. *P. Clément, S. Invernizzi, E. Mitidieri, and I. I. Vrabie,* Semigroup Theory and Applications
117. *J. Vinuesa,* Orthogonal Polynomials and Their Applications: Proceedings of the International Congress
118. *C. M. Dafermos, G. Ladas, and G. Papanicolaou,* Differential Equations: Proceedings of the EQUADIFF Conference
119. *E. O. Roxin,* Modern Optimal Control: A Conference in Honor of Solomon Lefschetz and Joseph P. Lasalle
120. *J. C. Díaz,* Mathematics for Large Scale Computing
121. *P. S. Milojević,* Nonlinear Functional Analysis
122. *C. Sadosky,* Analysis and Partial Differential Equations: A Collection of Papers Dedicated to Mischa Cotlar
123. *R. M. Shortt,* General Topology and Applications: Proceedings of the 1988 Northeast Conference
124. *R. Wong,* Asymptotic and Computational Analysis: Conference in Honor of Frank W. J. Olver's 65th Birthday
125. *D. V. Chudnovsky and R. D. Jenks,* Computers in Mathematics
126. *W. D. Wallis, H. Shen, W. Wei, and L. Zhu,* Combinatorial Designs and Applications
127. *S. Elaydi,* Differential Equations: Stability and Control
128. *G. Chen, E. B. Lee, W. Littman, and L. Markus,* Distributed Parameter Control Systems: New Trends and Applications
129. *W. N. Everitt,* Inequalities: Fifty Years On from Hardy, Littlewood and Pólya
130. *H. G. Kaper and M. Garbey,* Asymptotic Analysis and the Numerical Solution of Partial Differential Equations
131. *O. Arino, D. E. Axelrod, and M. Kimmel,* Mathematical Population Dynamics: Proceedings of the Second International Conference
132. *S. Coen,* Geometry and Complex Variables
133. *J. A. Goldstein, F. Kappel, and W. Schappacher,* Differential Equations with Applications in Biology, Physics, and Engineering
134. *S. J. Andima, R. Kopperman, P. R. Misra, J. Z. Reichman, and A. R. Todd,* General Topology and Applications
135. *P Clément, E. Mitidieri, B. de Pagter,* Semigroup Theory and Evolution Equations: The Second International Conference
136. *K. Jarosz,* Function Spaces
137. *J. M. Bayod, N. De Grande-De Kimpe, and J. Martínez-Maurica,* p-adic Functional Analysis
138. *G. A. Anastassiou,* Approximation Theory: Proceedings of the Sixth Southeastern Approximation Theorists Annual Conference
139. *R. S. Rees,* Graphs, Matrices, and Designs: Festschrift in Honor of Norman J. Pullman
140. *G. Abrams, J. Haefner, and K. M. Rangaswamy,* Methods in Module Theory

141. *G. L. Mullen and P. J.-S. Shiue*, Finite Fields, Coding Theory, and Advances in Communications and Computing
142. *M. C. Joshi and A. V. Balakrishnan*, Mathematical Theory of Control: Proceedings of the International Conference
143. *G. Komatsu and Y. Sakane*, Complex Geometry: Proceedings of the Osaka International Conference
144. *I. J. Bakelman*, Geometric Analysis and Nonlinear Partial Differential Equations
145. *T. Mabuchi and S. Mukai*, Einstein Metrics and Yang–Mills Connections: Proceedings of the 27th Taniguchi International Symposium
146. *L. Fuchs and R. Göbel*, Abelian Groups: Proceedings of the 1991 Curaçao Conference
147. *A. D. Pollington and W. Moran*, Number Theory with an Emphasis on the Markoff Spectrum
148. *G. Dore, A. Favini, E. Obrecht, and A. Venni*, Differential Equations in Banach Spaces
149. *T. West*, Continuum Theory and Dynamical Systems
150. *K. D. Bierstedt, A. Pietsch, W. Ruess, and D. Vogt*, Functional Analysis
151. *K. G. Fischer, P. Loustaunau, J. Shapiro, E. L. Green, and D. Farkas*, Computational Algebra
152. *K. D. Elworthy, W. N. Everitt, and E. B. Lee*, Differential Equations, Dynamical Systems, and Control Science
153. *P.-J. Cahen, D. L. Costa, M. Fontana, and S.-E. Kabbaj*, Commutative Ring Theory
154. *S. C. Cooper and W. J. Thron*, Continued Fractions and Orthogonal Functions: Theory and Applications
155. *P. Clément and G. Lumer*, Evolution Equations, Control Theory, and Biomathematics
156. *M. Gyllenberg and L. Persson*, Analysis, Algebra, and Computers in Mathematical Research: Proceedings of the Twenty-First Nordic Congress of Mathematicians
157. *W. O. Bray, P. S. Milojević, and Č. V. Stanojević*, Fourrier Analysis: Analytic and Geometric Aspects
158. *J. Bergen and S. Montgomery*, Advances in Hopf Algebras

Additional Volumes in Preparation

advances in Hopf algebras

edited by

Jeffrey Bergen
DePaul University
Chicago, Illinois

Susan Montgomery
University of Southern California
Los Angeles, California

CRC Press
Taylor & Francis Group
Boca Raton London New York

CRC Press is an imprint of the
Taylor & Francis Group, an **informa** business

First published 1994 by Marcel Dekker

Published 2018 by CRC Press
Taylor & Francis Group
6000 Broken Sound Parkway NW, Suite 300
Boca Raton, FL 33487-2742

First issued in hardback 2018

ISBN 13: 978-1-138-40180-8 (hbk)
ISBN 13: 978-0-8247-9065-3 (pbk)

Library of Congress Cataloging-in-Publication Data

Advances in Hopf algebras / edited by Jeffrey Bergen, Susan Montgomery.
p. cm. -- (Lecture notes in pure and applied mathematics ; v. 158)
Lectures from a conference held Aug. 10-14, 1992 at DePaul University in Chicago.
Includes bibliographical references
ISBN 0—8247—9065—0 (acid-free)
1. Hopf algebras--Congresses. I. Bergen, Jeffrey. II. Montgomery, Susan. III. Series.
QA613.8.A38 1994
510'.55--dc20 94-804
CIP

Preface

The NSF-CBMS conference Hopf Algebras and Their Actions on Rings was held at DePaul University in Chicago, Illinois. The conference featured a series of ten lectures by Susan Montgomery as well as nine supporting lectures by Miriam Cohen, Yukio Doi, Warren Nichols, Bodo Pareigis, Donald Passman, David Radford, Hans-Jurgen Schneider, Earl Taft, and Mitsuhiro Takeuchi. The conference, which served as a "summer school" for both experts and nonexperts in the field, attracted approximately 90 participants representing 11 countries.

This volume contains the expository lectures by the nine supporting lecturers as well as invited expository papers by Lindsay Childs and David Moss, Shahn Majid, and Akira Masuoka. The lectures by Susan Montgomery appear in the Conference Board of the Mathematical Sciences series published by the American Mathematical Society.

We would like to thank the National Science Foundation and the University Research Council at DePaul University for their financial support of this conference. We would also like to thank Maria Allegra of Marcel Dekker, Inc., for her assistance in putting this volume together. Finally, we thank all of our anonymous referees.

Jeffrey Bergen
Susan Montgomery

Contents

Contributors

LINDSAY CHILDS State University of New York, Albany, New York

M. COHEN Ben Gurion University, Beer Sheva, Israel

YUKIO DOI Fukui University, Fukui, Japan

SHAHN MAJID Cambridge University, Cambridge, England

AKIRA MASUOKA Shimane University, Matsue, Shimane, Japan

DAVID J. MOSS State University of New York, Albany, New York

WARREN D. NICHOLS Florida State University, Tallahassee, Florida

BODO PAREIGIS University of Munich, Munich, Germany

D. S. PASSMAN University of Wisconsin, Madison, Wisconsin

DAVID E. RADFORD University of Illinois—Chicago, Chicago, Illinois

H.-J. SCHNEIDER University of Munich, Munich, Germany

EARL J. TAFT Rutgers University, New Brunswick, New Jersey

MITSUHIRO TAKEUCHI University of Tsukuba, Tsukuba, Ibaraki, Japan

Hopf Algebras and Local Galois Module Theory

LINDSAY CHILDS State University of New York, Albany, New York
DAVID J. MOSS State University of New York, Albany, New York

Let $L \supset K$ be a finite Galois extension of algebraic number fields with Galois group G, and with rings of integers $S \supset R$. Galois module theory seeks to understand S as an RG-module. If L/K is tamely ramified, then S is a locally free RG-module by a classical theorem of E. Noether, and a rich theory has been developed to understand the obstructions to freeness: see, for example [$F83$] or a forthcoming book by B. Erez. However, if L/K is wildly ramified the situation is much less well-understood, for the local structure is unclear.

In 1959 Leopoldt [$L59$] showed that a useful approach to wild extensions is to view S as a module, not over RG, but over the larger order

$$\mathcal{A}(S) = \{\alpha \in KG | \alpha S \subseteq S\}.$$

He showed that if $K = \mathbf{Q}$ and G is abelian, then S is free over $\mathcal{A}(S)$. However, Leopoldt's theorem does not extend beyond $K = \mathbf{Q}$ and G abelian, and since the appearance of Leopoldt's paper, positive results on local freeness over the Leopoldt order have been scarce.

One of the first general positive results was in [$C87$], where it was shown that if G is abelian and $\mathcal{A}(S)$ is a Hopf order in KG, then S is locally free as an $\mathcal{A}(S)$-module. This introduced the theme of "taming wild extensions with Hopf algebras".

The purpose of this paper is to offer further positive results on local freeness, built around the Hopf algebra theme. We note that this theme, when it applies, opens the possibility of extending the global theory of tame extensions to certain classes of wild extensions. Such a program has been pursued in recent work of M. Taylor and his collaborators ([$ST90$], [$T88$], [$T90a$], [$T90b$]), and is nicely described in a recent survey paper by Taylor and Byott [$TB92$].

Taylor and Byott almost always assume that L/K is a Galois extension with group G; however, in view of the work of Greither and Pareigis [$GP87$], as well as the examples of section 3, below, it will be useful to assume only that L/K is a Hopf Galois extension.

For the remainder of this introduction, assume K is a local field, a finite extension of $\mathbf{Q}_p$, with valuation ring R.

In section 1, using a previously overlooked theorem of H.-J. Schneider, we show that with an appropriate notion of tameness Noether's theorem cited above generalizes to Hopf Galois extensions L/K with Hopf algebra A, where A is any finite dimensional cocommutative K-Hopf algebra. In particular, commutativity of A is not needed.

Let L/K be a Hopf Galois extension with Hopf algebra A. For an order S_0 over R in L (not necessarily the full integral closure of R in L), call

$$\mathcal{A}(S_0) = \{\alpha \text{ in } A | \alpha S_0 \subseteq S_0\}$$

the Leopoldt order of S_0. In section 2 we show that if $\mathcal{A}(S_0)$ is a Hopf order in A, then every Hopf order H over R in A containing $\mathcal{A}(S_0)$ is the Leopoldt order for some order S in L such that S is free over H.

In sections 3 and 4 we study Kummer extensions with respect to a formal group of dimension one. This is a class of extensions $S \supseteq R$ which are orders in Hopf Galois extensions $L \supseteq K$ and which are free over their Leopoldt orders. These extensions were introduced by Taylor [$T86$] and studied in special cases in [$T85$] and [$T87$]. They include a large collection of wildly ramified local Galois extensions L/K such that the Leopoldt order $\mathcal{A}(S)$ of the valuation ring S is Hopf and hence S is free over $\mathcal{A}(S)$. The freeness of S over $\mathcal{A}(S)$ follows from the fact that the algebras S under consideration are H-Galois objects (or in the terminology of [$TB92$], H-principal homogeneous spaces) where H is the representing Hopf algebra for a finite subgroup of the formal group. Then $\mathcal{A}(S)$ will be the dual of H; the main technical difficulty then becomes describing that dual, which we study in section 4, adapting techniques of Taylor.

In a brief final section we introduce the following question. Let L/K be an A-Hopf Galois extension, with valuation rings $S \supseteq R$. Let $\mathcal{A}(A, S)$ be the Leopoldt order of S. Does the structure of $\mathcal{A}(A, S)$ depend on A? This question is meaningful because, as Greither and Pareigis have shown ([$GP87$], c.f. [$C89$] and [$P90$]), a given extension L/K may be A-Hopf Galois for more than one Hopf algebra A. We give an example of an extension of degree 4 which is Hopf Galois for two Hopf algebras A_1 and A_2, such that only one of the corresponding Leopoldt orders of S is Hopf.

1. NOETHER'S THEOREM

The cornerstone of Galois module theory is Noether's theorem. Let L/K be a Galois extension of number fields with rings of integers S, R, respectively and Galois group G. Viewing S as an RG-module, we may ask if S is free over RG. This is the same as asking if S has a normal basis as a free R-module (or that L/K has a normal integral basis). Noether's theorem asserts that S is locally free over RG (where "local" means at the completion at any finite place of R) iff L/K is tamely ramified ("tame", for short), i.e. the ramification index of any non-zero prime ideal p of R is relatively prime to the characteristic of R/p.

Noether's theorem implies, in particular, that when L/K is wildly ramified (= "wild", i.e. not tame) there is no hope that S could be free over RG. To deal with this situation, Leopoldt introduced the idea of viewing S over the ring

$$\mathcal{A}(S) = \{\alpha \text{ in } KG | \alpha s \in S \text{ for all } s \text{ in } S\},$$

an order over R in KG which contains RG, and which we will call the Leopoldt order of S. Leopoldt [$Le59$] showed that if $K = \mathbf{Q}$ and G is abelian, then S is always free over $\mathcal{A}(S)$. However, subsequent examples showed that S need not be locally free over $\mathcal{A}(S)$ if $K \neq \mathbf{Q}$ or G is not abelian. See [$BF72a$].

In [$CH87$] (c.f. also [$W88$]), S. Hurley and the first author defined the notion of tame extension with respect to a Hopf algebra. Let S be a commutative R-algebra and an H-module algebra, where H is an R-Hopf algebra. Suppose S and H are both finitely generated projective R-modules of the same rank, and the fixed ring $S^H = R$. If I is the space of left integrals of H, then IS is contained in $S^H = R$. We called S/R tame if $IS = R$.

Assuming that H is commutative and cocommutative, we showed in [$CH86$] that S is locally isomorphic to H as an H-module if S is a tame H-extension of R.

This applies in the case where K is a number field with ring of integers R, L is a finite extension of K, S is an order contained in the ring of integers of L, and S is an H-module algebra, where H is a cocommutative R-Hopf algebra, finitely generated and projective as an R-module. If $A = K \otimes_R H$ and L is an A-Hopf Galois extension of K, then S is an H-tame extension of R if $IS = R$, where I is the space of left integrals of H. The result of [$CH86$] showed that, assuming H is also commutative, then S is locally free over H.

It turns out that the assumption of commutativity on H is not necessary, thanks to a deep result of H.-J. Schneider. In fact:

THEOREM 1.1. Let R be a complete discrete valuation ring of characteristic zero, with quotient field K. Let A be a cocommutative K-Hopf algebra, of finite rank as a K-module. Let L be a K-algebra which is a Hopf Galois extension of K with Hopf algebra A. Let H be an order over R in A with module of left integrals I and S be an order over R in L, such that S is an H- module algebra. If $IS = R$ (that is, the H-extension S/R is tame), then $S \cong H$ as left H-modules.

The proof of this is a matter of putting together two results.

One, found as Theorem 5.1 of [CH86], is that if $IS = R$ then S is H-projective. To sketch this generalization of a well-known result in representation theory of finite groups (c.f. [S77], Lemma 20, page 118): let ϑ generate the free rank one R-module I. Since $IS = R$, there is some z in S so that $\vartheta z = 1$. Now S is a free R- module, so $H \otimes_R S$, viewed as a left H-module via the H-action on H, is a projective left H-module, and the scalar multiplication map $\mu : H \otimes_R S \to S$ is a left H–module homomorphism. To show that S is H- projective, we find a left H-module splitting map ν for μ, namely $\nu : S \to H \otimes_R S$ by

$$\nu(s) = \sum_{(\vartheta)} \vartheta_{(1)} \otimes z \cdot (\vartheta_{(2)}^{\lambda} s)$$

(usual Sweedler notation, and with λ as the antipode of H). It is a technical exercise to verify that $\mu \circ \nu$ is the identity on S. One uses the fact that if ϑ is a left integral then for all h in H,

$$\sum_{(\vartheta)} h\vartheta_{(1)} \otimes \vartheta_{(2)}^{\lambda} = \sum_{(\vartheta)} \vartheta_{(1)} \otimes \vartheta_{(2)}^{\lambda} h$$

to verify that ν is a left H-module homomorphism. Technical details may be found in Theorem 5.1 of [CH86].

The other result is Schneider's. We have $K \otimes_R S \cong L$ and $K \otimes_R H \cong A$. Now L is an A-Hopf Galois extension of K, so by a result of Kreimer and Cook [KC76], $L \cong A$ as left A-modules. Hence S and H are two projective left H-modules so that $K \otimes_R S \cong K \otimes_R H$ as left $K \otimes_R H$-modules. But then Theorem 4.1 of [Sch77] applies to yield that $S \cong H$ as left H-modules. ∎

Theorem 1.1 extends Noether's theorem. For if L/K is a Galois extension of number fields with group G, and S is the integral closure of R in L, then S is tamely ramified iff the trace map $tr : S \to R$, $tr(s) = \sum_{\sigma \in G} \sigma(s)$, is onto. But $\sum_{\sigma \in G} \sigma$ generates the module of integrals I of RG. So L/K tame is equivalent to the condition $IS = R$. Schneider's theorem then plays the same role in Theorem 1.1 as Swan's theorem ([$Sw60$], Corollary 6.4, which Schneider's theorem extends) does in the proof of Noether's theorem (see [$CF67$], page 22).

This extension of Noether's theorem to Hopf orders has a nice interpretation involving the Leopoldt order. Note that S is any order over R in L, not necessarily the maximal order:

COROLLARY 1.2. With K, R, L, S and A as in Theorem 1.1, suppose the Leopoldt order $\mathcal{A}(S)$ of S in A, namely, $\mathcal{A}(S) = \{\alpha \text{ in } A | \alpha s \in S \text{ for all } s \text{ in } S\}$, is an R-Hopf algebra order in A. Then S is a free $\mathcal{A}(S)$-module.

PROOF. (c.f. [$C87$], Theorem 2.1). Since L/K is A-Galois, the fixed ring $L^A = \{s \text{ in } L | \text{ as} = \epsilon(a)s \text{ for all a in } A\} = K$. We have easily that $IS \subseteq S^H \subseteq L^A \cap S = R$, where I is the module of left integrals of H. Let ϕ be a generator of the one-dimensional K-space of left integrals of A. Since L is an A-Hopf Galois extension of K, $\phi L = K$ and ϕS is a fractional ideal of K. Thus $\phi S = aR$ for some a in K. But then $\vartheta = \phi/a$ is a left integral of A which maps S onto $R \subseteq S$. By definition of $H = \mathcal{A}(S)$, ϑ is in H, so is in I, and $IS = R$. The result then follows from Theorem 1.1. ▌

This result raises the question, given a Hopf Galois extension L/K of number fields with Hopf algebra A, under what conditions is the Leopoldt order

$$\mathcal{A}(S) = \{\alpha \text{ in } A | \alpha s \in S \text{ for all } s \text{ in } S\},$$

of the ring of integers S of L a Hopf order in A? This question was considered in [$C87$] for abelian extensions of $\mathbf{Q}$(i.e. $A = \mathbf{Q}G, G$ abelian) and for Kummer extensions of prime order.

Over $\mathbf{Q}$, it turns out that $\mathcal{A}(S)$ is Hopf iff the extension $L/\mathbf{Q}$ is tamely ramified at all odd primes, and the ramification group for $L/\mathbf{Q}$ at the prime 2 has order at most 2 ([$C87$], Theorem 5.1). By contrast, Leopoldt's main result in [$Le59$] is that S is free over $\mathcal{A}(S)$ for $A = \mathbf{Q}G$, G any finite abelian group.

In the case of Kummer extensions of a local field K of prime order p with ramification number t, $\mathcal{A}(S)$ is a Hopf order iff $t \equiv -1 (\text{mod} p)$; if $t < pe_0/(p-1) - 1$, where e_0 is the ramification index of K over $\mathbf{Q}_p$, then S is free over $\mathcal{A}(S)$ iff

$t \equiv a(\bmod p)$ and a divides $p-1$. The first result is a reformulation by Greither [$Gr92$] of the main result of [$C87$]; the second is due to Bertrandias and Ferton [$BF72a$]; c.f. [$BF72b$] for the case $t \geq pe_0/(p-1)-1$. Greither's reformulation, with a suitably generalized ramification number, holds for any totally ramified Hopf Galois extension L/K of order p ([$Gr92$], Theorem 2.7).

Greither also has necessary conditions on the ramification numbers of a cyclic Galois extension L/K of degree p^2 in order that $\mathcal{A}(S)$ be Hopf (see [$Gr92$], Theorem 3.2).

2. ORDERING ORDERS

Rather than starting with a wildly ramified Galois extension of number fields and asking if the Leopoldt order of its ring of integers is Hopf, a relatively successful strategy has been to begin with a number field K with ring of integers R and a finite abelian group G, consider all the Hopf algebra orders over R in KG, and, for a wild extension L/K with group G, see if any Hopf algebra order is the Leopoldt order of the ring of integers of L. This was essentially the strategy of [$Ch87$] and [Gr92]. The basic approach is that starting from a Hopf algebra order one can construct an order over R in L. More precisely, let L be a Hopf Galois extension of K, a local number field, with Hopf algebra A. Let R be the valuation ring of K, let S be the integral closure of R in L (we do not assume L is a field) and let H be a Hopf order over R in A. Then

$$\tilde{\mathcal{O}}(H) = \{s \text{ in } L | hs \in S \text{ for all } h \text{ in } H\}$$

is a lattice in L (i.e. an R-finitely generated submodule of L which contains a K-basis of L). Taylor has observed:

PROPOSITION 2.1. $\tilde{\mathcal{O}}(H)$ is an order over R in L.

PROOF. ([$T87$], Lemma 3.1). To see that $\tilde{\mathcal{O}}(H)$ is an R-lattice in L, observe that since 1 is in H, $\tilde{\mathcal{O}}(H) \subseteq S$; on the other hand, if $\{h_i\}$ is an R-basis of H and $\{s_j\}$ is an R-basis of S (for $i,j=1,\ldots,n$), then there is some r in R so that $r(h_i s_j)$ is in S for all i and j. So $rS \subseteq \tilde{\mathcal{O}}(H)$ and $\tilde{\mathcal{O}}(H)$ is a lattice. Now 1 is in $\tilde{\mathcal{O}}(H)$ because for all h in $H, h \cdot 1 = \epsilon(h) \cdot 1$ and $\epsilon(h)$ is in R, hence $h \cdot 1$ is in S for all h in H. If s,t are in $\tilde{\mathcal{O}}(H)$, then, for all h in $H, h(st) = \sum_{(h)} h_{(1)}(s) \cdot h_{(2)}(t)$ is in S. So st is in $\tilde{\mathcal{O}}(H)$. Thus $\tilde{\mathcal{O}}(H)$ is an order in L. ▌

Thus given a Hopf Galois extension L/K of number fields with Hopf algebra A, we have the map $\mathcal{A}$, from orders over R in L to orders over R in A, and the map $\tilde{\mathcal{O}}$, from orders over R in A to lattices over R in L. For an order S over R in L, sometimes $\mathcal{A}(S)$ is a Hopf order in A; if H is a Hopf order over R in A, $\tilde{\mathcal{O}}(H)$ is an order over R in L. It is not the case that $\tilde{\mathcal{O}}$ and $\mathcal{A}$ are always inverses of each other. The simplest example is to take a wildly ramified abelian extension $L/\mathbf{Q}$, with ring of integers S and Galois group G; then $\mathbf{Z}G$ acts on S, so, since S is the maximal order of L, $\tilde{\mathcal{O}}(\mathbf{Z}G) = S$. But $\mathcal{A}(S)$ is necessarily larger than $\mathbf{Z}G$, for since $L/\mathbf{Q}$ is wildly ramified, S cannot be projective over $\mathbf{Z}G$ by Noether's theorem, but Leopoldt's main theorem [Le59] is that S is free over $\mathcal{A}(S)$. Thus $\mathcal{A}\tilde{\mathcal{O}}(\mathbf{Z}G)$ is strictly larger than ZG.

The following results bear on the question of when $\mathcal{A}$ and $\tilde{\mathcal{O}}$ are inverses of each other.

PROPOSITION 2.2. Let K be a local field with valuation ring R. Let H be a commutative, cocommutative R-Hopf algebra, finitely generated and free as R-module, and $A = K \otimes_R H$. Let L be an A-Hopf Galois extension of K. Let S be an order over R in L such that S/R is a tame H-extension. Then S is a free rank one H-module and $H = \mathcal{A}(S)$. If S is an H-Galois extension of R or H is a local ring, then $S = \tilde{\mathcal{O}}(H)$, hence $H = \mathcal{A}(\tilde{\mathcal{O}}(H))$ and S is the unique order over R in L which is a tame H-extension.

The hypothesis that S is H-tame reflects a strategy often used in the theory: start with H, construct an S so that S is H-tame (a trace condition if A is a group ring), then apply this result.

PROOF. Since S/R is H-tame, by the extension of Noether's theorem, S is free of rank one.

To show $H = \mathcal{A}(S)$, first observe that since $\mathcal{A}(S) = \{a \text{ in } A \mid aS \subseteq S\}$, we have $H \subseteq \mathcal{A}(S)$. Let $S = Hw$, the free rank one H-module with basis w. Then $L = Aw$. If a is in $\mathcal{A}(S)$, then $aw \in S$, so $aw = hw$ for some h in $H \subseteq A$. But since L is A-free on w, $a = h$ in H. Hence $\mathcal{A}(S) \subseteq H$.

To show $S = \tilde{\mathcal{O}}(H)$, recall that

$$\tilde{\mathcal{O}}(H) = \{s \text{ in } L | Hs \subseteq \mathcal{O}_L\}$$

where $\mathcal{O}_L$ is the integral closure of R in L, and $HS \subseteq S \subseteq \mathcal{O}_L$, so $S \subseteq \tilde{\mathcal{O}}(H)$. First assume S is an H-Galois extension of R. The inclusion $S \subseteq \tilde{\mathcal{O}}(H)$ is an R-algebra, H-module homomorphism, hence induces an $S\#H$-module structure on $\tilde{\mathcal{O}}(H)$. But $S\#H \cong \mathrm{End}_R(S)$ since S is H- Galois, and we therefore have a Morita isomorphism $\tilde{\mathcal{O}}(H) \cong S \otimes_R \tilde{\mathcal{O}}(H)^H$ given by multiplication in $\tilde{\mathcal{O}}(H)$. But

$R \subseteq \tilde{\mathcal{O}}(H)^H \subseteq \tilde{\mathcal{O}}(H) \cap L^A \subseteq \mathcal{O}_L \cap K = R$, hence $\tilde{\mathcal{O}}(H)^H = R$ and $S = \tilde{\mathcal{O}}(H)$. Uniqueness of S follows.

If H is a local ring and $S \cong H$ as left H-module, then S is an H-Galois extension of R by [W92]. ▌

The following result says that if you find one Hopf order which is the Leopoldt order of some order in L, then the same is true for any larger Hopf order.

THEOREM 2.3. Let L/K be an A-Galois extension of local fields, and R be the valuation ring of K. Let H_0 be a Hopf order in A so that $\tilde{\mathcal{O}}(H_0)$ is H_0-tame. Then $H_0 = \mathcal{A}\tilde{\mathcal{O}}(H_0))$. If H is any Hopf order in A containing H_0, then $\tilde{\mathcal{O}}(H)$ is free over H and $\mathcal{A}\tilde{\mathcal{O}}(H)) = H$.

PROOF. That $H_0 = \mathcal{A}(\tilde{\mathcal{O}}(H_0))$ follows from Proposition 2.2.

Let ϑ_0 generate the module of left integrals of H_0. Since $S_0 = \tilde{\mathcal{O}}(H_0)$ is H_0-tame, there is a z_0 in S_0 so that $\vartheta_0 z_0 = 1$. Let ϑ generate the module of left integrals of H, then $\vartheta_0 = r\vartheta$ for some r in R, since $H_0 \subseteq H$. Let $z = rz_0$. Claim:

1) z is in $\tilde{\mathcal{O}}(H) = S$

2) $\vartheta z = 1$, hence S is H-tame.

Claim 2) is obvious: $\vartheta z = (\vartheta_0/r)(rz_0) = \vartheta_0 z_0 = 1$. To prove claim 1), first note that since $H_0 \subseteq H$, $H^* \subseteq H_0^*$ (linear duals over R). We have $H = H^* \cdot \vartheta$, so for any ξ in H, there exists f in H^* with $\xi = f \cdot \vartheta$. To show z is in S, we need to show that for any ξ in H, ξz is in $\mathcal{O}_L$, the valuation ring of L. But
$\xi z = (f \cdot \vartheta)z = (f \cdot (\vartheta_0/r))(rz_0) = (f \cdot \vartheta_0)z_0$. Now since f is in $H^* \subseteq H_0^*$, $f \cdot \vartheta_0$ is in H_0, and since z_0 is in $\tilde{\mathcal{O}}(H_0)$, $(f \cdot \vartheta_0)z_0$ is in $\mathcal{O}_L$. Thus ξz is in $\mathcal{O}_L$, and z is in $\tilde{\mathcal{O}}(H)$. ▌

COROLLARY 2.4. If L/K is a Galois extension of local fields with Galois group G and L/K is tamely ramified, then for every Hopf order H in KG, $\tilde{\mathcal{O}}(H)$ is free over H and $H = \mathcal{A}(\tilde{\mathcal{O}}(H))$.

This follows immediately from Theorem 2.3 and the fact that any Hopf order in KG contains RG (because the dual of any Hopf order in KG is contained in the maximal order of KG^*, namely RG^*). ▌

3. KUMMER THEORY OF FORMAL GROUPS

In this section we describe a large class of extensions of a local field K which have orders whose Leopoldt orders are Hopf.

The extensions are called Kummer extensions with respect to a formal group. Classical cyclic Kummer extensions of prime power order may be described from this point of view, as we will show.

Fix a prime p, and let K be a local field, a finite extension of $\mathbf{Q}_p$. Let R be the valuation ring of K, with maximal ideal m generated by π. Let $\bar{K}$ be an algebraic closure of K, and let $\bar{R}$ be the integral closure of R in $\bar{K}$, with maximal ideal $\overline{m}$. A formal group $F = F(x, y)$ of dimension one defined over R is a power series in two variables with coefficients in R so that the operation $\alpha +_F \beta = F(\alpha, \beta)$ for any α, β in $\overline{m}$ makes $\overline{m}$ into an abelian group with identity element 0. A homomorphism $f : F \to G$ from one formal group of dimension one to another is a power series $f = f(x)$ in $R[[x]]$ so that for any α, β in $\overline{m}$, $f(\alpha +_F \beta) = f(\alpha) +_G f(\beta)$. We denote $\overline{m}$ with operation $+_F$ by $F(\bar{K})$. For any extension L of K contained in $\bar{K}$, $F(L)$ is defined similarly.

Unreferenced notation and facts about formal groups are from Fröhlich [F68].

There is a map $[\,] = [\,]_F : \mathbf{Z} \to \mathrm{End(F)}$ given by $[0] = 0$, $[1](x) = x$, $[-1](x)$ is defined by $F(x, [-1](x)) = 0$, and for any n,

$$[n+1](x) = F([n](x), x) \qquad (n > 0)$$

$$[n-1](x) = F([n](x), [-1](x)) \qquad (n < 0).$$

The formal group F has finite height if the power series $[p](x)$ is non-zero modulo m.

Given formal groups F and G of dimension one and finite height defined over R, and a homomorphism $f : F \to G$, we may define an R- Hopf algebra H by $H = R[[x]]/(f(x))$. Here the counit map ϵ is the algebra homomorphism induced by sending x to 0; the antipode is the algebra homomorphism induced by sending x to $[-1]_F(x)$, and the comultiplication map Δ is the algebra map from H to $H \otimes_R H$ induced by sending x to $F(x \otimes 1,\ 1 \otimes x)$.

To see that Δ is well-defined, we define Δ in the same way from $R[[x]]$ to $R[[x]]\hat{\otimes}R[[x]]$ and show that $(f(x))$ is mapped to $(f(x)) \otimes R[[x]] + R[[x]] \otimes (f(x))$ (that is, $(f(x))$ is a coideal). Thus it suffices to show that $\Delta(f(x))$ is in the ideal generated by $f(x) \otimes 1 = f(x \otimes 1)$ and $1 \otimes f(x) = f(1 \otimes x)$. But if we write $x \otimes 1$ as y and $1 \otimes x$ as z, then $R[[x]]\hat{\otimes}R[[x]] \cong R[[y, z]$, and $\Delta(x) = F(y, z)$. We then have

$$\Delta(f(x)) = f(F(y, z)) = G(f(y), f(z)).$$

Since $G(y, z)$ has no constant term, $G(f(y), f(z))$ is in the ideal generated by $f(y)$ and $f(z)$, as we wished to show.

Let $A = K \otimes_R H$.

If f has height h, that is, Weierstrass degree $q = p^h$, then by the Weierstrass preparation theorem, $f = f_0 \cdot u$, where f_0 is a Weierstrass polynomial of degree q and u is an invertible power series. Then, since f has no multiple roots, ([F68], p.107-8) $H \cong R[x]/(f_0(x))$ is a free R-module of rank q and Γ, the set of roots of f_0 in $\overline{m}$, is a subgroup of $F(\bar{K})$ of order q.

Following Taylor [T86], we define the Kummer order

$$S_c = R[[z]]/(f(z) - c)$$

for any c in m . As with H, S_c is a free R-module of rank q. We make S_c into an H-comodule algebra by defining the R-algebra homomorphism

$$\alpha : S_c \to S_c \otimes H \cong R[[z, x]]/(f(z) - c, f(x))$$

to be the homomorphism induced by sending z to $F(z, x)$. Then α is well-defined, since

$$\alpha(f(z)) = f(F(z, x)) = G(f(z), f(x)) = G(c, 0) = c = \alpha(c).$$

THEOREM 3.1. For any c in m , S_c is an H-Galois object.

PROOF. It suffices to show that $T \otimes_R S_c$ is a $T \otimes_R H$-Galois object for some faithfully flat R-algebra T. For that, it suffices to find a faithfully flat R-algebra T so that $T \otimes_R S_c$ is isomorphic to $T \otimes_R H$ as $T \otimes_R H$-comodule algebras, for then $T \otimes_R S_c$ will be isomorphic as Galois object to the trivial $T \otimes_R H$-Galois object.

Let a in $\bar{K}$ be a root of $f(x) - c$, and let $L = K[a]$, T the valuation ring of L with maximal ideal m_T generated by π_T. Define an algebra homomorphism γ from $T \otimes_R H \cong T[[x]]/(f(x))$ to $T \otimes_R S_c \cong T[[t]]/(f(t) - c)$ induced by sending x to $t -_F a$. Then $0 = f(x)$ is sent by γ to

$$\begin{aligned} f(t -_F a) &= f(F(t, [-1]_F(a)) \\ &= G(f(t), [-1]_G(f(a)) \\ &= G(c, [-1]_G(c)) \\ &= 0. \end{aligned}$$

Thus γ is a well-defined T-algebra homomorphism. To show that γ is a $T \otimes H$-comodule homomorphism, we show $\alpha \circ \gamma = (\gamma \otimes 1) \circ \Delta$ as maps from $T \otimes_R H$

to $(T \otimes S_c) \otimes_T (T \otimes H)$. We write $T \otimes H$ as the image of $T[[x]]$ and $(T \otimes S_c) \otimes_T (T \otimes H)$ as the image of $T[[t,x]]$. Now

$$\alpha \circ \gamma(x) = \alpha(t -_F a) = \alpha(F(t, [-1]_F(a)))$$

$$= F(\alpha(t), [-1]_F(a))$$

$$= F(F(t,x), [-1]_F(a)),$$

while

$(\gamma \otimes 1) \circ \Delta(x) = (\gamma \otimes 1)F(x \otimes 1,\ 1 \otimes x)$ in $(\gamma \otimes 1)(R[[x \otimes 1,\ 1 \otimes x]])$. Now $(\gamma \otimes 1)(x \otimes 1)$ is the image in $(T \otimes S_c) \otimes_T (T \otimes H)$ of $t -_F a$ in $T[[t,x]$, and $(\gamma \otimes 1)(1 \otimes x)$ is the image of x. So we have

$$(\gamma \otimes 1)F(x \otimes 1,\ 1 \otimes x) = F(t -_F a, x)$$

$$= F(F(t, [-1]_F(a)), x))$$

which, using the associativity and commutativity of F, is the same as $\alpha \circ \gamma(x)$. Thus the map γ is a $T \otimes H$-comodule homomorphism. ∎

We can also use the map γ to show that S_c is isomorphic to $H^* = \mathrm{Hom}_R(H, R)$ as H^*-modules, and we give an explicit Galois generator for S_c:

COROLLARY 3.2. S_c is a free H^*-module on the image in S_c of t^{q-1} in $R[[t]]$.

PROOF. Let I be the free rank one R-module of integrals of H. Since R is local and H is commutative and cocommutative we know that H is isomorphic to H^* as H^*-modules, with $H = H^*j$ where j is any generator of I. However, since $H = R[[x]]/(f(x))$ and $\epsilon(x^k) = 0$ for all $k > 0$, an easy calculation shows that $f(x)/x$ is a generator of I.

Viewing the situation over the faithfully flat R-algebra T, we now see that $T \otimes_R H$ is a free $T \otimes_R H^*$-module with generator $f(x)/x$. Since γ is an isomorphism of $T \otimes_R H$-comodules (i.e. $T \otimes_R H^*$-modules), $T \otimes_R S_c$ is isomorphic to $T \otimes_R H^*$ as a $T \otimes_R H^*$-module and is generated by the image in $T \otimes_R S_c$ of $\gamma(f(x)/x)$ in $T[[t]]$.

Let $w(x) = f(x)/x$. Then $\gamma(f(x)/x) = w(\gamma(x)) = w(F(t, [-1](a))$. Since $f(x)$ has Weierstrass degree q, $w(x) \equiv x^{q-1} (\bmod \pi)$ and so $\gamma(f(x)/x) \equiv F(t, [-1](a))^{q-1} \equiv t^{q-1} (\bmod \pi_T)$.

Let $\psi = \gamma(f(x)/x)$ in $T[[t]]$. If $\{b_1, \ldots, b_q\}$ is a T-basis of $T \otimes_R H^*$, then $\{b_1\psi, \ldots, b_q\psi\}$ is a T-basis of $T \otimes_R S_c$. This also yields a $T/\pi_T T$-basis of $T \otimes_R S_c/\pi_T T \otimes_R S_c$. But then $\{b_1 t^{q-1}, \ldots, b_q t^{q-1}\}$ also is a set in $T \otimes_R S_c$which reduces modulo $\pi_T T$ to a $T/\pi_T T$-basis of $T \otimes_R S_c/\pi_T T \otimes_R S_c$. So by Nakayama's Lemma, $\{b_1 t^{q-1}, \ldots, b_q t^{q-1}\}$ is also a T-basis for $T \otimes_R S_c$. Hence $T \otimes_R S_c = T \otimes_R H^* t^{q-1}$, and since T is a faithfully flat R-algebra, $S_c = H^* t^{q-1}$. ▌

COROLLARY 3.3. $S_c = \tilde{O}(H^*)$ and $H^* = \mathcal{A}(S_c)$.

This follows from Theorem 2.2. ▌

If we apply Weierstrass preparation to $f(t) - c$, we may write $f(t) - c = g(t)v(t)$, $g(t)$ a Weierstrass polynomial of degree q, and $v(t)$ an invertible power series. Then $S_c \cong R[t]/(g(t))$ as R-algebras. This identification confuses the H-comodule structure, however.

Now we consider special cases.

♣ Suppose $g(t)$ is irreducible over K. Then L_c is a field extension of K. If Γ, the set of roots of $f(x)$ in $\overline{m}$, is contained in K, then L_c is a (classical) Galois extension of K with Galois group G isomorphic to Γ. This follows because of

PROPOSITION 3.4. If the roots Γ of $f(x)$ are in K, then $A = K[[x]]/(f(x)) \cong K^\Gamma$. Hence L_c is a Galois extension of K, where the Galois group $G \cong \Gamma$ acts on L_c by translating (under $+_F$) the generator t of L_c by elements of Γ.

PROOF. Since $f_0(x)$ splits in K, $A \cong K[x]/(f_0(x)) \cong K^G$ where G is a set in $1-1$ correspondence with the roots of $f_0(x)$, that is, with the elements of Γ, and the map $\varphi : A \to K^G$ is induced by $\varphi(x)(s_g) = g$ for $g \in \Gamma$ and s_g the element of G which corresponds to g. Then φ may be viewed as corresponding to a pairing

$$<>: G \times A \to K$$

by

$$s_g \times m(x) \to < s_g, m(x) > = m(g)$$

where $m(X)$ is a polynomial in $R[X]$. Then the comultiplication on A defines a multiplication on G by

$$\begin{aligned} < s_g s_h, x > &= < s_g \otimes s_h, \Delta(x) > \\ &= < s_g \otimes s_h, F(y, z) > \end{aligned}$$

(identifying $A \otimes A$ as the image of $R[[x]]\hat{\otimes}R[[x]] \cong R[[y,z]]$)

$$= F(< s_g, y >, < s_h, z >)$$
$$= g +_F h$$
$$=< s_{g+_F h}, x > .$$

Thus the multiplication on G is that induced on G from the formal group multiplication on $\Gamma \subseteq F(\bar{K})$.

In case $A \cong K^\Gamma$, the action of the Galois group G on L_c is induced by translating the generator t by elements of Γ. To see this, observe that since $L_c = K[[t]]/(f(t) - c)$ is a K^G-Galois object, then L_c is a Galois extension of K with group G. The action of G on L_c is induced from the coaction map

$\alpha : L_c \to L_c \otimes A$, where $A = K[[x]]/(f(x))$ and $\alpha(t) = F(t,x)$, by

$$s_g \cdot t = F(t, < s_g, x >) = F(t, g) = t +_F g$$

for g in Γ corresponding to s_g in G. Thus G acts on the generator t of L_c by translating t by the roots of $f(x)$. ▌

♣ If $c \in m_K, c \notin m_K^2$, then $S_c = \mathcal{O}_{L_c}$. For the Newton polygon $N(f(x) - c)$ of $f(x) - c$ and $N(g(x))$ agree to the left of $(q, 0)$. Since $N(f(x) - c)$ has a vertex at $(0, v(c))$, so does $N(g(x))$. But then $v(g(0)) = v(c)$, and so $g(0) \in m_K, \notin m_K^2$, and $g(x)$ is Eisenstein. Therefore $S_c = \mathcal{O}_{L_c}$. If π is a generator of m_K, then c is in m_K and not in m_K^2 iff $c = u\pi$ for some u in R^*.

The intersection of these special cases gives our main local Galois module result.

THEOREM 3.5. Let F, G be formal groups of dimension one, Γ a finite subgroup of $F(K)$, $f : F \to G$ a homomorphism with kernel $= \Gamma$. Let m_K be generated by π. Then for any unit u of $\mathcal{O}_K$, $L = K[[z]]/(f(z) - u\pi)$ is a Galois field extension of K with group $\cong \Gamma$, and $\mathcal{O}_L = R[[z]]/(f(z) - u\pi)$ is a free rank one module over its associated order $\mathcal{A} = \mathcal{A}(\mathcal{O}_L)$, where $\mathcal{A}^* \cong R[[x]]/(f(x))$. ▌

Adapting methods of Lubin [$Lu79$] (see Example 4.5 below), a large number of examples of Hopf algebras H of the form described in the theorem may be constructed from congruence-torsion subgroups of formal groups, as is shown in [$CZ93$].

To explain the terminology, "Kummer extension with respect to the formal group F", we conclude this section by specializing F to the multiplicative formal group $\mathbf{G}_m$.

PROPOSITION 3.6. Let $F = G = \mathbf{G}_m$, the multiplicative formal group defined as $\mathbf{G}_m(x, y) = x + y + xy$. Let $q = p^n$ and consider the endomorphism $[q] : \mathbf{G}_m \to \mathbf{G}_m$. Suppose K contains a primitive q th root of unity. Then the Kummer extensions of K with respect to $\mathbf{G}_m$ corresponding to $f = [q]$ are classical Kummer extensions with Galois group C_q cyclic of order q.

PROOF. We consider $H = R[[x]]/([q](x))$. It is easy to see by induction that for any $m > 0$, $[m](x) = (x+1)^m - 1$, so

$$\begin{aligned} H = R[x]/([p^n](x)) &= R[x]/((x+1)^q - 1) \\ &= R[y]/(y^q - 1) \\ &\cong RC_q, \end{aligned}$$

the group ring of the cyclic group of order q, as R-algebras, where $y = x + 1$. This last isomorphism is in fact as Hopf algebras, for

$$\begin{aligned} \Delta(y) &= \Delta(x+1) = \Delta(x) + \Delta(1) \\ &= (x \otimes 1 + 1 \otimes x + x \otimes x) + 1 \otimes 1 \\ &= (x+1) \otimes (x+1) \\ &= y \otimes y \end{aligned}$$

so the generator y of H is grouplike.

Given any c in m, $S_c = R[t]/([q](t) - c) = R[z]/(z^q - (1+c))$, where $z = t + 1$. Since $c \in m$, then $1 + c$ is a unit of R.

Suppose K contains a primitive q th root of unity ζ. Then

$$\Gamma = \{\zeta^r - 1 | r = 0, 1, \ldots, q-1\} \subseteq K$$

is the set of roots of $[q](x)$. So by Proposition 3.4, L_c is a Galois extension of K with group $G \cong \Gamma$, where if s_r in G corresponds to $\zeta^r - 1$ in Γ, then for the generator t of L_c,

$$\begin{aligned} s_r \cdot t &= \mathbf{G}_m(t, < s_r, x >) \\ &= \mathbf{G}_m(t, \zeta^r - 1) \end{aligned}$$

$$= t + \zeta^r - 1 + (\zeta^r - 1)t.$$

Hence

$$s_r \cdot z = s_r \cdot t + 1$$
$$= \zeta^r t + \zeta^r = \zeta^r z$$

and the Galois group G acts on the generator z by multiplication by q th roots of unity. Thus L_c is a Kummer extension of K with group $G = C_q$. ∎

4. DESCRIBING H^*

Let F be a formal group of dimension one and finite height defined over the valuation ring R of a local field $K \supseteq \mathbf{Q}_p$. Let m_K be the maximal ideal of R, $m_K = \pi R$ for some parameter π.

In the last section we showed that given a homomorphism f with domain F and an element c in m_K, the Kummer extension S_c is isomorphic to $H = R[[x]]/(f(x))$ as an H-comodule, hence $S_c \cong H^*$ as H^*-modules. Thus it is of interest to describe H^*. Taylor [T85], [T87] has found a basis of H^* as an R-module when H arises from a Lubin-Tate formal group. In this section we extend this description.

Let $G \subseteq m_K$ be a finite group under the action of F: that is, for g_1, g_2 in G, $g_1 +_F g_2 = F(g_1, g_2)$. Let F_1 be a formal group and $f : F \to F_1$ be a homomorphism of formal groups with $\ker(f) = G$, then $H = R[[x]]/(f(x))$ is a Hopf R-algebra with comultiplication induced by F, and f will have height h where $p^h = q = |G|$. The Weierstrass Preparation Theorem yields a factorization of $f(x)$ as $f(x) = h(x)u(x)$, where $h(x)$ is a Weierstrass polynomial of degree q and $u(x)$ is an invertible element of $R[[x]]$. Then

$$h(x) = \prod_{g \in G} (x - g) \text{ in } R[x] \text{ and } H \cong R[x]/(h(x)).$$

Let Γ be an abstract group isomorphic to G, and let $\chi : \Gamma \to G \subseteq K$ be an isomorphism. Then $A = K \otimes_R H \cong K[x]/(h(x)) \cong K^\Gamma$, via the map

$$\alpha : K[x]/(h(x)) \to K^\Gamma$$

induced by $\alpha(p(x))(\gamma) = p(\chi(\gamma))$ for $p(x)$ in $K[x]$. The standard duality pairing $K^\Gamma \times K\Gamma \to K$ becomes $A \times K\Gamma \to K$ given by:

$$< p(x), k_\gamma \gamma > = \sum_{\gamma \in \Gamma} k_\gamma p(\chi(\gamma)) = \sum_{g \in G} k_{\chi^{-1}(g)} p(g).$$

We wish to identify the dual of H.

We begin with Euler's formula: if G is the set of roots of $h(x)$, then

$$\frac{1}{h(x)} = \sum_{g \in G} \frac{1}{h'(g)(x-g)}$$

(To prove this one verifies that the polynomial

$$\sum_{g \in G} \left(\frac{\frac{h(x)}{x-g}}{h'(g)} \right)$$

of degree $\leq q-1$ has the value 1 on all q elements of G, hence by the uniqueness in the Chinese Remainder Theorem, must be the constant polynomial 1.)

Following Taylor ([T85], Section 2), set $x = 1/T$ in Euler's formula and expand both sides as power series in T. If

$$h(x) = x^q + b_{q-1}x^{q-1} + \ldots + b_1 x$$

with all b_j in m_K, then the left side of Euler's formula becomes

$$\frac{1}{h(1/T)} = T^q\left(\frac{1}{1 + b_{q-1}T + \ldots + b_1 T^{q-1}}\right) = T^q + c_{q+1}T^{q+1} + \ldots$$

with all $c_j$in πR, while the right side,

$$\sum_{g \in G} \frac{1}{h'(g)(\frac{1}{T} - g)} = \sum_{g \in G} \frac{T}{h'(g)}(1 + gT + g^2T^2 + \ldots)\,.$$

Equating coefficients of the powers of T, we get

$$\sum_{g \in G} \frac{g^i}{h'(g)} = \begin{cases} 0 & \text{if } 0 \leq i < q-1 \\ 1 & \text{if } i = q-1 \\ c_{i+1} & \text{if } i > q-1, \text{ where } c_{i+1} \in \pi R \end{cases}$$

(where $g^0 = 1$ for all g in G, including $g = 0$). Using this formula, we have

PROPOSITION 4.1. The dual U in $K\Gamma$ of H is the R-submodule of $K\Gamma$ with basis

$$\{\sum_{\gamma\in\Gamma} \frac{\chi(\gamma)^i\gamma}{h'(\chi(\gamma))} \mid i = 0, 1, \dots, q-1\} .$$

PROOF. Let $\{e_0, e_1, \dots, e_{q-1}\}$ be the dual basis in $K\Gamma$ of the basis $\{1, x, x^2, \dots, x^{q-1}\}$ of H. Then $U = \sum_{i=0}^{q-1} Re_i$. and $< e_i, x^j >= \delta_{i,j}$. Let

$$f_i = \sum_{\gamma\in\Gamma} \frac{\chi(\gamma)^i\gamma}{h'(\chi(\gamma))} .$$

Then

$$\begin{aligned} < f_i, x^j > &= \sum_{\gamma\in\Gamma} \frac{\chi(\gamma)^i}{h'(\chi(\gamma))} < \gamma, x^j > \\ &= \sum_{\gamma\in\Gamma} \frac{\chi(\gamma)^i}{h'(\chi(\gamma))} \chi(\gamma)^j \\ &= \sum_{g\in G} \frac{g^{i+j}}{h'(g)} \\ &= \begin{cases} 0 & \text{if } i+j < q-1 \\ 1 & \text{if } i+j = q-1 \\ c_{i+j+1} & \text{if } i+j > q-1. \end{cases} \end{aligned}$$

Then

$$\begin{aligned} f_i &= \sum_{j=0}^{q-1} \langle f_i, x^j \rangle e_j \\ &= e_{q-1-i} + \sum_{j=q-i}^{q-1} c_{i+j+1}\ e_j\ , \end{aligned}$$

or

$$(f_0, f_1, \dots, f_{q-1}) = (e_0, e_1, \dots, e_{q-1})M$$

where M is the $q \times q$ matrix

$$\begin{pmatrix} 0 & 0 & \dots & 0 & 0 & 1 \\ 0 & 0 & & 0 & 1 & c_{q+1} \\ 0 & 0 & & 1 & c_{q+1} & c_{q+2} \\ & & \dots & & & \\ 0 & 1 & & & & \\ 1 & c_{q+1} & \dots & & & c_{2q-1} \end{pmatrix} .$$

Since the matrix M is in $GL_q(R)$, $\{f_0, f_1, \dots, f_{q-1}\}$ is a basis of U. ∎

The next proposition recovers Taylor's description in $[T87]$. Let v be the valuation on K, normalized so that $v(\pi) = 1$.

PROPOSITION 4.2. Suppose $h(x)$ has the property

(4.3) $h'(0) = b$ with $v(b) = r$, and $h'(x) = \pi^r u(x)$ with $u(x)$ invertible in H.

Then $\{\sigma_0, \sigma_1, \ldots, \sigma_{q-1}\}$ is a basis of U, where for each $i = 0, \ldots, q-1$,

$$\sigma_i = \frac{1}{\pi^r} \sum_{\gamma \in \Gamma} \chi(\gamma)^i \gamma.$$

PROOF. Since $u(x)$ is invertible in H, we may choose as a basis of H the set $\{\frac{1}{u(x)}, \frac{x}{u(x)}, \ldots, \frac{x^{q-1}}{u(x)}\}$. Then

$$\begin{aligned}
\langle \frac{x^i}{u(x)}, \sigma_j \rangle &= \langle \frac{x^i}{u(x)}, \frac{1}{\pi^r} \sum_{\gamma \in \Gamma} \chi(\gamma)^j \gamma \rangle \\
&= \frac{1}{\pi^r} \sum_{\gamma \in \Gamma} \chi(\gamma)^j \frac{\chi(\gamma)^i}{u(\chi(\gamma))} \\
&= \sum_{\gamma \in \Gamma} \frac{\chi(\gamma)^{i+j}}{h'(\chi(\gamma))} \\
&= \sum_{g \in G} \frac{g^{i+j}}{h'(g)} \\
&= \langle f_i, x^j \rangle
\end{aligned}$$

So the matrix relating the dual basis of $\{\frac{x^i}{u(x)}\}$ with $\{\sigma_j\}$ is the invertible matrix M. Hence $\{\sigma_j | j = 0, \ldots, q-1\}$ is a basis for U. ▌

Suppose $H = R[[X]]/(f(X))$ where f is a homomorphism of formal groups from F to F_1, and $f(X) = h(X)u(X)$ where $h(X)$ is a Weierstrass polynomial of degree q and $u(X)$ is a unit in $R[[X]]$. Then $H \cong R[X]/(h(X))$. Let x be the image of X in H. When does $h(x)$ satisfy (4.3), namely, $h'(x) = h'(0)v(x), v(x)$ a unit in H? If $h(x) = h_1 x + h_2 x^2 + \ldots + h_{q-1} x^{q-1} + x^q$ and $h'(x) = h'(0)v(x)$ with $v(x)$ in $R[x]$, then, since $v(0) = 1$, $h_1 = h'(0)$ must divide q and rh_r for all r, $1 \leq r < q$. We conclude with three examples where (4.3) holds. The first is Taylor's $[T87]$.

EXAMPLE 4.4. Let F be a Lubin-Tate formal group defined over R which admits as an endomorphism $[\pi](x) = \pi x + x^q$, where $q = | R/\pi R |$. Then $R[x]/([\pi](x))$ is a Hopf R-algebra and $[\pi](x)$ clearly satisfies (4.3). Moreover, as Taylor points out and is easily seen by induction on n using the chain rule, $[\pi^n](x) = [\pi]([\pi^{n-1}](x)$ also satisfies (4.3).

On the other hand, if $f(x)$ and $g(x)$ are power series of finite heights whose corresponding Weierstrass polynomials satisfy (4.3), it need not follow that $(g \circ f)(x)$ has a Weierstrass polynomial which satisfies (4.3). (For an example, take $p = 3, f(x) = 3x + x^3 + x^4, g(x) = 3x + x^3$.)

EXAMPLE 4.5. Let F_t be a standard generic formal group of height h. This is a formal group defined over $\mathbf{Z}_p[[t_1, \ldots, t_{h-1}]]$ such that

$$[p](x) = pxu_0(x) + t_1x^pu_1(x) + \ldots + t_{h-1}x^{p^{h-1}}u_{h-1}(x) + x^{p^h}u_h(x)$$

where for each $i < h$, $u_i(x)$ is a unit in $\mathbf{Z}_p[[t_1, \ldots, t_i]][[x]]$, and $u_h(x)$ is a unit in $\mathbf{Z}_p[[t_1, \ldots, t_{h-1}]]$. See ([$Lu79$], p. 105). We may specialize F_t to a formal group F_a over R by replacing t_i by a_i in m_K for all $i = 1, \ldots, h-1$.

If we choose the a_i so that $v(a_i) \geq v(a_1)$ for all $i \geq 1$, then F_a will have height h and $[p]_{F_a}(x) = \sum b_i x^i$ with $v(a_1) = v(b_1) \leq v(b_i)$ for all i, $1 \leq i < p^h$. If $[p](x) = h(x)u(x)$ where $u(x)$ is a unit of $R[[x]]$ and $h(x) = \sum h_i x^i$ is a Weierstrass polynomial of degree p^h then $v(h_1) = v(b_1)$ and $v(h_i) \geq v(h_1)$ for $1 \leq i < p^h$, as is easily seen by writing $h(x) = [p](x)u^{-1}(x)$ and successively comparing coefficients of $1, x, \ldots, x^{p^{h-1}}$. Hence $R[x]/(h(x))$ is a Hopf R-algebra and $h(x)$ satisfies (4.3). Thus if $h(x)$ splits in K then H^* has a basis of the type described in Proposition 4.2.

EXAMPLE 4.6. Let F be a formal group of height h defined over $R \supseteq \mathbf{Z}_p$, and suppose the Newton polygon of $[p]_F$, $N([p])$, has a vertex at p. (By appropriate specialization of the generic formal group F_t of Example 4.5, such an F is easily constructed.)

Now by Lubin's Lemma (Lemma 4.1.2 of [$Lu64$], c.f. [$Z88$], p. 27), there exists an invertible power series $u(x)$ in $R[[x]]$ so that $u(F(u^{-1}(x), u^{-1}(y))) = F^u(x, y)$ has $[m]_{F^u}(x) = [m]_F^u(x) = u([m](u^{-1}(x))$ for all m in $\mathbf{Z}_p$, and $[\zeta]_{F^u}(x) = \zeta x$ for all ζ in the group μ_{p-1} of $p-1$st roots of unity in $\mathbf{Z}_p$. If $[p]_F(a) = 0$ for a in $m_{\bar{K}}$, then $[p]_{F^u}(u(a)) = 0$, and, since $u(x)$ is invertible in $R[[x]]$, the elements a and $u(a)$ have the same valuation. It follows that the Newton polygons of $[p]_F$ and of $[p]_{F^u}$ agree to the left of the abscissa p^h, since the slopes of the edges of the Newton polygon of $[p]_F$ to the left of p^h are the negatives of the valuations of the roots of $[p]_F$. In particular, the Newton polygon of $[p]_{F^u}$ will have a vertex at p iff it is so for $[p]_F$. So, without loss of generality, we shall assume that F has the property that $[\zeta]_F(x) = \zeta x$ for all ζ in μ_{p-1}.

By Lubin's Local Factorization Principle ([$Lu79$], p. 106), there exists a factorization $[p](x) = h(x)g(x)$ in $R[[x]]$ where $h(x)$ is a Weierstrass polynomial of degree p whose roots are 0 and the $p-1$ roots of $[p]$ in $\bar{K}$ whose valuation is equal to $-m$ where m is the slope of the edge joining $(1, v(p))$ and the vertex at p in the Newton polygon of $[p]$. In fact, $h(x)$ arises as a factor via the Weierstrass Preparation Theorem of a homomorphism $f : F \to F_1$ of formal groups, where F_1 is some formal group defined over R (as is f) and $\ker f =$ roots of $h(x)$ ([F68], Theorem 4, p. 112).

Now if $a \in m_{\bar{K}}$ is in $\ker f$, so is $[\zeta](a) = \zeta a$ for any ζ in μ_{p-1}, and $v(\zeta a) = v(a)$. Thus if a is a root of $h(x)$, then in $\bar{K}[x], h(x) = x \prod_{\zeta \in \mu_{p-1}} (x - \zeta a)$, hence $a^{p-1} = b$ in R and $h(x) = x^p - bx$. Then $H = R[[x]]/(f(x))$ is a Hopf R-algebra, and since $f(x) = h(x)u(x)$ for some invertible power series by Weierstrass preparation, $H \cong R[x]/(h(x))$ and $h(x) = x^p - bx$ satisfies (4.3). Thus if $h(x)$ splits in K, then $U = H^*$ has a basis as in Proposition 4.2.

5. HOPF GALOIS STRUCTURES

C. Greither and B. Pareigis ([GP87, p.245; [P90], p.84) have shown that the non-normal extension $\mathbf{Q}(2^{1/4})/\mathbf{Q}$ is a Hopf Galois extension for two different $\mathbf{Q}$-Hopf algebras. In this section we anticipate future research in local Galois module theory by elaborating on this example. We work locally, over $\mathbf{Q}_2$. Since $x^4 - 2$ is an Eisenstein polynomial, letting ω be a root of $x^4 - 2$, the valuation ring S of $L = \mathbf{Q}_2(\omega)$ is $S = \mathbf{Z}_2[\omega]$.

EXAMPLE 5.1. Let A_1, A_2 be the two $\mathbf{Q}_2$-Hopf algebras acting on L, and let $\mathcal{A}_i(S)$ be the Leopoldt order of S in $A_i, i = 1, 2$. Then one $\mathcal{A}_i$ is a $\mathbf{Z}_2$-Hopf order and the other is not.

As Pareigis observes ([P90], p.85), field extensions L/K with more than one Hopf Galois structure are very common. For example, if L/K is a Galois extension with group $C_q, q = p^n$ with p an odd prime, then L/K has a unique Hopf Galois structure iff $n = 1$ (c.f. [$Ch89$] and [$P90$], section 5). Example 5.1 shows that choosing which Galois module structure to use on L/K relates to the attractiveness of the resulting local Galois module structure for L/K.

PROOF. The Hopf algebra $A_1 = \mathbf{Q}_2[c, s]/(c^2 + s^2 - 1, cs)$ with comultiplication $\Delta(c) = c \otimes c - s \otimes s$, $\Delta(s) = c \otimes s + s \otimes c$. One sees that A_1 is contained in $\mathbf{Q}_2[i]C_4$, where C_4 is the cyclic group of order 4 generated by σ, as follows:

$$c = (\sigma + \sigma^3)/2, \quad s = i(\sigma - \sigma^3)/2.$$

Note that $\mathbf{Q}_2[i]C_4$ is split and A_1 contains the pairwise orthogonal minimal idempotents

$$e_1 = (c^2 + c)/2, \quad e_{-1} = (c^2 - c)/2,$$

$$e_i = (s^2 + s)/2 \qquad e_{-i} = (s^2 - s)/2.$$

Thus $A_1 \cong (\mathbf{Q}_2 C_4)^*$, the dual of the group ring.

Now A_1 acts on S as follows:

	1	ω	ω^2	ω^3
c	1	0	$-\omega^2$	0
s	0	$-\omega$	0	ω^3
c^2	1	0	ω^2	0
s^2	0	ω	0	ω^3

Hence we have

	1	ω	ω^2	ω^3
e_1	1	0	0	0
e_i	0	0	0	ω^3
e_{-1}	0	0	ω^2	0
e_{-i}	0	ω	0	0

Let $\mathcal{A}_1 = \{\alpha \text{ in } A_1 | \alpha S \subseteq S\}$. Then $\mathcal{A}_1$ is contained in the maximal order of A_1, namely the $\mathbf{Z}_2$-algebra generated by the idempotents e_1, e_i, e_{-1} and e_{-i}. But each of the idempotents is in $\mathcal{A}_1$. So $\mathcal{A}_1$ is the maximal order of A_1, and is isomorphic to $\mathcal{A}(\mathbf{Z}_2 C_4)^*$, hence is a $\mathbf{Z}_2$-Hopf algebra.

The other Hopf algebra acting on L is

$$A_2 = \mathbf{Q}_2[c, s]/(s^2 - 2c^2 + 2, cs)$$

with $\mathbf{Q}_2$-basis $1, c, s, c^2$ and comultiplication given by

$$\Delta(c) = c \otimes c - (s \otimes s)/2$$

$$\Delta(s) = c \otimes s + s \otimes c.$$

The action of A_2 on L is given by

	1	ω	ω^2	ω^3
1	1	ω	ω^2	ω^3
c	1	0	$-\omega^2$	0
s	0	ω^3	0	-2ω
c^2	1	0	ω^2	0

Let $\alpha = a + bc + ds + ec^2$ be a typical element of A_2. We find all a, b, d, e in $\mathbf{Q}_2$ so that $\alpha(s)$ is in S for any s in S.

We have

$$\alpha(1) = (a + b + e)1$$
$$\alpha(\omega) = a\omega + d\omega^3$$
$$\alpha(\omega^2) = (a - b + e)\omega^2$$
$$\alpha(\omega^3) = a\omega^3 - 2d\omega.$$

So α is in $\mathcal{A}_2$ iff all coefficients are in $\mathbf{Z}_2$, iff $a \in \mathbf{Z}_2, d \in \mathbf{Z}_2,\ \ b+e \in \mathbf{Z}_2$, and $b-e \in \mathbf{Z}_2$. Clearly $1, c, s, c^2$ are in $\mathcal{A}_2$. For b, e in $\mathbf{Q}_2$, $bc + ec^2$ is in $\mathcal{A}_2$ iff $b+e$ and $b-e$ are in $\mathbf{Z}_2$. Set $b = e = 1/2$, then we get $(c+c^2)/2$ in $\mathcal{A}_2$. If $b+e$ and $b-e$ are in $\mathbf{Z}_2$, then $2b = b'$ and $2e = e'$ are in $\mathbf{Z}_2$, and $b - e = (b' - e')/2$ is in $\mathbf{Z}_2$, so $b' = e' + 2f$ with f in $\mathbf{Z}_2$.

Hence if $bc + ec^2$ is in $\mathcal{A}_2$, then $bc + ec^2 = (b'c + e'c^2)/2 = e'(c + c^2)/2 + fc$, where e' and f are in $\mathbf{Z}_2$. It follows that $\mathcal{A}_2$ has a basis $\{1, c, s, (c^2 + c)/2\}$.

But $\Delta(c) = c \otimes c - (s \otimes s)/2$, which is not in $\mathcal{A}_2 \otimes \mathcal{A}_2$. So $\mathcal{A}_2$ is not a Hopf algebra. ∎

One easily verifies that $S = \mathcal{A}_2(1 + \omega + \omega^2)$ is a free $\mathcal{A}_2$-module, even though $\mathcal{A}_2$ is not a Hopf order in A_2.

REFERENCES

[BF72a] F. Bertrandias, M.-J. Ferton, Sur l'anneau des entiers d'une extension cyclique de degre premier d'un corps local, C. R. Acad. Sci. Paris 274 (1972), 1330-1333.

[BF72b] F. Bertrandias, J.-P. Bertrandias, M.-J. Ferton, Sur l'anneau des entiers d'une extension cyclique de degre premier d'un corps local, C. R. Acad. Sci. Paris 274 (1972), 1388-1391.

[CS66] S. Chase, M. Sweedler, *Hopf algebras and Galois theory*, Springer LNM 97 (1966).

[CH86] L. Childs, S. Hurley, Tameness and local normal bases for objects of finite Hopf algebras, Trans. Amer. Math. Soc. 298 (1986), 763-778.

[C87] L. Childs, Taming wild extensions with Hopf algebras, Trans. Amer. Math. Soc. 304 (1987), 111-140.

[C89] L. Childs, On the Hopf Galois theory for separable field extensions, Comm. Algebra 17 (1989), 809-825.

[CZ93] L. Childs, K. Zimmermann, Congruence-torsion subgroups of dimension one formal groups, J. Algebra (to appear).

[F62] A. Fröhlich, The module structure of Kummer extensions over Dedekind domains, J. reine angew. Math. 209 (1962), 39-53.

[F67] A. Fröhlich, Local Fields, in *Algebraic Number Theory* (J. W. S. Cassels and A. Fröhlich, eds.), Thompson, Washington, D. C., 1967.

[F68] A. Fröhlich, *Formal Groups*, Springer LNM 74 (1968).

[F83] A. Fröhlich, *Galois Module Structure of Algebraic Integers*, Springer-Verlag, 1983

[Gr92] C. Greither, Extensions of finite group schemes, and Hopf Galois theory over a complete discrete valuation ring, Math. Z. 210 (1992), 37-67.

[GP87] C. Greither, B. Pareigis, Hopf Galois theory for separable field extensions, J. Algebra 106 (1987), 239-258.

[KC76] H. F. Kreimer, P. Cook, Galois theories and normal bases, J. Algebra 43 (1976), 115-121.

[L59] H. Leopoldt, Über die Hauptordnung der ganzen Elemente eines abelschen Zahlkorpers, J. reine angew. Math. 201 (1959), 119-149.

[Lu64] J. Lubin, One-parameter formal Lie groups over p-adic integer rings, Annals of Math. 80 (1964), 464-484.

[Lu79] J. Lubin, Canonical subgroups of formal groups, Trans. Amer. Math. Soc. 251 (1979), 103-127.

[P90] B. Pareigis, Forms of Hopf algebras and Galois theory, Topics in Algebra, Banach Center Publ. 26 (1990), 75-93.

[Sch77] H.-J. Schneider, Cartan matrix of liftable finite group schemes, Comm. Algebra 5 (1977), 795-819.

[S77] J.-P. Serre, *Linear Representations of Finite Groups*, Springer-Verlag, 1977.

[Sw60] R. G. Swan, Induced representations and projective modules, Annls of Math. 71 (1960), 552-578.

[ST90] A. Srivastav, M. J. Taylor, Elliptic curves with complex multiplication and Galois module structure, Invent. Math. 99 (1990), 165-184.

[T85] M. J. Taylor, Formal groups and the Galois module structure of local rings of integers, J. reine angew. Math. 358 (1985), 97-103.

[T86] M. J. Taylor, A note on Galois modules and group schemes, informal manuscript, 1986.

[T87] M..J. Taylor, Hopf structure and the Kummer theory of formal groups, J. reine angew Math. 375/376 (1987), 1-11.

[T88] M. J. Taylor, Mordell-Weil groups and the Galois module structure of rings of integers, Illinois J. Math. 32 (1988), 428-452.

[T90a] M. J. Taylor, Resolvendes et espaces homogenes principaux de schémas en groupe, Sem. de Théorie des Nombres, Bordeaux, 2 (1990), 255-271.

[T90b] M. J. Taylor, The Galois module structure of certain arithmetic principal homogeneous spaces, J. Algebra 153 (1992), 203-214.

[TB92] M. J. Taylor, N. P. Byott, Hopf orders and Galois module structure, DMV Seminar 18, Birkhäuser Verlag, Basel, 1992, 154-210.

[W88] W. C. Waterhouse, Tame objects for finite commutative Hopf algebras, Proc. Amer. Math. Soc. 103 (1988), 354-356.

[W92] W. C. Waterhouse, Normal basis implies Galois for coconnected Hopf algebras, preprint, 1992.

[Z88] K. Zimmermann, Points of order p of generic formal groups, Ann. Inst. Fourier, Grenoble 38, 4 (1988), 17-32.

State University of New York at Albany

Partially supported by NSA grant #MDA90492 – $H3025$ and NSF grant #DMS90 – 01722.

Quantum Commutativity and Central Invariants

M. COHEN Ben Gurion University, Beer Sheva, Israel

0. INTRODUCTION

Let us start by considering the following:

1. *Graded commutative superalgebras.* That is, $A = A_0 \oplus A_1$, a Z_2-graded algebra over a field k (of characteristic $\neq 2$) where $ab = (-1)^{|a||b|}ba$, for all homogeneous elements a, b.
2. *The Quantum plane* $\mathbb{C}_q[X,Y]$. That is $\mathbb{C}[X,Y]$ where $xy = q^{-1}yx$.
3. *The Quantum superplane* (or q-Grassman algebra) $\mathbb{C}_q[\xi,\eta]$. That is $\mathbb{C}[\xi,\eta]$ with, $\xi^2 = \eta^2 = 0$, $\xi\eta = -qn\xi$.

What do the above have in common? They evidently describe a situation of "almost" commtutativity.

This research was supported by the Basic Research Foundation administered by the Israel Academy of Sciences and Humanities.

The first section of this paper deals with a concept which encompasses these examples. We show that they are all instances of what we term "quantum-commutative" H-module (or comodule) algebras A, where H is a (co)quasitriangular Hopf algebra. *The basic philosophy is that quantum-commutative algebras are to quasitriangular Hopf algebras what commutative algebras are to cocommutative Hopf algebras.* For example, left $A\#H$-modules are also right A-modules, and A^H is contained in the center of A.

In section 2 we describe more such properties of A and ${}_{A\#H}\mathrm{mod}$. In section 3 we single out the property that a Hopf algebra acts on an algebra A with $A^H \subset Z(A)$. This property leads to interesting connections between the ideal structures of A, A^H and$A\#H$.

Having central invariants and its affect on A was realized already in 1961 by Herstein [Her] where he conjectured:

Conjecture [Her] Let R be a ring and let $G = \langle g \rangle$ be a cyclic group of automorphisms of R of prime period p. If R^G is central in R then the commutator ideal of R is nil.

This conjecture is still open. There are several positive answers in certain cases. For example, it was proved by Osterburg:

Theorem [0] If R, R^G and G are as in the conjecture and R is semiprime then R is commutative.

§1 QUANTUM-COMMUTATIVITY

Let H be a Hopf algebra over a field k, (notation as in [S]),and ${}_H\mathrm{mod}$ the category of left H-modules. One of the important properties of this category is that for any $V, W \in {}_H\mathrm{mod}$ also $V \otimes_k W \in {}_H\mathrm{mod}$ via:

$$h \cdot (v \otimes w) = \Sigma h_1 \cdot v \otimes h_2 \cdot w$$

This makes ${}_H\mathrm{mod}$ into a so called monodial category [EK, Mac]. If $A \in {}_H\mathrm{mod}$, and A is moreover an algebra with multiplication μ, then being an H-module algebra [S] is just saying thac μ is a morphism in the category, that is:

$$\mu(h \cdot (v \otimes w)) = h \cdot (\mu(v \otimes w)),$$

and so A is an algebra in the category. When is A a commutative algebra in the category?

Well, in order to start answering such a question one needs to first establish what is meant by an "interchange" of two elements. The most naive interchange is the usual twist maps τ, which is defined for any k vector space V, W by

$$\tau_{V,W} : v \otimes w \to w \otimes v$$

thus saying that A is a commutative algebra means that $\mu\tau = \mu$. But is τ a morphism in the category ${}_H\text{mod}$? Well, usually not. There is no reason for

$$\sum h_2 \cdot w \otimes h_1 \cdot v = \sum h_1 \cdot w \otimes h_2 \cdot v$$

Obviously, if H is cocommutative then the above holds, and so in this case saying that $A \in {}_H\text{mod}$ is commutative in the category (with τ as the interchange maps) is the same as saying that A is a commutative H-module algebra.

What can be said failing the fact that τ is an H-module homomorphism? One should be able to replace τ by another interchange map which is a morphism in the category. One of the important features of (co) quasitriangular Hopf algebras is that such a map $\psi_{V,W}$ exists. This interchange lies in the basis of the connection between quantum Yang Baxter equations and braids and knots, it gives rise to actions of the symmetric group, or more generally, the braid group on $V^{\otimes n}$ for any $V \in {}_H\text{mod}$.

If $\psi^2 = id$ (as does τ), then ψ is suggestively called a symmetry and ${}_H\text{mod}$ is a symmetric (tensorial) category [EK,G, Mac, Ma, JS].

Returning to the question of commutativity in the category ${}_H\text{mod}$ it is now evident that we mean by it an algebra A with multiplication μ so that:

$$\mu\psi = \mu$$

Explicitly, let (H, R) be a *quasitriangular Hopf* algebra [D]. That is, H is a Hopf algebra over k and $R = \sum R^{(1)} \otimes R^{(2)}$ is an invertible element of $H \otimes H$ satisfying the following (with $r = R$):

1. $\sum \Delta(R^{(1)}) \otimes R^{(2)} = \sum R^{(1)} \otimes r^{(1)} \otimes R^{(2)}r^{(2)}$.

2. $\sum R^{(1)} \otimes \Delta(R^{(2)}) = \sum R^{(1)}r^{(1)} \otimes r^{(2)} \otimes R^{(2)}$

3. $\Delta^{cop}(h) = R\Delta(h)R^{-1}$, all $h \in H$ (where $\Delta^{cop}(h) = \sum h_2 \otimes h_1$)

(H, R) is called triangular if $R^{-1} = \sum R^{(2)} \otimes R^{(1)}$.

Let $V, W \in {}_H\text{mod}$ and define the interchange:

$$\psi_{V,W} : V \otimes W \to W \otimes V \text{ by:}$$

$$v \otimes w \to \sum R^{(2)} \cdot w \otimes R^{(1)} \cdot v$$

all $v \in V, w \in W$.

Then $\psi_{V,W}$ is an H-module isomorphism by property (3). It is a symmetry exactly when (H, R) is triangular.

Now, if A is an H-module algebra then we call A *quantum-commutative* with respect to (H, R) if it is commutative in the category ${}_H\text{mod}$ with the interchange map ψ as above. That is:

$$ab = \sum (R^{(2)} \cdot b)(R^{(1)} \cdot a), \text{ all } a, b \in A$$

Dually, let $(H, \langle \mid \rangle)$ be a coquasitriangular (braided) Hopf algebra (see [Hay, LT, Maj1] for the definition) and let ${}_H\text{comod}$ denote the category of left H-comodules. Denote for each $V \in {}_H\text{comod}$ its structure map
$\rho_V : V \to H \otimes V$ by $\rho(v) = \sum v_1 \otimes v_2$.

There exist an interchange map here as well

$$\psi_{V,W} : V \otimes W \to W \otimes V \text{ by}$$

$$v \otimes w \to \sum \langle v_1 | w_1 \rangle w_2 \otimes v_2$$

all $v \in V,\ w \in W$.

Thus saying that a comodule algebra $A \in {}_H\text{comod}$ is quantum-commutative with respect to $(H, \langle \mid \rangle)$ means that

$$ab = \sum \langle a_1 | b_1 \rangle b_2 a_2$$

all $a, b \in A$.

Returning now to the examples in the introduction we show next that they are all commutative algebras in an appropriate category. There might of course be several categories for which the same algebra is a commutative object, and other categories for which it is not.

Example 1. Graded-commutative super algebras

Let $A = A_{\bar{0}} \oplus A_{\bar{1}}$ be a Z_2-graded algebra over a field k of characteristic not 2, assume $A_i A_j \subset A_{i+j}$ and $ab = (-1)^{|a|\,|b|} ba$ for all homogeneous elements a, b.

A being Z_2-graded is equivalent to its being a $(kZ_2)^*$-module algebra [CM]. Moreover, as for any cyclic group, $(kZ_2)^*$ is isomorphic to kZ_2 by:

$$\phi : kZ_2 \to (kZ_2)^* \text{ via}$$

$$1 \to p_1 + p_g$$

$$g \to p_1 - p_g$$

(where $Z_2 = \{1, g\}$ and $\{p_1, p_g\}$ are dual bases of kZ_2 and $(kZ_2)^*$ respectively).

This isomorphism induces an action of kZ_2 on A by:

$$g \cdot a = (p_1 - p_g) \cdot a = (-1)^{|a|} a$$

(we write the elements of Z_2 as $\{1, g\}$ rather than $\bar{0}$, $\bar{1}$ to avoid confusion with addition in kZ_2).

Let $H = kZ_2$, since H is cocommutative, it has a trivial triangular structure, ($R = 1 \otimes 1$) but then, since A is not commutative, it would not be quantum-commutative with respect to $(H, 1 \otimes 1)$. Giving H however the following non-trivial triangular structure [Fi, Maj2]:

$$R = \frac{1}{2}(1 \otimes 1 + 1 \otimes g + g \otimes 1 - g \otimes g)$$

yields the desired property on A.

Indeed, for example if $a, b \in A_{\bar{1}}$, then

$$\begin{aligned} \sum R^{(2)} \cdot b)(R^{(1)} \cdot a) &= \tfrac{1}{2}[ba + (g \cdot b)a + b(g \cdot a) - (g \cdot b)(g \cdot a)] = \\ &= \tfrac{1}{2}[ba - ba - ba - ba] = -ba = ab \end{aligned}$$

The rest are checked similarly.

Conversely, let (H, R) be as above and let $A \in {}_H\text{mod}$ be an algebra in the category, then A is Z_2-graded using the induced $(kZ_2)^*$-module structure. That is,

$$A_{\bar{0}} = p_1 \cdot A = \frac{1+g}{2} \cdot A$$

$$A_{\bar{1}} = p_g \cdot A = \frac{1-g}{2} \cdot A$$

It is now easy to see that A is quantum-commutative with respect to (H, R) if and only if it is graded-commutative.

Example 2 The quantum plane

let $\mathbb{C}_q[X,Y]$ be the complex algebra generated by X, Y subject to: $xy = q^{-1}yx$, q an nth root of 1. As we shall see for the quantum superplane, $\mathbb{C}_q[X,Y]$ is in a sense a "universal" quantum-commutative algebra with respect to a coaction. However, following [CW2] we show now that it is quantum-commutative with respect to an action as well.

Denote $A = \mathbb{C}_q[X,Y]$, then A has a natural $Z \times Z$ grading by degrees of x and y, however, since x^n and y^n are central in A, this grading induces a grading of A by $G = Z_n \times Z_n$. Explicitly:

$$A = \sum_{(i,j) \in G} A_1 x^i y^j, \text{ with } A_1 = \mathbb{C}[x^n, y^n].$$

Thus by [CM] A is an $H = (kG)^*$-module algebra. Again, as in example 1, since H is cocommutative and A is not commutative, A is not commutative in the category if $R = 1 \otimes 1$. However, if we let:

$$R = \sum_{i,j,k,l=0}^{n-1} q^{jk-il} p_{i,j} \otimes p_{k,l}$$

(where $\{(i,j)\}$ and $\{p_{i,j}\}$ are dual bases of $\mathbb{C}G$ and $(\mathbb{C}G)^*$ respectively), then (H, R) is quasitriangular. But now since x, y have degrees $(1,0), (0,1)$ respectively, we have:

$$\sum (R^{(2)} \cdot y)(R^{(1)} \cdot x) = \sum q^{jk-il}(p_{k,l} \cdot y)(p_{i,j} \cdot x)$$

$$= q^{-1}yx = xy$$

Thus A is quantum-commutative with respect to (H, R).

Moreover, since

$$\phi : \mathbb{C}\, G \to (\mathbb{C}\, G)^*,$$

defined by

$$(s,t) \to \sum_{i,j,k,l} q^{jk-il} \langle p_{k,l}, (s,t) \rangle p_{i,j} = \sum_{i,j} q^{js-it} p_{i,j}$$

is a Hopf algebra isomorphism, this induces an action of $\mathbb{C}\, G$ on A by:

$$(s,t) \cdot x = q^{-t} x$$

$$(s,t) \cdot y = q^{s} y$$

This isomorphism induces a nontrivial quasitriangular structure on $\mathbb{C}\, G$ with:

$$r = \frac{1}{n^2} \sum_{a,b,s,t} q^{sb-ta} (s,t) \otimes (a,b)$$

Under these, A is then obviously quantum-commutative with respect to $(\mathbb{C}\, G, r)$.

Before considering example 3 let us explain what we mean by a "universal" construction of a quantum-commutative object.

Let V be a vector space, $T(V) = \sum V^{\otimes n}$ the tensor algebra, and I the ideal of $T(V)$ generated by:

$$\{ v \otimes w - w \otimes v \; : \; v, w \in V \}$$

Then the well-known symmetric algebra $S(V) = T(V)/I$ is a "universal" commutative algebra. If H is a Hopf algebra and $V \in {}_H\text{mod}$, then $T(V)$ is an H-module algebra. If H is cocommutative then I is H-stable and so $S(V)$ is an H-module, as a matter of fact it is a "universal" commutative H-module algebra. In general though, I need not be H-stable and so $S(V)$ is not an H-module. Following the philosophy that replacing the twist map by another interchange yields an analogous construction, an analogue of $S(V)$ for quasitriangular (H, R) was first defined by Gurevich [G, section2] who studied so called "q-Hecke symmetries", which in our notation is:

A quasitriangular (H, R) with $(q\,id - \psi)(id + \psi) = 0$, some $q \in k$.

This analogue was later emphasised by Manin [Ma] in the triangular case (i.e. $q = 1$). The analogous construction is done by replacing I by I_R, the ideal of $T(V)$ generated by

$$\{ v \otimes w - \sum R^{(2)} \cdot w \otimes R^{(1)} \cdot v \; : \; v, w \in V \}$$

and $S(V)$ by

$$S_R(V) = T(V)/I_R$$

The ideal I_R is H-stable for:

$$\begin{aligned}&h\cdot(v\otimes w-\textstyle\sum R^{(2)}\cdot w\otimes R^{(1)}\cdot v)=\\&\textstyle\sum h_1\cdot v\otimes h_2\cdot w-h_1R^{(2)}\cdot w\otimes h_2R^{(1)}\cdot v= \quad (\text{ by property (3)})\\&\textstyle\sum h_1\cdot v\otimes h_2\cdot w-R^{(2)}h_2\cdot w\otimes R^{(1)}h_1\cdot v\in I_R\end{aligned}$$

Hence $S_R(V)$ is an H-module algebra which is "universal" with respect to being quantum-commutative with respect to (H,R). There is one problem though, as conveyed to us by Majid. That is: unlike $S(V)$, there is no guarantee that $S_R(V)$ is not trivial. This cannot happen if (H,R) is triangular, or more generally, as shown in [G] if (H,R) is a q-Hecke symmetry.

The dual notion of $S_R(V)$ for coquasitriangular Hopf algebras (bialgebras) and their comodules is the same as above where one uses the previously mentioned interchange ψ in the definition of I. This point of view is taken in the next example [Ma]. Since it requires a considerable background we just hint at the details, for a thorough exposition see [LT, CW1, Maj1].

Example 3 The quantum superplane

Let $\mathbb{C}_q[\xi,\eta]$ be the algebra generated over $\mathbb{C}$ by ξ and η subject to $\xi^2=\eta^2=0$, $\xi\eta=-q\eta\xi$, $q^2\neq-1$. Let $B=-q\begin{pmatrix}q&0&0&0\\0&0&1&0\\0&1&q-q^{-1}&0\\0&0&0&q\end{pmatrix}$ index B by pairs (ij,kl) $i,j,k,l=1,2$.

Let V be a complex vector space with basis $\{f^1,f^2\}$ and define an operator B on $V\otimes V$ by:

$$B(f^1\otimes f^2)=\sum B_{(ij,kl)}f^k\otimes f^l$$

Then following [FRT] one forms a coquasitriangular bialgebra $(A(B),\langle\,|\,\rangle_B)$ [LT] for which V becomes an $A(B)$-comodule, and so does $T(V)$.

Let I_B be the ideal of $T(V)$ generated by:

$$\{f^i\otimes f^j-B(f^i\otimes f^j)\}$$

then $S_B(V)=T(V)/I_B$ is exactly $\mathbb{C}_q[\xi,\eta]$ upon identifying $\xi=\bar f^1$ and $\eta=\bar f^2$. Indeed, for example:

$$\begin{aligned}B(f^1\otimes f^1)&=\textstyle\sum B_{11,kl}f^k\otimes f^l\\&=-q^2(f^1\otimes f^1)\end{aligned}$$

(for $-q^2$ is the only non-zero entry in the 11-row).

Hence

$$\begin{aligned}\bar 0&=\overline{f^1\otimes f^1-Bf^1\otimes f^1}=\\&=\overline{f^1\otimes f^1+q^2f^1\otimes f^1}\end{aligned}$$

That is $(1+q^2)\overline{f^1\otimes f^1}=\bar 0$ in $S_B(V)$ so $\overline{f^1\otimes f^1}=\bar 0$, that is $\xi^2=0$.

We end this section by exhibiting a large class of quantum commutative actions which was introduced in [CW1].

Let H be a finite-dimensional cocommutative Hopf algebra and let H^{*cop} be the Hopf algebra resulting from H^* by using Δ^{cop} for comultiplication, and S^{-1} for the antipode. Define an action of H on H^{*cop} by:

$$h \cdot p = \sum \langle h_1, p_3 \rangle \langle Sh_2, p_1 \rangle p_2,$$

all $h \in H$, $p \in H^*$.

Form $D(H) = H^{*cop} \# H$, which is a particular case of the Drienfeld double [Maj1, R] and as such has a quasitriangular structure via:

$$R = \sum h_i \otimes h_i^*$$

(where $\{h_i\}$ and $\{h_i^*\}$ are dual bases of H and H^* respectively).

Now, any twisted product of H, $k_\sigma[H]$, where σ is an invertible cocycle [BCM], was shown in [CW1] to be a $D(H)$-module algebra via:

$$h \cdot (1 \#_\sigma l) = \sum \gamma(h_1)(1 \#_\sigma l)\gamma^{-1}(h_2)$$

(where $\gamma : H \to k_\sigma[H]$ via $\gamma(h) = 1 \#_\sigma h$ and γ^{-1} is its inverse in the convolution algebra $Hom(H, k_\sigma[H])$,

$$p \cdot (1 \#_\sigma l) = 1 \#_\sigma \langle Sp, h_1 \rangle h_2$$

all $h \in H$, $p \in H^{*cop}$.

Under these we proved:

THEOREM 1.1[CW1, 1.5] *Let H be a finite-dimensional cocommutative Hopf algebra, then $k_\sigma[H]$ is quantum-commutative with respect to $(D(H), R)$.*

An immediate yet open question is

Question 1 Can theorem 1.1 be generalized to non-cocommutative finite-dimensional Hopf algebras? We do not know at present even how to make such H into a $D(H)$-module algebra.

§2.PROPERTIES OF QUANTUM-COMMUTATIVE ACTIONS

We summarize some properties of quantum-commutative actions:

THEOREM 2.1 *Let (H, R) be a quasitriangular Hopf algebra and A quantum-commutative with respect to (H, R) then*

1. *[CW1, 2.2] Every H-stable left (right) ideal of A is two-sided, and its right and left annihilators in A coincide.*

2. *[CW1] $A \subset Z(A)$ (=the center of A).*

3. *[C1, 2.6] If (H, R) is triangular and A, B are both quantum-commutative with respect to (H, R) then so is $A \otimes B$ (where the product is defined using the interchange, i.e. $(a \otimes b)(c \otimes d) = \sum a(R^{(2)} \cdot c) \otimes (R^{(1)} \cdot b)d$*

4. *[CW2, 1.8] $Q^r(A)$ (defined below) is quantum-commutative with respect to (H, R) and it moreover equals $Q^s(A)$.*

By $Q^r(A)$ we mean the right H-quotient ring of A, that is: Let $\mathcal{F} = \{ I : I$ is an H-stable ideal of A with 0 left and right annihilators$\}$, and let

$$Q^r = \lim_{I \in \mathcal{F}} Hom(I_A, A_A)$$

with addition and mulitplication defined as usual. For $q \in Q^r$, $q : I \to A$ and for $h \in H$ define an action $h \cdot q : I \to A$ by $(h \cdot q)(y) = \sum h_1 \cdot [qSh_2 \cdot y]$ all $y \in I$. This extends the action on A. By Q^s, the H-symmetric ring of quotients we mean $\{ q \in Q^r : Iq \subset A$ and $qI \subset A$, some $I \in \mathcal{F}\}$ Considering the category ${}_{A\#H}mod$, the following shows that it resembles the category of modules over commutative rings. In particular, as known [Kos] for modules over graded-commutative algebras A (example 1), every left A-module is naturally a right A-module.

THEOREM 2.2 *Let (H, R) be a quasitriangular Hopf algebra acting on A and assume A is quantum commutative with respect to (H, R). Let M be a left $A\#H$-module and define a right action of A on M by*

$$m \leftarrow a = \sum (R^{(2)} \cdot a)(R^{(1)} \cdot m)$$

then:

1. *Under the above M is an $A-A$-bimodule.*

2. *If $M, N \in_{A\#H} mod$, then*
$$M \otimes_A N \in_{A\#H} mod$$
(where $(a\#h) \cdot (m \otimes n) = \sum ah_1 \cdot m \otimes h_2 \cdot n$ all $a \in A, h \in H$, $m \in M, n \in N$).

3. *For any* $M, N \in_{A\#H}$ *mod, we have*

$$K = Hom_A(M_A, N_A) \in_{A\#H} mod$$

(via: $((a\#h)\cdot f)(m) = a\sum h_1 \cdot f(Sh_2 \cdot m)$*, all* $a \in A,\ h \in H, f \in K, m \in M$*), and moreover,*

$$K^H = Hom_{A\#H}({}_AM, {}_AN).$$

4. *If* Φ *is the standard*

$$\Phi : (M \otimes_A N) \otimes_A L \to M \otimes_A (N \otimes_A L),$$

then $({}_{A\#H}mod, \otimes_A, \Phi, A)$ *is a monodial category.*

As a corollary of Theorem 2.1 we can deduce various connections between the ideal structure is of A and $A\#H$ for example.

COROLLARY 2.3 *If* (H, R) *is finite dimensional and* A *is quantum-commutation with respect to it then the following are equivalent.*

1. $A\#H$ *is simple and* A *has finite Goldie rank.*

2. $A\#H$ *is simple artinian.*

3. A *is* H*-simple (in particular* A^H *is a field) and*

$$\dim_{A^H} A = \dim_k H.$$

In the next section we isolate the fact: $A^H \subset Z(A)$ and show that much can be said under this condition.

§3. CENTRAL INVARIANTS

Let H be a Hopf algebra acting on A and assume that $A^H \subset Z(A)$. If non-zero H-stable ideals of A intersect A^H nontrivially then there exists a strong connection between A^H, A and $A\#H$. This condition is satisfied when H is finite-dimensional and A/A^H is H^*-Galois [BCM], that is, the map $[,] : A \otimes_{A^H} A \to A\#H$, defined by: $[a,b] = (a\#t)(b\#1)$ (where $t \neq 0$ a left integral of H) is surjective. In this case the connection is strongest.

THEOREM 3.1[CW2, 0.5] *Let H be a finite-dimensional Hopf algebra acting on A so that $A^H \subset Z(A)$ and A/A^H is H^*-Galois then*

1. *For every ideal I of $A\#H$:*

$$I = (I \cap A)\#H = (I \cap A^H)A\#H$$

2. *The map $(,) : A \otimes_{A\#H} A \to A^H$ defined by $(a,b) = t \cdot (ab)$ is surjective. Hence $A\#H$ and A^H are Morita equivalent.*

An example of the hypothesis of the theorem is a localization of the quantum plane.

Example 2-revisited

Let $A = \mathbb{C}_q[x,y]$, $q^n = 1$ which as seen before is $G = Z_n \times Z_n$-graded and quantum commutative with respect to $((\mathbb{C}\,G)^*, R)$. If we consider now the right $(\mathbb{C}\,G)^*$-quotient ring $Q^r(A)$, then it is quantum-commutative with respect to $(\mathbb{C}\,G)^*, R)$ as well (theorem 2.1) (in particular it is G-graded). Denote by x^{-1} and y^{-1} the right A-module homomorphism

$$x^{-1} : xA \to A, \text{ given by } xa \to a$$

$$y^{-1} : yA \to A, \text{ given by } ya \to a$$

all $a \in A$. Then x^{-1} and y^{-1} are homogeneous of degree $(n-1,0)$ and $(0,n-1)$ respectively. Let $B = A[x^{-1},y^{-1}] \subset Q^r$ then B is quantum commutative with respect to $((\mathbb{C}\,G)^*, R)$, and since $x^iy^j \in B_{i,j}$ and $x^{-i}y^{-j} \subset B_{n-i,n-j}$, for all $(i,j) \in G$, we have $B_{i,j}B_{n-i,n-j} = B_{0,0}$. That is B is strongly graded, which is equivalent to saying that B/B^H is H^*-Galois, where $H = (CG)^*$ ■

An extension A/A^H being H^*-Galois implies that A is a faithful left $A\#H$-module [CFM]. There are certain situations in which faithful actions can never occur though $A^H \subset Z(A)$.

THEOREM 3.2 *[CW2, 0.11] Let H be a finite-dimensional Hopf algebra acting on a field. If $S^2 \neq id$ then the action is not faithful.*

We ask here:

Question 2 Let H be a non-cocommutative Hopf algebra acting on a commutative algebra. Can such an action be faithful?

A consequence of 3.1 is

COROLLARY 3.3 *Under the hypothesis of 3.1:*

1. *A is H-prime* $\Leftrightarrow A^H$ *is prime* $\Leftrightarrow A\#H$ *is prime and then* $A\#H$ *satisfies a polynomial identity of degree* $\leq 2dim_k H$.

2. *A is H-simple* $\Leftrightarrow A^H$ *is simple* $\Leftrightarrow A\#H$ *is simple* $\Leftrightarrow A\#H$ *is simple artinian.*

If the Galois condition is relaxed, much can still be said; for example:

THEOREM 3.4 [CW2, 1.3] *Let H be a Hopf algebra acting on A so that* $A^H \subset Z(A)$. *If A is left H-artinian and H-semiprime, and every non-zero left H-stable ideal of A intersects* A^H *non-trivially then:*

1. A^H *is a finite direct sum of fields.*

2. *Every H-stable left ideal of A is of the form eA where* $e \in A^H$ *is an idempotent. (Hence all H-stable left ideals are two-sided).*

3. *A is a finite direct sum of H-simple algebras.*

Summarizing then, quantum-commutative actions are abundant and tangible, it is worthwhile to find out more about how close they are in their behaviour to commutative algebras acted upon by cocommutative Hopf algebras. For example we ask if an analogue of [FS] holds here.

Question 3 Let A be quantum-commutative with respect to a finite-dimensional (quasi-triangular Hopf algebra, is A/A^H an integral extension?

A much more elaborate plan [Ma] is to develope a non-commutative algebraic geometry for the category of quantum commutative algebras. A step in this direction was taken by [Ha], though for $H = D(kG)$ only, and by [B] and Berevkin [Ma], in the more general setting.

REFERENCES

[**B**] J.C. Baez, "R-commutative differential geometry and Poisson algebras", Adv. Math., to appear.

[**BCM**] R.J. Blattner, M. Cohen and S. Montgomery, "Crossed products and inner actions of Hopf algebras", AMS Trans. 298 (1986), 671-711.

[**C**] M. Cohen, "Smash products, inner actions and quotient rings", Pacific. J. Math. 125 (1986), 45-66.

[**CFM**] M. Cohen, D. Fischman and S. Montgomery, "Hopf Galois extensions, smash products and Morita equivalence", J. Alg. 113, (1990), 351-372.

[**CM**] M. Cohen and S. Montgomery, "Group graded rings, smash products and group actions" AMS. Trans., 282 (184), 237-258.

[**CW1**] M. Cohen and S. Westreich, "From supersymmetry to Quantum-commutativity", J. of Algebra, to appear.

[**CW2**] M. Cohen and S. Westreich, "Central invariants of H-module algebras", Comm. in Algebra, to appear.

[**D**] V.G. Drinfeld, "Quantum groups", Proc. of the Berkely, CA (1987), 789-820.

[**EK**] S.Eilenberg and G.M. Kelly, "Closed categories", Proc. Cong. Categorical Algebra (La Jolla, 1965), Springer, Berlin (1966), 421-562.

[**Fi**] D. Fischman, "Schur's double centralizer theorem: a Hopf algebra approach", J. of Algebra, to appear.

[**FRT**] L.D. Faddeev, N.Y. Reshetikin and L.A. Takjtajan, "Quantization of Lie groups and Lie algebras", Algebraic Anal. 1 (1988), 129-140.

[**FS**] W. Ferrer Santos, "Invariants of finite-dimensional Hopf algebra", preprint.

[**G**] D.I. Gurevich, "Algebraic aspects of the quantum-Yang-Baxter equation", Leningrad Math. J. Vol. 2, (1991), 801-828.

[**Ha**] S. Haran, "An invitation to dyslectic geometry", J. of Algebra, to appear.

[**Hay**] T. Hayashi, "Quantum groups and quantum determinants", J. of Algebra 152 (1992), 146-165.

[**Her**] I.N. Herstein, "Rings admitting ceratin automorphism, Istanbul Univ. Fen. Fac. Mecm. Ser A. 26 (1961), 45-49.

[**JS**] A. Joyal and R. Street, "Braided tensor categories", Adv in Math., to appear.

[**Kos**] B. Kostant, "Graded manifolds, graded Lie theory and prequantization", Lecture notes inMath. 570 (1976), 177-306.

[**LT**] R.G. Larson and J. Towber, "Braided Hopf algebras, a dual concept to quasitriangularity", to appear.

[**Mac**] S. MacLane, "Natural associativity and commtuativity", Rice Unive. Stud. Vol 49, Rice Unive. Houston, TX (1963), 28-46.

[**Ma**] Y. Manin, "Quantum groups and non-commutative geometry", U. of Montreal lectures (1988).

[**Maj1**] S. Majid "Quasitriangular Hopf algebras and Yang-Baxter equations", Int. J. Modern Physics, A, 5 (1) (1990), 1-91.

[**Maj2**] S. Majid, "Examples of braided groups and braided matrices", J. Math. Phys. 32 (1991), 3246-3253.

[**O**] J. Osterburg, "Central fixed rings", J. London Math. Soc. (2) (1981), 246-248.

[**R**] D.E. Radford, "Minimal quasitriangular Hopf algebras", J. of Algebra, to appear.

[**S**] M. Sweedler, "Hopf algebras", Benjamin, N.Y. (1969).

Generalized Smash Products and Morita Contexts for Arbitrary Hopf Algebras

YUKIO DOI Fukui University, Fukui, Japan

Introduction From the viewpoint of Morita theory we study the relationship between an algebra B and its coinvariants subalgebra $C = B^{coA} = \{b \in B \mid \rho(b) = b \otimes 1_A\}$, under the right coaction ρ of a fixed bialgebra A over a commutative ring k. We use in an essential way the generalized smash product $\#(A,B) = (\mathrm{Hom}(A,B), \cdot)$ and its right ideal

$$Q := \{\lambda \in \#(A,B) \mid \Sigma \lambda(a_2)_0 \otimes \lambda(a_2)_1 a_1 = \lambda(a) \otimes 1_A,\ \forall a \in A\}.$$

We show that $\#(A,B)$ and C are always connected via a Morita context, using B and Q as the connecting bimodules. Along the way we reconsider the concept of a Hopf Galois extension and the role of "so called" total integrals, and give more transparent proofs of several well known resuls.

More generally, for any left A-module coalgebra D we can also construct a smash product $\#(D,B)$. Whenever we fix a grouplike element x in D, $\#(D,B)$ and $C' := \{b \in B \mid \Sigma b_0 \otimes (b_1 \rightharpoonup x) = b \otimes x\}$ are also connected via a Morita context, using B and

$$Q' := \{\lambda \in \#(D,B) \mid \Sigma \lambda(d_2)_0 \otimes (\lambda(d_2)_1 \rightharpoonup d_1) = \lambda(d) \otimes x,\ \forall d \in D\}.$$

Some results still hold in this more general context.

Our approach to the subject was inspired primarily by the works of Chase-Sweedler [CS] and Cohen-Fischman-Montgomery [CFM]. But our context is very close to [CS], rather than [CFM].

1 Preliminaries

Throughout, A denotes a bialgebra over a commutative ring k, and B denotes a right A-comodule algebra; thus B is a (k-)algebra which is a right A-comodule such that its comodule structure map $\rho_B: B \to B\otimes A,\ b \mapsto \Sigma b_0 \otimes b_1$ is an algebra map. C denotes the A-coinvariants subalgebra of B;

$$C = B^{coA} = \{b\in B \mid \rho_B(b) = b\otimes 1_A\}.$$

1.1 We form the (generalized) <u>smash product</u> $\#(A,B)$ ([D2]) as follows; $\#(A,B) = \mathrm{Hom}(A,B)$ as a k-module, and the multiplication is defined by

$$(f\cdot g)(a) = \Sigma f(g(a_2)_1 a_1) g(a_2)_0,\quad f,g\in \mathrm{Hom}(A,B),\ a\in A.$$

Then $\#(A,B)$ becomes an (associative) k-algebra with identity $u_B\circ\varepsilon_A$, where $u_B: k \to B$ is the unit map of B and $\varepsilon_A: A \to k$ is the counit map of A. We always view B as a subalgebra of $\#(A,B)$ via

$$B \subsetneq \#(A,B) : b \mapsto (a \mapsto \varepsilon(a)b).$$

We then get the following equations in $\#(A,B)$: for any $f\in \#(A,B)$,

(1.1a) $(b\cdot f)(a) = bf(a)$ (so, let us write bf for $b\cdot f$).

(1.1b) $(f\cdot b)(a) = \Sigma f(b_1 a) b_0$.

Note that if A is finitely generated projective over k, then our smash product $\#(A,B)$ is isomorphic to $B\# A^*$, the usual smash product associated from the left A^*-module algebra structure on B.

1.2 <u>Bimodules B and Q</u>. The left regular B-module B can be extended to a left $\#(A,B)$-module by the following rule

(1.2a) $f \rightharpoonup b := \Sigma f(b_1) b_0$ $(= (f\cdot b)(1_A)$ by 1.1b).

Further if $c\in C$ then $f\rightharpoonup (bc) = \Sigma f(b_1)b_0 c = (f\rightharpoonup b)c$. Thus B is a $(\#(A,B), C)$-bimodule. We set

$$Q := \{\lambda\in \#(A,B) \mid \Sigma \lambda(a_2)_0 \otimes \lambda(a_2)_1 a_1 = \lambda(a)\otimes 1_A,\ \forall a\in A\}.$$

It plays an important role in this paper. We have

(1.2b) $f\cdot\lambda = f(1)\lambda$, for all $\lambda\in Q$ and $f\in \#(A,B)$,

since $(f\cdot\lambda)(a) = \Sigma f(\underline{\lambda(a_2)_1 a_1})\underline{\lambda(a_2)_0} = f(\underline{1})\underline{\lambda(a)}$ (by $\lambda\in Q$). Also, the following calculation shows that Q is a right ideal of $\#(A,B)$: for $\lambda\in Q$ and $g\in \#(A,B)$, $\Sigma(\lambda\cdot g)(a_2)_0\otimes(\lambda\cdot g)(a_2)_1 a_1$

$$= \Sigma \underline{\lambda(g(a_3)_2 a_2)_0} g(a_3)_0 \otimes \underline{\lambda(g(a_3)_2 a_2)_1 g(a_3)_1 a_1}$$

$$= \Sigma \underline{\lambda(g(a_2)_1 a_1)} g(a_2)_0 \otimes \underline{1} = (\lambda\cdot g)(a)\otimes 1.$$

Furthermore, if $c\in C$ then $c\lambda\in Q$. Hence Q is a $(C, \#(A,B))$-bimodule.

1.3 EXAMPLE Let A = kG a group algebra. In this case, as is well-known, B is a G-graded algebra $B = \bigoplus_{\sigma\in G} B_\sigma$ with $B_1 = C$. Identifying $\#(A,B)$ with Map(G,B), we have for all f, $g\in$ Map(G,B), $\tau\in G$,

$$(f\cdot g)(\tau) = \Sigma_{\sigma\in G} f(\sigma\tau) g(\tau)_\sigma \text{ and } Q = \{\lambda\in \mathrm{Map}(G,B) \mid \lambda(\tau)\in B_{\tau^{-1}},\ \forall\tau\in G\}$$

where $g(\tau)_\sigma$ denotes the σ-component of $g(\tau)$.

1.4 <u>Bimodule maps F and G</u>. We define a $\#(A,B)$-bimodule map

(1.4a) $F: B\otimes_C Q \to \#(A,B)$, $F(b\otimes\lambda) = b\lambda$ $(= b\cdot\lambda)$.

We also define a C-bimodule map

(1.4b) $G: Q\otimes_{\#(A,B)} B \to C$, $G(\lambda\otimes b) = \lambda\rightharpoonup b$ $(= \Sigma\lambda(b_1)b_0)$.

Note that Im $G \subset C$, since $\rho(\lambda\rightharpoonup b) = \rho(\Sigma\lambda(b_1)b_0) = \Sigma\underline{\lambda(b_2)_0}b_0\otimes\underline{\lambda(b_2)_1 b_1}$ $= \Sigma\underline{\lambda(b_1)}b_0\otimes\underline{1} = (\lambda\rightharpoonup b)\otimes 1$.

We then have the following associativity relations:

(1.4c) $(b\lambda)\rightharpoonup b' = b(\lambda\rightharpoonup b')$ for $b, b'\in B$, $\lambda\in Q$,

(1.4d) $(\lambda\rightharpoonup b)\lambda' = \lambda\cdot(b\lambda')$ for $b\in B$, $\lambda,\lambda'\in Q$.

(The second relation follows from (1.2a) and (1.2b).) Thus we have shown:

THEOREM $(\#(A,B), C, {}_{\#(A,B)}B_C, {}_CQ_{\#(A,B)}, F, G)$ forms a Morita context.

1.5 More general contexts. Let D be any left A-module coalgebra; thus D is a k-coalgebra which is a left A-module such that its module action $A\otimes D \to D$, $a\otimes d \mapsto a\rightharpoonup d$, is a coalgebra map. We may then form the smash product $\#(D,B) = \mathrm{Hom}(D,B)$ ([K]) with multiplication

$$(f\cdot g)(d) = \Sigma f(g(d_2)_1\rightharpoonup d_1)g(d_2)_0,\ f,g\in\#(D,B),\ d\in D.$$

Consider $B\subset\#(D,B)$ via $b\mapsto (d\mapsto \varepsilon(d)b)$. Then for $b\in B$, $f\in\#(D,B)$, $d\in D$,

$$(b\cdot f)(d) = bf(d) \text{ and } (f\cdot b)(d) = \Sigma f(b_1\rightharpoonup d)b_0.$$

If we fix a grouplike element x in D, i.e. $\Delta_D(x) = x\otimes x$ and $\varepsilon(x) = 1$, then B is a left $\#(D,B)$-module via

(1.5a) $f\rightharpoonup b := \Sigma f(b_1\rightharpoonup x)b_0$ $(= (f\cdot b)(x))$, $f\in\#(D,B)$, $b\in B$.

(A fundamental example of such a D is a quotient coalgebra and a quotient left A-module of A. Here we take $x = p(1_A)$, where $p: A\to D$ is the canonical projection. In this case $\#(D,B)$ can be embedded in $\#(A,B)$ via $f\mapsto f\circ p$.) If we set

(1.5b) $C' := \{b\in B \mid \Sigma b_0\otimes(b_1\rightharpoonup x) = b\otimes x\}$,

then it is a subalgebra of B with $C\subset C'$, and B is a $(\#(D,B),C')$-bimodule. Define

(1.5c) $Q' := \{\lambda\in\#(D,B) \mid \Sigma\lambda(d_2)_0\otimes(\lambda(d_2)_1\rightharpoonup d_1) = \lambda(d)\otimes x,\ d\in D\}$,

which is a right ideal of $\#(D,B)$. Moreover Q' is a $(C',\#(D,B))$-bimodule. Thus we have:

THEOREM $(\#(D,B), C', B, Q', F', G')$ is a Morita context, where

(1.5d) $F': B\otimes_{C'} Q' \to \#(D,B)$, $b\otimes\lambda\mapsto b\lambda$,

(1.5e) $G': Q'\otimes_{\#(D,B)} B\to C'$, $\lambda\otimes b\mapsto\lambda\rightharpoonup b$ $(= \Sigma\lambda(b_1\rightharpoonup x)b_0)$.

1.6 x-invariants M^x. For any left $\#(D,B)$-module M we define the subspace of x-invariants by

(1.6a) $M^x := \{m\in M \mid f\cdot m = f(x)m,\ \forall f\in\#(D,B)\} \cong \mathrm{Hom}_{\#(D,B)}(B,M)$,

In particular $B^x = \{b\in B \mid f\rightharpoonup b = f(x)b,\ \forall f\in\#(D,B)\}$ is a subalgebra of B, and any M is a left B^x-module. Obviously one sees that

(1.6b) $C'\subset B^x$ and $Q'\subset\#(D,B)^x$.

1.7 Hopf modules. We denote by ${}_B\mathbf{M}^D$ the category of (D,B)-Hopf modules [D3]: Whose objects are both left B-modules and right D-comodules M with

(1.7a) $\qquad \rho_M(bm) = \Sigma\, b_0 m_0 \otimes (b_1 \rightharpoonup m_1), \quad \forall b \in B,\ m \in M,$

where ρ_M: $M \to M \otimes D$, $m \mapsto \Sigma\, m_0 \otimes m_1$ is the D-comodule structure map on M. Whose morphisms are D-comodule maps which are B-linear.

For any $M \in {}_B\mathbf{M}^D$, the subspace of (x-)coinvariants is the set

(1.7b) $\qquad M^{coD} = \{m \in M \mid \rho_M(m) = m \otimes x\},$

which is a left C'-module. Note that $B^{coD} = C'$ if we consider B as an object of ${}_B\mathbf{M}^D$ via $b \mapsto \Sigma\, b_0 \otimes (b_1 \rightharpoonup x)$.

If $M \in {}_B\mathbf{M}^D$ then M can be viewed as a left # (D,B)-module via

(1.7c) $\qquad f \cdot m = \Sigma\, f(m_1) m_0, \quad$ for all $f \in \#(D,B)$, $m \in M$.

Then we have

(1.7d) $\qquad M^{coD} \subset M^x \quad$ for all $M \in {}_B\mathbf{M}^D$.

1.8 The adjunctions Φ_V and Ψ_M. For any left C'-module V, the induced left # (D,B)-module $B \otimes_{C'} V$ is an object of ${}_B\mathbf{M}^D$ via

$$b'(b \otimes v) = b'b \otimes v \text{ and } \rho(b \otimes v) = \Sigma\, b_0 \otimes v \otimes (b_1 \rightharpoonup x).$$

This defines a functor $B \otimes_{C'} (-)$: ${}_{C'}\mathbf{M} \to {}_B\mathbf{M}^D$, $V \mapsto B \otimes_{C'} V$, which is left adjoint to the functor $(\ )^{coD}$: ${}_B\mathbf{M}^D \to {}_{C'}\mathbf{M}$, $M \mapsto M^{coD}$. The adjunctions are as follows:

(1.8a) $\qquad \Phi_V: V \to (B \otimes_{C'} V)^{coD}, \quad v \mapsto 1 \otimes v,$

(1.8b) $\qquad \Psi_M: B \otimes_{C'} M^{coD} \to M, \quad b \otimes n \mapsto bn.$

If Ψ_M is an isomorphism for all $M \in {}_B\mathbf{M}^D$, we say that ${}_B\mathbf{M}^D$ satisfies the weak structure theorem. If, in addition, Φ_V is an isomorphism for all $V \in {}_{C'}\mathbf{M}$, we say that ${}_B\mathbf{M}^D$ satisfies the strong structure theorem. This means that $B \otimes_{C'} (\ )$ is an equivalence of categories.

If W is a left B-module, then $W \otimes D$ is an object of ${}_B\mathbf{M}^D$ via

$$b(w \otimes d) = \Sigma\, b_0 w \otimes (b_1 \rightharpoonup d) \text{ and } \rho(w \otimes d) = w \otimes \Delta(d).$$

Then $(W \otimes D)^{coD} \cong W$ (via $\Sigma\, w_i \otimes d_i \mapsto \Sigma\, w_i \varepsilon(d_i)$ and $w \otimes x \leftarrow\!\shortmid w$), and the adjunction Ψ_M for $M := W \otimes D$ can be identified with the map

(1.8c) $\qquad \gamma_W: B \otimes_{C'} W \to W \otimes D, \quad b \otimes w \mapsto \Sigma\, b_0 w \otimes (b_1 \rightharpoonup x).$

In particular, if ${}_B\mathbf{M}^D$ satisfies the weak structure theorem, then the map

(1.8d) $\qquad \gamma'(=\gamma_B): B \otimes_{C'} B \to B \otimes D, \quad b \otimes b' \mapsto \Sigma\, b_0 b' \otimes (b_1 \rightharpoonup x)$

is bijective.

Recall that the extension B/C is called A-Galois if the map

(1.8e) $\qquad \beta: B \otimes_C B \to B \otimes A, \quad b' \otimes b \mapsto \Sigma\, b' b_0 \otimes b_1$

is bijective. We say that B/C is A-anti-Galois if the map

(1.8f) $\qquad \gamma: B \otimes_C B \to B \otimes A, \quad b \otimes b' \mapsto \Sigma\, b_0 b' \otimes b_1,$

is bijective. This means that B^{op}/C^{op} is A^{op}-Galois. If ${}_B\mathbf{M}^A$ (take $D = A$ and $x = 1_A$) satisfies the weak structure theorem then B/C is A-anti-Galois. Thus we will use the map γ mainly. Remark that when A has a bijective antipode, β is surjective (resp. bijective) if and only if so is γ (cf. [D2, (2.3)]).

2 The surjectivity of G'

In this section we consider when the map G' (1.5e) is surjective.

2.1 THEOREM In the general Morita context (1.5), the following (a)-(c) are equivalent:

(a) $G'\colon Q'\otimes_{\#(D,B)} B \to C'$ is surjective (bijective).

(b) There exists $\chi \in Q'$ such that $\chi(x) = 1_B$.

(c) For any left $\#(D,B)$-module M, $\xi_M\colon Q'\otimes_{\#(D,B)} M \to M^x$, $\lambda\otimes m \mapsto \lambda\cdot m$, is a C'-module isomorphism.

If these conditions hold, then we have:

(d) $M^x = M^{coD}$ for all $M \in {}_B\mathbf{M}^D$.

(e) χ as in (b) is an idempotent in $\#(D,B)$ and $\chi\cdot\#(D,B)\cdot\chi = C'\chi \cong C'$ as algebras.

(f) For any left C'-module V, Φ_V (1.8a) is an isomorphism.

(g) C' is a right C'-direct summand of B.

(h) B and Q' are generators as C'-modules.

(i) B and Q' are finitely generated projective as $\#(D,B)$-modules.

(j) The map F induces bimodule isomorphisms

$$B \cong \mathrm{Hom}_{-\#(D,B)}(Q', \#(D,B)),\ b \mapsto (\lambda \mapsto b\lambda),$$

$$Q' \cong \mathrm{Hom}_{\#(D,B)-}(B, \#(D,B)),\ \lambda \mapsto (b \mapsto b\lambda).$$

(k) The algebra maps induced by the bimodule structures,

$$C' \to \mathrm{End}_{-\#(D,B)}(Q'),\ c \mapsto (\lambda \mapsto c\lambda),$$

$$C' \to (\mathrm{End}_{\#(D,B)-}(B))^{op},\ c \mapsto (b \mapsto bc),$$

are isomorphisms.

Proof. (a)⇒(b): Assume G' is surjective, so there exists $\Sigma\lambda_i\otimes b_i \in Q'\otimes_{\#(D,B)} B$ such that $\Sigma\lambda_i \rightharpoonup b_i = 1_B$. Set $\chi = \Sigma\lambda_i\cdot b_i$. Then $\chi\in Q'$ since Q' is a right ideal of $\#(D,B)$. Moreover $\chi(x) = \Sigma(\lambda_i\cdot b_i)(x) = \Sigma\lambda_i \rightharpoonup b_i = 1_B$.

(b)⇒(c): Let $\chi \in Q'$ with $\chi(x) = 1$. For any left $\#(D,B)$-module M, define $\eta_M\colon M^x \to Q'\otimes_{\#(D,B)} M$ by $\eta_M(n) = \chi\otimes n$ ($n\in M^x$). Then $\xi_M(\eta_M(n)) = \chi\cdot n = \chi(x)n = n$ and $\eta_M(\xi_M(\lambda\otimes m)) = \chi\otimes\lambda\cdot m = \chi\cdot\lambda\otimes m = \chi(x)\lambda\otimes m$ (by (1.6b)) $= \lambda\otimes m$. Hence ξ_M is bijective.

(c)⇒(a): Take M = B. Then G' $(=\xi_B)$ is bijective with $C' = B^x$.

(d): It is enough to show that $M^x \subset M^{coD}$ by (1.7d). Let $m\in M^x$. Then $m = \chi(x)m = \chi\cdot m = \Sigma\chi(m_1)m_0$, and so $\rho_M(m) = \Sigma\chi(m_2)_0 m_0 \otimes (\chi(m_2)_1 \rightharpoonup m_1)$ (by (1.7a)) $= \Sigma\chi(m_1)m_0\otimes x$ (by $\chi\in Q'$) $= m\otimes x$. Hence $m \in M^{coD}$.

(e): Clearly χ is an idempotent. Next, for $f \in \#(D,B)$, $\chi\cdot f\cdot\chi = (\chi\cdot f)(x)\chi$ (by 1.2b)) $= (\chi\rightharpoonup f(x))\chi \in C'\chi$. In particular, for any $c\in C'$, $\chi\cdot c\cdot\chi = (\chi\rightharpoonup c)\chi = c\chi$. This shows $\chi\cdot\#(D,B)\cdot\chi = C'\chi$. It is easily verified that $C' \to C'\chi$, $c \mapsto c\chi$ is an algebra isomorphism.

(f): Φ_V is the composition of the following canonical isomorphisms:

$$V \cong C'\otimes_{C'} V \cong Q'\otimes_{\#(D,B)} B\otimes_{C'} V \quad \text{(since G' is an isomorphism)}$$

$$\cong (B\otimes_{C'} V)^x \quad (\text{by } \xi_M \text{ for } M = B\otimes_{C'} V) = (B\otimes_{C'} V)^{coD} \quad (\text{by (d)}).$$

(g): The map tr: $B \to C'$, $b \mapsto \chi \rightharpoonup b$ $(= \sum \chi(b_1)b_0)$, is right C'-linear and $tr(c) = c$ for all $c \in C'$. Hence C' is a right C'-direct summand of B.

(h)-(k) follow from the standard argument of Morita theory (cf. [B]). □

2.2 PROPOSITION Assume that there exists a left D-comodule map $q: D \to A$ (i.e. $\sum (q(d)_1 \rightharpoonup x) \otimes q(d)_2 = \sum d_1 \otimes q(d_2)$, $\forall d \in D$). Then for any $\lambda \in Q$, $\lambda \circ q$ is an element in Q'. In particular, if $q(x) = 1_A$ and $G: Q \otimes_{\#(A,B)} B \to C$ is surjective, then $G': Q' \otimes_{\#(D,B)} B \to C'$ is surjective.

Proof. For any $d \in D$ we have $\sum (\lambda \circ q)(d_2)_0 \otimes ((\lambda \circ q)(d_2)_1 \rightharpoonup d_1)$
$= \sum \lambda(q(d)_2)_0 \otimes (\lambda(q(d)_2)_1 q(d)_1 \rightharpoonup x)$ (by the assumption on q)
$= \sum (\lambda \circ q)(d) \otimes x$ (by $\lambda \in Q$), and so $\lambda \circ q \in Q'$. □

Recall that the convolution inverse of id_A is called an antipode for A which will be denoted by S. A bialgebra with antipode is called a Hopf algebra. On the other hand the twist-convolution inverse of id_A is called a pode for A which will be denoted by $\overline{S}$. Thus,

$$\sum a_2 \overline{S}(a_1) = \varepsilon(a) 1_A = \sum \overline{S}(a_2) a_1, \text{ for all } a \in A.$$

A bialgebra with pode is called an anti-Hopf algebra. S and $\overline{S}$ are algebra and coalgebra antimorphisms. A bialgebra has both an antipode and a pode if and only if it is a Hopf algebra with bijective antipode (in this case, $\overline{S}$ is the composite-inverse of S) (see [DT1, Prop.7]).

The next result gives a relation between elements in Q' (or Q) and integrals from D (or A) into B.

2.3 PROPOSITION

(a) If A has an antipode S (i.e. A is Hopf algebra), then

$$Q' = \{\lambda \in \#(D,B) \mid \sum \lambda(d_2) \otimes d_1 = \sum \lambda(d)_0 \otimes (S(\lambda(d)_1) \rightharpoonup x),\ \forall d \in D\}.$$

(i.e. $\lambda \in Q'$ means that λ is a left D-comodule map from D to B, where B is a left D-comodule via $b \mapsto \sum (S(b_1) \rightharpoonup x) \otimes b_0$.) In particular,

$$Q = \{\lambda \in \#(A,B) \mid \sum \lambda(a_2) \otimes a_1 = \sum \lambda(a)_0 \otimes S(\lambda(a)_1),\ \forall a \in A\}.$$

(b) Assume that A is a Hopf algebra. If there is $\chi \in Q$ with $\chi(1_A) = 1_B$, then $\phi := \chi \circ S: A \to B$ is a total integral (i.e. a right A-comodule map with $\phi(1_A) = 1_B$), and all $M \in {}_B\mathbf{M}^A$ and $N \in \mathbf{M}_B^{\hat{A}}$ are relative injective as A-comodules.

(c) Assume that A is an anti-Hopf algebra. If $\phi: A \to B$ is a total integral then $\chi := \phi \circ \overline{S}$ is in Q with $\chi(1_A) = 1_B$. In particular, when A is a Hopf algebra with bijective antipode, there is a total integral $\phi: A \to B$ if and only if $G: Q \otimes_{\#(A,B)} B \to C$ is surjective.

Proof. (a): If $\lambda \in Q'$ then $\sum \lambda(d_2)_0 \otimes \lambda(d_2)_1 \otimes (\lambda(d_2)_2 \rightharpoonup d_1)$
$= \sum \lambda(d)_0 \otimes \lambda(d)_1 \otimes x$, and so, $\sum \lambda(d_2)_0 \otimes (S(\lambda(d_2)_1)\lambda(d_2)_2 \rightharpoonup d_1)$
$= \sum \lambda(d)_0 \otimes (S(\lambda(d)_1) \rightharpoonup x)$. Hence $\sum \lambda(d_2) \otimes d_1 = \sum \lambda(d)_0 \otimes (S(\lambda(d)_1) \rightharpoonup x)$.

Conversely, if $\sum \lambda(d_2)\otimes d_1 = \sum \lambda(d)_0 \otimes (S(\lambda(d)_1) \rightharpoonup x)$, then

$$\sum \lambda(d_2)_0\otimes\lambda(d_2)_1\otimes d_1 = \sum \lambda(d)_0\otimes\lambda(d)_1\otimes(S(\lambda(d)_2)\rightharpoonup x),$$

and so $\sum\lambda(d_2)_0\otimes(\lambda(d_2)_1\rightharpoonup d_1) = \sum\lambda(d)_0\otimes(\lambda(d)_1 S(\lambda(d)_2)\rightharpoonup x)$.
Hence, $\sum\lambda(d_2)_0\otimes(\lambda(d_2)_1\rightharpoonup d_1) = \lambda(d)\otimes x$.

(b): We compute $\sum\phi(a)_0\otimes\phi(a)_1 = \sum \underline{\chi(S(a_1)_0}\otimes\underline{\chi(S(a_1))_1 S(a_2)}a_3$
$= \sum\chi(S(a_1))\otimes a_2$ (by $\lambda\in Q$) $= \sum\phi(a_1)\otimes a_2$, so ϕ is an integral.
Clearly $\phi(1_A) = 1_B$. For $M\in {}_B\mathbf{M}^A$ and $N\in\mathbf{M}_B^{\hat{A}}$ we define

$$q_M\colon M\otimes A\to M,\ m\otimes a\mapsto \sum\chi(m_1 S(a))m_0,$$

$$q_N\colon N\otimes A\to N,\ n\otimes a\mapsto\sum n_0\chi(S(S(n_1)a)).$$

Then q_M (resp. q_N) is an A-comodule section of ρ_M (resp. ρ_N). This shows that M and N are relative injective as A-comodules ([D2, (1.4)]).

(c): We have $\sum\chi(a_2)_0\otimes\chi(a_2)_1 a_1 = \sum\phi(\bar{S}(a_2))_0\otimes\phi(\bar{S}(a_2))_1 a_1$
$= \sum\phi(\bar{S}(a_3))\otimes\bar{S}(a_2)a_1 = \chi(a)\otimes 1$. Thus $\lambda\in Q$. The last statement follows from 2.1. □

In the next 2.4 and 2.5, A need not have an antipode.

2.4 LEMMA (1) Let $\chi\in\#(D,B)$ be invertible with respect to twist-convolution X, i.e. there is a unique $\chi^-\in\#(D,B)$ such that $\sum\chi(d_2)\chi^-(d_1) = \varepsilon(d)1_B = \sum\chi^-(d_2)\chi(d_1)$ for all $d\in D$. Then the following (a)-(c) are equivalent:

(a) $\chi\in Q'$.

(b) $\sum\chi(d_3)_0\chi^-(d_1)\otimes(\chi(d_3)_1\rightharpoonup d_2) = \varepsilon(d)1_B\otimes x,\ \forall d\in D$.

(c) $\chi^-\colon D\to B$ is a right D-comodule map, i.e.

$$\sum\chi^-(d)_0\otimes(\chi^-(d)_1\rightharpoonup x) = \sum\chi^-(d_1)\otimes d_2,\ \forall d\in D.$$

(2) If there is an X-invertible $\chi\in Q'$ then G' is surjective.

Proof. (1) (a)$\Rightarrow$(b): Let $\chi\in Q'$. Then $\sum\underline{\chi(d_3)_0}\chi^-(d_1)\otimes(\underline{\chi(d_3)_1\rightharpoonup d_2})$
$= \sum\underline{\chi(d_2)}\chi^-(d_1)\otimes\underline{x} = \varepsilon(d)1_B\otimes x$ for $d\in D$..

(b)$\Rightarrow$(c): $\sum\chi^-(d)_0\otimes(\chi^-(d)_1\rightharpoonup x) = \sum\chi^-(d_2)_0\underline{\varepsilon(d_1)1_B}\otimes(\chi^-(d_2)_1\rightharpoonup\underline{x})$
$= \sum\chi^-(d_4)_0\underline{\chi(d_3)_0\chi^-(d_1)}\otimes(\chi^-(d_4)_1\underline{\chi(d_3)_1\rightharpoonup d_2})$ (by (b))
$= \sum(\chi^-(d_4)\chi(d_3))_0\chi^-(d_1)\otimes((\chi^-(d_4)\chi(d_3))_1\rightharpoonup d_2)$
$= \sum\varepsilon(d_3)\chi^-(d_1)\otimes d_2 = \sum\chi^-(d_1)\otimes d_2$.

(c)$\Rightarrow$(a): $\sum\chi(d_2)_0\otimes(\chi(d_2)_1\rightharpoonup d_1) = \sum\chi(d_4)_0\underline{\chi^-(d_2)}\chi(d_1)\otimes(\chi(d_4)_1\rightharpoonup\underline{d_3})$
$= \sum\chi(d_3)_0\underline{\chi^-(d_2)_0}\chi(d_1)\otimes(\chi(d_3)_1\underline{\chi^-(d_2)_1\rightharpoonup} x)$ (by (c))
$= \sum(\chi(d_3)\chi^-(d_2))_0\chi(d_1)\otimes((\chi(d_3)\chi^-(d_2))_1\rightharpoonup x)$
$= \sum\varepsilon(d_2)\chi(d_1)\otimes x = \chi(d)\otimes x$.

(2): Let $\chi' := \chi^-(x)\chi$. Since $\chi^-(x)\in C'$ by the above (c), χ' is an element in Q' with $\chi'(x) = 1$. Hence, by 2.1(b), G' is surjective. □

B is said to be anti-cleft for D ([DT1]) if there exists an X-invertible right D-comodule map from D to B, or equivalently, there exists an X-invertible element χ in Q'(by 2.4). The next theorem is a "left" version of [MD, Theorem 1.5], which says anti-cleftness implies the "very" strong structure theorem:

2.5 THEOREM The following are equivalent:

(a) B is anti-cleft for D.

(b) ${}_B\mathbf{M}^D$ satisfies the strong structure theorem, and B has the right normal basis property in the sense that: There is a right C'-module and a right D-comodule isomorphism from $C'\otimes D$ to B, where $C'\otimes D$ is a right C'-module by $(c\otimes d)c' = cc'\otimes d$ and a right D-comodule by $c\otimes d \mapsto c\otimes\Delta_D(d)$.

(c) The map γ': $B\otimes_{C'} B \to B\otimes D$, $b\otimes b' \mapsto \Sigma\, b_0 b'\otimes(b_1 \rightharpoonup x)$, is bijective, and B has the right normal basis property.

(d) The canonical algebra map

$$\pi:\ \#(D,B) \to \mathrm{End}_{-C'}(B),\ \pi(f)(b) = f\rightharpoonup b\ (=\Sigma f(b_1\rightharpoonup x)b_0),$$

is bijective, and B has the right normal basis property.

Proof. (a)⇒(b): Choose an X-invertible $\chi\in Q'$. Let $M\in {}_B\mathbf{M}^D$. Since $\chi\bullet m = \Sigma\,\chi(m_1)m_0\in M^{coD}$ and $\chi\bullet(bn) = (\chi\rightharpoonup b)n$ for $m\in M$, $b\in B$ and $n\in M^{coD}$, one sees that the map $M \to B\otimes_{C'} M^{coD}$, $m\mapsto \Sigma\,\chi^-(m_1)\otimes\chi\bullet m_0$, is the inverse of Ψ_M: $B\otimes_{C'} M^{coD} \to M$, $b\otimes n\mapsto bn$. That ${}_B\mathbf{M}^D$ satisfies the strong structure theorem is then an immediate consequence of Theorem 2.1(f) and Lemma 2.4(2).

Observe that the map

$$\sigma:\ M \to M^{coD}\otimes D,\ m\mapsto \Sigma\,(\chi\bullet m_0)\otimes m_1$$

is a D-comodule isomorphism (the inverse is given by $\sigma^{-1}(n\otimes d) = \chi^-(d)n$). In fact, for any $m\in M$, $\Sigma\,\chi^-(m_1)\chi\bullet m_0 = \Sigma\,\chi^-(m_2)\chi(m_1)m_0 = \varepsilon(m_1)m_0 = m$, and for any $n\in M^{coD}$, $d\in D$,

$\sigma(\chi^-(d)n) = \Sigma\,\chi\bullet(\chi^-(d)_0 n)\otimes\chi^-(d)_1\rightharpoonup x$ (since $n\in M^{coD}$)

$= \Sigma\,\chi\bullet(\chi^-(d_1)n)\otimes d_2$ (by (2.4c)) $= \Sigma\,\chi(\chi^-(d_1)_1\rightharpoonup x)\chi^-(d)_0 n\otimes d_2$

$= \Sigma\,\chi(d_2)\chi^-(d_1)n\otimes d_3$ (by (2.4c)) $= \Sigma\,\varepsilon(d_1)n\otimes d_2 = n\otimes d$.

In particular when $M = B$ we get a D-comodule isomorphism:

$$\sigma_1:\ B\cong C'\otimes D,\ b\mapsto \Sigma\,(\chi\rightharpoonup b_0)\otimes(b_1\rightharpoonup x)\ \text{and}\ \chi^-(d)c \leftarrow\!\shortmid c\otimes d.$$

This is also a right C'-module map. So, (b) is proved.

(b)⇒(c) follows from (1.8d).

(c)⇒(d): Note that γ' is a right B-isomorphism. Then π: $\#(D,B) \to \mathrm{End}_{-C'}(B)$ is the compositions of the following canonical isomorphism

$$\#(D,B)\cong \mathrm{Hom}_{-B}(B\otimes D,B)\cong \mathrm{Hom}_{-B}(B\otimes_{C'} B,B)\cong \mathrm{End}_{-C'}(B).$$

(d)⇒(a): Let θ: $C'\otimes D \to B$ be a right C'-module isomorphism which is also a right D-comodule map. Define ϕ: $D\to B$ by $\phi(d) = \theta(1\otimes d)$. Clearly ϕ is a right D-comodule map. Define g: $B\to C'$ by $g := (id\otimes\varepsilon)\theta^{-1}$. Then g is a right C'-module map. Note that $\theta^{-1}(b) = \Sigma\, g(b_0)\otimes(b_1\rightharpoonup x)$ for all $b\in B$. Since π is bijective there exists $\chi\in\#(D,B)$ such that $\pi(\chi) = g$. We claim that χ is the X-inverse of ϕ. First, we have

$(\chi X\phi)(d) = \Sigma\,\chi(d_2)\phi(d_1) = \Sigma\,\chi(\phi(d)_1\rightharpoonup x)\phi(d)_0$ (since ϕ is D-colinear)

$= \chi\rightharpoonup(\phi(d)) = g(\phi(d)) = (id\otimes\varepsilon)\theta^{-1}(\theta(1\otimes d)) = \varepsilon(d)1_B$.

On the other hand, we have

$\pi(\phi X\chi)(b) = (\phi X\chi)\rightharpoonup b = \Sigma\,(\phi X\chi)(b_1\rightharpoonup x)b_0 = \Sigma\,\phi(b_2\rightharpoonup x)\chi(b_1\rightharpoonup x)b_0$

$= \Sigma\,\theta(1\otimes(b_1\rightharpoonup x))g(b_0) = \Sigma\,\theta(g(b_0)\otimes(b_1\rightharpoonup x))$ (since $g(b_0)\in C'$)

$= \theta(\theta^{-1}(b)) = b$, and so, by the bijectivity of π, we have $\phi X\chi = u\circ\varepsilon$. □

3 The image of F'

In general it seems impossible to consider the situation when F' (1.5d) is surjective, because $\#(D,B)$ is too big for $B\otimes_{C'} Q'$. We shall consider a weaken condition "Im F' $= \#(D,B)^{rat}$" instead of the surjectivity of F'. To do this we assume in this section that D is projective over k.

Remark that the notion of a rational D^*-module by Sweedler makes sense for D projective over k.

3.1 We consider the question of which modules in ${}_{\#(D,B)}\mathbf{M}$ arise from Hopf modules in ${}_B\mathbf{M}^D$ as in (1.7c). Let $M\in{}_{\#(D,B)}\mathbf{M}$. Define

$$\rho': M \to \mathrm{Hom}_{B-}(\#(D,B),\ M) \text{ by } \rho'(m)(f) = f\cdot m,$$
$$\rho: M \to \mathrm{Hom}(D^*,M) \text{ by } \rho(m)(d^*) = \iota(d^*)\cdot m,$$

where $\iota: D^* \to \#(D,B)$ denotes a canonical algebra map (i.e. $\iota(d^*)(d) = \langle d^*,d\rangle 1_B$). Observe that the canonical maps

$$M\otimes D \to \mathrm{Hom}_{B-}(\#(D,B),\ M),\ m\otimes d \mapsto (f \mapsto f(d)m),$$
$$M\otimes D \to \mathrm{Hom}(D^*,M),\ m\otimes d \mapsto (d^* \mapsto \langle d^*,d\rangle m),$$

are injective since D is k-projective, by a routine direct sum argument. Note that this implies $Q' = \#(D,B)^x$, the x-invariants of $\#(D,B)$, and $M^{coD} = M^x$ whenever $M\in{}_B\mathbf{M}^D$.

Regard the above embeddings as inclusions, so that

$$M\otimes D \subset \mathrm{Hom}_{B-}(\#(D,B),\ M) \text{ and } M\otimes D \subset \mathrm{Hom}(D^*,M).$$

The left $\#(D,B)$-module M is called $\#$-rational (resp. D^*-rational) in case $\rho'(M)\subset M\otimes D$ (resp. $\rho(M)\subset M\otimes D$). (When D is finitely generated projective over k, $M\otimes D = \mathrm{Hom}(D^*,M) = \mathrm{Hom}_{B-}(\#(D,B),\ M)$, and so any $M\in{}_{\#(D,B)}\mathbf{M}$ is $\#$- and D^*-rational.)

PROPOSITION Suppose M is a left $\#(D,B)$-module.

(a) If M is $\#$-rational then it is D^*-rational.

(b) If M is D^*-rational then it is an object in ${}_B\mathbf{M}^D$.

(c) If M is D^*-rational and $\Psi_M: B\otimes_{C'} M^{coD} \to M,\ b\otimes n \mapsto bn$, is surjective, then M is $\#$-rational.

Proof: (a) is clear.

(b): One easily checks that the above ρ makes M a right D-comodule. To see that M is an object in ${}_B\mathbf{M}^D$, we first observe that for $b\in B$, $d^*\in D^*$,

$\iota(d^*)\cdot b = \Sigma\, b_0\cdot \iota(d^*\leftharpoonup b_1)$ in $\#(D,B)$, where $\langle d^*\leftharpoonup b_1,d\rangle = \langle d^*,b_1\rightharpoonup d\rangle$.

If we write $\rho(m) = \Sigma\, m_0\otimes m_1$ for $m\in M$, then $(\iota(d^*)\cdot b)\cdot m$

$= \Sigma\, b_0(\iota(d^*\leftharpoonup b_1)\cdot m) = \Sigma\, b_0\langle d^*\leftharpoonup b_1,m_1\rangle m_0$ (since M is D^*-rational)

$= \Sigma\langle d^*,b_1\rightharpoonup m_1\rangle b_0m_0$. On the other hand, $(\iota(d^*)\cdot b)\cdot m = \iota(d^*)\cdot(bm)$

$= \Sigma\langle d^*,(bm)_1\rangle(bm)_0$. Hence $\rho(bm) = \Sigma\, b_0m_0\otimes b_1\rightharpoonup m_1$. Thus $M \in{}_B\mathbf{M}^D$.

(c): Let $m = bn$, where $b\in B$, $n\in M^{coD}$. Then for any $f\in\#(D,B)$ we have $f\cdot m = f\cdot(bn) = (f\cdot b)\cdot n = (f\cdot b)(x)n$ (by $n\in M^{coD} = M^x$) $= \Sigma\, f(b_1\rightharpoonup x)b_0n$ $= \Sigma\, f(m_1)m_0$ (since $M \in{}_B\mathbf{M}^D$). This shows that M is $\#$-rational. □

Remark that any $M \in {}_B\mathbf{M}^D$ defines a $\#$-rational module structure on M by (1.7c). Thus ${}_B\mathbf{M}^D$ is naturally equivalent to the category of $\#$-rational left $\#(D,B)$-modules, which is a full subcategory of ${}_{\#(D,B)}\mathbf{M}$. If D is finitely generated projective over k then ${}_B\mathbf{M}^D = {}_{\#(D,B)}\mathbf{M}$.

3.2 THEOREM (a variation of [CFM, Theorem 1.2])
Assume that D is finitely generated projective over k. In the Morita context $(\#(D,B), C', B, Q', F', G')$, the following (1)-(5) are equivalent:

(1) $F': B \otimes_{C'} Q' \to \#(D,B)$, $b \otimes \lambda \mapsto b\lambda$ is surjective (bijective).

(2) ${}_B\mathbf{M}^D$ satisfies the weak structure theorem.

(3) B is a left $\#(D,B)$-generator.

(4)(a) B is a finitely generated projective right C'-module.

(b) $\pi: \#(D,B) \to \mathrm{End}_{-C'}(B)$, $\pi(f)(b) = f \rightharpoonup b$, is an algebra isomorphism.

(5)(c) Q' is a finitely generated projective left C'-module.

(d) $B \cong \mathrm{Hom}_{C'-}(Q', C')$ $(b \mapsto (\lambda \mapsto \lambda \rightharpoonup b))$ as bimodules.

(e) For $f \in \#(D,B)$, $Q' \cdot f = 0$ implies $f = 0$.

If these conditions hold, then we have:

(f) $Q' \cong \mathrm{Hom}_{-C'}(B, C')$ $(\lambda \mapsto (b \mapsto \lambda \rightharpoonup b))$ as bimodules.

(g) $\#(D,B) \cong (\mathrm{End}_{C'-}(Q'))^{op}$ $(f \mapsto (\lambda \mapsto \lambda \cdot f))$ as algebras.

(h) The map $\gamma': B \otimes_{C'} B \to B \otimes D$, $b \otimes b' \mapsto \Sigma\, b_0 b' \otimes (b_1 \rightharpoonup x)$, is bijective.

Proof. We know that (1) implies (3), (4), (5), (f) and (g), by the standard argument of Morita theory (cf. [B, Thm (3.4)]). These hold without the assumption that D is finitely generated projective. Note that (e) follows from the injectivity of (f).

(1)⇒(2): The hypothesis on F means that there exist $b_1, \cdots, b_s \in B$ and $\lambda_1, \cdots, \lambda_s \in Q'$ such that $\Sigma\, b_i \lambda_i = u_B \circ \varepsilon_D$ in $\#(D,B)$. For any $M \in {}_B\mathbf{M}^D$ we may define $M \to B \otimes_{C'} M^{coD}$ by $m \mapsto \Sigma\, b_i \otimes \lambda_i \cdot m$. It is easy to see that this map is the inverse of Ψ_M.

(2)⇒(1): Take $M = \#(D,B) \in {}_B\mathbf{M}^D$. Then $\Psi_M = F'$.

(3)⇒(1): Since $Q' = \#(D,B)^{\pi} \cong \mathrm{Hom}_{\#(D,B)}(B, \#(D,B))$, Im F' coincides with the trace ideal of the left $\#(D,B)$-module B. So F' is surjective if and only if B is a left $\#(D,B)$-generator.

(4)⇒(1): Assuming (a), there are $b_1, \cdots, b_s \in B$ and $p_1, \cdots, p_s \in \mathrm{Hom}_{-C'}(B, C')$ such that for all $b \in B$, $b = \Sigma\, b_i p_i(b)$. Now by (b) there are $\lambda_1, \cdots, \lambda_s \in \#(D,B)$ such that $\pi(\lambda_i) = p_i$, $i = 1, \cdots, s$. We claim that $\lambda_i \in Q'$. If $f \in \#(D,B)$ then $\pi(f \cdot \lambda_i)(b) = (f \cdot \lambda_i) \rightharpoonup b = f \rightharpoonup (\lambda_i \rightharpoonup b) = f \rightharpoonup p_i(b) = f(x) p_i(b)$ (by $p_i(b) \in C'$) $= \pi(f(x)\lambda_i)(b)$. Thus we have $f \cdot \lambda_i = f(x)\lambda_i$ (all i), since π is bijective. Hence $\lambda_i \in Q'$ with $\Sigma\, b_i \lambda_i = u_B \circ \varepsilon_D$, and so F' is surjective.

(5)⇒(1): Assuming (c), there are $\lambda_1, \cdots, \lambda_s \in Q'$, $p_1, \cdots, p_s \in \mathrm{Hom}_{C'-}(Q', C')$ such that for all $\lambda \in Q'$, $\lambda = \Sigma\, p_i(\lambda)\lambda_i$. Now by (d) there exist $b_1, \cdots, b_s \in B$ such that $p_i(\lambda) = \lambda \rightharpoonup b_i$ (all $\lambda \in Q'$). Then we have $\lambda \cdot (\Sigma\, b_i \lambda_i) = \Sigma\, (\lambda \rightharpoonup b_i)\lambda_i = \Sigma\, p_i(\lambda)\lambda_i = \lambda$. This implies $\Sigma\, b_i \lambda_i = u_B \circ \varepsilon_D$ by (e).

Finally, (h) follows from (2) (see (1.8d)). □

3.3 PROPOSITION If $F: B\otimes_C Q \to \#(A,B)$ is surjective and there exists a left D-comodule map $q: D \to A$ with $\varepsilon_A \circ q = \varepsilon_D$, then $F': B\otimes_{C'} Q' \to \#(D,B)$ is surjective.

Proof. The hypothesis on F means that there exist $b_1,\cdots,b_s \in B$ and $\lambda_1,\cdots,\lambda_s \in Q$ such that $\Sigma b_i\lambda_i = u_B\circ\varepsilon_A$ in $\#(A,B)$. Then

$$\Sigma b_i(\lambda_i\circ q) = u_B\circ\varepsilon_A\circ q = u_B\circ\varepsilon_D, \text{ with } \lambda_i\circ q \in Q' \text{ (by 2.2).}$$

This shows that $F': B\otimes_{C'} Q' \to \#(D,B)$ is surjective. □

3.4 PROPOSITION Let M be any left $\#(D,B)$-module. Define

$$M^{rat} := \rho'^{-1}(M\otimes D) \quad \text{and} \quad M^{D^*-rat} = \rho^{-1}(M\otimes D),$$

where ρ' and ρ are as in (3.1). Then:

(a) M^{rat} is a unique maximal $\#$-rational submodule of M, and M^{D^*-rat} is a unique maximal D^*-rational submodule of M. Moreover $M^{rat} \subset M^{D^*-rat}$.

(b) If $\#(D,B)$ is a left $\#(D,B)$-module by left multiplication, then

$$\#(D,B)^{rat} = \#(D,B)^{D^*-rat}.$$

(c) $\#(D,B)^{rat}$ is a two-sided ideal of $\#(D,B)$. Moreover

$$\mathrm{Im}\, F' \subset \#(D,B)^{rat} \text{ and } (\#(D,B)^{rat})^{coD} = Q'.$$

Proof. (a): For $m\in M^{rat}$, let $\rho'(m) = \Sigma_i m_i\otimes d_i \in M\otimes D$. Thus,

$$f\cdot m = \Sigma f(d_i)m_i \text{ for all } f\in \#(D,B).$$

For any $g\in \#(D,B)$, we have $\rho'(f\cdot m)(g) = g\cdot(f\cdot m) = (g\cdot f)\cdot m = \Sigma(g\cdot f)(d_i)m_i$ $= \Sigma g(f(d_{i(2)})_{(1)} \rightharpoonup d_{i(1)})f(d_{i(2)})_{(0)}m_i$, so

$$\rho'(f\cdot m) = \Sigma f(d_{i(2)})_{(0)}m_i \otimes f(d_{i(2)})_{(1)} \rightharpoonup d_{i(1)} \in M\otimes D.$$

Thus $f\cdot m\in M^{rat}$, so that M^{rat} is a left $\#(D,B)$-submodule of M. Next, we will show $\rho'(M^{rat}) \subset M^{rat}\otimes D$. Choose $\{d_i\}_{i\in I} \subset D$ and $\{p_i\}_{i\in I} \subset D^*$ such that for $d\in D$, $\langle p_i,d\rangle = 0$ for almost all i and $d = \Sigma_{i\in I}\langle p_i,d\rangle d_i$ (dual basis lemma). If $m\in M^{rat}$ with $\rho'(m) = \Sigma_{i\in I} m_i\otimes d_i$, then $\iota(p_j)\cdot m\in M^{rat}$ $(\forall j)$ and $\Sigma_{j\in I}\iota(p_j)\cdot m\otimes d_j = \Sigma_{i,j\in I}\langle p_j,d_i\rangle m_i\otimes d_j = \Sigma_{i\in I} m_i\otimes(\Sigma_{j\in I}\langle p_j,d_i\rangle d_j) =$ $\Sigma_{i\in I} m_i\otimes d_i$, and hence $\rho'(m)\in M^{rat}\otimes D$. Thus M^{rat} is $\#$-rational.

The maximality of M^{rat} is clear. A similar argument shows the same for M^{D^*-rat} (see [Sw]). It is clear that $M^{rat} \subset M^{D^*-rat}$.

(b): It is enough to show that $\#(D,B)^{D^*-rat} \subset \#(D,B)^{rat}$. Let $g\in$ $\#(D,B)^{D^*-rat}$. Then by definition there are $g_1,\cdots,g_s\in \#(D,B)$ and $d_1,\cdots,d_s$ $\in D$ such that $d^*\cdot g = \Sigma\langle d^*,d_i\rangle g_i$, for all $d^*\in D^*$. Hence we have

$$\Sigma\langle d^*, g(d_{(2)})_{(1)} \rightharpoonup d_{(1)}\rangle g(d_{(2)})_{(0)} = \Sigma\langle d^*,d_i\rangle g_i(d), \text{ for all } d\in D.$$

It follows that $\Sigma g(d_{(2)})_{(0)}\otimes(g(d_{(2)})_{(1)} \rightharpoonup d_{(1)}) = \Sigma g_i(d)\otimes d_i$ in $B\otimes D$. Thus $\Sigma f(g(d_{(2)})_{(1)} \rightharpoonup d_{(1)})g(d_{(2)})_{(0)} = \Sigma f(d_i)g_i(d)$ for all $f\in \#(D,B)$, and so $f\cdot g = \Sigma f(d_i)g_i$. Thid shows that $g\in \#(D,B)^{rat}$.

(c) is easy. □

3.5 THEOREM Assume that $G': Q'\otimes_{\#(D,B)} B \to C'$ is surjective. Then ${}_B\mathbf{M}^D$ satisfies the strong structure theorem if and only if the following two conditions hold:

(i) $\mathrm{Im}\, F' = \#(D,B)^{rat}$,

(ii) ω_M: $\#(D,B)^{rat} \otimes_{\#(D,B)} M \to M$, $g \otimes m \mapsto g \cdot m \ (= \Sigma g(m_1)m_0)$, is surjective for any $M \in {}_B\mathbf{M}^D$.

In this case, F' induces an isomorphism $B \otimes_{C'} Q' \cong \#(D,B)^{rat}$ and ω_M is bijective.

Proof. ("if" part): For any $M \in {}_B\mathbf{M}^D$ we have the following commutative diagram (note $M^{coD} = M^*$):

$$\begin{array}{ccc} B\otimes\xi_M: B\otimes_{C'} Q'\otimes_{\#(D,B)} M & \to & B\otimes_{C'} M^{coD} \\ \downarrow F'\otimes M & & \downarrow \Psi_M \\ \omega_M: \#(D,B)^{rat}\otimes_{\#(D,B)} M & \to & M. \end{array}$$

Therefore Ψ_M is surjective. To see that Ψ_M is injective, let N be the kernel of Ψ_M. Since Ψ_M is a morphism in ${}_B\mathbf{M}^D$, N is an object in ${}_B\mathbf{M}^D$, and so Ψ_N: $B \otimes_{C'} N^{coD} \to N$ is surjective. But, since $N^{coD} = N^* = Q' \otimes_{\#(D,B)} N = Q' \otimes_{\#(D,B)} (\mathrm{Ker}\Psi_M) = \mathrm{Ker}(Q' \otimes_{\#(D,B)} \Psi_M)$ (by (2.1i)) $= \mathrm{Ker}(id_{M^*}) = 0$, we have $N = 0$ by the surjectivity of Ψ_N. Hence Ψ_M is bijective. It follows, together with (2.1f), that ${}_B\mathbf{M}^D$ satisfies the strong structure theorem.

("only if" part): Since F' is the adjunction Ψ_M for $M := \#(D,B)^{rat}$, we have an isomorphism F': $B \otimes_{C'} Q' \cong \#(D,B)^{rat}$. Hence ω_M is also a bijection, using the above diagram. □

We remark that in case $M = B$, the map ω_B: $\#(D,B)^{rat} \otimes_{\#(D,B)} B \to B$, $g \otimes b \mapsto g \rightharpoonup b \ (= \Sigma g(b_1 \rightharpoonup x)b_0)$, is surjective if and only if there exists $h \in \#(D,B)^{rat}$ with $h(x) = 1_B$. In fact, assume ω_B is surjective, then there exists $\Sigma g_i \otimes b_i \in \#(D,B)^{rat} \otimes B$ with $\Sigma g_i \rightharpoonup b_i = 1_B$. Set $h = \Sigma g_i \cdot b_i$. Then $h \in \#(D,B)^{rat}$ since $\#(D,B)^{rat}$ is a ideal. Moreover $h(x) = \Sigma (g_i \cdot b_i)(x) = \Sigma g_i \rightharpoonup b_i = 1_B$. Conversely, let $h \in \#(D,B)^{rat}$ with $h(x) = 1_B$. Then $(bh) \rightharpoonup 1_B = b$ for $b \in B$. Hence ω_B is surjective.

We also remark that if ω_B is surjective and Ψ_M is surjective for $M \in {}_B\mathbf{M}^D$, then ω_M is surjective, since

$$\omega_M \circ (\#(D,B)^{rat} \otimes_{\#(D,B)} \Psi_M) = \Psi_M \circ (\omega_B \otimes_{C'} M^*).$$

In case k is a field, there is another sufficient condition for ${}_B\mathbf{M}^D$ to satisfy the strong structure theorem (by using cotensor products):

3.6 REMARK Assume A is an anti-Hopf algebra over a field k, and B is a flat (resp. faithfully flat) right C-module such that the map γ (1.8f) is bijective. Assume A is faithfully coflat as a left D-comodule, where A is a left D-comodule via $a \mapsto \Sigma (a_1 \rightharpoonup x) \otimes a_2$. Then ${}_B\mathbf{M}^D$ satisfies the weak (resp. strong) structure theorem.

Proof. Apply Theorem (2.3) in [D3] for B^{op}/C^{op} (as A^{op}-Galois extension) and D (as right A^{op}-module coalgebra). □

4 The structure of $\#(A,B)^{rat}$

Throughout this section A is a Hopf algebra which is projective over k. We study the structure of $\#(A,B)^{rat}$ (see (3.4) where $A = D$ and $x = 1_A$).

For $a \in A$, $f \in \#(A,B)$, we define $a \rightharpoonup f$ and $f \leftharpoondown a \in \#(A,B)$ by

$$(a \rightharpoonup f)(a') = f(a'a) \text{ and } (f \leftharpoondown a)(a') = f(a'S(a)),\ a' \in A.$$

Since the antipode S is an algebra anti-morphism, $\leftharpoondown$ is a right A-module structure on $\#(A,B)$. Let ρ: $\#(A,B)^{rat} \to \#(A,B)^{rat} \otimes A$ be the comodule structure and as usual write $\Sigma g_0 \otimes g_1$ for $\rho(g)$, whenever $g \in \#(A,B)^{rat}$. We then have that $f \cdot g = \Sigma f(g_1)g_0$ for all $f \in \#(A,B)$.

4.1 THEOREM

(1) $\#(A,B)^{rat}$ is a right A-submodule of $\#(A,B)$ under $\leftharpoondown$, and $\leftharpoondown$ and ρ make $\#(A,B)^{rat}$ an A-Hopf module. In particular, there is an isomorphism of A-Hopf modules

$$\text{(4.1a)} \qquad Q \otimes A \cong \#(A,B)^{rat}, \quad \lambda \otimes a \mapsto \lambda \leftharpoondown a, \quad \Sigma (g_0 \leftharpoondown S(g_1)) \otimes g_2 \leftarrow\!\shortmid g.$$

(2) Q is a left B-module by the following rule

$$\text{(4.1b)} \qquad b \rhd \lambda := \Sigma b_0(\lambda \leftharpoondown S(b_1)), \quad \text{for } b \in B,\ \lambda \in Q.$$

Moreover the map (4.1a) is also a left B-module isomorphism, where the B-module structure on $Q \otimes A$ is $b \cdot (\lambda \otimes a) = \Sigma (b_0 \rhd \lambda) \otimes b_1 a$.

(3) For $m \in M \in {}_B\mathbf{M}^A$ and $g \in \#(A,B)^{rat}$, we have

$$\text{(4.1c)} \qquad \rho_M(g \cdot m) = \Sigma g_0 \cdot m \otimes g_1.$$

(4) If there exists $h \in \#(A,B)^{rat}$ with $h(1_A) = 1_B$ (for example, if G: $Q \otimes_{\#(A,B)} B \to C$, is surjective), then for any $M \in {}_B\mathbf{M}^A$

$$\omega_M: \#(A,B)^{rat} \otimes_{\#(A,B)} M \to M,\ g \otimes m \to g \cdot m\ (= \Sigma g(m_1)m_0),$$

is an isomorphism and the inverse is given by $m \mapsto \Sigma (h \leftharpoondown m_1) \otimes m_0$.

Proof. (1): The proof is similar in spirit to [Sw, 5.1.2]. It is enough to establish for all $f \in \#(A,B)$, $g \in \#(A,B)^{rat}$, $a \in A$ that

$$\text{(4.1d)} \qquad f \cdot (g \leftharpoondown a) = \Sigma f(g_1 a_2)(g_0 \leftharpoondown a_1).$$

For suppose this holds. Then this implies

$$\text{(4.1e)} \qquad \rho(g \leftharpoondown a) = \Sigma (g_0 \leftharpoondown a_1) \otimes g_1 a_2.$$

Thus $g \leftharpoondown a \in \rho^{-1}(\#(A,B) \otimes A) = \#(A,B)^{rat}$. Also (4.1e) is exactly the A-Hopf module coherence condition. Now one easily checks that for every $a \in A$, f, $g \in \#(A,B)$

$$\text{(4.1f)} \qquad f \cdot (g \leftharpoondown a) = \Sigma ((a_2 \rightharpoonup f) \cdot g) \leftharpoondown a_1.$$

Hence, since $g \in \#(A,B)^{rat}$, we have $f \cdot (g \leftharpoondown a) = \Sigma ((a_2 \rightharpoonup f)(g_1)g_0) \leftharpoondown a_1$ $= \Sigma (f(g_1 a_2)g_0) \leftharpoondown a_1 = \Sigma f(g_1 a_2)(g_0 \leftharpoondown a_1)$. The last statement is just the fundamental theorem of Hopf modules [Sw. 4.1.1].

(2): We first show that $b \rhd \lambda$ is in Q: For any $f \in \#(A,B)$ we have

$$f \cdot (b \rhd \lambda) = (f \cdot b_0) \cdot (\lambda \leftharpoondown S(b_1))$$
$$= \Sigma (f \cdot b_0)(S(b_1)(\lambda \leftharpoondown S(b_2)) \quad \text{(by (4.1d) and } \rho(\lambda) = \lambda \otimes 1)$$

$= \sum f(b_1 S(b_2)) b_0 (\lambda \leftharpoondown S(b_3))$ (by (1.1b)) $= f(1_A)(b \triangleright \lambda)$.
This shows $b \triangleright \lambda \in Q$, since $Q = \#(A,B)^x$. For $b, b' \in B$ we have that
$b \triangleright (b' \triangleright \lambda) = \sum b \triangleright (b'_0 (\lambda \leftharpoondown S(b'_1))) = \sum b_0 b'_0 (\lambda \leftharpoondown S(b'_1) S(b_1)) = (bb') \triangleright \lambda$.

(3) For any $a^* \in A^*$, we have $(id \otimes a^*) \rho_M (g \cdot m) = a^* \cdot (g \cdot m) = (a^* \cdot g) \cdot m$
$= \sum \langle a^*, g_1 \rangle g_0 \cdot m = (id \otimes a^*)(\sum g_0 \cdot m \otimes g_1)$.

(4) We have $\sum (h \leftharpoondown m_1) \rightharpoonup m_0 = \sum (h \leftharpoondown m_2)(m_1) m_0 = \sum h(m_1 S(m_2)) m_0 = h(1_A) m$
$= m$ for any $m \in M$.

Conversely, $\sum (h \leftharpoondown (g \cdot m)_1) \otimes (g \cdot m)_0 = \sum (h \leftharpoondown g_1) \otimes g_0 \cdot m$ (by (3))
$= \sum (h \leftharpoondown g_1) \cdot g_0 \otimes m = \sum (h \leftharpoondown g_2)(g_1) g_0 \otimes m$ (by $g_0 \in \#(A,B)^{rat}$)
$= \sum h(g_1 S(g_2)) g_0 \otimes m = h(1_A) g \otimes m = g \otimes m$. □

The next is the main result in this section.

4.2 THEOREM If $\gamma: B \otimes_C B \to B \otimes A$, $b \otimes b' \mapsto \sum b_0 b' \otimes b_1$, is surjective, then $\mathrm{Im}\, F = \#(A,B)^{rat}$.

Proof. Consider the map $\gamma_Q: B \otimes_C Q \to Q \otimes A$, $b \otimes \lambda \to \sum (b_0 \triangleright \lambda) \otimes b_1$ (see (1.8c)). This is surjective since we may identify $\gamma_Q = \gamma \otimes_B Q$. We have the commutative diagram

$$\begin{array}{ccc} \gamma_Q: B \otimes_C Q & \to & Q \otimes A \\ & F \searrow & \downarrow (4.1b) \\ & & \#(A,B)^{rat}. \end{array}$$

Since the map (4.1b) is an isomorphism, it follows immediately that $\mathrm{Im}\, F = \#(A,B)^{rat}$. □

The next corollary is an essencially Kreimer-Takeuchi's fundamental theorem [KT, (1.7)] for any finite Hopf algebra.

4.3 COROLLARY Assume that A is a Hopf algebra which is finitely generated projective over k. Then the following are equivalent:

(a) $F: B \otimes_C Q \to \#(A,B)$ is surjective.

(b) $\gamma: B \otimes_C B \to B \otimes A$, $b \otimes b' \mapsto \sum b_0 b' \otimes b_1$, is surjective.

If these conditions hold, then β is bijective, i.e. B/C is an A-Galois extension. Moerover all statements as in Theorem 3.2 hold (replacing D, $\#(D,B)$, C', Q',F' and G' by A, $\#(A,B)$, C, Q, F and G).

Proof. (a)⇒(b): By Theorem 3.2(h), γ is bijective. In paticular since the antipode S is bijective, we have β is bijective.

(b)⇒(a): follows from 4.2, since $\#(A,B)^{rat} = \#(A,B)$. □

For the general case, we have the next result:

4.4 COROLLARY Let A be a Hopf algebra which is k-projective.

(1) Assume that $\mathrm{Im}\, F = \#(A,B)^{rat}$ and $G: Q \otimes_{\#(A,B)} B \to C$ is surjective. (In this case, we say that our Morita context $(\#(A,B), C, B, Q, F, G)$ is rationally strict). Then ${}_B\mathbf{M}^A$ satisfies the strong structure theorem. In

particular γ is bijective(i.e. B/C is A-anti-Galois) and B is a faithfully flat right C-module.

(2) When the antipode S is bijective, our Morita context is rationally strict if and only if γ(or β) is surjective and there is a total integral ϕ: A → B. In this case, B/C is a faithfully flat A-Galois extension. (Compare with [Sc, Theorem 3.5].)

Proof. (1): By 3.5 and 4.1(4), ${}_B\mathbf{M}^A$ satisfies the strong structure theorem. In particular γ is bijective. Furthermore, since $B\otimes_C(\ ): {}_C\mathbf{M} \to {}_B\mathbf{M}^A$ is an equivalence of categories, B is a faithfully flat right C-module.

(2) follows from the above (1), 2.3(c) and 4.2. ∎

ACKNOWLEDGEMENT

The auther would like to thank Akira Masuoka for many fruitful discussions.

REFERENCES

[B] H.Bass, Algebraic K-theory, Benjamin, New York 1968.

[CFM] M.Cohen, D.Fischman and S.Montgomery, Hopf Galois extensions, smash products, and Morita equivalence, J. Algebra 133 (1990), 351-372.

[CSw] S.U.Chase and M.E.Sweedler, Hopf algebras and Galois Theory, Lecture Notes in Math. vol 97, Springer 1969.

[D1] Y.Doi, Cleft comodule algebras and Hopf modules, Comm. Algebra 12 (1984), 1155-1169.

[D2] Y.Doi, Algebras with total integrals, Comm. Algebra (1985), 2137-2159.

[D3] Y.Doi, Unifying Hopf modules, J.Algebra 153 (1992).

[DT1] Y.Doi and M.Takeuchi, Cleft comodule algebras for a bialgebra, Comm. Algebra 14 (1986), 801-818.

[DT2] Y.Doi and M.Takeuchi, Hopf-Galois extensions of algebras, the Miyashita-Ulbrich Action, and Azumaya algebras, J. Algebra 121 (1989), 488-516.

[K] M.Koppinen, Variations on the smash product with applications to group-graded rings, preprint.

[KT] H.F.Kreimer and M.Takeuchi, Hopf algebras and Galois extensions of an algebra, Indiana Univ. M. J. 30 (1981), 675-692.

[MD] A.Masuoka and Y.Doi, Generalization of cleft comodule algebras, Comm. Algebra (to appear).

[Sc] Hans-J.Schneider, Principal homogeneous spaces for arbitrary Hopf algebras, Israel J. Math. 72 (1990), 167-195.

[Sw] M.E.Sweedler, Hopf algebras, Benjamin, New York 1969.

Algebras and Hopf Algebras in Braided Categories

SHAHN MAJID Cambridge University, Cambridge, England

ABSTRACT

This is an introduction for algebraists to the theory of algebras and Hopf algebras in braided categories. Such objects generalise super-algebras and super-Hopf algebras, as well as colour-Lie algebras. Basic facts about braided categories $\mathcal{C}$ are recalled, the modules and comodules of Hopf algebras *in* such categories are studied, the notion of 'braided-commutative' or 'braided-cocommutative' Hopf algebras (braided groups) is reviewed and a fully diagrammatic proof of the reconstruction theorem for a braided group $\mathrm{Aut}(\mathcal{C})$ is given. The theory has important implications for the theory of quasitriangular Hopf algebras (quantum groups). It also includes important examples such as the degenerate Sklyanin algebra and the quantum plane.

One of the main motivations of the theory of Hopf algebras is that they provide a generalization of groups. Hopf algebras of functions on groups provide examples of commutative Hopf algebras, but it turns out that many group-theoretical constructions work just as well when the Hopf algebra is allowed to be non-commutative. This is the philosophy associated to some kind of non-commutative (or so-called quantum) algebraic geometry. In a Hopf algebra context one can say the same thing in a dual way: group algebras and enveloping algebras are cocommutative but many constructions are not tied to this. This point of view has been highly successful in recent years, especially in regard to the quasitriangular Hopf algebras of Drinfeld[11]. These are non-cocommutative but the non-cocommutativity is controlled by a quasitriangular structure $\mathcal{R}$. Such objects are commonly called quantum groups. Coming out of physics, notably associated to solutions of the Quantum Yang-Baxter Equations (QYBE) is a rich supply of quantum groups.

Here we want to describe some kind of rival or variant of these quantum groups, which we call braided groups[44]–[52]. These are motivated by an earlier revolution that was very popular some decades ago in mathematics and physics, namely the theory of super or $\mathbb{Z}_2$-graded algebras and Hopf algebras. Rather than make the algebras non-commutative

[1]1991 Mathematics Subject Classification 18D10, 18D35, 16W30, 57M25, 81R50, 17B37
This paper is in final form and no version of it will be submitted for publication elsewhere

[2]SERC Fellow and Fellow of Pembroke College, Cambridge

etc one makes the notion of tensor product $\otimes$ non-commutative. The algebras remain commutative with respect to this new tensor product (they are super-commutative). Under this point of view one has super-groups, super-manifolds and super-differential geometry. In many ways this line of development was somewhat easier than the notion of quantum geometry because it is conceptually easier to make an entire shift of category from vector spaces to super-vector spaces. One can study Hopf algebras in such categories also (super-quantum groups).

In this second line of development an obvious (and easy) step was to generalise such constructions to the case of symmetric tensor categories[27]. These have a tensor product $\otimes$ and a collection of isomorphisms Ψ generalizing the transposition or super-transposition map but retaining its general properties. In particular, one keeps $\Psi^2 = \mathrm{id}$ so that these generalized transpositions still generate a representation of the symmetric group. Since only such general properties are used in most algebraic constructions, such as Hopf algebras and Lie algebras, these notions immediately (and obviously) generalise to this setting. See for example Gurevich [19], Pareigis [61], Scheunert [67] and numerous other authors. On the other hand, the theory is *not* fundamentally different from the super-case.

Rather more interesting is the further generalization to relax the condition that $\Psi^2 = \mathrm{id}$. Now Ψ and Ψ^{-1} must be distinguished and are more conveniently represented by braid-crossings rather than by permutations. They generate an action of the Artin braid group on tensor products. Such quasitensor or braided-tensor categories have been formally introduced into category theory in [24] and also arise in the representation theory of quantum groups. The study of algebras and Hopf algebras in such categories is rather more non-trivial than in the symmetric case. It is this theory that we wish to describe here. It has been introduced by the author under the heading 'braided groups' as mentioned above. Introduced were the relevant notions (not all of them obvious), the basic lemmas (such as a braided-tensor product analogous to the super-tensor product of super-algebras) and a construction leading to a rich supply of examples.

On the mathematical level this project of 'braiding' all of mathematics is, I believe, a deep one (provided one goes from the symmetric to the truly braided case). Much of mathematics consists of manipulating symbols, making transpositions etc. The situation appears to be that in many constructions the role of permutation group can (with care) be played equally well by the braid group. Not only the algebras and braided groups to be described here, but also braided differential calculus, braided-binomial theorems and braided-exponentials are known[57] as well as braided-Lie algebras[58]. Much more can be expected. Ultimately we would like some kind of braided geometry comparable to the high-level of development in the super case.

Apart from this long-term philosophical motivation, one can ask what are the more immediate applications of this kind of braided geometry? I would like to mention five of them.

1. Many algebras of interest in physics such as the degenerate Sklyanin algebra, quantum planes and exchange algebras are not naturally quantum groups but turn out to be braided ones[52][56]. There are braided-matrices $B(R)$ and braided-vectors $V(R')$ associated to R-matrices.

2. The category of Hopf algebras is not closed under quotients in a good sense. For example, if $H \subset H_1$ is covered by a Hopf algebra projection then $H_1 \cong B \rtimes H$ where

B is a braided-Hopf algebra. This is the right setting for Radford's theorem as we have discovered and explained in detail in [52].

3. Braided groups are best handled by means of braid diagrams in which algebraic operations 'flow' along strings. This means deep connections with knot theory and is also useful even for ordinary Hopf algebras. For example, you can dualise theorems geometrically by turning the diagram-proof up-side-down and flip conventions by viewing in a mirror.

4. A useful tool in the theory of quasitriangular Hopf algebras (quantum groups) via a process of *transmutation*. By encoding their non-cocommutativity as braiding in a braided category they appear 'cocommutative'. Likewise, dual quasitriangular Hopf algebras are rendered 'commutative' by this process [45][49].

5. In particular, properties of the quantum groups $\mathcal{O}_q(G)$ and $U_q(g)$ are most easily understood in terms of their braided versions $B_q(G)$ and $BU_q(g)$. This includes an Ad-invariant 'Lie algebra-like' subspace $\mathcal{L} \subset U_q(g)$ and an isomorphism $B_q(G) \cong U_q(g)$ [50][52].

An outline of the paper is the following. In Section 1 we recall the basic notions of braided tensor categories and how to work in them, and some examples. We recall basic facts about quasitriangular and dual quasitriangular Hopf algebras and the braided categories they generate. In Section 2 we do diagrammatic Hopf-algebra theory in this setting. In Section 3 we give a new diagrammatic proof of our generalised Tannaka-Krein-type reconstruction theorem. In Section 4 we explain the results about ordinary quantum groups obtained from this braided theory. In Section 5 we end with basic examples of braided matrices etc associated to an R-matrix. Although subsequently of interest in physics, the braided matrices arose quite literally from the Tannaka-Krein theorem mentioned above. This is an example of pure mathematics feeding back into physics rather than the other way around (for a change).

Our work on braided groups (or Hopf algebras in braided categories) was presented to the Hopf algebra community at the Euler Institute in Leningrad, October 1990 and at the Biannual Meeting of the American Maths Society in San Francisco, January 1991 and published in [48][49]. The result presented at these meetings was the introduction of Hopf algebras living *in* the braided category of comodules of a dual quasitriangular Hopf algebra. The connection between crossed modules (also called Drinfeld-Yetter categories) and the quantum double as well as the connection with Radford's theorem were introduced in [38] in early 1990. The braided interpretation of Radford's theorem was introduced in detail in [52] and circulated at the start of 1992. Dual quasitriangular (or coquasitriangular) Hopf algebras themselves were developed in connection with Tannaka-Krein ideas in [36, Sec. 4][48][49, Appendix] (and earlier in other equivalent forms). A related Tannaka-Krein theorem in the quasi-associative dual quasitriangular setting was obtained in [43] at the Amherst conference and circulated in final form in the Fall of 1990.

It is a pleasure to see that some of these ideas have subsequently proven of interest in Hopf algebra circles (directly or indirectly). I would also like to mention some constructions of Lyubashenko[30][31] relating to our joint work[32]. Also in joint work with Gurevich[21] the transmutation construction is related to Drinfeld's process of twisting[13]. Several other papers can be mentioned here. On the whole I have resisted the temptation to give a full survey of all results obtained so far. Instead, the aim here is a more pedagogical exposition of the more elementary results, with proofs.

Throughout this paper we assume familiarity with usual techniques of Hopf algebras such as in the book of Sweedler[69]. In this sense the style (and also the motivation) is somewhat different from our braided-groups review article for physicists[42]. We work over a field k. With more care one can work here with a ring just as well. When working with matrix or tensor components we will use the convention of summing over repeated indices. Some of the elementary quantum groups material should appear in more detail in my forthcoming book.

1 Braided Categories

Here we develop the braided categories within which we intend to work, namely those coming from (co)modules of quantum groups. In fact, the theory in Sections 2,3 is not tied to quantum groups and works in any braided category. The material in the present section is perfectly standard by now.

1.1 Definition and General Constructions

Symmetric monoidal (=tensor) categories have been known for some time and we refer to [27] for details. The model is the category of k-modules. The notion of braided monoidal (=braided tensor=quasitensor) category is a small generalization if this.

Briefly, a *monoidal category* means $(\mathcal{C},\otimes,\Phi_{V,W,Z},\underline{1},l,r)$ where $\mathcal{C}$ is a category with objects V,W,Z etc, $\otimes:\mathcal{C}\times\mathcal{C}\to\mathcal{C}$ is a functor and Φ is a natural transformation between the two functors $\otimes(\ \otimes\ )$ and $(\ \otimes\ )\otimes$ from $\mathcal{C}\times\mathcal{C}\times\mathcal{C}\to\mathcal{C}$. This means a functorial collection of isomorphisms $\Phi_{V,W,Z}:V\otimes(W\otimes Z)\to(V\otimes W)\otimes Z$. These are in addition required to obey the 'pentagon' coherence condition of Mac Lane. This expresses equality of two ways to go via Φ from $U\otimes(V\otimes(W\otimes Z))\to((U\otimes V)\otimes W)\otimes Z$. Once this is assumed Mac Lane's theorem ensures that all other re-bracketing operations are consistent. In practice this means we can forget Φ and brackets entirely. We also assume a unit object $\underline{1}$ for the tensor product and associated functorial isomorphisms $l_V:V\to\underline{1}\otimes V, r_V:V\to V\otimes\underline{1}$ for all objects V, which we likewise suppress.

A monoidal category $\mathcal{C}$ is *rigid* (=has left duals) if for each object V, there is an object V^* and morphisms $\mathrm{ev}_V:V^*\otimes V\to\underline{1}$, $\mathrm{coev}_V:\underline{1}\to V\otimes V^*$ such that

$$V\stackrel{\mathrm{coev}}{\to}(V\otimes V^*)\otimes V\cong V\otimes(V^*\otimes V)\stackrel{\mathrm{ev}}{\to}V \tag{1}$$

$$V^*\stackrel{\mathrm{coev}}{\to}V^*\otimes(V\otimes V^*)\cong(V^*\otimes V)\otimes V^*\stackrel{\mathrm{ev}}{\to}V^* \tag{2}$$

compose to id_V and id_{V^*} respectively. A single object has a left dual if $V^*,\mathrm{ev}_V,\mathrm{coev}_V$ exist. The model is that of a finite-dimensional vector space (or finitely generated projective module when k is a ring).

Finally, the monoidal category is *braided* if it has a *quasisymmetry* or 'braiding' Ψ given as a natural transformation between the two functors $\otimes$ and $\otimes^{\mathrm{op}}$ (with opposite product) from $\mathcal{C}\times\mathcal{C}\to\mathcal{C}$. This is a collection of functorial isomorphisms $\Psi_{V,W}:V\otimes W\to W\otimes V$ obeying two 'hexagon' coherence identities. In our suppressed notation these are

$$\Psi_{V\otimes W,Z}=\Psi_{V,Z}\circ\Psi_{W,Z},\qquad \Psi_{V,W\otimes Z}=\Psi_{V,Z}\circ\Psi_{V,W} \tag{3}$$

while identities such as

$$\Psi_{V,\underline{1}} = \mathrm{id}_V = \Psi_{\underline{1},V} \tag{4}$$

can be deduced. If $\Psi^2 = \mathrm{id}$ then one of the hexagons is superfluous and we have an ordinary symmetric monoidal category.

Let us recall that the functoriality of maps such as those above means that they commute in a certain sense with morphisms in the category. For example, functoriality of Ψ means

$$\Psi_{Z,W}(\phi\otimes \mathrm{id}) = (\mathrm{id}\otimes\phi)\Psi_{V,W}\ \forall \phi \overset{V}{\underset{Z}{\downarrow}}, \qquad \Psi_{V,Z}(\mathrm{id}\otimes\phi) = (\phi\otimes\mathrm{id})\Psi_{V,W}\ \forall\phi \overset{W}{\underset{Z}{\downarrow}}. \tag{5}$$

These conditions (3)-(5) are just the obvious properties that we take for granted when transposing ordinary vector spaces or super-vector spaces. In these cases Ψ is the twist map $\Psi_{V,W}(v\otimes w) = w\otimes v$ or the supertwist

$$\Psi_{V,W}(v\otimes w) = (-1)^{|v||w|} w\otimes v \tag{6}$$

on homogeneous elements of degree $|v|, |w|$. The form of Ψ in these familiar cases does not depend directly on the spaces V, W so we often forget this. But in principle there is a different map $\Psi_{V,W}$ for each V, W and they all connect together as explained.

In particular, note that for any two V, W we have two morphisms $\Psi_{V,W}, \Psi^{-1}_{W,V} : V\otimes W \to W\otimes V$ and in the truly braided case these can be distinct. A convenient notation in this case is to write them not as permutations but as braid crossings. Thus we write morphisms pointing downwards (say) and instead of a usual arrow, we use the shorthand

$\Psi_{V,W} =$  V W / W V $\qquad \Psi^{-1}_{W,V} =$ V W / W V.  (7)

In this notation the hexagons (3) appear as

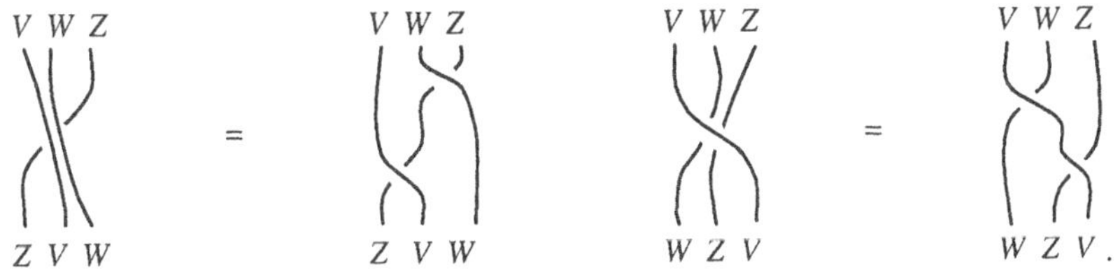

V W Z / Z V W = V W Z / Z V W $\qquad$ V W Z / W Z V = V W Z / W Z V. (8)

The doubled lines refer to the composite objects $V\otimes W$ and $W\otimes Z$ in a convenient extension of the notation. The coherence theorem for braided categories can be stated very simply in this notation: if two series of morphisms built from Ψ, Φ correspond to the same braid then they compose to the same morphism. The proof is just the same as Mac Lane's proof in the symmetric case with the action of the symmetric group replaced by that of the Artin braid group.

This notation is a powerful one. We can augment it further by writing any other morphisms as nodes on a string connecting the inputs down to the outputs. Functoriality

(5) then says that a morphism $\phi : V \to Z$ say can be pulled through braid crossings,

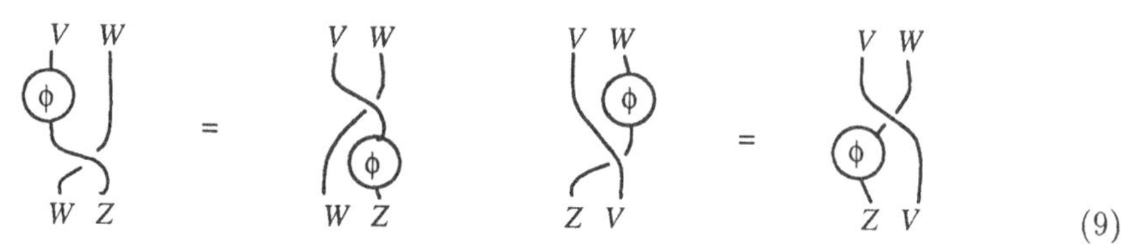

(9)

Similarly for Ψ^{-1} with inverse braid crossings. An easy lemma using this notation is that for any braided category $\mathcal{C}$ there is another mirror-reversed braided monoidal category $\bar{\mathcal{C}}$ with the same monoidal structure but with braiding

$$\bar{\Psi}_{V,W} = \Psi^{-1}_{W,V} \tag{10}$$

in place of $\Psi_{V,W}$, i.e with the interpretation of braid crossings and inverse braid crossings interchanged.

Finally, because of (4) we can suppress the unit object entirely so the evaluation and co-evaluation appear simply as ev $= \cup$ and coev $= \cap$. Then (1)-(2) appear as

$$\mathrm{ev}_V = \underset{}{\overset{V^*\ V}{\cup}} \qquad \mathrm{coev}_V = \underset{V\ V^*}{\cap} \qquad \ldots \tag{11}$$

There is a similar notion of right duals $V^{\vee}$ and $\bar{\mathrm{ev}}_V, \bar{\mathrm{coev}}_V$ for which the mirror-reflected double-bend here can be likewise straightened.

Example 1.1 *Let $R \in M_n(k) \otimes M_n(k)$ be invertible and obey the QYBE*

$$R_{12}R_{13}R_{23} = R_{23}R_{13}R_{12}$$

then the monoidal category $\mathcal{C}(V, R)$ generated by tensor products of $V = \mathbb{C}^n$ is braided.

Proof This is an elementary exercise (and extremely well-known). The notation is $R_{12} = R \otimes \mathrm{id}$ and $R_{23} = \mathrm{id} \otimes R$ in $M_n^{\otimes 3}$. The braiding on basis vectors $\{e_i\}$ is

$$\Psi(e_i \otimes e_j) = e_b \otimes e_a R^a{}_i{}^b{}_j \tag{12}$$

extended to tensor products according to (3). The morphisms in the category are linear maps such that Ψ is functorial with respect to them in the sense of (5). The associativity Φ is the usual one on vector spaces. □

If R obeys further conditions then $\mathcal{C}(V, V^*, R)$ generated by V, V^* is rigid. One says that such an R is *dualizable.* For this there should exist among other things a 'second-inverse'

$$\tilde{R} = ((R^{t_2})^{-1})^{t_2} \tag{13}$$

where t_2 is transposition in the second M_n factor. This defines one of the mixed terms in the braiding

$$\Psi_{V^*,V^*}(f^i \otimes f^j) = R^i{}_a{}^j{}_b f^b \otimes f^a \tag{14}$$

$$\Psi_{V,V^*}(e_i \otimes f^j) = \widetilde{R}^a{}_i{}^j{}_b f^b \otimes e_a \tag{15}$$

$$\Psi_{V^*,V}(f^i \otimes e_j) = e_a \otimes f^b R^{-1}{}^i{}_b{}^a{}_j \tag{16}$$

where $V^* = \{f^i\}$ is a dual basis. The evaluation and coevaluation are given by the usual morphisms

$$\mathrm{ev}_V(f^i \otimes e_j) = \delta^i{}_j, \quad \mathrm{coev}_V(1) = \sum_i e_i \otimes f^i. \tag{17}$$

One needs also the second-inverse $\widetilde{R^{-1}}$ for Ψ to be invertible. In this way one translates the various axioms into a linear space setting. We see in particular that the QYBE are nothing other than the braid relations in matrix form.

We turn now to some general categorical constructions. One construction in [37][40] is based on the idea that a pair of monoidal categories $\mathcal{C} \to \mathcal{V}$ connected by a functor behaves in many ways like a bialgebra with $\otimes$ in $\mathcal{C}$ something like the product. In some cases this is actually true as we shall see in Section 3 in the form of a Tannaka-Krein-type reconstruction theorem, but we can keep it in general as motivation. Motivated by this we showed that for every pair $\mathcal{C} \to \mathcal{V}$ of monoidal categories there is a dual one $\mathcal{C}^\circ \to \mathcal{V}$ where $\mathcal{C}^\circ$ is the *Pontryagin dual monoidal category*[37]. This generalised the usual duality for Abelian groups and bialgebras to the setting of monoidal categories. We also proved such things as a canonical functor

$$\mathcal{C} \to {}^\circ(\mathcal{C}^\circ). \tag{18}$$

Of special interest to us now is the case $\mathcal{C} \to \mathcal{C}$ where the functor is the identity one. So associated to every monoidal category $\mathcal{C}$ is another monoidal category $\mathcal{C}^\circ$ of 'representations' of $\otimes$. This special case can also be denoted by $\mathcal{C}^\circ = \mathcal{Z}(\mathcal{C})$ the 'center' or 'inner double' of $\mathcal{C}$ for reasons that we shall explain shortly. This case was found independently by V.G. Drinfeld who pointed out that it is braided.

Proposition 1.2 *[37][10] Let $\mathcal{C}$ be a monoidal category. There is a braided monoidal category $\mathcal{C}^\circ = \mathcal{Z}(\mathcal{C})$ defined as follows. Objects are pairs (V, λ_V) where V is an object of $\mathcal{C}$ and λ_V is a natural isomorphism in $\mathrm{Nat}(V \otimes \mathrm{id}, \mathrm{id} \otimes V)$ such that*

$$\lambda_{V,\underline{1}} = \mathrm{id}, \qquad (\mathrm{id} \otimes \lambda_{V,Z})(\lambda_{V,W} \otimes \mathrm{id}) = \lambda_{V,W \otimes Z}.$$

and morphisms are $\phi : V \to W$ such that the modules are intertwined in the form

$$(\mathrm{id} \otimes \phi)\lambda_{V,Z} = \lambda_{W,Z}(\phi \otimes \mathrm{id}), \qquad \forall Z \text{ in } \mathcal{C}.$$

The monoidal product and braiding are

$$(V, \lambda_V) \otimes (W, \lambda_W) = (V \otimes W, \lambda_{V \otimes W}), \ \lambda_{V \otimes W,Z} = (\lambda_{V,Z} \otimes \mathrm{id})(\mathrm{id} \otimes \lambda_{W,Z})$$

$$\Psi_{(V,\lambda_V),(W,\lambda_W)} = \lambda_{V,W}.$$

Proof The monoidal structure was found in the author's paper [37] where full proofs were also given. We refer to this for details. Its preprint was circulated in the Fall of 1989. The braiding was pointed out by Drinfeld[10] who had considered the construction from a very different and independent point of view to our duality one, namely in connection with the double of a Hopf algebra as we shall explain below. Another claim to the construction is from the direction of tortile categories[25]. See also [40] for further work from the duality point of view. □

The 'double' point of view for this construction is based on the following example cf[10].

Example 1.3 *Let H be a bialgebra over k and $\mathcal{C} = {}_H\mathcal{M}$ the monoidal category of H-modules. Then an object of $\mathcal{Z}(\mathcal{C})$ is a vector space V which is both a left H-module and an invertible left H-comodule such that*

$$\sum h_{(1)}v^{(\bar{1})} \otimes h_{(2)}{\triangleright}v^{(\bar{2})} = \sum (h_{(1)}{\triangleright}v)^{(\bar{1})}h_{(2)} \otimes (h_{(1)}{\triangleright}v)^{(\bar{2})}, \qquad \forall h \in H,\ v \in V.$$

In this form $\mathcal{Z}({}_H\mathcal{M})$ coincides with the category ${}^H_H\mathcal{M}$ of H-crossed modules[72] with an additional invertibility condition. The braiding is

$$\Psi_{V,W}(v \otimes w) = \sum v^{(\bar{1})}{\triangleright}w \otimes v^{(\bar{2})}.$$

The invertibility condition on the comodules ensures that Ψ^{-1} exists, and is automatic if the bialgebra H has a skew-antipode.

Proof The proof is standard from the point of view of Tannaka-Krein reconstruction methods (which we shall come to later). From $\mathcal{C}$ we can reconstruct H as the representing object for a certain functor. This establishes a bijection $\mathrm{Lin}(V, H \otimes V) \cong \mathrm{Nat}(V \otimes \mathrm{id}, \mathrm{id} \otimes V)$ under which λ_V corresponds to a map $V \to H \otimes V$. That λ_V represents $\otimes$ corresponds then to the comodule property of this map. That λ_V is a collection of morphisms corresponds to the stated compatibility condition between this coaction and the action on V as an object in $\mathcal{C}$. To see this in detail let H_L denote H as an object in $\mathcal{C}$ under the left action. Given λ_V a natural transformation we define

$$\sum v^{(\bar{1})} \otimes v^{(\bar{2})} = \lambda_{V,H_L}(v \otimes 1) \tag{19}$$

and check

$$\begin{aligned}(\mathrm{id} \otimes \lambda_{V,H_L})(\lambda_{V,H_L} \otimes \mathrm{id})(v \otimes 1 \otimes 1) &= \lambda_{V,H_L \otimes H_L}(v \otimes (1 \otimes 1)) \\ &= \lambda_{V,H_L \otimes H_L}(v \otimes \Delta(1)) = (\Delta \otimes \mathrm{id}) \circ \lambda_{V,H_L}(v \otimes 1)\end{aligned}$$

where the first equality is the fact that λ_V 'represents' $\otimes$ and the last is that λ_V is functorial under the morphism $\Delta : H_L \to H_L \otimes H_L$. The left hand side is the map $V \to H \otimes V$ in (19) applied twice so we see that this map is a left coaction. Moreover,

$$\begin{aligned}\sum h_{(1)}v^{(\bar{1})} \otimes h_{(2)}{\triangleright}v^{(\bar{2})} &= h{\triangleright}\lambda_{V,H_L}(v \otimes 1) = \lambda_{V,H_L}(h{\triangleright}(v \otimes 1)) \\ &= \sum \lambda_{V,H_L}(h_{(1)}{\triangleright}v \otimes R_{h_{(2)}}(1)) = \sum \left(\lambda_{V,H_L}(h_{(1)}{\triangleright}v \otimes 1)\right)(h_{(2)} \otimes 1)\end{aligned}$$

where the first equality is the definition (19) and the action of H on $H_L \otimes V$. The second equality is that λ_{V,H_L} is a morphism in $\mathcal{C}$. The final equality uses functoriality under the morphism $R_{h_{(2)}} : H_L \to H_L$ given by right-multiplication to obtain the right hand side of the compatibility condition. The converse directions are easier. Given a coaction $V \to H \otimes V$ define $\lambda_{V,W}(v \otimes w) = \sum v^{(\bar{1})} \triangleright w \otimes v^{(\bar{2})}$. This also implies at once the braiding $\Psi = \lambda$ as stated.

Finally we note that in Proposition 1.2 the definition assumes that the λ_V are invertible. If we were to relax this then we would have a monoidal category which is just that of crossed modules as in [72], but then Ψ would not necessarily be invertible and hence would not be a true braiding. The invertible λ_V correspond to left comodules which are *invertible* in the following sense: there exists a linear map $V \to V \otimes H$ sending v to $\sum v^{[2]} \otimes v^{[1]}$ say, such that

$$\sum v^{[2](\bar{1})} v^{[1]} \otimes v^{[2](\bar{2})} = 1 \otimes v = \sum v^{(\bar{2})[1]} v^{(\bar{1})} \otimes v^{(\bar{2})[2]}, \qquad \forall v \in V. \tag{20}$$

One can see that if such an 'inverse' exists, it is unique and a right comodule. Moreover, it is easy to see that the invertible comodules are closed under tensor products. They correspond to λ_V^{-1} in a similar way to (19) and with $\lambda_{V,W}^{-1}(w \otimes v) = \sum v^{[2]} \otimes v^{[1]} \triangleright w$ for the converse direction. In the finite-dimensional case they provide left duals V^* with left coaction $\beta_{V^*}(f)(v) = \sum v^{[1]} f(v^{[2]})$. If the bialgebra H has a skew-antipode then every left comodule is invertible by composing with the skew-antipode. So in this case the condition becomes empty.

From the categorical point of view in Proposition 1.2, if $\mathcal{C}$ has right duals then every $\lambda_{V,W}$ is invertible, cf[37]. The inverse is the right-adjoint of λ_{V,W^*}, namely $\lambda_{V,W}^{-1} = (\bar{\mathrm{ev}}_W \otimes \mathrm{id}) \circ \lambda_{V,W^*} \circ (\mathrm{id} \otimes \bar{\mathrm{coev}}_W)$. When $\mathcal{C} = {}_H\mathcal{M}$ then the finite-dimensional left modules have right duals if the bialgebra H has a skew-antipode, so in this case the invertibility of λ_V is automatic. On the other hand, we do not need to make these suppositions here.

This completes our computation of $\mathcal{Z}({}_H\mathcal{M})$. Apart from the invertibility restriction we see that it consists of compatible module-comodule structures as stated. □

Note that the notion of a crossed module is an immediate generalisation of the notion of a crossed G-module[71] with $H = kG$, the group algebra of a finite group G. In this case the category of crossed G-modules is well-known to be braided[18]. Moreover, because the objects can be identified with underlying vector spaces, we know by the Tannaka-Krein reconstruction theorem[66] that there must exist a bialgebra $coD(H)$ such that our braided-category is equivalent to that of right $coD(H)$-comodules

$$\mathcal{M}_{f.d.}^{coD(H)} = {}_H^H\mathcal{M}_{f.d.} \tag{21}$$

Here we take the modules to be finite-dimensional as a sufficient (but not necessary) condition for a Tannaka-Krein reconstruction theorem to apply and the co-double $coD(H)$ to exist. In the nicest case the category is also ${}_{D(H)}\mathcal{M}_{f.d.}$ for some $D(H)$. This is an abstract definition of Drinfeld's quantum double and works for a bialgebra.

If it happens that H is a Hopf algebra with invertible antipode then one can see from the above that ${}_H^H\mathcal{M}_{f.d.}$ is rigid and so $coD(H)$ and $D(H)$ will be Hopf algebras. The categorical reason is that ${}_H\mathcal{M}_{f.d.}$ is rigid and this duality extends to $\mathcal{Z}(\mathcal{C})$ with the dual of λ_V defined by the left-adjoint of $\lambda_{V,W}^{-1}$, namely $\lambda_{V^*,W} = (\mathrm{ev}_V \otimes \mathrm{id}) \circ \lambda_{V,W}^{-1} \circ (\mathrm{id} \otimes \mathrm{coev}_V)$. We will study details about categories of modules and comodules and the reconstruction theorems later in this section and in Section 3. The point is that these categorical methods are very powerful.

Proposition 1.4 *[33]cf[11] If H is a finite-dimensional Hopf algebra then $D(H)$ (the quantum double Hopf algebra of H) is built on $H^*\otimes H$ as a coalgebra with the product*

$$(a\otimes h)(b\otimes g)=\sum b_{(2)}a\otimes h_{(2)}g < Sh_{(1)},b_{(1)} >< h_{(3)},b_{(3)} >,\ \ h,g\in H,\ a,b\in H^*$$

where $< \ , \ >$ denotes evaluation.

Proof The quantum double $D(H)$ was introduced by Drinfeld[11] as a system of generators and relations built from the structure constants of H. The formula stated on $H^*\otimes H$ is easily obtained from this as done in [33]. We have used here the conventions introduced in [38] that avoid the use of the inverse of the antipode. Also in [38] we showed that the modules of the double were precisely the crossed modules category as required. To see this simply note that H and $H^{*\mathrm{op}}$ are sub-Hopf algebras and hence a left $D(H)$-module is a left H-module and a suitably-compatible right H^*-module. The latter is equally well a left H-comodule compatible as in Example 1.3. See [38] for details. □

In [33] we introduced a further characterization of the quantum double as a member of a class of *double cross product* Hopf algebras $H_1\bowtie H_2$ (in which H_i are mutually acting on each other). Thus, $D(H)=H^{*\mathrm{op}}\bowtie H$ where the actions are mutual coadjoint actions. In this form it is clear that the role of H^* can be played by H° in the infinite dimensional Hopf algebra case. We will not need this further here.

1.2 Quasitriangular Hopf Algebras

We have already described one source of braided categories, namely as modules of the double $D(H)$ (or comodules of the codouble) of a bialgebra. Abstracting from this one has the notion, due to Drinfeld, of a quasitriangular Hopf algebra. These are such that their category of modules is braided.

Definition 1.5 *[11] A* quasitriangular bialgebra or Hopf algebra *is a pair $(H,\mathcal{R})$ where H is a bialgebra or Hopf algebra and $\mathcal{R}\in H\otimes H$ is invertible and obeys*

$$(\Delta\otimes\mathrm{id})\mathcal{R}=\mathcal{R}_{13}\mathcal{R}_{23},\quad (\mathrm{id}\otimes\Delta)\mathcal{R}=\mathcal{R}_{13}\mathcal{R}_{12}. \tag{22}$$

$$\tau\circ\Delta h=\mathcal{R}(\Delta h)\mathcal{R}^{-1},\ \forall h\in H. \tag{23}$$

Here $\mathcal{R}_{12}=\mathcal{R}\otimes 1$ and $\mathcal{R}_{23}=1\otimes\mathcal{R}$ etc, and τ is the usual twist map.

Thus these Hopf algebras are like cocommutative enveloping algebras or group algebras but are cocommutative now only up to an isomorphism implemented by conjugation by an element $\mathcal{R}$. Some elementary (but important) properties are

Lemma 1.6 *[12] If $(H,\mathcal{R})$ is a quasitriangular bialgebra then $\mathcal{R}$ as an element of $H\otimes H$ obeys*

$$(\epsilon\otimes\mathrm{id})\mathcal{R}=(\mathrm{id}\otimes\epsilon)\mathcal{R}=1. \tag{24}$$

$$\mathcal{R}_{12}\mathcal{R}_{13}\mathcal{R}_{23}=\mathcal{R}_{23}\mathcal{R}_{13}\mathcal{R}_{12} \tag{25}$$

If H is a Hopf algebra then one also has

$$(S\otimes\mathrm{id})\mathcal{R}=\mathcal{R}^{-1},\qquad (\mathrm{id}\otimes S)\mathcal{R}^{-1}=\mathcal{R},\quad (S\otimes S)\mathcal{R}=\mathcal{R} \tag{26}$$

$$\exists S^{-1},u,v,\quad S^2(h)=uhu^{-1},\quad S^{-2}(h)=vhv^{-1}\quad \forall h\in H \tag{27}$$

Proof For (24) apply ϵ to (22), thus $(\epsilon\otimes\mathrm{id}\otimes\mathrm{id})(\Delta\otimes\mathrm{id})\mathcal{R}=\mathcal{R}_{23}=(\epsilon\otimes\mathrm{id}\otimes\mathrm{id})\mathcal{R}_{13}\mathcal{R}_{23}$ so that (since $\mathcal{R}_{23}$ is invertible) we have $(\epsilon\otimes\mathrm{id})\mathcal{R}=1$. Similarly for the other side. For (25) compute $(\mathrm{id}\otimes\tau\circ\Delta)\mathcal{R}$ in two ways: using the second of axioms (22) directly or using axiom (23), and then the second of (22). For (26) consider $\sum\mathcal{R}^{(1)}{}_{(1)}S\mathcal{R}^{(1)}{}_{(2)}\otimes\mathcal{R}^{(2)}=1$ by the property of the antipode and equation (24) already proven, but equals $\mathcal{R}(S\otimes\mathrm{id})\mathcal{R}$ by axiom (22). Similarly for the other side, hence $(S\otimes\mathrm{id})\mathcal{R}=\mathcal{R}^{-1}$. Similarly for $(\mathrm{id}\otimes S)\mathcal{R}^{-1}=\mathcal{R}$ once we appreciate that $(\Delta\otimes\mathrm{id})(\mathcal{R}^{-1})=(\mathcal{R}_{13}\mathcal{R}_{23})^{-1}=\mathcal{R}_{23}^{-1}\mathcal{R}_{13}^{-1}$ etc, since Δ is an algebra homomorphism. For (27) the relevant expressions are

$$u=\sum(S\mathcal{R}^{(2)})\mathcal{R}^{(1)},\quad u^{-1}=\sum\mathcal{R}^{(2)}S^2\mathcal{R}^{(1)},\quad v=Su$$

which one can verify to have the right properties. In addition one can see that $\Delta u=(\mathcal{R}_{21}\mathcal{R}_{12})^{-1}(u\otimes u)$ and similarly for v so that uv^{-1} is group-like (and implements S^4). For details of the computations see [12] or reviews by the author. □

Here (25) is the reason that Physicists call $\mathcal{R}$ the 'universal R-matrix' (compare Example 1.1). Indeed, in any finite-dimensional representation the image of $\mathcal{R}$ is such an R-matrix. There are well-known examples such as $U_q(sl_2)$ and $U_q(g)$[11][23]. Here we give perhaps the simplest known quasitriangular Hopf algebras

Example 1.7 *[47] Let $\mathbb{Z}_n=\mathbb{Z}/n\mathbb{Z}$ be the finite cyclic group of order n and $k\mathbb{Z}_n$ its group algebra with generator g. Let q be a primitive n-th root of unity. Then there is a quasitriangular Hopf algebra $\mathbb{Z}_n'$ consisting of this group algebra and*

$$\Delta g=g\otimes g,\quad \epsilon g=1,\quad Sg=g^{-1},\quad \mathcal{R}=n^{-1}\sum_{a,b=0}^{n-1}q^{-ab}g^a\otimes g^b. \tag{28}$$

Proof We assume that k is of suitable characteristic. To verify the non-trivial quasitriangular structure we use that $n^{-1}\sum_{b=0}^{n-1}q^{ab}=\delta_{a,0}$. Then $\mathcal{R}_{13}\mathcal{R}_{23}=n^{-2}\sum q^{-(ab+cd)}g^a\otimes g^c\otimes g^{b+d}$ $=n^{-2}\sum q^{-b(a-c)}q^{-cb'}g^a\otimes g^c\otimes g^{b'}=n^{-1}\sum q^{-ab'}g^a\otimes g^a\otimes g^{b'}=(\Delta\otimes\mathrm{id})\mathcal{R}$ where $b'=b+d$ was a change of variables. Similarly for the second of (22). The remaining axiom (23) is automatic because the Hopf algebra is both commutative and cocommutative. □

Example 1.8 *[2][54] Let G be a finite Abelian group and $k(G)$ its function Hopf algebra. Then a quasitriangular structure on $k(G)$ means a function $\mathcal{R}\in H\otimes H$ obeying*

$$\mathcal{R}(gh,f)=\mathcal{R}(g,f)\mathcal{R}(h,f),\quad \mathcal{R}(g,hf)=\mathcal{R}(g,h)\mathcal{R}(g,f),\quad \mathcal{R}(g,e)=1=\mathcal{R}(e,g)$$

for all g,h,f in G and e the identity element. I.e., a quasitriangular structure on $k(G)$ means precisely a bicharacter of G.

Proof We identify $k(G)\otimes k(G)$ with functions on $G\times G$, with pointwise multiplication. Using the comultiplication given by multiplication in G we have at once that (22) corresponds to the first two displayed equations. Axiom (23) becomes $hg\mathcal{R}(g,h)=\mathcal{R}(g,h)gh$ and so is automatic because the group is Abelian. Given these first two of the stated conditions, the latter two hold iff $\mathcal{R}$ is invertible. □

The $\mathbb{Z}_n'$ example here also has an immediate generalization to the group algebra kG of a finite Abelian group equipped with a bicharacter on $\hat{G}$. This just coincides with the last example applied to $k(\hat{G})=kG$. Finally, we return to our basic construction,

Example 1.9 *[11] Let H be a finite-dimensional bialgebra. Then $D(H)$ is quasitriangular. In the Hopf algebra case the quasitriangular structure is $\mathcal{R} = \sum_a (f^a \otimes 1) \otimes (1 \otimes e_a)$ where $H = \{e_a\}$ is a basis and $\{f^a\}$ a dual basis.*

Proof The result is due to Drinfeld. A direct proof in the abstract Hopf algebra setting appeared in [33]. The easiest way to show that $\mathcal{R}$ is invertible is to verify in view of (26) that $(S \otimes \mathrm{id})(\mathcal{R})$ is the inverse. □

Theorem 1.10 *e.g.[35] Let $(H, \mathcal{R})$ be a quasitriangular bialgebra. Then the category ${}_H\mathcal{M}$ of modules is braided. In the Hopf algebra case the finite-dimensional modules are rigid. The braiding and the action on duals are*

$$\Psi_{V,W}(v \otimes w) = \sum \mathcal{R}^{(2)} \triangleright w \otimes \mathcal{R}^{(1)} \triangleright v, \quad h \triangleright f = f((Sh) \triangleright (\ ))$$

with ev, coev *as in (17).*

Proof $h \triangleright \Psi(v \otimes w) = (\Delta h) \triangleright \tau(\mathcal{R} \triangleright (v \otimes w)) = \tau((\Delta^{\mathrm{op}} h) \mathcal{R} \triangleright (v \otimes w)) = \tau(\mathcal{R}(\Delta h) \triangleright (v \otimes w)) = \Psi(h \triangleright (v \otimes w))$ in virtue (23). It is easy to see that (22) likewise just correspond to the hexagons (3) or (8). Functoriality is also easily shown. For an early treatment of this topic see [35, Sec. 7]. Note that if $\mathcal{R}_{21} = \mathcal{R}^{-1}$ (the triangular rather than quasitriangular case) we have Ψ symmetric rather than braided. This was the case treated in [11] though surely the general quasitriangular case was also known to some experts at the time or shortly thereafter. □

Proposition 1.11 *[50, Sec. 6] Let $H = \mathbb{Z}_2'$ denote the quantum group in Example 1.7 with $n = 2$. Then $\mathcal{C} = {}_{\mathbb{Z}_2'}\mathcal{M} =$ SuperVec the category of super-vector spaces.*

Proof One can easily check that this $\mathbb{Z}_2'$ is indeed a quasitriangular (in fact, triangular) Hopf algebra. Hence we have a (symmetric) tensor category of representations. Writing $p = \frac{1-g}{2}$ we have $p^2 = p$ hence any representation V splits as $V_0 \oplus V_1$ according to the eigenvalue of p. We can also write $\mathcal{R} = 1 - 2p \otimes p$ and hence from Theorem 1.10 we compute $\Psi(v \otimes w) = \tau(\mathcal{R} \triangleright (v \otimes w)) = (1 - 2p \otimes p)(w \otimes v) = (1 - 2|v||w|)w \otimes v = (-1)^{|v||w|} w \otimes v$ as in (6). □

So this non-standard quasitriangular Hopf algebra $\mathbb{Z}_2'$ (non-standard because of its non-trivial $\mathcal{R}$) recovers the category of super-spaces with its correct symmetry Ψ. In just the same way the category $\mathcal{C}_n = {}_{\mathbb{Z}_n'}\mathcal{M}$ consists of vector spaces that split as $V = \oplus_{a=0}^{n-1} V_a$ with the degree of an element defined by the action $g \triangleright v = q^{|v|} v$ where q is a primitive n-th root of unity. From Theorem 1.10 and (28) we find

$$\Psi_{V,W}(v \otimes w) = q^{|v||w|} w \otimes v. \tag{29}$$

Thus we call $\mathcal{C}_n$ the category of *anyonic vector spaces* of fractional statistics $\frac{1}{n}$, because just such a braiding is encountered in anyonic physics. For $n > 2$ the category is strictly braided in the sense that $\Psi \neq \Psi^{-1}$. There are natural anyonic traces and anyonic dimensions generalizing the super-case[47]

$$\underline{\dim}(V) = \sum_{a=0}^{n-1} q^{-a^2} \dim V_a, \qquad \underline{\mathrm{Tr}}(f) = \sum_{a=0}^{n-1} q^{-a^2} \mathrm{Tr}\, f|_{V_a}. \tag{30}$$

We see that this anyonic category is generated by the quantum group $\mathbb{Z}'_n$.

Obviously we can take this idea for generalising super-symmetry to the further case of Example 1.8. In this case a $k(G)$-module just means a G-graded space where $f{\triangleright}v = f(|v|)v$ on homogeneous elements of degree $|v| \in G$. This is well-known to Hopf algebraists for some time: the new ingredient is that a bicharacter gives our G-graded spaces a natural braided-transposition Ψ. We have given plenty of other less obvious examples of braided categories generated in this way from quasitriangular Hopf algebras [50, Sec. 6][45]. Our idea in this work is not to use Hopf algebras in connection with deformations (the usual setting) but rather as the 'generator' of a category within which we shall later make algebraic constructions. This is how quantum groups are naturally used to generalise supersymmetry. In this context they are typically discrete.

1.3 Dual Quasitriangular Structures

In this section we describe the dual results to those above. If a quasitriangular Hopf algebra is almost cocommutative up to conjugation then its dual Hopf algebra should be almost commutative up to 'conjugation' in the convolution algebra. The relevant axioms are obtained by dualizing in the standard way by writing out the axioms of a quasitriangular Hopf algebra as diagrams and then reversing all the arrows (and a left-right reversal). Obviously it is the axioms that are being dualised and not any specific Hopf algebra. This is important because in the infinite-dimensional case the dual axioms are weaker. This is a rigorous way to work with the standard quantum groups over a field as appreciated in [36] among other places.

We will always denote our dual quasitriangular bialgebras and Hopf algebras by A (to avoid confusion). These are equipped now with a map $\mathcal{R} : A \otimes A \to k$ which should be invertible in $\mathrm{Hom}(A \otimes A, k)$ in the sense that there exists a map $\mathcal{R}^{-1} : A \otimes A \to k$ such that

$$\sum \mathcal{R}^{-1}(a_{(1)} \otimes b_{(1)})\mathcal{R}(a_{(2)} \otimes b_{(2)}) = \epsilon(a)\epsilon(b) = \sum \mathcal{R}(a_{(1)} \otimes b_{(1)})\mathcal{R}^{-1}(a_{(2)} \otimes b_{(2)}).$$

Keeping such considerations in mind, it is easy to dualize the remainder of Drinfeld's axioms to obtain the following definition.

Definition 1.12 *A dual quasitriangular bialgebra or Hopf algebra $(A, \mathcal{R})$ is a bialgebra or Hopf algebra A and a convolution-invertible map $\mathcal{R} : A \otimes A \to k$ such that*

$$\mathcal{R}(ab \otimes c) = \sum \mathcal{R}(a \otimes c_{(1)})\mathcal{R}(b \otimes c_{(2)}), \quad \mathcal{R}(a \otimes bc) = \sum \mathcal{R}(a_{(1)} \otimes c)\mathcal{R}(a_{(2)} \otimes b) \tag{31}$$

$$\sum b_{(1)}a_{(1)}\mathcal{R}(a_{(2)} \otimes b_{(2)}) = \sum \mathcal{R}(a_{(1)} \otimes b_{(1)})a_{(2)}b_{(2)} \tag{32}$$

for all $a, b, c \in A$.

This looks a little unfamiliar but is in fact obtained by replacing the multiplication in Definition 1.5 by the convolution product and the comultiplication by the multiplication in A. Axiom (32) is the dual of (23) and says, as promised, that A is almost commutative - up to $\mathcal{R}$. Axioms (31) are the dual of (22) and say that $\mathcal{R}$ is a 'bialgebra bicharacter'. They should be compared with Example 1.14 below. We also have analogues of the various results in Section 1.2. Again, the new language is perhaps unfamiliar so we give some of the proofs in this dual form in detail.

Lemma 1.13 *[49] If $(A, \mathcal{R})$ is a dual quasitriangular bialgebra then*

$$\mathcal{R}(a \otimes 1) = \epsilon(a) = \mathcal{R}(1 \otimes a). \tag{33}$$

$$\sum \mathcal{R}(a_{(1)} \otimes b_{(1)})\mathcal{R}(a_{(2)} \otimes c_{(1)})\mathcal{R}(b_{(2)} \otimes c_{(2)}) = \sum \mathcal{R}(b_{(1)} \otimes c_{(1)})\mathcal{R}(a_{(1)} \otimes c_{(2)})\mathcal{R}(a_{(2)} \otimes b_{(2)}) \tag{34}$$

for all a, b, c in A. If A is a Hopf algebra then in addition,

$$\mathcal{R}(Sa \otimes b) = \mathcal{R}^{-1}(a \otimes b), \qquad \mathcal{R}^{-1}(a \otimes Sb) = \mathcal{R}(a \otimes b), \quad \mathcal{R}(Sa \otimes Sb) = \mathcal{R}(a \otimes b) \tag{35}$$

$$\exists S^{-1}, \quad v(a) = \sum \mathcal{R}(a_{(1)} \otimes Sa_{(2)}), \qquad \sum a_{(1)}v(a_{(2)}) = \sum v(a_{(1)})S^2 a_{(2)} \tag{36}$$

Proof Using (31) we have

$$\begin{aligned}\mathcal{R}(a \otimes 1) &= \sum (\mathcal{R}^{-1}(a_{(1)} \otimes 1)\mathcal{R}(a_{(2)} \otimes 1))\mathcal{R}(a_{(3)} \otimes 1)\\ &= \sum \mathcal{R}^{-1}(a_{(1)} \otimes 1)(\mathcal{R}(a_{(2)} \otimes 1)\mathcal{R}(a_{(3)} \otimes 1)) = \sum \mathcal{R}^{-1}(a_{(1)} \otimes 1)\mathcal{R}(a_{(2)} \otimes 1.1) = \epsilon(a)\end{aligned}$$

as in (33). Likewise on the other side. Also, if $\mathcal{R}^{-1}$ exists it is unique. Hence for A a Hopf algebra it is given by $\mathcal{R}^{-1}(a \otimes b) = \mathcal{R}(Sa \otimes b)$ (use axioms (31)). In this case $a \otimes b \mapsto \mathcal{R}(Sa \otimes Sb)$ is convolution inverse to $\mathcal{R}^{-1}$ because $\sum \mathcal{R}(Sa_{(1)} \otimes Sb_{(1)})\ \mathcal{R}(Sa_{(2)} \otimes b_{(2)}) = \sum \mathcal{R}(Sa \otimes (Sb_{(1)})b_{(2)}) = \mathcal{R}(Sa \otimes 1)\epsilon(b) = \epsilon(a)\epsilon(b)$ etc. Hence $\mathcal{R}(Sa \otimes Sb) = \mathcal{R}(a \otimes b)$, proving the other side and the third part of (35). For (34) we apply the second of (31), (32) and the second of (31) again,

$$\begin{aligned}\sum \left(\mathcal{R}(a_{(1)} \otimes b_{(1)})\mathcal{R}(a_{(2)} \otimes c_{(1)})\right) \mathcal{R}(b_{(2)} \otimes c_{(2)}) &= \sum \mathcal{R}(a \otimes c_{(1)}b_{(1)})\mathcal{R}(b_{(2)} \otimes c_{(2)})\\ = \sum \mathcal{R}(b_{(1)} \otimes c_{(1)})\mathcal{R}(a \otimes b_{(2)}c_{(2)}) &= \sum \mathcal{R}(b_{(1)} \otimes c_{(1)}) \left(\mathcal{R}(a_{(1)} \otimes c_{(2)})\mathcal{R}(a_{(2)} \otimes b_{(2)})\right).\end{aligned}$$

For (36) one defines $v : A \to k$ as shown (and similarly a map $u : A \to k$) and checks the relevant facts analogous to Lemma 1.6. This is done in complete detail in [49, Appendix] to which we refer the reader. □

Note that if $\mathcal{R}$ is a linear map obeying (33) and (31) and if A is a Hopf algebra, we can use (35) as a definition of $\mathcal{R}^{-1}$. Some authors in defining similar notions have made (33) an axiom in the bialgebra case (this is the case in [22] and in first versions of some other works). For this and other reasons we stick to our original terminology from[36][43][48][49] with axioms and properties as above.

Example 1.14 *Let G be an Abelian group and kG its group algebra. This is dual quasitriangular iff there is a function $\mathcal{R} : G \times G \to k$ obeying the bicharacter conditions in Example 1.8.*

Proof In the group algebra we can work with group-like elements (these form a basis). On such elements the axioms in (31)-(32) simplify: simply drop the $_{(1)}, _{(2)}$ suffixes! This immediately reduces to the bimultiplicativity while invertibility corresponds once again (given this) to (33). □

This example is clearly identical in content to Example 1.8. Because kG is dual to $k(G)$ by evaluation, it is obvious that a dual quasitriangular structure on kG is just the same thing as a quasitriangular structure on $k(G)$, namely as we see, a bicharacter. The result is however, more transparent from the dual quasitriangular point of view and slightly more general. This example is the reason that we called $\mathcal{R}$ obeying (31) a bialgebra bicharacter[36]. For a concrete example, one can take $G = \mathbb{Z}_n$ and $\mathcal{R}(a,b) = q^{ab}$ with q a primitive n-th root of unity to give a non-standard dual quasitriangular structure on $k\mathbb{Z}_n$. It is just Example 1.7 after a $\mathbb{Z}_n$-Fourier transform.

Now let R be an invertible matrix solution of the QYBE as in Example 1.1. There is a by-now standard bialgebra $A(R)$[11][14] defined by generators 1 and $\mathbf{t} = \{t^i{}_j\}$ (regarded as an $n \times n$ matrix) and the relations, comultiplication and counit

$$R^i{}_a{}^j{}_b t^a{}_k t^b{}_l = t^j{}_b t^i{}_a R^a{}_k{}^b{}_l,\ \Delta t^i{}_j = t^i{}_a \otimes t^a{}_j,\ \epsilon t^i{}_j = \delta^i{}_j,\ \text{i.e. } R\mathbf{t}_1\mathbf{t}_2 = \mathbf{t}_2\mathbf{t}_1 R,\ \Delta\mathbf{t} = \mathbf{t}\otimes\mathbf{t},\ \epsilon\mathbf{t} = \mathrm{id} \tag{37}$$

where $\mathbf{t}_1$ and $\mathbf{t}_2$ denote $\mathbf{t}\otimes\mathrm{id}$ and $\mathrm{id}\otimes\mathbf{t}$ in $M_n \otimes M_n$ with values in $A(R)$.

The power of this matrix notation lies in the fact that $\otimes$ is used only to refer to the abstract tensor product of copies of the algebra (as in defining the axioms of a Hopf algebra etc). The matrix tensor product as in $M_n \otimes M_n$ is suppressed and its role is replaced by the suffices $_{1,2}$ etc when needed. Thus $R = R_{12}$ (with the indices suppressed when there are only two M_n in the picture) while $\mathbf{t}_1$ means that the $\mathbf{t}$ is viewed as a matrix in the first M_n (with values in $A(R)$). The rules of the notation are that matrices are understood as multiplied in the usual order, independently in the $_{1,2}$ etc copies of M_n. Using this notation (or directly with indices) one can see at once that $A(R)$ has two fundamental representations $\rho^\pm$ in M_n defined by

$$\rho^+(t^i{}_j)^k{}_l = R^i{}_j{}^k{}_l,\quad \rho^-(t^i{}_j)^k{}_l = R^{-1}{}^k{}_l{}^i{}_j,\quad \text{i.e.}\quad \rho_2^+(\mathbf{t}_1) = R_{12},\quad \rho_2^-(\mathbf{t}_1) = R_{21}^{-1}. \tag{38}$$

In the compact notation the proof reads

$$\begin{aligned}
\rho_3^+(R_{12}\mathbf{t}_1\mathbf{t}_2) &= R_{12}\rho_3^+(\mathbf{t}_1)\rho_3^+(\mathbf{t}_2) = R_{12}R_{13}R_{23}\\
&= R_{23}R_{13}R_{12} = \rho_3^+(\mathbf{t}_2)\rho_3^+(\mathbf{t}_1)R_{12} = \rho_3^+(\mathbf{t}_2\mathbf{t}_1R_{12})\\
\rho_3^-(R_{12}\mathbf{t}_1\mathbf{t}_2) &= R_{12}\rho_3^-(\mathbf{t}_1)\rho_3^-(\mathbf{t}_2) = R_{12}R_{31}^{-1}R_{32}^{-1}\\
&= R_{32}^{-1}R_{31}^{-1}R_{12} = \rho_3^-(\mathbf{t}_2)\rho_3^-(\mathbf{t}_1)R_{12} = \rho_3^-(\mathbf{t}_2\mathbf{t}_1R_{12}).
\end{aligned}$$

Note that for ρ^- we need $R_{12}R_{31}^{-1}R_{32}^{-1} = R_{32}^{-1}R_{31}^{-1}R_{12}$, i.e. $R_{31}R_{32}R_{12} = R_{12}R_{32}R_{31}$ which is again the QYBE after a relabeling of the positions in $M_n^{\otimes 3}$.

This bialgebra $A(R)$ is important because for the standard R-matrices one has a convenient construction of the quantum function algebras $\mathcal{O}_q(G)$ deforming the ring of representative functions on compact simple group G[14]. One has to quotient the bialgebra by suitable further relations (or localise a determinant) to obtain a Hopf algebra. In the standard case of course one knew that the result was dual quasitriangular because of Drinfeld's result that $U_q(g)$ was quasitriangular (over formal power-series). The question for a general R-matrix was not so clear at the time and was resolved in [35][34] where we showed that there is always some kind of quasitriangular structure in the form of a map $\mathcal{R} : A(R) \to A(R)^{**}$. In the more modern setting our result reads as follows. One can also see subsequent works such as [28] but I retain here the strategy (which comes from physics) of my original proofs, this time with full pedagogical details.

Proposition 1.15 *Let R be an invertible solution of the QYBE in $M_n \otimes M_n$. Then the associated matrix bialgebra $A(R)$ is dual quasitriangular with $\mathcal{R} : A(R) \otimes A(R) \to k$ given by $\mathcal{R}(\mathbf{t} \otimes 1) = \mathrm{id} = \mathcal{R}(1 \otimes \mathbf{t})$ and $\mathcal{R}(\mathbf{t}_1 \otimes \mathbf{t}_2) = R$ extended as a bialgebra bicharacter according to (31). Explicitly,*

$$
\begin{aligned}
&\mathcal{R}(t^{i_1}{}_{j_1} t^{i_2}{}_{j_2} \cdots t^{i_M}{}_{j_M} \otimes t^{k_N}{}_{l_N} t^{k_{N-1}}{}_{l_{N-1}} \cdots t^{k_1}{}_{l_1}) \\
&= R^{i_1}{}_{m_{11}}{}^{k_1}{}_{n_{11}} \; R^{m_{11}}{}_{m_{12}}{}^{k_2}{}_{n_{21}} \; \cdots \; R^{m_{1N-1}}{}_{j_1}{}^{k_N}{}_{n_{N1}} \\
&\quad R^{i_2}{}_{m_{21}}{}^{n_{11}}{}_{n_{12}} R^{m_{21}}{}_{m_{22}}{}^{n_{21}}{}_{n_{22}} \\
&\quad \vdots \qquad\qquad\qquad\qquad \vdots \\
&\quad R^{i_M}{}_{m_{M1}}{}^{n_{1M-1}}{}_{l_1} \; \cdots \; \cdots \; R^{m_{MN-1}}{}_{j_M}{}^{n_{NM-1}}{}_{l_N} = Z_R(\, {}_I \overset{K}{\underset{L}{\square}} {}_J \,)
\end{aligned}
$$

where the last notation is as a partition function[35, Sec. 5.2]. Here $I = (i_1, \cdots, i_M)$ and $K = (k_1, \cdots k_N)$ etc. There is a similar expression for $\mathcal{R}^{-1}$. If we adopt the notation $\bar{K} = (k_N, \cdots k_1)$ and $t^{i_1}{}_{j_1} \cdots t^{i_M}{}_{j_M} = t^I{}_J$ then

$$
\mathcal{R}(t^I{}_J \otimes t^{\bar{K}}{}_{\bar{L}}) = Z_R(\, {}_I \overset{K}{\underset{L}{\square}} {}_J \,), \quad \mathcal{R}^{-1}(t^{\bar{I}}{}_J \otimes t^K{}_L) = Z_{R^{-1}}(\, {}_I \overset{K}{\underset{L}{\square}} {}_J \,)
$$

Proof Note that $\mathcal{R}(\mathbf{t}_1 \otimes \mathbf{t}_2) = \rho_2^+(\mathbf{t}_1)$ and the proof that this extends in its first input as a bialgebra bicharacter is exactly the proof above that ρ^+ extends to products as a representation. Thus we have $\mathcal{R}(a \otimes \mathbf{t}) = \rho^+(a)$ for all a and $\mathcal{R}(ab \otimes \mathbf{t}) = \mathcal{R}(a \otimes \mathbf{t})\mathcal{R}(b \otimes \mathbf{t})$ as we require for the first of (31). In particular,

$$
\mathcal{R}(\mathbf{t}_1 \mathbf{t}_2 \cdots \mathbf{t}_M \otimes \mathbf{t}_{M+1}) = \rho_{M+1}^+(\mathbf{t}_1 \mathbf{t}_2 \cdots \mathbf{t}_M) = R_{1M+1} \cdots R_{MM+1}
$$

is well-defined. Next the tensor product of representations is also a representation (because $A(R)$ is a bialgebra), hence there is a well-defined algebra map $\rho^{+\otimes N} : A(R) \to M_n^{\otimes N}$ given by $\rho^{+\otimes N}(a) = (\rho_1^+ \otimes \rho_2^+ \otimes \cdots \otimes \rho_N^+) \circ \Delta^{N-1}(a)$. In particular,

$$
\begin{aligned}
&R_{1M+1} R_{1M+2} \cdots R_{1M+N} \\
&R_{2M+1} R_{2M+2} \cdots R_{2M+N} = \rho_{M+1}^+(\mathbf{t}_1 \cdots \mathbf{t}_M) \cdots \rho_{M+N}^+(\mathbf{t}_1 \cdots \mathbf{t}_M) = (\rho^+)^{\otimes N}(\mathbf{t}_1 \cdots \mathbf{t}_M) \\
&\cdots \quad \cdots \\
&R_{MM+1} \cdots \quad \cdots R_{MM+N}
\end{aligned}
$$

also depends only on $\mathbf{t}_1 \mathbf{t}_2 \cdots \mathbf{t}_M$ as an element of $A(R)$. The array on the left can be read (and multiplied up) column after column (so that the first equality is clear) or row after row (like reading a book). The two are the same when we bear in mind that R living in distinct copies of $M_n \otimes M_n$ commute. The expression is just the array Z_R in our compact notation. If we define $\mathcal{R}(\mathbf{t}_1 \mathbf{t}_2 \cdots \mathbf{t}_M \otimes \mathbf{t}_{M+N} \cdots \mathbf{t}_{M+1})$ as this array, we know that the second of (31) will hold and that $\mathcal{R}$ is well defined in its first input.

Now we repeat the steps above for the second input of $\mathcal{R}$. Thus $\mathcal{R}(\mathbf{t}_1 \otimes \mathbf{t}_2) = R = \bar{\rho}_1^+(\mathbf{t}_2)$ extends in its second input as a bialgebra bicharacter since this $\bar{\rho}^+$ extends as an antirepresentation $A(R) \to M_n$ (proof similar to that for ρ^+). Thus we define $\mathcal{R}(\mathbf{t} \otimes a) = \bar{\rho}^+(a)$ and in particular,

$$
\mathcal{R}(\mathbf{t}_M \otimes \mathbf{t}_{M+N} \cdots \mathbf{t}_{M+2} \mathbf{t}_{M+1}) = \bar{\rho}_M^+(\mathbf{t}_{M+N} \cdots \mathbf{t}_{M+2} \mathbf{t}_{M+1}) = R_{MM+1} R_{MM+2} \cdots R_{MM+N}
$$

is well-defined. Likewise, we can take tensor products of $\bar{\rho}^+$ and will again have well-defined anti-representations. Hence

$$\begin{array}{l} R_{1M+1}R_{1M+2}\cdots R_{1M+N} \\ R_{2M+1}R_{2M+2}\cdots R_{2M+N} = \bar{\rho}_1^+(\mathbf{t}_{M+N}\cdots\mathbf{t}_{M+1})\cdots\bar{\rho}_M^+(\mathbf{t}_{M+N}\cdots\mathbf{t}_{M+1}) = (\bar{\rho}^+)^{\otimes M}(\mathbf{t}_{M+N}\cdots\mathbf{t}_{M+1}) \\ \cdots \quad \cdots \\ R_{MM+1}\cdots \quad \cdots R_{MM+N} \end{array}$$

depends only on $\mathbf{t}_{M+N}\cdots\mathbf{t}_{M+2}\mathbf{t}_{M+1}$ as an element of $A(R)$. The first equality comes from writing out the $\bar{\rho}^+$ and rearranging the resulting array (bearing in mind that copies of R in distinct M_n tensor factors commute). The resulting array of matrices then coincides with that above, which we have already defined as $\mathcal{R}(\mathbf{t}_1\cdots\mathbf{t}_M\otimes\mathbf{t}_{M+N}\cdots\mathbf{t}_{M+1})$. We see that this array then is well defined as a map $A(R)\otimes A(R)\to k$, in its fist input (for fixed $\mathbf{t}_{M+N},\cdots,\mathbf{t}_{M+1}$) by its realization as a tensor power of ρ^+ and in its second input (for fixed $\mathbf{t}_1,\cdots,\mathbf{t}_M$) by its realization as a tensor power of $\bar{\rho}^+$. By its construction, it obeys (31).

Next, we note that when R is invertible there is a similar construction for $\mathcal{R}^{-1}$ to that for $\mathcal{R}$ above. Here $\mathcal{R}^{-1}$ obeys equations similar to (31) but with its second input multiplicative and its first input antimultiplicative. We use R^{-1} in the role of R, for example, $\mathcal{R}^{-1}(\mathbf{t}\otimes a) = \rho^-(a)$ extends as a representation, while $\mathcal{R}^{-1}(a\otimes\mathbf{t})$ extends as an antirepresentation. The steps are entirely analogous to those above, and we arrive at the partition function $Z_{R^{-1}}$. We have to show that $\mathcal{R},\mathcal{R}^{-1}$ are inverse in the convolution algebra of maps $A(R)\otimes A(R)\to k$. Explicitly, we need,

$$\mathcal{R}(t^I{}_A\otimes t^{\bar{K}}{}_{\bar{B}})\mathcal{R}^{-1}(t^{\bar{A}}{}_J\otimes t^B{}_L) = \delta^I{}_J\delta^K{}_L, \quad \text{i.e.} \quad Z_R(\,{}_I\overset{\bar{K}}{\underset{\bar{B}}{\Box}}{}_A)Z_{R^{-1}}(\,{}_{\bar{A}}\overset{B}{\underset{L}{\Box}}{}_{\bar{J}}) = \delta^I{}_J\delta^K{}_L \tag{39}$$

and similarly on the other side. Writing the arrays in our compact notation we have

$$\begin{array}{l} R_{1M+N}\cdots R_{1M+1} \\ \cdots \quad \cdots \\ R_{M-1M+N}\cdots R_{M-1M+1} \\ R_{MM+N}\cdots R_{MM+1}R^{-1}_{MM+1}\cdots R^{-1}_{MM+N} \\ \qquad\qquad\qquad R^{-1}_{M-1M+1}\cdots R^{-1}_{M-1M+N} \\ \qquad\qquad\qquad \cdots \quad \cdots \\ \qquad\qquad\qquad R^{-1}_{1M+1}\cdots R^{-1}_{1M+N} \end{array}$$

Here the copies of M_n numbered $1\cdots M$ on the left correspond to the index I, on the right to $\bar{J}$ (they occur reversed). The copies of M_n numbered $M+1\cdots M+N$ correspond on the top to $\bar{K}$ (they occur reversed) and on the bottom to L. In between they are matrix-multiplied as indicated, corresponding to the sum over A, B. The overlapping line here collapses after cancellation of inverses ending in id in the copy of M_n numbered M, and results in a similar picture with one row less. Repeating this, the whole thing collapses to the identity in all the copies of M_n.

Finally, we check (32), which now takes the form

$$t^{\bar{K}}{}_{\bar{B}}t^I{}_A Z_R(\,{}_A\overset{B}{\underset{L}{\Box}}{}_J) = Z_R(\,{}_I\overset{K}{\underset{B}{\Box}}{}_A)t^A{}_J t^{\bar{B}}{}_{\bar{L}}. \tag{40}$$

In the compact notation we compute

$$\begin{aligned}
&\mathbf{t}_{M+N}\cdots\mathbf{t}_{M+1}\mathbf{t}_1\cdots\mathbf{t}_M R_{1M+1}\cdots R_{1M+N}\\
&\qquad\qquad\qquad\qquad\cdots\qquad\cdots\\
&\qquad\qquad\qquad\qquad R_{MM+1}\cdots R_{MM+N}\\
&= R_{1M+1}\cdots R_{1M+N}\mathbf{t}_1\mathbf{t}_{M+N}\cdots\mathbf{t}_{M+1}\mathbf{t}_2\cdots\mathbf{t}_M R_{2M+1}\cdots R_{2M+N}\\
&\qquad\qquad\qquad\qquad\qquad\qquad\qquad\qquad\cdots\qquad\cdots\\
&\qquad\qquad\qquad\qquad\qquad\qquad\qquad\qquad R_{MM+1}\cdots R_{MM+N}\\
&=\\
&\vdots\\
&= R_{1M+1}R_{1M+2}\cdots R_{1M+N}\\
&\quad R_{2M+1}R_{2M+2}\cdots R_{2M+N}\\
&\qquad\cdots\qquad\cdots\\
&\quad R_{MM+1}\cdots\qquad\cdots R_{MM+N}\mathbf{t}_1\mathbf{t}_2\cdots\mathbf{t}_M\mathbf{t}_{M+N}\cdots\mathbf{t}_{M+1}.
\end{aligned}$$

Here the copies of M_n numbered $1,\cdots,M$ on the left correspond to the index I, and the copies of M_n numbered $M+1,\cdots,M+N$ correspond on the top to $\bar{K}$ (they occur reversed), etc. The first equality makes repeated use of the relations (37) of $A(R)$ to give

$$\mathbf{t}_{M+N}\cdots\mathbf{t}_{M+1}\mathbf{t}_1 R_{1M+1}\cdots R_{1M+N} = R_{1M+1}\cdots R_{1M+N}\,\mathbf{t}_1\mathbf{t}_{M+N}\cdots\mathbf{t}_{M+1}.$$

The $\mathbf{t}_2\cdots\mathbf{t}_M$ move past the $R_{1M+1}\cdots R_{1M+N}$ freely since they live in different matrix spaces. This argument for the first equality is then applied to move $\mathbf{t}_{M+N}\cdots\mathbf{t}_{M+1}\mathbf{t}_2$, and so on. The arguments in this proof may appear complicated, but in fact this kind of repeated matrix multiplication (multiplication of entire rows or columns of matrices) is quite routine in the context of exactly solvable statistical mechanics (where the QYBE originated). □

Let us note that while the algebra relations (37) of $A(R)$ do not depend on the normalization of R, the dual quasitriangular structure does. The elements $t^{i_1}{}_{j_1}\cdots t^{i_M}{}_{j_M}$ of $A(R)$ have a well-defined degree $|t^{i_1}{}_{j_1}\cdots t^{i_M}{}_{j_M}| = M$ (the algebra is graded), and if $R' = \lambda R$ is a non-zero rescaling of our solution R then the corresponding dual quasitriangular structure is changed to

$$\mathcal{R}'(a\otimes b) = \lambda^{|a||b|}\mathcal{R}(a\otimes b) \tag{41}$$

on homogeneous elements. This is evident from the expression in terms of Z_R that we have obtained in the last proposition. Finally, in view of the reasons that we passed to the dual setting it is obvious that

Theorem 1.16 *Let $(A,\mathcal{R})$ be a dual quasitriangular bialgebra. Then $\mathcal{M}^A$ the category of right A-comodules is braided. In the Hopf algebra case the finite-dimensional comodules are rigid,*

$$\Psi_{V,W}(v\otimes w) = \sum w^{(\bar{1})}\otimes v^{(\bar{1})}\mathcal{R}(v^{(\bar{2})}\otimes w^{(\bar{2})}),\quad \beta_{V^*}(f) = (f\otimes S)\circ\beta_V$$

where $\beta_{V^*}(f)\in V^*\otimes A$ is given as a map $V\to A$.

Proof This is an entirely trivial dualization of the proof of Theorem 1.10 above. For example, Ψ is an intertwiner because

$$\begin{aligned}\Psi_{V,W}(v\otimes w) \ \mapsto\ & \sum w^{(\bar 1)(\bar 1)}\otimes v^{(\bar 1)(\bar 1)}\otimes w^{(\bar 1)(\bar 2)}v^{(\bar 1)(\bar 2)}\mathcal{R}(v^{(\bar 2)}\otimes w^{(\bar 2)})\\ &= \sum w^{(\bar 1)}\otimes v^{(\bar 1)}\otimes w^{(\bar 2)}{}_{(1)}v^{(\bar 2)}{}_{(1)}\mathcal{R}(v^{(\bar 2)}{}_{(2)}\otimes w^{(\bar 2)}{}_{(2)})\\ &= \sum w^{(\bar 1)}\otimes v^{(\bar 1)}\otimes \mathcal{R}(v^{(\bar 2)}{}_{(1)}\otimes w^{(\bar 2)}{}_{(1)})v^{(\bar 2)}{}_{(2)}w^{(\bar 2)}{}_{(2)}\\ &= \sum w^{(\bar 1)(\bar 1)}\otimes v^{(\bar 1)(\bar 1)}\otimes \mathcal{R}(v^{(\bar 1)(\bar 2)}\otimes w^{(\bar 1)(\bar 2)})v^{(\bar 2)}w^{(\bar 2)}\end{aligned}$$

where the arrow is the tensor product $W\otimes V$ coaction. We used (32). The result is Ψ applied to the result of the tensor product coaction $V\otimes W$. The hexagons (3) correspond in a similarly trivial way to (31). □

Note that $\mathcal{C}(V,R)$ in Example 1.1 forms a subcategory of $\mathcal{M}^{A(R)}$. Moreover, in the dualizable case there is a Hopf algebra $GL(R)\supset A(R)$ such that $\mathcal{C}(V,V^*,R)$ is a subcategory of $\mathcal{M}^{GL(R)}$. The relevant coactions are

$$e_i\mapsto e_j\otimes t^j{}_i,\qquad f^i\mapsto f^j\otimes St^i{}_j \tag{42}$$

and we recover from Theorem 1.16 the braidings quoted. Also, in [36][43][48][49] we regarded this proposition as a starting point and set out to prove something further, namely its converse. If $\mathcal{M}^A$ is braided then A has induced on it by Tannaka-Krein reconstruction a dual quasitriangular structure. We will see this in Section 3. [43] generalised the Tannaka-Krein theorem to the setting of dual quasi-Hopf algebras (associative up to an isomorphism cf[13]) while [48][49] generalised it to the braided setting. It is more or less the *sine qua non* for the work here.

2 Braided Tensor Product Algebra and Braided Hopf Algebras

So far we have described braided monoidal or quasitensor categories and ways to obtain them. Now we begin our main task and study algebraic structures living in such categories. For this we use the diagrammatic notation of Section 1.1. Detailed knowledge of quantum groups etc is not required in this section.

The idea of an algebra B in a braided category is just the usual one. Thus there should be product and unit morphisms

$$\cdot : B\otimes B\to B,\qquad \eta:\underline{1}\to B \tag{43}$$

obeying the usual associativity and unity axioms but now as morphisms in the category. Note that the term 'algebra' is being used loosely since we have not discussed direct sums and linearity under a field or ring. These notions are perfectly compatible with what follows but do not play any particular role in our general constructions.

The fundamental lemma for the theory we need is the generalization to this setting of the usual $\mathbb{Z}_2$-graded or super-tensor product of superalgebras:

Lemma 2.1 *[44]-[50] Let B,C be two algebras in a braided category. There is a* braided tensor product algebra *$B\underline{\otimes}C$, also living in the braided category. It has product $(\cdot_B\otimes\cdot_C)\circ(\mathrm{id}\otimes\Psi_{C,B}\otimes\mathrm{id})$ and tensor product unit morphism.*

Proof We repeat here the diagrammatic proof[41]. The box is the braided tensor product multiplication,

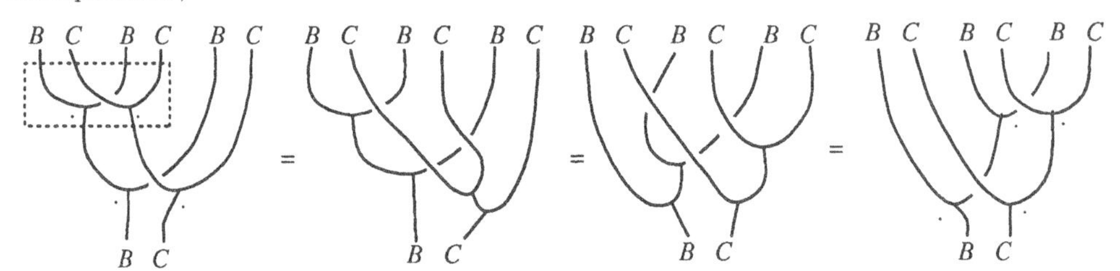

The first step uses functoriality as in (9) to pull the product morphism through the braid crossing. The second equality uses associativity of the products in B, C and the third equality uses functoriality again in reverse. The product is manifestly a morphism in the category because it is built out of morphisms. Finally, the unit is the tensor product one because the braiding is trivial on $\underline{1}$. □

In the concrete case the braided tensor product is generated by $B = B\otimes 1$ and $C = 1\otimes C$ and an exchange law between the two factors given by Ψ. This is because $(b\otimes 1)(1\otimes c) = (b\otimes c)$ while $(1\otimes c)(b\otimes 1) = \Psi(c\otimes b)$. Another notation is to label the elements of the second copy in the braided tensor product by $'$. Thus $b\equiv(b\otimes 1)$ and $c'\equiv(1\otimes c)$. Then if $\Psi(c\otimes b) = \sum b_k\otimes c_k$ say, we have the braided-tensor product relations

$$c'b \equiv (1\otimes c)(b\otimes 1) = \Psi(c\otimes b) = \sum b_k\otimes c_k \equiv \sum b_k c'_k. \tag{44}$$

This makes clear why the lemma generalizes the notion of $\mathbb{Z}_2$-graded or super-tensor product. Note also that there is an equally good *opposite braided tensor product* with the inverse braid crossing in Lemma 2.1. This is simply the braided tensor product algebra constructed in the mirror-reversed category $\bar{\mathcal{C}}$ but with the result viewed in our original category.

2.1 Braided Hopf Algebras

Armed with the braided tensor product of algebras in a braided category we can formulate the notion of Hopf algebra.

Definition 2.2 *[44]-[50] A Hopf algebra in a braided category or* braided-Hopf algebra *is (B, Δ, ϵ, S) where B is an algebra in the category and $\Delta : B \to B\underline{\otimes}B$, $\epsilon : B \to \underline{1}$ are algebra homomorphisms where $B\underline{\otimes}B$ has the braided tensor product algebra structure. In addition, Δ, ϵ obey the usual coassociativity and counity axioms to form a coalgebra in the category, and $S : B \to B$ obeys the usual axioms of an antipode. If there is no antipode then we speak of a braided-bialgebra or bialgebra in a braided category.*

In diagrammatic form the algebra homomorphism and braided-antipode axioms read

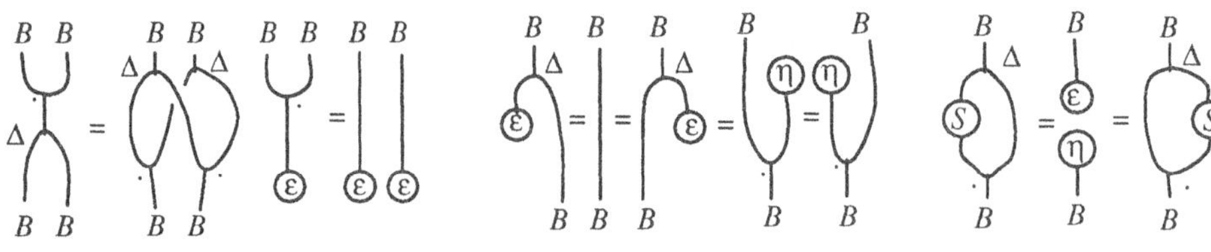

(45)

One then proceeds to develop the usual elementary theory for these Hopf algebras. For example, recall that the usual antipode is an antialgebra map.

Lemma 2.3 *For a braided-Hopf algebra B, the braided-antipode obeys $S(b{\cdot}c) = \cdot\Psi(Sb \otimes Sc)$ and $S(1) = 1$, or more abstractly, $S \circ \cdot = \cdot \circ \Psi_{B,B} \circ (S \otimes S)$ and $S \circ \eta = \eta$.*

Proof In diagrammatic form the proof is[45]

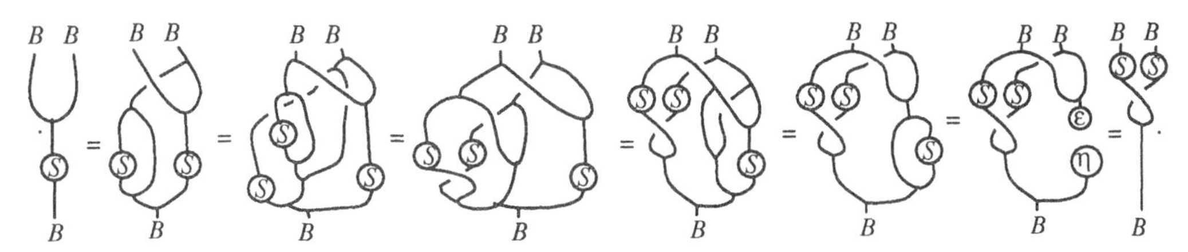

In the first two equalities we have grafted on some circles containing the antipode, knowing they are trivial from (45). We then use the coherence theorem to lift the second S over to the left, and associativity and coassociativity to reorganise the branches. The fifth equality uses the axioms (45) for Δ. □

Here we want to mention a powerful *input-output symmetry* of these axioms. Namely, turn the pages of this book upside down and look again at these diagrams. The axioms of a braided-Hopf algebra (45) are unchanged except that the roles of product/coproduct and unit/counit morphisms are interchanged. The proof of Lemma 2.1 becomes the proof of a new lemma expressing coassociativity of the braided-tensor product of two coalgebras. Meanwhile the proof of Lemma 2.3 reads as the proof of a new lemma that the braided-antipode is a braided-anti-coalgebra map.

This applies therefore to all results that we prove about bialgebras or Hopf algebras in braided categories provided all notions are suitably turned up-side-down. This is completely rigorous and nothing to do with finite-dimensionality or individual dual objects. In addition, there is a *left-right symmetry* of the axioms consisting of reflecting in mirror about a vertical axis combined with reversal of all braid crossings. These symmetries of the axioms can be taken together so that we obtain precisely four theorems for the price of one when we use the diagrammatic method.

An endemic problem for those working in Hopf algebras is that every time something is proven one has to laboriously figure out its input-output reversed version or its version with opposite left-right conventions. This problem is entirely solved by reflecting in a mirror or turning up-side-down.

2.2 Dual Braided Hopf Algebras

Suppose now that the category has dual objects (is rigid) in the sense explained in Section 1.1. In this case the input-output symmetry of the axioms of a Hopf algebra becomes realised concretely as the construction of a dual Hopf algebra.

Proposition 2.4 *If B is a braided-Hopf algebra, then its left-dual B^* is also a braided-Hopf algebra with product, coproduct, antipode, counit and unit given by*

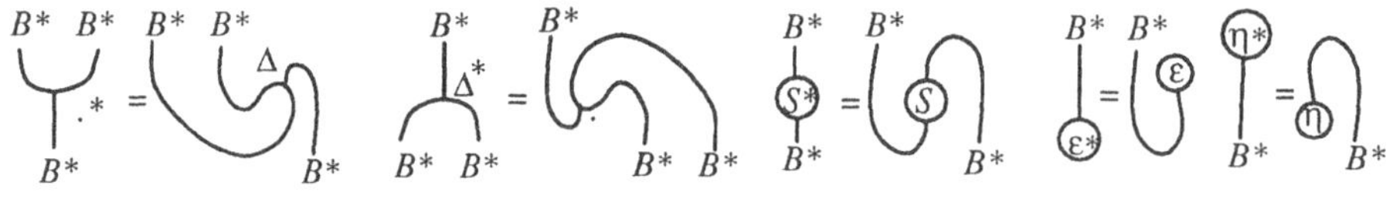

Proof Associativity and coassociativity follow at once from coassociativity and associativity of B. Their crucial compatibility property comes out as

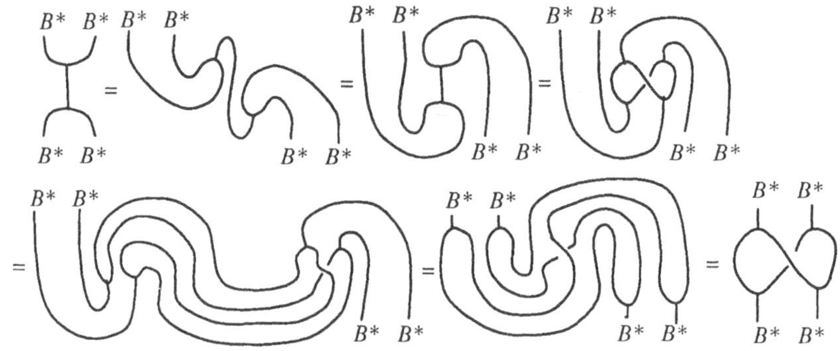

where we use the double-bend axiom (11) for dual objects. The antipode property comes out just as easily. □

We see that in diagrammatic form the dual bialgebra or Hopf algebra is obtained by rotating the desired structure map in an anticlockwise motion and without cutting any of the attaching strings. For right duals the motion should be clockwise. Let us stress that this dual-Hopf algebra construction of an individual object should not be confused with the rather more powerful input-output symmetry for the axioms introduced above.

2.3 Braided Actions and Coactions

Another routine construction is the notion of module and its input-output-reversed notion of comodule. These are just the obvious ones but now as morphisms in the category. Let us check that the tensor product of modules of a bialgebra is a module. If $V, \alpha_V : B \otimes V \to V$ and $W, \alpha_W : B \otimes W \to W$ are two left modules then

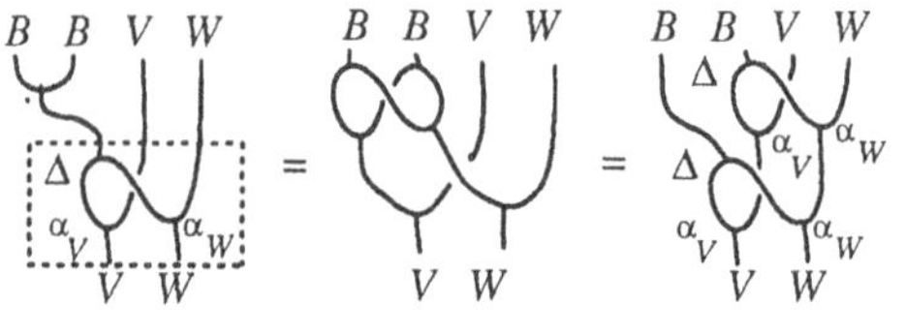

(46)

is the definition (in the box) of tensor product module and proof that it is indeed a module. The first equality is the homomorphism property of Δ. Likewise for right modules by left-right reflecting the proof in a mirror and also reversing all braid crossings.

As for usual Hopf algebras, the left-right symmetry can be concretely realised via the antipode. Thus if $V, \alpha^R : V \otimes B \to V$ is a right module then

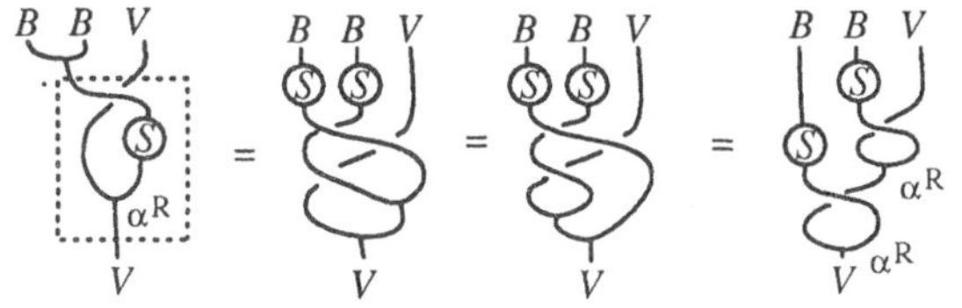

(47)

shows the construction of the corresponding left module. This is shown in the box.

Also if the left-module V has a left-dual V^* then this becomes a right-module with

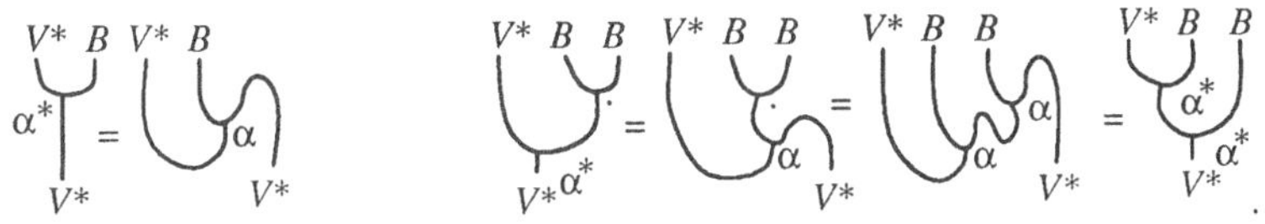

(48)

Combining these two constructions we conclude (with the obvious definition of intertwiners or morphisms between braided modules):

Proposition 2.5 *[49] Let B be a bialgebra in the braided category $\mathcal{C}$. Then the category ${}_B\mathcal{C}$ of braided left-modules is a monoidal category. If B is a Hopf algebra in $\mathcal{C}$ and $\mathcal{C}$ is rigid then ${}_B\mathcal{C}$ is rigid.*

Proof For the second part we feed the result of (48) into (47). A more traditional-style proof with commuting diagrams is in [49] for comparison. It should convince the reader of the power of the diagrammatic method. □

Naturally, a braided left B-module algebra is by definition an algebra living in the category ${}_B\mathcal{C}$. This means an algebra C such that

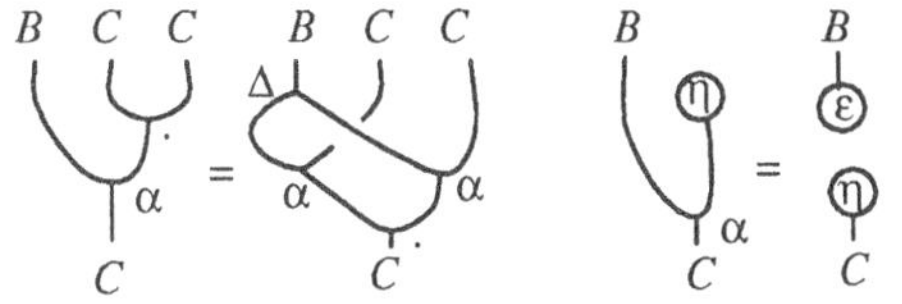

B–Module Algebra (49)

Likewise for other constructions familiar for actions of bialgebras or Hopf algebras. For example, a coalgebra C in the category ${}_B\mathcal{C}$ is a coalgebra such that

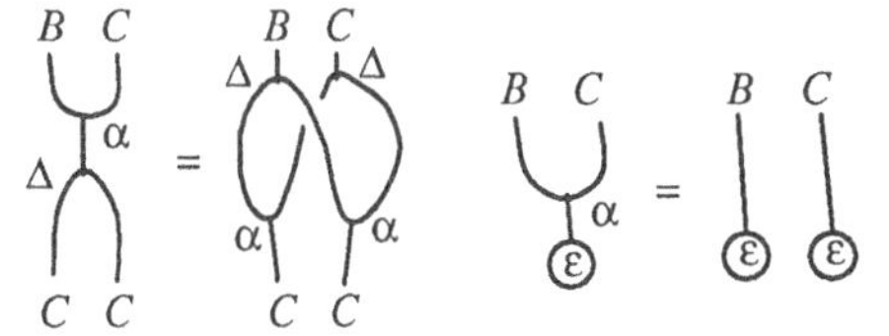

B–Module Coalgebra. (50)

For the right-handed theory reflect the above in a mirror and reverse all braid crossings. This gives the notion of right B-module algebras etc. Next, by turning the pages of this book upside down we have all the corresponding results for comodules in place of modules. Thus the category $\mathcal{C}^B$ of right-comodules had a tensor product and in the Hopf algebra and rigid case is also rigid. Likewise for left comodules.

Proposition 2.6 *[46] If C is a left B-module algebra then there is a braided* cross product *or semidirect product algebra $C{>\!\!\!\triangleleft}B$ built on the object $C\otimes B$. Likewise for right B-module algebras, left B-module coalgebras and right B-comodule coalgebras. The semidirect (co)product maps are*

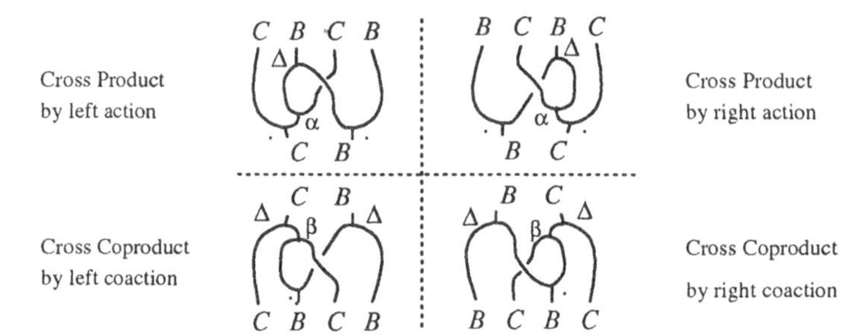

Proof We only need to prove one of these by our diagrammatic means to conclude all four. Full details are in [46]. □

Example 2.7 *[51, Appendix] Let B be a braided-Hopf algebra. Then B is a left B-module algebra by the* braided adjoint *action* $\cdot^2\circ(\mathrm{id}\otimes\Psi_{B,B})\circ(\mathrm{id}\otimes S\otimes\mathrm{id})\circ(\Delta\otimes\mathrm{id})$.

Example 2.8 *[58] Let B be a braided-Hopf algebra with left dual B^*. Then B is a right B^*-module algebra by a* braided right regular *action* $\mathrm{ev}_B\circ(\mathrm{id}\otimes S\otimes\mathrm{id})\circ(\mathrm{id}\otimes\Delta)\circ\Psi_{B,B^*}$

The verification of these examples is a nice demonstration of the techniques above. In diagrammatic form they read

Ad =   Reg =   (51)

The adjoint action leads to a notion of braided Lie algebra[58] among other applications, while the right regular action corresponds to the action of fundamental vector fields. It can also be used to construct a braided Weyl algebra cf [57].

2.4 Braided-(Co)-Commutativity

Next we come to the question of commutativity or cocommutativity in a braided category. Again, we only have to work with one of these and turn our diagrams up-side-down for the other. The main problem is that the naive opposite-coproduct

$$\bar{\Delta}=\Psi^{-1}_{B,B}\circ\Delta \tag{52}$$

does *not* make B into a bialgebra in our original braided category $\mathcal{C}$, but rather gives a bialgebra in the mirror-reversed category $\bar{\mathcal{C}}$. Thus there is a notion of opposite bialgebra (and if B is a Hopf algebra with invertible antipode then S^{-1} provides an antipode) but it forces us to leave the category.

Hence there is no way to consider bialgebras or Hopf algebras that are cocommutative in the sense that they coincide with their opposite. There is so far no intrinsic notion of braided-cocommutative Hopf algebras for this reason. On the other hand we have introduced in [44] the notion of a braided-cocommutativity with respect to a module. This is a property of a module on which B acts.

Definition 2.9 *[44] A braided left module* (V, α_V) *is* braided-cocommutative *(or B is braided-cocommutative with respect to V) if*

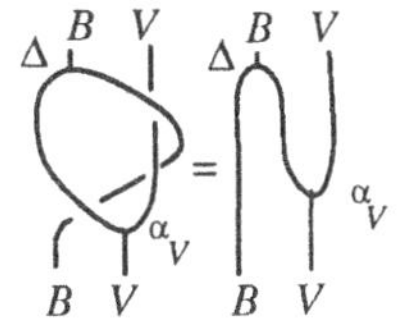

Braided – Cocommutativity

To understand this notion suppose that the category is symmetric not strictly braided. In this case $\Psi^2 = \mathrm{id}$ and we see that the condition is implied by $\bar{\Delta} = \Delta$. But in a general braided category we cannot disentangle V and must work with this weaker notion. Moreover, as far as such modules are concerned the bialgebra B has all the usual representation-theoretic features of usual cocommutative Hopf algebras. One of these is that their tensor product is symmetric under the usual transposition of the underlying vector spaces of modules. The parallel of this is

Proposition 2.10 *[45] Let B be a bialgebra in a braided category and define $\mathcal{O}(B) \subset {}_B\mathcal{C}$ the subcategory of braided-cocommutative modules. Then $\mathcal{O}(B)$ is closed under $\otimes$. Moreover, the tensor product in $\mathcal{O}$ is braided with braiding induced by the braiding in $\mathcal{C}$,*

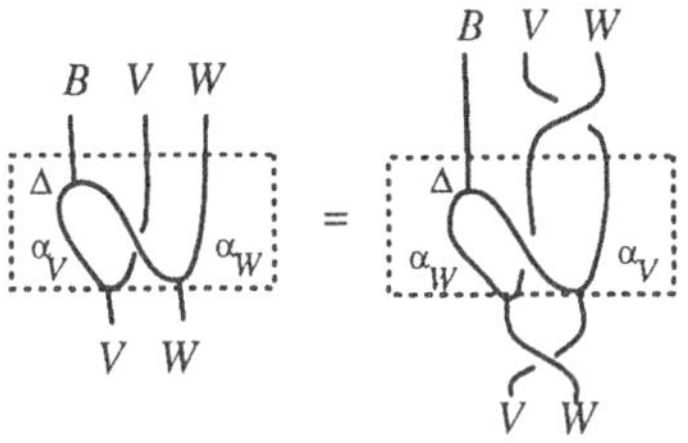

Braided – Commutativity of Product of Modules

Proof The first part is given in detail in [45] in a slightly more general context. The second part follows from Definition 2.9 by adding an action on W to both sides. □

The trivial representation is always braided-cocommutative. In many examples the adjoint representation in Example 2.7 is also braided-cocommutative. In this case one can formulate properties like those of an enveloping algebra of a braided-Lie algebra[58]. One can formally define a *braided group* as a pair consisting of a Hopf algebra in a braided category and a class of braided-cocommutative modules. This turns out to be a useful notion because in many situations it is only this weak notion of cocommutativity that is needed. For example

Theorem 2.11 *[46] Let B, C be Hopf algebras in a braided category and C a braided-cocommutative-B-module algebra and coalgebra. Then $C {>\!\!\!\triangleleft} B$ forms a Hopf algebra in the braided category with the braided tensor product coalgebra structure.*

2.5 Quantum-Braided Groups

We can go further and consider bialgebras that are quasi-cocommutative in some sense, analogous to the idea of a quasitriangular bialgebra in Section 1.2. To do this we require a second coproduct which we denote $\Delta^{\rm op}: B \to B\otimes B$ also making B into a bialgebra. In this case we have the notion of a braided B-module with respect to which $\Delta^{\rm op}$ behaves like an opposite coproduct. This is just as in Definition 2.9 but with the left hand Δ replaced by $\Delta^{\rm op}$. The class of such B-modules is denoted $\mathcal{O}(B,\Delta^{\rm op})$.

The second ingredient that we need is a quasitriangular structure which is understood now as a convolution-invertible morphism $\mathcal{R}: \underline{1} \to B\otimes B$. With these ingredients the analogue of Definition 1.5 is[45]

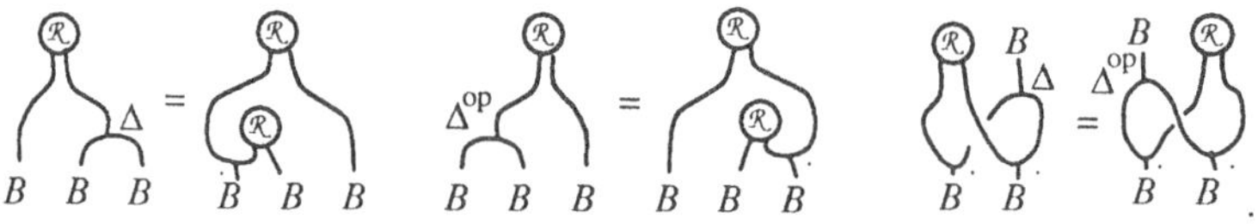

$$(53)$$

The braided analogue of Theorem 1.10 is then

Theorem 2.12 *[45] Let $(B,\Delta^{\rm op},\mathcal{R})$ be a quasitriangular bialgebra in a braided category. Then $\mathcal{O}(B,\Delta^{\rm op}) \subset {}_B\mathcal{C}$ is a braided monoidal category with braiding $\Psi^{\mathcal{O}}_{V,W} = \Psi_{V,W}\circ(\alpha_V\otimes\alpha_W)\circ\Psi_{B,V}\circ(\mathcal{R}\otimes{\rm id})$.*

Proof Diagrammatic proofs are in [45, Sec. 3]. □

The dual theory with comodules and an opposite product is developed in [48][49]. We mention here only that turning (53) up-side-down and then setting the category to be the usual one of vector spaces returns not the axioms of a dual quasitriangular structure as in Section 1.3 but its inverse. This reversal is due to the fact that the categorical dualization in Section 2.2 yields in the vector space category the opposite coproduct and product to the usual dualization.

3 Reconstruction Theorem

In this section we give a construction (not the only one) for bialgebras and Hopf algebras in braided categories. There is such a bialgebra associated to a pair of braided categories $\mathcal{C}\to\mathcal{V}$ or even to a single braided category $\mathcal{C}$. The idea behind this is the theory of Tannaka-Krein reconstruction generalised to the braided setting.

The Tannaka-Krein reconstruction theorems should be viewed as a generalization of the simple notion of Fourier Transform. The idea is that the right notion of representation of an algebraic structure should itself have enough structure to reconstruct the original algebraic object. On the other hand many constructions may appear very simple in terms of the representation theory and highly non-trivial in terms of the original algebraic object, and vice versa.

In the present setting we know that quantum groups give rise to braided categories as their representations, while conversely we will see that the representations or endomorphisms of a category $\mathcal{C}$ in a category $\mathcal{V}$ gives rise to a quantum group in $\mathcal{V}$.

3.1 Usual Tannaka-Krein Theorem

The usual Tannaka-Krein theorem for Hopf algebras says that a monoidal category $\mathcal{C}$ equipped with a functor to the category of vector spaces (i.e. whose objects can be identified in a strict way with vector spaces) is equivalent to that of the comodules of a certain bialgebra A reconstructed from $\mathcal{C}$. All our categories $\mathcal{C}$ are assumed equivalent to small ones.

Theorem 3.1 *Let $F : \mathcal{C} \to \mathrm{Vec}$ be a monoidal functor to the category of vector spaces with finite-dimensional image. Then there exists a bialgebra A uniquely determined as universal with the property that F factors through $\mathcal{M}^A$. If $\mathcal{C}$ is braided then A is dual quasitriangular. If $\mathcal{C}$ is rigid then A has an antipode.*

Proof We defer this to Theorem 3.11 below. Just set $\mathcal{V} = \mathrm{Vec}$ there. □

An early treatment of the bialgebra case is in [66]. See also [9]. The part concerning the antipode was shown in [70]. That a symmetric category gives a dual-triangular structure was pointed out in the modules setting in [11]. See also [29]. It is a trivial step to go from there to the braided case in which case the result is dual quasitriangular. This has been done by the author, while at the same time (in order to say something new) generalising in two directions. One is to the quasi-Hopf algebra setting[35][43] and the other to the braided-Hopf algebra setting[39][49].

This theorem tells us that (dual)quasitriangular Hopf algebras are rather more prevalent in mathematics (and physics) than we might have otherwise suspected. It also gives us a useful perspective on any Hopf algebra construction, by allowing us to go backwards and forwards between representations and the algebra itself. For example, if we are already given a bialgebra A then coming out of the reconstruction theorem one has associated to any subcategory

$$\mathcal{O} \subset \mathcal{M}^A \tag{54}$$

closed under tensor product, a sub-bialgebra

$$A_{\mathcal{O}} = \cup_{(V,\beta_V)\in\mathcal{O}} \mathrm{image}(\beta_V) \subset A, \qquad \mathrm{image}(\beta_V) = \{(f \otimes \mathrm{id}) \circ \beta_V(v); v \in V,\ f \in V^*\}. \tag{55}$$

If the sub-category is braided then $A_{\mathcal{O}}$ is dual quasitriangular etc. So this is a concrete form of the reconstruction theorem in the case where $\mathcal{O}$ is already in the context of a bialgebra.

For example if $A = A(R)$ and $\mathcal{O} = \mathcal{C}(V, R)$ in Section 1 then $A_{\mathcal{O}} = A(R)$ again. This is because the image of tensor powers of V for the coaction in (42) is clearly any monomial in the generators $\mathbf{t}$ of $A(R)$. Hence in this case the subcategory reconstructs all of $A(R)$. The result is due to Lyubashenko though the proof in [29] is different (and stated in the triangular case).

For another example let A be a bialgebra and $\mathcal{O}$ the category of comodules which are commutative in the sense $\sum v^{(\bar{1})} \otimes av^{(\bar{2})} = \sum v^{(\bar{1})} \otimes v^{(\bar{2})}a$ for all $a \in A$ and v in the comodule. Then $A_{\mathcal{O}}$ is a bialgebra contained in the center of A. This is therefore a canonically associated 'bialgebra centre' construction.

There are analogous results to these for modules. At the level of Theorem 3.1 the module theory is less powerful only if one functors (as usual) into familiar finite-dimensional vector spaces.

3.2 Braided Reconstruction Theorem

In this section we come to the fully-fledged braided Tannaka-Krein-type reconstruction theorem. We follow for pedagogical reasons the original module version[39][45], mainly because the comodule version was already given in complete detail in [49] and we do not want to repeat it. Also, we give here for the first time a fully diagrammatic proof.

Throughout this section we fix $F : \mathcal{C} \to \mathcal{V}$ a monoidal functor between monoidal categories. At least $\mathcal{V}$ should be braided. In this case there is an induced functor $V \mapsto \mathrm{Nat}(V \otimes F, F)$. We suppose that this functor is representable. So there is an object $B \in \mathcal{V}$ such that $\mathrm{Nat}(V \otimes F, F) \cong \mathrm{Hom}_{\mathcal{V}}(V, B)$ by functorial bijections. Let $\{\alpha_X : B \otimes F(X) \to F(X);\ X \in \mathcal{C}\}$ be the natural transformation corresponding to the identity morphism $B \to B$. Then using α and the braiding we get an induced map

$$\mathrm{Hom}_{\mathcal{V}}(V, B^{\otimes n}) \to \mathrm{Nat}(V \otimes F^n, F^n) \tag{56}$$

and we assume that these are likewise bijections. This is the *representability assumption for modules* and we assume it in what follows.

Theorem 3.2 *[45] Let $F : \mathcal{C} \to \mathcal{V}$ obey the representability assumption for modules. Then B is a bialgebra in $\mathcal{V}$, uniquely determined as universal with the property that F factors through ${}_B\mathcal{V}$. If $\mathcal{C}$ is braided then B is quasitriangular in the braided category with $\mathcal{R}$ given by the ratio of the braidings in $\mathcal{C}$ and $\mathcal{V}$. If $\mathcal{C}$ is rigid then B has a braided-antipode.*

We will give the proof in diagrammatic form. The bijections (56) and the structure maps in the theorem are characterized by

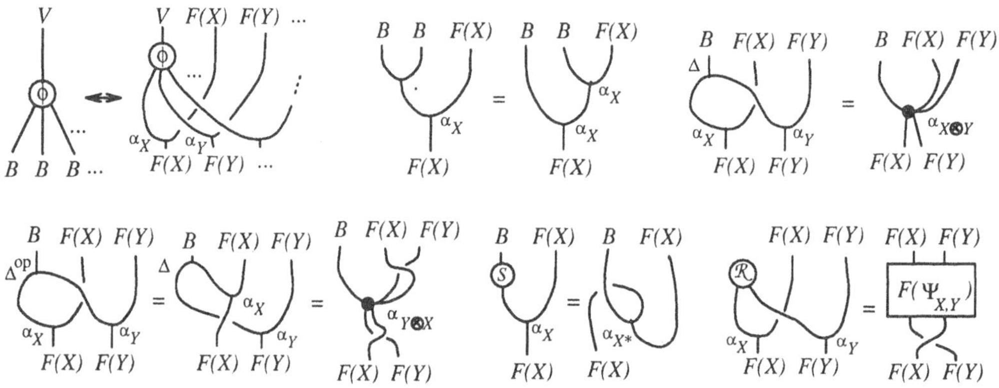

(57)

Here the assumption that F is monoidal means that there are functorial isomorphisms $F(X \otimes Y) \cong F(X) \otimes F(Y)$ and in the rigid case $F(X^*) \cong F(X)^*$. The latter follow from the uniqueness of duals up to isomorphism (for example one can define $F(X)^* = F(X^*)$ etc. and any other dual is isomorphic). These isomorphisms are used freely and suppressed in the notation. The solid node $\alpha_{X \otimes Y}$ is α on the composite object $X \otimes Y$ but viewed via the first of these isomorphisms as a morphism $B \otimes F(X) \otimes F(Y) \to F(X) \otimes F(Y)$. Similarly for α_{X^*}. In this way all diagrams refer to morphisms in $\mathcal{V}$. The unit $\underline{1} \to B$ corresponds to the identity natural transformation and the counit to $\alpha_{\underline{1}}$. Their proofs are suppressed.

Lemma 3.3 *The product on B defined in (57) is associative.*

Proof We use the definition of $\cdot$ twice in terms of its corresponding natural transformations and then in reverse

Hence the natural transformations corresponding to the two morphisms $B \otimes B \otimes B \to B$ coincide and we have an algebra in the category. □

Lemma 3.4 *The coproduct Δ on B defined in (57) is coassociative.*

Proof We use the definition of Δ twice and then in reverse, using in the middle that F is monoidal and hence compatible with the (suppressed) associativity in the two categories

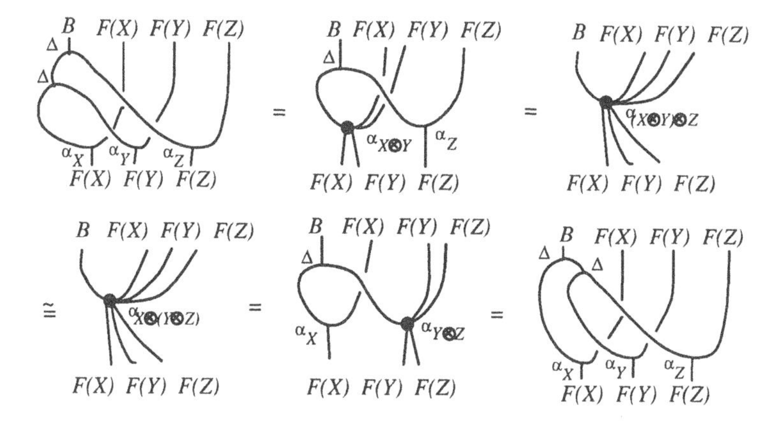

The key step is the third equality which follows from functoriality of α under the associativity morphism $X \otimes (Y \otimes Z) \to (X \otimes Y) \otimes Z$ and that F is monoidal. If F is not monoidal but merely multiplicative one has here a quasi-associative coproduct as explained in [35][43]. □

Proposition 3.5 *The product and coproduct in the last two lemmas fit together to form a bialgebra in $\mathcal{V}$.*

Proof We use the definitions of $\cdot$ and Δ

□

Lemma 3.6 *The second coproduct $\Delta^{\rm op}$ on B defined in (57) is coassociative.*

Proof This is similar to the proof of Lemma 3.4

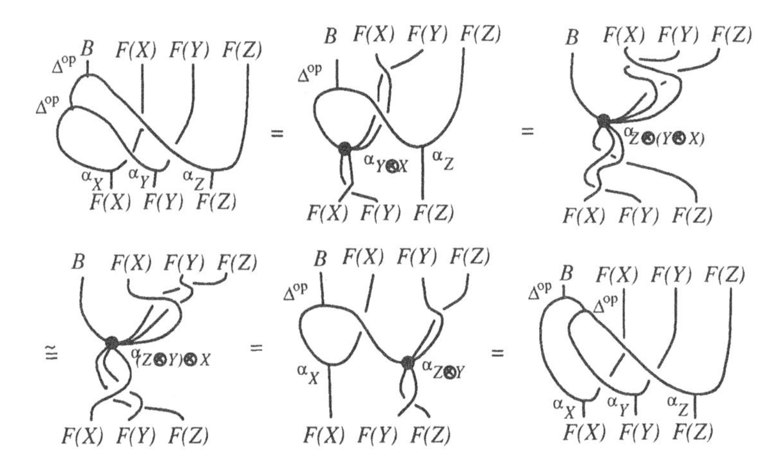

□

Proposition 3.7 *The product and the second coproduct fit together to form a second bialgebra in $\mathcal{V}$.*

Proof This is similar to the proof of Proposition 3.5

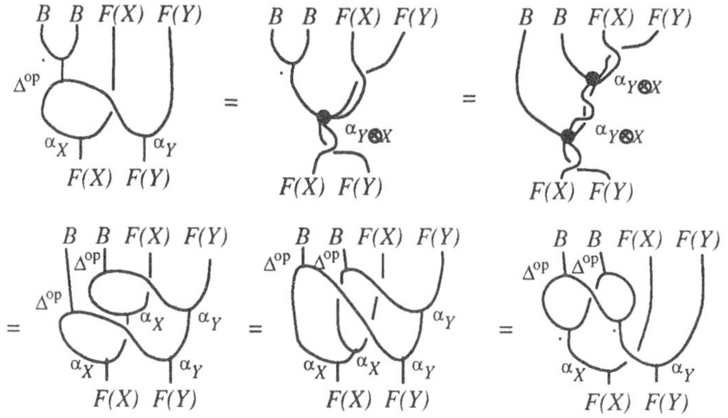

□

Proposition 3.8 *If $\mathcal{C}$ is rigid then S defined in (57) is an antipode for the coproduct Δ.*

Proof The first, second and fourth equalities are the definitions of $\cdot, S, \Delta$. The fifth uses functoriality of α under the evaluation $X^* \otimes X \to \underline{1}$

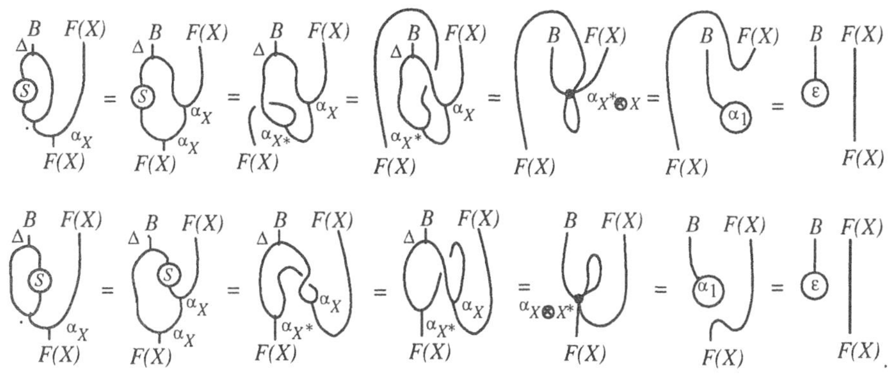

The result is the natural transformation corresponding to $\eta \circ \epsilon$. Similarly for the second line using functoriality under the coevaluation morphism $\underline{1} \to X \otimes X^*$. □

Proposition 3.9 *If $\mathcal{C}$ is braided then $\mathcal{R}$ defined in (57) makes B into a quasitriangular bialgebra.*

Proof To prove the first of (53) we evaluate the definitions and use F applied to the hexagon identity in $\mathcal{C}$ for the fifth equality, and then in reverse.

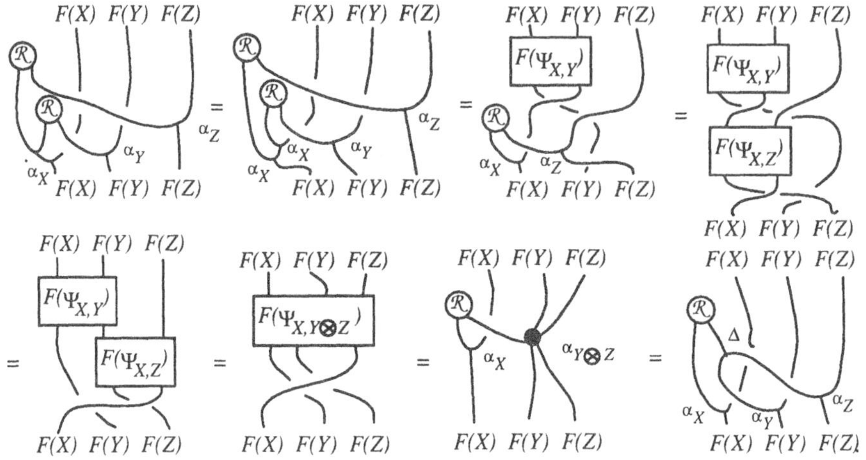

The same strategy works for the second of (53)

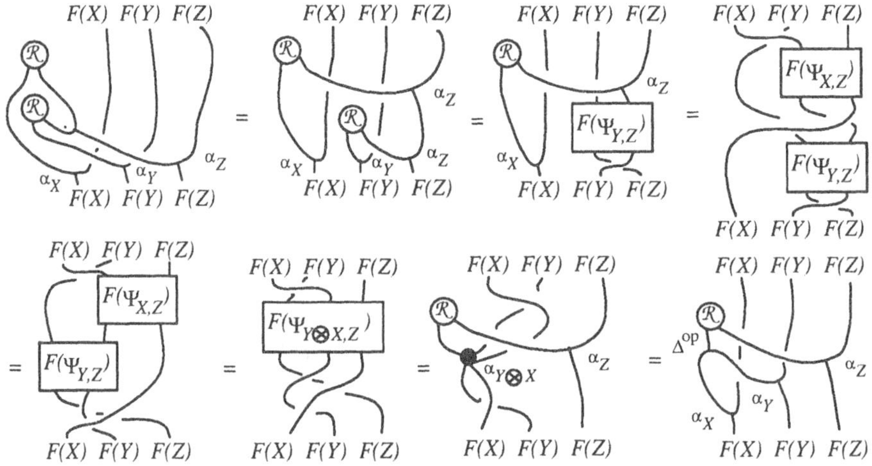

Finally, to prove the last of (53) we use in the third equality the functoriality of α under the morphism $\Psi_{X,Y}$:

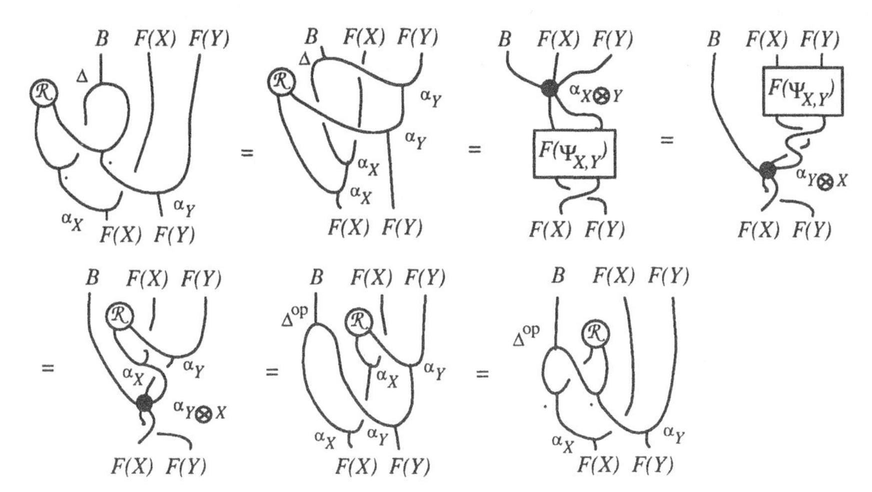

The construction of $\mathcal{R}^{-1}$ is based in the inverse natural transformation to that for $\mathcal{R}$ and the proof that this then is inverse in the convolution algebra $\underline{1} \to B \otimes B$ is straightforward using the same techniques. □

Clearly the definition of the product in (57) is such that α_X become modules. So we have a functor $\mathcal{C} \to {}_B\mathcal{V}$. The universal property of B follows easily from its role as representing object for natural transformations (56).

Corollary 3.10 *If $\mathcal{C}$ is braided and F is a tensor functor in the sense that the braiding of $\mathcal{C}$ is mapped on to the braiding of $\mathcal{V}$ then $\Delta^{\rm op} = \Delta$, $\mathcal{R}$ is trivial and B is a braided group (braided-cocommutative) with respect to the image of the functor $\mathcal{C} \to {}_B\mathcal{V}$.*

Proof This follows at once from the form of $\mathcal{R}, \Delta^{\rm op}$ in (57). □

For example if $F = \mathrm{id}$ (or the canonical functor into a suitable completion of $\mathcal{C}$) then to every rigid braided $\mathcal{C}$ we have an associated braided Hopf algebra B and a large class of braided-cocommutative modules $\{\alpha_X\}$. The ratio of the braidings is trivial and this is why we have from this point of view some kind of braided group rather than braided quantum group.

Finally, given a monoidal functor $F : \mathcal{C} \to \mathcal{V}$ we can equally well require representability of the functor $V \mapsto \mathrm{Nat}(F, F \otimes V)$ and its higher order products i.e., bijections

$$\mathrm{Hom}_{\mathcal{V}}(B^{\otimes n}, V) \to \mathrm{Nat}(F^n, F^n \otimes V) \tag{58}$$

This is the *representability assumption for comodules* and is always satisfied if $\mathcal{V}$ is co-complete and if the image of F is rigid. In this case one can write B as a coend $B = \int^X F(X)^* \otimes F(X)$.

Theorem 3.11 *[49] Let $F : \mathcal{C} \to \mathcal{V}$ obey the representability assumption for comodules. Then B is a bialgebra in $\mathcal{V}$, uniquely determined as universal with the property that F factors through $\mathcal{V}^B$. If $\mathcal{C}$ is braided then B is dual quasitriangular in the braided category with $\mathcal{R}$ given by the ratio of the braidings in $\mathcal{C}$ and $\mathcal{V}$. If $\mathcal{C}$ is rigid then B has a braided-antipode.*

Proof Literally turn the above proofs up-side-down. [49] has more traditional proofs. Note also the slightly different conventions there which are chosen so as to ensure that the dual quasitriangular structure reduces for $\mathcal{V} = \mathrm{Vec}$ to the usual notion as in Section 1.3 rather than its convolution-inverse. □

By taking the identity functor to a cocompletion we obtain a canonical braided-Hopf algebra $B = \mathrm{Aut}\,(\mathcal{C})$ associated to a rigid braided monoidal category $\mathcal{C}$[49]. By turning Corollary 3.10 up-side-down the braided-Hopf algebra this time is braided-commutative with respect to a class of comodules. In this sense $\mathrm{Aut}\,(\mathcal{C})$ is a braided group of function algebra type.

4 Applications to Ordinary Hopf Algebras

In this section we give some applications of the above braided theory to ordinary Hopf algebras. In this case there is either a background quasitriangular bialgebra or Hopf algebra H and we work in the braided category ${}_H\mathcal{M}$ or a background dual quasitriangular Hopf algebra A and we work in the braided category $\mathcal{M}^A$. See Sections 1.2 and 1.3 respectively. The latter theory is the dual theory to the former and given by turning the diagram-proofs in this book upside down. There need be no relationship between H or A since we are dualizing the theory and not any specific Hopf algebra.

In this context the content of Lemma 2.1 is immediate: it says that two H-module algebras or A-comodule algebras have a braided tensor product which is also an H-module algebra or A-comodule algebra respectively. From (44) and Theorem 1.10 or Theorem 1.16 we have obviously

$$B, C, B\underline{\otimes}C \in {}_H\mathcal{M}: \qquad (a\otimes c)(b\otimes d) = \sum a(\mathcal{R}^{(2)}\triangleright b)\otimes(\mathcal{R}^{(1)}\triangleright c)d \tag{59}$$

$$B, C, B\underline{\otimes}C \in \mathcal{M}^A: \qquad (a\otimes c)(b\otimes d) = \sum ab^{(\bar{1})}\otimes c^{(\bar{1})}d\mathcal{R}(c^{(\bar{2})}\otimes b^{(\bar{2})}) \tag{60}$$

for all $a, b \in B,\ c, d \in C$. We have introduced this construction in [44]–[50] and explained there that it is exactly a generalization of the $\mathbb{Z}_2$-graded or super-tensor product of super-algebras. This in turn has specific applications and spin-offs. An amusing one is

Proposition 4.1 *[1] Let H be a Hopf algebra. Then the usual n-fold tensor product $H^{\otimes^n}$ is an H-module algebra under the action*

$$h\triangleright(b_1\otimes\cdots\otimes b_n) = \sum h_{(1)}b_1Sh_{(2n)}\otimes h_{(2)}b_2Sh_{(2n-1)}\otimes\cdots\otimes h_{(n-1)}b_{n-1}Sh_{(n+2)}\otimes h_{(n)}b_nSh_{(n+1)}.$$

Proof This arises from the general theorems below as the n-fold braided tensor product of H as an H-module algebra and in the case where H is quasitriangular. This n-fold braided tensor product turns out to be isomorphic to the usual tensor product in a non-trivial way using the quasitriangular structure. Computing the resulting action through this isomorphism and using the axioms (23) etc one finds the stated action. On the other hand the resulting formula does not require any quasitriangular structure at all and can then be checked directly to work for any Hopf algebra as stated. □

This is a typical example in that the braided theory leads one to unexpected formulas (as far as I know the last proposition is unexpected) which can then be verified directly. We will come to another such spin-off in Section 4.3.

4.1 Transmutation

In this section we shall show how to obtain non-trivial examples of bialgebras and Hopf algebras in braided categories. The key construction is one that we have called *transmutation* because it asserts that an ordinary Hopf algebra can be turned by this process into a braided one. This is achieved as an application of the generalised reconstruction theorem in Section 3 and is actually part of a rather general principle: by viewing an algebraic structure in terms of its representations, and targeting these by means of a functor into some new category, we can reconstruct our algebraic object in this new category. In this way the category in which an algebraic structure lives can be changed. Moreover, this 'mathematical alchemy' can be useful in that the structure may look more natural and have better properties after transmutation to the new category.

In the present setting the data for our transmutation is a pair $H_1 \xrightarrow{f} H$ where H_1 is a quasitriangular Hopf algebra, H is at least a bialgebra, and f a bialgebra map.

Theorem 4.2 *[44][45] H can be viewed equivalently as a bialgebra $B(H_1, H)$ living in the braided category ${}_{H_1}\mathcal{M}$ by the adjoint action induced by f. Here*

$$B(H_1, H) = \begin{cases} H & \text{as an algebra} \\ \underline{\Delta},\ \underline{S},\ \underline{\mathcal{R}} & \text{modified coproduct, antipode, quasitriangular structure} \end{cases}$$

where B has a braided-antipode if H has an antipode and a braided-quasitriangular structure if H is quasitriangular.

Proof We let $\mathcal{C} = {}_H\mathcal{M}$ and $\mathcal{V} = {}_{H_1}\mathcal{M}$ and F the functor by pull-back along f. Then Theorem 3.2 tells us there is a braided-Hopf algebra B. □

Explcit formulae for the transmuted structure are

$$\underline{\Delta} b = \sum b_{(1)} f(S\mathcal{R}_1{}^{(2)}) \otimes \mathcal{R}_1{}^{(1)} \triangleright b_{(2)}, \quad \underline{S} b = \sum f(\mathcal{R}_1{}^{(2)}) S(\mathcal{R}_1{}^{(1)} \triangleright b) \tag{61}$$

$$\underline{\mathcal{R}} = \sum \rho^{(1)} f(S\mathcal{R}_1{}^{(2)}) \otimes \mathcal{R}_1{}^{(1)} \triangleright \rho^{(2)}, \qquad \rho = f(\mathcal{R}_1^{-1})\mathcal{R}. \tag{62}$$

There is also an opposite coproduct characterised by

$$\sum \Psi(b_{\underline{(1)}_{\rm op}} \otimes Q_1{}^{(1)} \triangleright b_{\underline{(2)}_{\rm op}}) f(Q_1{}^{(2)}) = \sum b_{\underline{(1)}} \otimes b_{\underline{(2)}} \tag{63}$$

where $Q_1 = (\mathcal{R}_1)_{21}(\mathcal{R}_1)_{12}$ and $f(Q_1{}^{(2)})$ right-multiplies the second tensor factor of the output of Ψ. The underlines in the superscripts are to remind us that we intend here the braided-coproducts $\underline{\Delta}$ and $\underline{\Delta}^{\rm op}$. The equation can also be inverted to give an explicit formula for $\underline{\Delta}^{\rm op}$. That these formulae obey the axioms (45) and (53) of Section 2 is verified explicitly in [45].

Corollary 4.3 *[47] Let H be a Hopf algebra containing a group-like element g of order n. Then H has a corresponding anyonic version B. It has the same algebra and*

$$\underline{\Delta} b = \sum b_{(1)} g^{-|b_{(2)}|} \otimes b_{(2)}, \quad \underline{\epsilon} b = \epsilon b, \quad \underline{S} b = g^{|b|} S b$$

$$\underline{\Delta}^{\rm op} b = \sum b_{(2)} g^{-2|b_{(1)}|} \otimes g^{-|b_{(2)}|} b_{(1)}, \quad \underline{\mathcal{R}} = \mathcal{R}_{\mathbb{Z}'_n}^{-1} \sum \mathcal{R}^{(1)} g^{-|\mathcal{R}^{(2)}|} \otimes \mathcal{R}^{(2)}$$

Proof We apply the transmutation theorem, Theorem 4.2 and compute the form of $B = B(\mathbb{Z}'_n, H)$. Here $\mathbb{Z}'_n$ is the non-standard quasitriangular Hopf algebra in Example 1.7 with quasitriangular structure $\mathcal{R}_{\mathbb{Z}'_n}$. The action of g on H is in the adjoint representation $g{\triangleright}b = gbg^{-1}$ for $b \in B$ and defines the degree of homogeneous elements by $g{\triangleright}b = q^{|b|}b$. □

Corollary 4.4 *Let H be a quasitriangular Hopf algebra containing a group-like element g of order 2. Then H has a corresponding super-version B.*

Proof The formulae are as in Corollary 4.3 with $n = 2$. The commutation relations with g define the grading of B. □

The first corollary was applied, for example to $H = u_q(g)$ at a root of unity to simplify its structure. It led to a new simpler form for its quasitriangular structure by finding its anyonic quasitriangular structure and working back[47]. The second corollary was usefully applied in [59] to superise the non-standard quantum group associated to the Alexander-Conway polynomial. In these examples, a sub-quasitriangular Hopf algebra is used to generate the braided category in which the entire quasitriangular Hopf algebra is then viewed by transmutation. In the process its quasitriangular structure becomes reduced because the part from the sub-Hopf algebra is divided out. This means that the part corresponding to the sub-Hopf algebra is made in some sense cocommutative.

Corollary 4.5 *[44] Every quasitriangular Hopf algebra H has a braided-group analogue $B(H, H)$ which is braided-cocommutative in the sense that $\underline{\mathcal{R}} = 1 \otimes 1$ and $\underline{\Delta}^{\rm op} = \underline{\Delta}$. The latter means*

$$\sum \Psi(b_{\underline{(1)}} \otimes Q^{(1)}{\triangleright}b_{\underline{(2)}})Q^{(2)} = \sum b_{\underline{(1)}} \otimes b_{\underline{(2)}}.$$

We call $B(H, H)$ the braided group of enveloping algebra type *associated to H. It is also denoted by $\underline{H}$.*

Proof Here we take the transmutation principle to its logical extreme and view any quasitriangular Hopf algebra H in its own braided category ${}_H\mathcal{M}$, by $H \subseteq H$. This is a bit like using a metric to determine geodesic co-ordinates. In that co-ordinate system the metric looks locally linear. Likewise, in its own category (as a braided group) our original quasitriangular Hopf algebra looks braided-cocommutative. From Corollary 3.10 we know that $\Delta^{\rm op} = \Delta$ and this gives the formula stated. □

This completely shifts then from one point of view (quantum=non-cocommutative object in the usual category of vector spaces) to another (classical='cocommutative' but braided object), and means that the theory of ordinary quasitriangular Hopf algebras is contained in the theory of braided-groups.

The braided-Hopf algebra B in Theorem 4.2 is equivalent to the original one in that spaces and algebras etc on which H act also become transmuted to corresponding ones for B. Partly, this is obvious since $B = H$ as an algebra, so any H-module V of H is also a braided B-module. The key point is that V is also acted upon by H_1 through the mapping $H_1 \to H$. So the action of H is used in two ways, both to define the corresponding action of B and to define the 'grading' of V as an object in a braided category ${}_{H_1}\mathcal{M}$. This extends the process of transmutation to modules.

Proposition 4.6 *[46, Prop 3.2] If C is an H-module algebra then its transmutation is a braided $B(H,H)$-module algebra in the sense of (49). Similarly an H-module coalgebra becomes a braided $B(H,H)$-module-coalgebra. Here the transmutation does not change the action, but simply views it in the braided category.*

Proof An elementary computation from the form of (61) and the braiding in Theorem 1.10. □

For example, the adjoint action of H on itself transmutes to the braided-adjoint action of $B = B(H,H)$ on itself in Example 2.7. Moreover, it means that $B(H,H)$ is braided-cocommutative with respect to Ad. Indeed

Proposition 4.7 *[44][45] For $B(H_1,H)$ the $\Delta^{\rm op}$ behaves like an opposite coproduct on all $B(H_1,H)$-modules that arise from transmutation. In particular, $B(H,H)$ is cocommutative in the sense of Definition 2.9 with respect to all braided-modules that arise from transmutation.*

Proof Writing the braids in Definition 2.9 in terms of the quasitriangular structure as explained in Theorem 1.10, we see that the condition for all V is implied by (and essentially equivalent to) the intrinsic braided-cocommutativity formula in Corollary 4.5. □

Finally, we mention a different aspect of this transmutation theory, namely a result underlying the direct proof that $B(H,H)$ is a braided-Hopf algebra (if one does not like the categorical one).

Lemma 4.8 *Let H be quasitriangular and C be an algebra in ${}_H\mathcal{M}$. Then there is an algebra isomorphism*

$$\theta_{H,C} : H\otimes C \to B(H,H)\underline{\otimes} C$$

where $\underline{\otimes}$ is the braided-tensor-product in Lemma 2.1.

Proof This is provided by $\theta_{H,C}(h\otimes c) = \sum hS\mathcal{R}^{(2)}\otimes\mathcal{R}^{(1)}\triangleright c$ and shows at once that $\underline{\Delta} = \theta_{H,H}\circ\Delta$ in (61) is an algebra homomorphism. The proof that θ is an algebra homomorphism is

$$\begin{aligned}
\theta_{H,C}(h\otimes c)\theta_{H,C}(g\otimes d) &= \sum h(S\mathcal{R}^{(2)})(\mathcal{R}''^{(2)}\triangleright(gS\mathcal{R}'^{(2)}))\otimes(\mathcal{R}''^{(1)}\mathcal{R}^{(1)}\triangleright c)(\mathcal{R}'^{(1)}\triangleright d)\\
&= \sum h(S\mathcal{R}^{(2)})\mathcal{R}'''^{(2)}g(S\mathcal{R}'^{(2)})S\mathcal{R}''^{(2)}\otimes(\mathcal{R}''^{(1)}\mathcal{R}'''^{(1)}\mathcal{R}^{(1)}\triangleright c)(\mathcal{R}'^{(1)}\triangleright d)\\
&= \sum hgS(\mathcal{R}''^{(2)}\mathcal{R}'^{(2)})\otimes(\mathcal{R}''^{(1)}\triangleright c)(\mathcal{R}'^{(1)}\triangleright d) = \theta_{H,C}(hg\otimes cd)
\end{aligned}$$

using the definition (59) and the axioms (22). □

As an immediate example one has that $H^{\otimes n}\cong B(H,H)^{\underline{\otimes}^n}$ for n-fold tensor products (iterate the lemma). Since $B(H,H)^{\underline{\otimes}^n}$ lives in ${}_H\mathcal{M}$ as an algebra (an H-module algebra) via the adjoint action, it follows that $H^{\otimes^n}$ does also. Computing this gives Proposition 4.1 as an amusing spin-off.

The transmutation theory obviously has a dual version for $A\xrightarrow{f}A_1$ a bialgebra map where A_1 is a dual quasitriangular Hopf algebra and A is at least a bialgebra.

Theorem 4.9 *[48][49] A can be viewed equivalently as a bialgebra $B(A, A_1)$ living in the braided category $\mathcal{M}^{A_1}$ by the right adjoint coaction induced by f. Here*

$$B(A, A_1) = \begin{cases} A & \text{as a coalgebra} \\ \underline{\cdot},\ \underline{S},\ \underline{\mathcal{R}} & \text{modified product, antipode, dual quasitriangular structure} \end{cases}$$

where B has a braided-antipode if A has an antipode and a braided-dual quasitriangular structure if A is dual quasitriangular.

Proof We let $\mathcal{C} = \mathcal{M}^A$ and $\mathcal{V} = \mathcal{M}^{A_1}$ and F the functor by push-out along f. Then Theorem 3.11 tells us there is a braided-Hopf algebra B. The exact conventions for $\mathcal{R}$ most useful here are in [49]. □

Explicit formulae for the transmuted structure are[49][48]

$$a\underline{\cdot}b = \sum a_{(2)}b_{(2)}\mathcal{R}((Sa_{(1)})a_{(3)} \otimes Sb_{(1)}), \quad \underline{S}a = \sum Sa_{(2)}\mathcal{R}((S^2a_{(3)})Sa_{(1)} \otimes a_{(4)}) \tag{64}$$

where for simplicity we concentrate on the case where f is the identity (the formulae in the general case are similar). The analogue of Corollary 4.5 is that $B(A, A)$ is braided-commutative in the sense of Definition 2.9 turned up-side-down, for all comodules V that come from transmutation of comodules of A. This reduces to an intrinsic form of commutativity dual to Corollary 4.5. This comes out explicitly as

$$b\underline{\cdot}a = \sum a_{(3)}\underline{\cdot}b_{(3)}\mathcal{R}(Sa_{(2)} \otimes b_{(1)})\mathcal{R}(a_{(4)} \otimes b_{(2)})\mathcal{R}(b_{(5)} \otimes Sa_{(1)})\mathcal{R}(b_{(4)} \otimes a_{(5)}). \tag{65}$$

We call $\underline{A} = B(A, A)$ the *braided group of function algebra type associated to* A. A direct proof that these formulae define a braided-Hopf algebra as in (45) appears in [49, Appendix].

Finally, we suppose that A is actually dual to H in a suitable sense (for example, one can suppose they are finite dimensional). Until now we have not assumed anything like this. Then from the two theorems above we have two Hopf algebras in braided categories, and moreover the two categories can be identified in the usual way. Thus a right A-comodule defines a left H-comodule and viewing everything in this way in ${}_H\mathcal{M}$ we have two Hopf algebras $B(H, H)$ and $B(A, A)$ in the same braided category.

Proposition 4.10 *If A is dual to H then the corresponding braided groups $B(A, A)$ and $B(H, H)$ are dual in the braided category, $B(A, A)^* = B(H, H)$.*

Proof Explicitly, the duality is given by $b \in B(H, H)$ mapping to a linear functional $< Sb, (\ ) >$ on $B(A, A)$, where S is the usual antipode of H. See [53] for full details. □

Thus the usual duality if it exists becomes the categorical duality as in Section 2.2. This is to be expected. More remarkable is the fact, also verified explicitly[52] that there is a canonical homomorphism of Hopf algebras in the braided category

$$Q : B(A, A) \to B(H, H), \qquad Q(a) = (a \otimes \mathrm{id})(\mathcal{R}_{21}\mathcal{R}_{12}) \tag{66}$$

given by evaluation against $\mathcal{R}_{21}\mathcal{R}_{12}$. In the standard examples $H = U_q(g)$ one has a formal expansion $\mathcal{R}_{21}\mathcal{R}_{12} = 1 + 2\hbar K^{-1} + O(\hbar^2)$ where K^{-1} is the inverse Killing form $g^* \to g$. So (66) is a version for braided groups of this linear map provided by Q. This point of view has already been developed at the level of linear maps $A \to H$ in [64] where the Hopf algebra is called *factorizable* if the map Q is a linear isomorphism. What we have in (66) is a much stronger statement: in the factorizable case $B(A, A) \cong B(H, H)$ as Hopf algebras in a braided category. Since the first of these is the braided version of the quantum function algebra $\mathcal{O}_q(G)$ and the second of the enveloping algebra $U_q(g)$, the isomorphism of their braided versions is remarkable.

4.2 Bosonization

In this section we prove results going the other way, turning any Hopf algebra in the braided category ${}_H\mathcal{M}$ or $\mathcal{M}^A$ into an ordinary Hopf algebra. This process has been introduced by the author under the heading *bosonization.* The origin of this term is from physics where $\mathbb{Z}_2$-graded algebras etc are called 'fermionic' while ordinary ungraded ones are called 'bosonic'. Not all braided groups are of the type coming from the transmutation in the last section, so bosonization is not simply transmutation in reverse.

Theorem 4.11 *[46, Thm 4.1] Suppose that B is a Hopf algebra living in a braided category of the form ${}_H\mathcal{M}$. Then there is an ordinary Hopf algebra* $\mathrm{bos}(B) = B{>\!\!\!\triangleleft} H$.

Proof The abstract way that the result arose in [46] is as follows (but once the result is known a direct proof is also easy). Since B lives in ${}_H\mathcal{M}$ it is in particular an H-module algebra. From Proposition 4.6 we see that the same linear map makes B a braided $B(H,H)$-module algebra. Likewise it is a braided module coalgebra and from Proposition 4.7 this braided-module structure is braided-cocommutative. Hence from Theorem 2.11 we have a semidirect product Hopf algebra $B{>\!\!\!\triangleleft} B(H,H)$. This contains $B(H,H)$ and it is easy to see that this is indeed the transmutation of $H \to H_2$ where H_2 is some ordinary Hopf algebra. It computes explicitly as follows. As an algebra it is the semidirect or smash product by the action of H on B. So $(1\otimes h)(b\otimes 1) = \sum h_{(1)}{\triangleright} b\otimes h_{(2)}$ where $\triangleright$ is the action of H. As a coalgebra it is[46]

$$\Delta(b\otimes h) = \sum b_{\underline{(1)}}\otimes \mathcal{R}^{(2)}h_{(1)}\otimes \mathcal{R}^{(1)}{\triangleright} b_{\underline{(2)}}\otimes h_{(2)}. \tag{67}$$

□

Once the result and formula (67) is known it is not hard to verify it directly. The key lemma for this direct verification is

Lemma 4.12 *[38] Let H be a quasitriangular bialgebra or Hopf algebra and B a left H-module with action $\triangleright$. Then*

$$\beta(b) = \sum \mathcal{R}^{(2)}\otimes \mathcal{R}^{(1)}{\triangleright} b$$

makes B into a left H-comodule. Moreover, it is compatible with $\triangleright$ in the sense of Example 1.3 and invertible so $B\in {}^H_H\mathcal{M}$.

Proof Using the axioms of a quasitriangular bialgebra one sees at once that this defines a coaction and this is compatible in the sense of Example 1.3. In the case when H is only a bialgebra we have to check invertibility in the sense of (20). The required inverse is provided by $\mathcal{R}^{-1}$ in place of $\mathcal{R}$ in the definition of β. □

As we explained in [38], this defines a functor ${}_H\mathcal{M} \to {}^H_H\mathcal{M} = \mathcal{Z}({}_H\mathcal{M})$. Since the functor takes morphisms to morphisms (or by direct computation) it is easy to see that if B is an H-module (co)algebra then it becomes in this way an H-comodule (co)algebra. It is completely clear then that $\mathrm{bos}(B) = B{>\!\!\!\triangleleft} H$ has the structure of a semidirect (co)product both as an algebra by $\triangleright$, and as a coalgebra by the coaction β from Lemma 4.12. This is the direct interpretation of (67). Simultaneous semidirect products and coproducts have been studied in [62] but the present construction of examples of them is of course new and due to the author.

One thing that we learn from the categorical point of view is that this ordinary Hopf algebra bos(B) is equivalent to the original B in the sense that its ordinary representations correspond to the braided-representations of B[46]. This applies as much to super-Hopf algebras as to Hopf algebras on other categories, so we recover a result known to experts working with super-Lie algebras and super-Hopf algebras that they can be reduced to ordinary ones. See [15] for an example of this strategy.

Corollary 4.13 *[46, Cor. 4.3] Any super-quasitriangular super-Hopf algebra can be bosonised to an equivalent ordinary quasitriangular Hopf algebra. It consists of adjoining an element g with relations $g^2 = 1$, $gb = (-1)^{|b|}bg$ and*

$$\Delta g = g \otimes g,\ \Delta b = \sum b_{\underline{(1)}} g^{|b_{\underline{(2)}}|} \otimes b_{\underline{(2)}}, \quad Sb = g^{-|b|}\underline{S}b, \quad \mathcal{R} = \mathcal{R}_{\mathbb{Z}'_2} \sum \underline{\mathcal{R}}^{(1)} g^{|\underline{\mathcal{R}}^{(2)}|} \otimes \underline{\mathcal{R}}^{(2)}.$$

Proof We have seen in Proposition 1.11 that the category of super-vector spaces is of the required form, with $H = \mathbb{Z}'_2$. Here the $\mathbb{Z}_2$-graded modules of the original super-Hopf algebra are in one-to-one correspondence with the usual representations of the bosonised ordinary Hopf algebra. We suppose that we work over characteristic not 2. Also, we have written the formulae in a way that works in the anyonic case with 2 replaced by n and (-1) by a primitive n-th root of unity for the Hopf algebra structure. □

This means that the theory of super-Lie algebras (and likewise for colour-Lie algebras[67], anyonic quantum groups etc) is in a certain sense redundant – we could have worked with their bosonized ordinary Hopf algebras. This is especially true in the super or colour case where there is no braiding to complicate the picture.

As usual the above theory has exactly a version for a Hopf algebra living in the braided category $\mathcal{M}^A$ where A is dual quasitriangular.

Theorem 4.14 *Suppose that B is a Hopf algebra living in a braided category of the form $\mathcal{M}^A$. Then there is an ordinary Hopf algebra* cobos$(B) = A{\triangleright\!\!\!<}B$.

Proof We turn our diagram-proofs in the theory leading to Theorem 4.11 up-side-down. This time the coproduct is the semidirect one by the coaction whereby B is an object in $\mathcal{M}^A$, and this also defines a right action $\triangleright$ (the dual version of Lemma 4.12) with respect to which we have a semidirect product algebra on $A \otimes B$,

$$b{\triangleleft}a = \sum b^{(\bar{1})} \mathcal{R}(b^{(\bar{2})} \otimes a), \quad (1 \otimes b)(a \otimes 1) = \sum a_{(1)} \otimes b{\triangleleft}a_{(2)}. \tag{68}$$

Just as one has a direct proof of Theorem 4.11, one can verify directly that this cobos(B) is a Hopf algebra. □

It is rather hard to consider this result and its attendant lemma as new results since they are exactly the dual construction (by turning proofs as diagrams up-side-down) of our bosonization Theorem 4.11. Equally well one could reflect in a mirror about a horizontal axis (or simply reverse the arrows in the more conventional commutative diagrams). In this case left modules/comodules become left comodules/modules and the Hopf algebra is $\bar{\text{cobos}}(B) = B{>\!\!\!\triangleleft}A$ by a left handed semidirect product and coproduct. Equally well we could reflect in a vertical axis turning left modules to right modules etc in Theorem 4.11 and giving a right-handed version $\bar{\text{bos}}(B) = H{\triangleright\!\!\!<}B$.

As before there is no suggestion here that A is dual to H since it is the construction that is being reversed and not any specific Hopf algebra. But if A is dual to H (say finite-dimensional) then we can make both constructions. Indeed, if B^* is the categorical dual as in Section 2.2 then

$$\mathrm{cobos}(B^*) = A{\bowtie\!\!<}B^*{\cong}(B{>\!\!\triangleleft}H)^* = \mathrm{bos}(B)^*. \tag{69}$$

Details and (more importantly) an application may be found in [53]. Another application of the bosonization theorem in its original form and in the dual form can be found in [7] and [16], to prove nice double-centraliser theorems for various kinds of Lie algebras in symmetric categories.

4.3 Radford's Theorem

In [52, Appendix] we have explained in detail how the above ideas provide a new braided interpretation of Radford's theorem[62] about Hopf algebras with projections. This theorem asserts that if H, H_1 are ordinary Hopf algebras and if

$$H_1 \underset{i}{\overset{p}{\rightleftarrows}} H \tag{70}$$

are bialgebra maps with $p \circ i = \mathrm{id}$ (a Hopf algebra projection), then there is an algebra and coalgebra B such that $H_1{\cong}B{>\!\!\triangleleft}H$ as a simultaneous semidirect product and semidirect coproduct. Radford called such simultaneous semidirect (co)products where the result is a Hopf algebra 'biproducts' and showed that they correspond to projections. We have already introduced some examples in the last section (arising from the bosonization process) but now we consider the general situation.

At the time of [62] the notion of braided categories was yet to be invented. Because of this the algebra and coalgebra B in Radford's theorem was simply some exotic object where the algebra and coalgebra did not form an ordinary Hopf algebra. We have pointed out in [52] and cf.[38] that B is in fact nothing other than a Hopf algebra in the braided category ${}^H_H\mathcal{M} = {}_{D(H)}\mathcal{M}$ of Example 1.3 (we stressed the latter in [52] for pedagogical reasons but explained that the former was more useful in the infinite-dimensional case). Thus we arrived at the following interpretation of Radford's theorem.

Proposition 4.15 *[52, Prop. A.2] Let $H_1 \underset{i}{\overset{p}{\rightleftarrows}} H$ be a Hopf algebra projection and let H have invertible antipode. Then there is a Hopf algebra B living in the braided category ${}^H_H\mathcal{M}$ such that $B{>\!\!\triangleleft}H{\cong}H_1$.*

Proof Explicitly, B is a subalgebra of H_1 and in ${}^H_H\mathcal{M}$ by action $\triangleright$ and coaction β,

$$B = \{b \in H_1 \mid \sum b_{(1)} \otimes p(b_{(2)}) = b \otimes 1\}, \quad h{\triangleright}b = \sum i(h_{(1)})bSoi(h_{(2)}), \quad \beta(b) = p(b_{(1)}) \otimes b_{(2)} \tag{71}$$

where $h \in H$. The braided-coproduct, braided-antipode and braiding of B are

$$\underline{\Delta}b = \sum b_{(1)}Soiop(b_{(2)}) \otimes b_{(3)}, \quad \underline{S}b = \sum iop(b_{(1)})Sb_{(2)}, \quad \Psi_{B,B}(b \otimes c) = \sum p(b_{(1)}){\triangleright}c \otimes b_{(2)}. \tag{72}$$

The isomorphism $\theta : B{>\!\!\triangleleft}H \to H_1$ is $\theta(b \otimes h) = bi(h)$, with inverse $\theta^{-1}(a) = \sum a_{(1)}S \circ i \circ p(a_{(2)}) \otimes p(a_{(3)})$ for $a \in H_1$. The only new part beyond [62] is the identification of the 'twisted Hopf algebra' B now as a Hopf algebra living in a braided category, and some slightly more explicit formulae for its structure. The set B coincides with the image of the projection $\Pi : H_1 \to H_1$ defined by $\Pi(a) = \sum a_{(1)} S \circ i \circ p(a_{(2)})$ in [62], while the pushed-out left adjoint coaction of H on B then reduces to the left coaction as stated. The braiding is from Example 1.3. The axioms of a Hopf algebra in a braided category require that $\underline{\Delta} : B \to B \otimes B$ is an algebra homomorphism with respect to the braided tensor product algebra structure on $B \otimes B$. Writing $\underline{\Delta} b = \sum b_{\underline{(1)}} \otimes b_{\underline{(2)}}$, this reads

$$\underline{\Delta}(bc) = \sum b_{\underline{(1)}} \Psi(b_{\underline{(2)}} \otimes c_{\underline{(1)}}) c_{\underline{(2)}} = \sum b_{\underline{(1)}} ({b_{\underline{(2)}}}^{(\bar{1})} \triangleright c_{\underline{(1)}}) \otimes {b_{\underline{(2)}}}^{(\bar{2})} c_{\underline{(2)}} \tag{73}$$

which indeed derives the condition in [62]. The structure of $B{>\!\!\triangleleft} H$ is the standard left-handed semidirect one by the action and coaction stated. Applying θ to these structures and evaluating further at once gives θ as a Hopf algebra isomorphism. Of course if H_1 is only a bialgebra then B is only a bialgebra in a braided category. In this case one can use the convolution inverse $i \circ S$ in the above. Also, the restriction to invertible antipode on H is needed only to ensure that Ψ is invertible as explained in Example 1.3. It is part of our interpretation of B as a braided-Hopf algebra rather than part of Radford's theorem itself. □

Such Hopf algebra projections have a geometrical interpretation as examples of trivial quantum principal bundles [5] and at the same time as quantum mechanics[53]. These papers also make some limited contact with more established ideas of non-commututive geometry as in [8]. Also note that in the above the Hopf algebra H need not be quasitriangular or dual quasitriangular. If it is then the above construction becomes related to the bosonization theorems of the last subsection, as mentioned there.

5 Braided Linear Algebra

In this section we describe some general constructions for examples of bialgebras and Hopf algebras in braided categories associated to a general matrix solution of the QYBE as in Example 1.1. This includes some interesting algebras for ring theorists. The first of these, $B(R)$, has a matrix of generators with matrix coproduct and includes a degenerate form of the Sklyanin algebra for the usual $GL_q(2)$ R-matrix. The second has a vector or covector of generators with linear coproduct, and includes the famous quantum plane $yx = qxy$ for this R-matrix. Thus the quantum plane does have a linear addition law provided we work in a braided category. Finally, we mention some recent developments such as a notion of braided-Lie algebra.

5.1 Braided Matrices

Let R be an invertible matrix solution of the QYBE and $A(R)$ the associated dual quasitriangular bialgebra as in Section 1.3. In the nicest case we can quotient $A(R)$ to obtain a dual quasitriangular Hopf algebra A. We say in this case that R is *regular*. This is true for the standard R matrices and one obtains $A = \mathcal{O}_q(G)$ as shown in [14]. Another way to obtain a dual quasitriangular Hopf algebra is if R is dualizable. In this case we have an associated dual quasitriangular Hopf algebra $A = GL(R)$. Either way we have a canonical bialgebra map $A(R) \to A$ and can apply the transmutation theorem, Theorem 4.9 to obtain a bialgebra $B(R) = B(A(R), A)$ in the category $\mathcal{M}^A$. This gives the following construction, which we verify directly.

Proposition 5.1 *[48][50] Let R be a bi-invertible solution of the QYBE with $v^i{}_j = \tilde{R}^i{}_a{}^a{}_j$ also invertible. Then there is a bialgebra $B(R)$ in the braided category of A-comodules, with matrix generators $\mathbf{u} = \{u^i{}_j\}$ and relations, braiding and coalgebra*

$$R^k{}_a{}^i{}_b u^b{}_c R^c{}_j{}^a{}_d u^d{}_l = u^k{}_a R^a{}_b{}^i{}_c u^c{}_d R^d{}_j{}^b{}_l, \quad \text{i.e.} \quad R_{21}\mathbf{u}_1 R_{12}\mathbf{u}_2 = \mathbf{u}_2 R_{21}\mathbf{u}_1 R_{12}.$$

$$\Psi(u^i{}_j \otimes u^k{}_l) = u^p{}_q \otimes u^m{}_n R^i{}_a{}^d{}_p R^{-1a}{}_m{}^q{}_b R^n{}_c{}^b{}_l \tilde{R}^c{}_j{}^k{}_d, \quad \text{i.e.} \quad \Psi(R^{-1}\mathbf{u}_1 \otimes R\mathbf{u}_2) = \mathbf{u}_2 R^{-1} \otimes \mathbf{u}_1 R$$

$$\underline{\Delta} u^i{}_j = u^i{}_a \otimes u^a{}_j, \ \underline{\epsilon} u^i{}_j = \delta^i{}_j \quad \text{i.e.} \quad \underline{\Delta}\mathbf{u} = \mathbf{u} \otimes \mathbf{u}, \quad \underline{\epsilon}\mathbf{u} = \mathrm{id}$$

Proof The formulae are obtained from Theorem 4.9 and then verified directly. Bi-invertible means R^{-1} and the second-inverse $\tilde{R}$ exist and is all that is needed to verify the braided-bialgebra axioms. The existence of v^{-1} is needed in defining Ψ^{-1} and is equivalent to demanding that R is dualizable. The commutation relations come from (65) using Proposition 1.15 to evaluate the formulae in matrix form. This gives a matrix equation of the form $uu = uuR^{-1}RR\tilde{R}$ with appropriate indices, see [48][50]. Putting two of the R's to the left (or rearranging (65)) gives the formula stated. The same applies to the braiding which comes out as a product of 4 R-matrices as shown but is also conveniently written in the compact form stated. Its extension to products is by definition in such a way that the product is a morphism in the category generated by this braiding Ψ and one can verify that this extension is consistent with the relations. To verify that the result is indeed a braided bialgebra an even more compact notation is useful. Namely as in (44) we can label the second copy of $B(R)$ in the braided tensor product by a prime, and then suppress Ψ. Then the commutation relations in the braided tensor product $B(R)\underline{\otimes}B(R)$ are $R^{-1}\mathbf{u}'_1 R\mathbf{u}_2 = \mathbf{u}_2 R^{-1}\mathbf{u}'_1 R$ and we compute

$$\begin{aligned} R_{21}\mathbf{u}_1\mathbf{u}'_1 R\mathbf{u}_2\mathbf{u}'_2 &= R_{21}\mathbf{u}_1 R(R^{-1}\mathbf{u}'_1 R\mathbf{u}_2)\mathbf{u}'_2 = (R_{21}\mathbf{u}_1 R\mathbf{u}_2)R^{-1}R_{21}^{-1}(R_{21}\mathbf{u}'_1 R\mathbf{u}'_2) \\ &= \mathbf{u}_2 R_{21}(\mathbf{u}_1 R_{21}^{-1}\mathbf{u}'_2 R_{21})\mathbf{u}'_1 R = \mathbf{u}_2 R_{21} R_{21}^{-1}\mathbf{u}'_2 R_{21}\mathbf{u}_1\mathbf{u}'_1 R = \mathbf{u}_2\mathbf{u}'_2 R_{21}\mathbf{u}_1\mathbf{u}'_1 R \end{aligned}$$

as required for $\underline{\Delta}$ to extend to $B(R)$ as a bialgebra in a braided category. In each expression, the brackets indicate how to apply the relevant relation to obtain the next expression. □

Note that the braided category here is generated by the matrix in $M_{n^2} \otimes M_{n^2}$ corresponding to Ψ as stated. On the other hand in the regular or dualizable case we can identify this category as contained in that of A-comodules, under the induced adjoint coaction in Theorem 4.9. In the present setting this is

$$\beta(u^i{}_j) = u^m{}_n \otimes (St^i{}_m)t^n{}_j, \qquad \text{i.e.} \quad \mathbf{u} \to \mathbf{t}^{-1}\mathbf{u}\mathbf{t}. \tag{74}$$

Note also that if $R_{21}R_{12} = 1$ (the triangular case) then the braiding and the commutation relations co-incide so that $\cdot = \Psi^{-1} \circ \cdot$ which is the naive notion of braided-commutativity. In general however this will not do and instead the braided-commutativity relations are different from the braiding itself.

Finally, it is good to know that the algebra $B(R)$ has at least one canonical representation. This is provided by

$$\rho(u^i{}_j)^k{}_l = Q^i{}_j{}^k{}_l, \quad \text{i.e.} \quad \rho_2(\mathbf{u}_1) = Q_{12}; \quad Q = R_{21}R_{12} \tag{75}$$

In the compact notation the proof reads

$$\begin{aligned}\rho_3(R_{21}\mathbf{u}_1R_{12}\mathbf{u}_2) &= R_{21}\rho_3(\mathbf{u}_1)R_{12}\rho_3(\mathbf{u}_2)\\ &= R_{21}Q_{13}R_{12}Q_{23} = Q_{23}R_{21}Q_{13}R_{12}\\ &= \rho_3(\mathbf{u}_2)R_{21}\rho_3(\mathbf{u}_1)R_{12} = \rho_3(\mathbf{u}_2R_{21}\mathbf{u}_1R_{12}).\end{aligned}$$

The middle equality follows from repeated use of the QYBE. Note that this representation is trivial in the triangular case. Other useful representations can be built from this.

From the theory above we obtain also that these braided matrices are related to usual quantum matrices by transmutation. The formula for the modified product comes out from (64) as [51]

$$\begin{aligned}u^i{}_j &= t^i{}_j\\ u^i{}_ju^k{}_l &= t^a{}_bt^d{}_lR^i{}_a{}^c{}_d\widetilde{R}^b{}_j{}^k{}_c\\ u^i{}_ju^k{}_lu^m{}_n &= t^d{}_bt^s{}_ut^z{}_nR^i{}_a{}^p{}_qR^a{}_d{}^w{}_y\widetilde{R}^b{}_c{}^v{}_w\widetilde{R}^c{}_j{}^k{}_pR^q{}_s{}^y{}_z\widetilde{R}^u{}_l{}^m{}_v\end{aligned} \tag{76}$$

etc. Here the products on the left are in $B(R)$ and are related by transmutation to the products on the right which are in $A(R)$. If we write some or all of the R-matrices over to the left hand side we have equally well the compact matrix form [51],

$$\begin{aligned}\mathbf{u} &= \mathbf{t}\\ R_{12}^{-1}\mathbf{u}_1R_{12}\mathbf{u}_2 &= \mathbf{t}_1\mathbf{t}_2\\ R_{23}^{-1}R_{13}^{-1}R_{12}^{-1}\mathbf{u}_1R_{12}\mathbf{u}_2R_{13}R_{23}\mathbf{u}_3 &= \mathbf{t}_1\mathbf{t}_2\mathbf{t}_3\end{aligned} \tag{77}$$

etc. This is just a rearrangement of (76) or our universal formula (64). For the transmuted product of multiple strings the universal formula from (64) involves a kind of partition function made from products of R to transmute the bosonic $A(R)$ to the braided $B(R)$ [51].

We believe these braided matrices deserve more study as certain well-behaved quadratic algebras. For the standard R-matrices they have quotients giving the braided versions $B_q(G)$ say of the quantum function algebras $\mathcal{O}_q(G)$, and are at the same time isomorphic via (66) and for generic q to the braided versions of the corresponding quantum enveloping algebras $U_q(g)$. This is related to constructions in physics[65]. On the other hand these $B(R)$ are interesting even at the quadratic level. The case of $BM_q(2)$ for the $GL_q(2)$ R-matrix was studied in [50] and shown in [52] to be a degenerate form of the Sklyanin algebra. Some remarkable homological properties of these braided matrices have recently been obtained in [3].

5.2 Braided Planes

If the algebras $B(R)$ are like 'braided matrices' because they have a matrix of generators with matrix coproduct, one can complete the picture with some notion of 'braided vectors' and 'braided covectors'. The usual algebra suggested here from physics is the Zamalodchikov or 'exchange algebra' of the form $\mathbf{x}_1\mathbf{x}_2 = \lambda\mathbf{x}_2\mathbf{x}_1R$ where $\mathbf{x}$ say is a row vector of generators and R is a matrix obeying the QYBE, and λ is a normalization constant. This is an interesting algebra in the Hecke case[20] but does not seem so interesting for a general solution of the QYBE. Coming out of the theory of Hopf algebras in braided categories it would seem rather more natural to use R for the braiding and some slightly different matrix R' say for the commutation relations of the algebra. This strategy works and gives the following construction.

Proposition 5.2 *[56] Let $R \in M_n \otimes M_n$ be an invertible solution of the QYBE and $R' \in M_n \otimes M_n$ an invertible matrix obeying*

$$R_{12}R_{13}R'_{23} = R'_{23}R_{13}R_{12}, \quad R_{23}R_{13}R'_{12} = R'_{12}R_{13}R_{23}, \quad (PR+1)(PR'-1) = 0$$

where P is the permutation matrix. Then there are braided-bialgebras $V\check{}(R')$ and $V(R')$ with row and column vectors of generators $\mathbf{x} = \{x_i\}$ and $\mathbf{v} = \{v^i\}$ respectively and with relations and braiding

$$x_i x_j = x_b x_a R'^a{}_i{}^b{}_j, \ \Psi(x_i \otimes x_j) = x_b \otimes x_a R^a{}_i{}^b{}_j, \quad \text{i.e. } \mathbf{x}_1\mathbf{x}_2 = \mathbf{x}_2\mathbf{x}_1 R', \ \Psi(\mathbf{x}_1 \otimes \mathbf{x}_2) = \mathbf{x}_2 \otimes \mathbf{x}_1 R$$

$$v^i v^j = R'^i{}_a{}^j{}_b v^b v^a, \ \Psi(v^i \otimes v^j) = R^i{}_a{}^j{}_b v^b \otimes v^a, \quad \text{i.e. } \mathbf{v}_1\mathbf{v}_2 = R'\mathbf{v}_2\mathbf{v}_1, \ \Psi(\mathbf{v}_1 \otimes \mathbf{v}_2) = R\mathbf{v}_2 \otimes \mathbf{v}_1$$

and linear coalgebra

$$\underline{\Delta}\mathbf{x} = \mathbf{x} \otimes 1 + 1 \otimes \mathbf{x}, \quad \underline{\epsilon}\mathbf{x} = 0, \quad \underline{\Delta}\mathbf{v} = \mathbf{v} \otimes 1 + 1 \otimes \mathbf{v}, \quad \underline{\epsilon}\mathbf{v} = 0.$$

If $R_{21}R'_{12} = R'_{21}R_{12}$ then these are braided-Hopf algebras with $\underline{S}\mathbf{x} = -\mathbf{x}$ and $\underline{S}\mathbf{v} = -\mathbf{v}$

Proof One can see that Ψ extends to products of generators in such a way that the product is a morphism in the braided category generated by R. For details see [56]. To see that the result forms a bialgebra we use the compact notation as in the proof of Proposition 5.1, $(\mathbf{x}_1+\mathbf{x}'_1)(\mathbf{x}_2+\mathbf{x}'_2) = \mathbf{x}_1\mathbf{x}_2+\mathbf{x}'_1\mathbf{x}_2+\mathbf{x}_1\mathbf{x}'_2+\mathbf{x}'_1\mathbf{x}'_2 = \mathbf{x}_2\mathbf{x}_1R'_{12}+\mathbf{x}_1\mathbf{x}'_2(PR_{12}+1)+\mathbf{x}'_2\mathbf{x}'_1R'_{12}$ while $(\mathbf{x}_2+\mathbf{x}'_2)(\mathbf{x}_1+\mathbf{x}'_1)R'_{12}$ has the same outer terms and the cross terms $\mathbf{x}'_2\mathbf{x}_1R'_{12} + \mathbf{x}_2\mathbf{x}'_1R'_{12} = \mathbf{x}_1\mathbf{x}'_2(R_{21}+P)R'$. These are equal since $PR+1 = PRPR' + PR'$ from the assumption on R'. Similarly for the other details. □

There are lots of ways to satisfy the auxiliary equation for the matrix R'. If R is a Hecke symmetry we can simply take $R' = \lambda R$ where λ is a suitable normalization. This is the familiar case. For example the standard $GL_q(n)$ R-matrix gives the algebra

$$x_i x_j = q x_j x_i \quad \text{if} \quad i > j \tag{78}$$

which is the n-dimensional quantum plane. Thus we see that it forms a Hopf algebra in the braided category generated by R. There is another normalization λ giving another quantum plane algebra and again a braided-Hopf algebra.

Another solution is $R' = P$ the permutation matrix. In this case the relations of the algebra are free (no relations). The R-matrix still enters into the braiding. This free-braided plane is therefore canonically associated to any invertible matrix solution of the QYBE. The simplest member of this family is the *braided line*. This is $B = k < x >$ (one generator) and

$$\Psi(x \otimes x) = qx \otimes x, \quad \underline{\Delta}x = x \otimes 1 + 1 \otimes x, \quad \underline{\epsilon}x = 0, \quad \underline{S}x = -x. \tag{79}$$

Finally, any R will obey an equation of the form $\prod_i(PR - \lambda_i) = 0$ and for each λ_i we can rescale R so that one of these factors become $PR + 1$. Then $PR' - 1$ defined as a multiple of the remaining factors gives us an R' obeying the required equations. Thus there are canonical braided Hopf algebras $V\check{}(R')$ and $V(R')$ for each eigenvalue λ_i in the decomposition. In the Hecke case there are by definition two such eigenvalues corresponding to the two normalizations mentioned, but in general R' will not be a multiple of R and need not obey the QYBE.

The algebra $V(R')$ can be called the *braided vector* algebra. Likewise the algebra $V\check{}(R')$ can be called the *braided covector* algebra. Another notation is $V^*(R')$ but we have avoided this here in order to not suggest that it is the left dual of the braided vector algebra in the sense of Section 2.2. In fact, the braiding is such that in the dualizable case the vector space generating the vector algebra is the left dual of that generating the covector algebra (compare with Section 1.1). Together with the braided matrices $B(R)$, these algebras form a kind of braided linear algebra. We refer to [51] for more details.

5.3 New Directions

In the above we have described the basic theory of Hopf algebras in braided categories and some canonical canonical constructions for them which have been established for the most part in the period 1989 – 1991. Some more recent directions are as follows.

One direction is that of *braided differential calculus.* The idea is that just as the differential structure on a group is obtained by making an infinitesimal group translation, now we can use our braided-coproducts on the braided-planes and braided-matrices to obtain corresponding differential operators.

Proposition 5.3 *[57] The operators $\partial^i : V\check{}(R') \to V\check{}(R')$ defined by*

$$\partial^i x_{i_1} \cdots x_{i_m} = \delta^i{}_{j_1} x_{j_2} \cdots x_{j_m} [m; R]^{j_1 \cdots j_m}_{i_1 \cdots i_m}$$

$$[m; R] = 1 + (PR)_{12} + (PR)_{12}(PR)_{23} + \cdots + (PR)_{12} \cdots (PR)_{m-1,m}$$

obey the relations of $V(R')$ and the braided-Leibniz rule

$$\partial^i(ab) = (\partial^i a)b + \cdot\Psi^{-1}(\partial^i \otimes a)b, \qquad \forall a, b \in V\check{}(R').$$

The result says that the braided-covector algebra (which is like the algebra of co-ordinate functions on some kind of non-commutative algebraic variety) is a left $V(R')$-module algebra in the category with reversed braiding. For the braided-line we obtain the usual Jackson q-derivative $(\partial f)(x) = \frac{f(qx)-f(x)}{(q-1)x}$. For the famous quantum plane we recover its well-known two dimensional differential calculus obtained usually by other means. We have also introduced in this context a braided-binomial theorem for the 'counting' of braided partitions. This is achieved by means of the *braided integers* $[m, R]$ in the proposition. One can also define a braided exponential map $\exp_R(\mathbf{x}|\mathbf{v})$ and prove a braided Taylor's theorem at least in the free case where $R' = P$. We refer to [57] for details.

Likewise, one has some natural right-handed differential operators on the braided matrices obtained from Example 2.8 applied to $B(R)$ and its braided-Hopf algebra quotients (the result lifts to the bialgebra setting). These are computed in detail in [58].

Related to this, one can formalise at least one notion of *braided Lie algebra*[58]. In a symmetric (not braided) category the notion of a Lie algebra and its enveloping algebra are just the obvious ones, see[19]. But as soon as one tries to make this work in the braided setting there are problems with the naive approach. One solution is based on the notion of braided-cocommutativity in Definition 2.9 with respect to the braided-adjoint action. Based on its properties in this case one can extract the axioms of a braided-Lie algebra as the following: a coalgebra $(\mathcal{L}, \Delta, \epsilon)$ in the braided category and a bracket $[\ ,\ ] : \mathcal{L} \otimes \mathcal{L} \to \mathcal{L}$

obeying

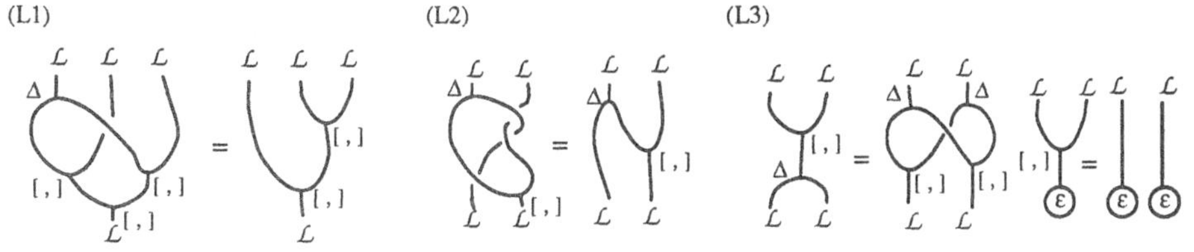

(L1) (L2) (L3)

$$(80)$$

We show that such a braided-Lie algebra has an enveloping bialgebra living in the category.

The braided matrices $B(R)$ (e.g. the degenerate Sklyanin algebra) double up in this role as braided-enveloping bialgebras. Here $\mathcal{L} = \mathrm{span}\{u^i{}_j\}$ with the matrix coproduct as in Section 5.1. These generators are a mixture of 'group-like' and 'primitive-like' generators. If one wants something more classical one can work equally well with the space $\mathcal{X} = \mathrm{span}\{\chi^i{}_j\}$ where $\chi^i{}_j = u^i{}_j - \delta^i{}_j$. In these terms the braiding has the same form as between the **u** generators, while the relations and coproduct become

$$R_{21}\chi_1 R_{12}\chi_2 - \chi_2 R_{21}\chi_1 R_{12} = Q_{12}\chi_2 - \chi_2 Q_{12}, \quad \underline{\Delta}\chi = \chi\otimes 1 + 1\otimes\chi + \chi\otimes\chi \tag{81}$$

in the compact notation. This is more like a 'Lie algebra' and for the standard R matrices, as the deformation parameter $q \to 1$ the matrix $Q = R_{21}R_{12} \to \mathrm{id}$ and the commutator on the right hand side vanishes. This means that it is the rescaled generators $\bar{\chi} = \hbar^{-1}\chi$ that tend to a usual Lie algebra and indeed the extra non-primitive term in the coproduct of the $\bar{\chi}$ now tends to zero. For details see [58]. We recall also that for the standard R matrices the algebras $B(R)$ have a quotient which is $U_q(g)$ at least formally, so these are understood as (the quotient of) the braided-enveloping algebra of a braided-Lie algebra.

We have not tried here to survey a number of more specific applications of this braided work. These include [4][21][32][55] as well as more physically-based applications. In addition among many relevant and interesting recent works by other mathematicians and that I have not had a chance to touch upon, I would like to at least mention [17][26][60][68] in the categorical direction and [6][63] as well as [2][7][16][3] already mentioned for works in an algebraic direction. What we have shown is that a number of general mathematical constructions can be braided. There is clearly plenty of scope for further work in this programme. Some potential areas are applications to knot theory, a general braided-combinatorics (based on the braided-integers above) and some kind of braided-analysis.

References

[1] W.K. Baskerville and S. Majid. Braided and unbraided harmonic oscillators. In M. Olmo et al, eds, *Anales de Física, Monografías*, 1:61-64. CIEMAT/RSEF, Madrid, 1993; The braided Heisenberg group. *J. Math. Phys.* 34:3588–3606, 1993.

[2] A. Borowiec, W. Marcinek, and Z. Oziewicz. On multigraded differential calculus. In R. Gierelak et al, eds, *Proc. of 1st Max Born Symposium, Poland, 1991*, pages 103–114. Kluwer.

[3] L. Le Bruyn. Homological properties of braided matrices. 1993, to appear in *J. Algebra*.

[4] T. Brzeziński and S. Majid. A class of bicovariant differential calculi on Hopf algebras. *Lett. Math. Phys*, 26:67–78, 1992.

[5] T. Brzeziński and S. Majid. Quantum group gauge theory on quantum spaces. May 1992, to appear in *Comm. Math. Phys.*

[6] M. Cohen and S. Westreich. From supersymmetry to quantum commutativity. 1991, to appear in *J. Algebra.*

[7] M. Cohen, D. Fischman and S. Westreich. Schur's double centralizer theorem for triangular Hopf algebras. 1992, to appear in Proc. AMS.

[8] A. Connes. C^* algebres et géométrie différentielle. *C.R. Acad. Sc. Paris*, 290:599–604, 1980.

[9] P. Deligne and J.S. Milne. Tannakian categories. *Springer Lec. Notes in Math*, 900, 1982.

[10] V.G. Drinfeld. Private communication, February 28th, 1990 (letter in response to preprint of [37]).

[11] V.G. Drinfeld. Quantum groups. In A. Gleason, editor, *Proceedings of the ICM*, pages 798–820, Rhode Island, 1987. AMS.

[12] V.G. Drinfeld. On almost-cocommutative Hopf algebras. *Algebra i Analiz*, 1(2):30–46, 1989. In Russian. Translation in *Leningrad Math. J.*

[13] V.G. Drinfeld. Quasi-Hopf algebras. *Algebra i Analiz*, 1(6):2, 1989. In Russian. Translation in *Leningrad Math J.*

[14] L.D. Faddeev, N.Yu. Reshetikhin, and L.A. Takhtajan. Quantization of Lie groups and Lie algebras. Translation in *Leningrad Math J.*, 1:193–225, 1990.

[15] D. Fischman. Schur's double centralizer theorem: A Hopf algebra approach. 1991, to appear in *J. Algebra.*

[16] D. Fischman and S. Montgomery. A Schur double centralizer theorem for cotriangular Hopf algebras and generalized Lie algebras. 1992 and 1993, to appear in *J. Algebra.*

[17] D. Freed. Higher algebraic structures and quantization. Preprint, 1992.

[18] P. Freyd and D. Yetter. Braided compact closed categories with applications to low dimensional topology. *Adv. Math.*, 77:156–182, 1989.

[19] D.I. Gurevich. The Yang-Baxter equations and a generalization of formal Lie theory. *Sov. Math. Dokl.*, 33:758–762, 1986.

[20] D.I. Gurevich. Algebraic aspects of the quantum Yang-Baxter equation. Translation in *Leningrad Math. J.*, 2:801–828, 1991.

[21] D.I. Gurevich and S. Majid. Braided groups of Hopf algebras obtained by twisting. 1991, to appear in *Pac. J. Math.*

[22] T. Hayashi. Quantum groups and quantum determinants. *J. Algebra*, 152:146–165, 1992.

[23] M. Jimbo. A q-difference analog of $U(g)$ and the Yang-Baxter equation. *Lett. Math. Phys.*, 10:63–69, 1985.

[24] A. Joyal and R. Street. Braided monoidal categories. Mathematics Reports 86008, Macquarie University, 1986.

[25] A. Joyal and R. Street. Tortile Yang-Baxter operators in tensor categories. *J. Pure Applied Algebra*, 71:43–51, 1991.

[26] M.M. Kapranov and V.A. Voevodsky. Braided monoidal 2-categories and Zamalodchikov tetrahedra equations. Mathematics report, Harvard and Northwestern Universities, 1991.

[27] S. Mac Lane. *Categories for the Working Mathematician.* Springer, 1974. GTM vol. 5.

[28] R.G. Larson and J. Towber. Two dual classes of bialgebras related to the concepts of 'quantum group' and 'quantum Lie algebra'. *Comm. Algebra*, 19(12):3295–3345, 1991.

[29] V.V. Lyubashenko. Hopf algebras and vector symmetries. *Usp. Math. Nauk.*, 41:185–186, 1986.

[30] V.V. Lyubashenko. Tangles and Hopf algebras in braided categories. 1991, to appear in *J. Pure and Applied Algebra.*

[31] V.V. Lyubashenko. Modular transformations for tensor categories. 1992, to appear in *J. Pure and Applied Algebra.*

[32] V.V. Lyubashenko and S. Majid. Braided groups and quantum Fourier transform. 1991, to appear in *J. Algebra.*

[33] S. Majid. Physics for algebraists: Non-commutative and non-cocommutative Hopf algebras by a bicrossproduct construction. *J. Algebra*, 130:17–64, 1990. From PhD Thesis, Harvard, 1988.

[34] S. Majid. More examples of bicrossproduct and double cross product Hopf algebras. *Isr. J. Math*, 72:133–148, 1990.

[35] S. Majid. Quasitriangular Hopf algebras and Yang-Baxter equations. *Int. J. Modern Physics A*, 5(1):1–91, 1990.

[36] S. Majid. Quantum groups and quantum probability. In *Quantum Probability and Related Topics VI (Proc. Trento, 1989)*, pages 333–358. World Sci.

[37] S. Majid. Representations, duals and quantum doubles of monoidal categories (in Proc. Winter School Geom. Phys., Srni, January 1990). *Suppl. Rend. Circ. Mat. Palermo, Ser. II*, 26:197–206, 1991.

[38] S. Majid. Doubles of quasitriangular Hopf algebras. *Comm. Algebra*, 19(11):3061–3073, 1991.

[39] S. Majid. Reconstruction theorems and rational conformal field theories. *Int. J. Mod. Phys. A*, 6(24):4359–4374, 1991.

[40] S. Majid. Braided groups and duals of monoidal categories. *Canad. Math. Soc. Conf. Proc.*, 13:329–343, 1992.

[41] S. Majid. Braided groups and braid statistics. In L. Accardi et al, eds, *Quantum Probability and Related Topics VIII (Proc. Delhi, 1990)*. World Sci.

[42] S. Majid. Beyond supersymmetry and quantum symmetry (an introduction to braided groups and braided matrices). In M-L. Ge and H.J. de Vega, eds, *Quantum Groups, Integrable Statistical Models and Knot Theory (Proc. Nankai, June 1992)*, pages 231–282. World Sci.

[43] S. Majid. Tannaka-Krein theorem for quasiHopf algebras and other results (in Proc. Amherst, June 1990). *Contemp. Maths*, 134:219–232, 1992.

[44] S. Majid. Braided groups and algebraic quantum field theories. *Lett. Math. Phys.*, 22:167–176, 1991.

[45] S. Majid. Transmutation theory and rank for quantum braided groups. *Math. Proc. Camb. Phil. Soc.*, 113:45–70, 1993.

[46] S. Majid. Cross products by braided groups and bosonization. March 1991, to appear in *J. Algebra*.

[47] S. Majid. Anyonic quantum groups. In Z. Oziewicz et al, eds, *Spinors, Twistors, Clifford Algebras and Quantum Deformations (Proc. Wroclaw, 1992)*, pages 327–336. Kluwer.

[48] S. Majid. Rank of quantum groups and braided groups in dual form. In *Proc. of the Euler Institute, St. Petersberg (1990), Springer Lec. Notes in Math*, 1510:79–89.

[49] S. Majid. Braided groups. *J. Pure and Applied Algebra*, 86:187–221, 1993.

[50] S. Majid. Examples of braided groups and braided matrices. *J. Math. Phys.*, 32:3246–3253, 1991.

[51] S. Majid. Quantum and braided linear algebra. *J. Math. Phys.*, 34:1176–1196, 1993.

[52] S. Majid. Braided matrix structure of the Sklyanin algebra and of the quantum Lorentz group. January 1992, to appear in *Comm. Math. Phys.*

[53] S. Majid. The quantum double as quantum mechanics. September 1992, to appear in *J. Geom. Phys.*

[54] S. Majid. C-statistical quantum groups and Weyl algebras. *J. Math. Phys.*, 33:3431–3444, 1992.

[55] S. Majid. Infinite braided tensor products and 2D quantum gravity. In *Proc. of XXI DGM, Tianjin, China 1992*. World Sci.

[56] S. Majid. Braided momentum in the q-Poincaré group. *J. Math. Phys.*, 34:2045–2058, 1993.

[57] S. Majid. Free braided differential calculus, braided binomial theorem and the braided exponential map. January 1993, to appear in *J. Math. Phys.*

[58] S. Majid. Quantum and braided Lie algebras. March 1993, to appear in *J. Geom. Phys.*

[59] S. Majid and M.J. Rodriguez-Plaza. Universal R-matrix for non-standard quantum group and superization. Preprint, DAMTP/91-47; Quantum and super-quantum group related to the Alexander-Conway polynomial. *J. Geom. Phys.* 11:437–443, 1993.

[60] J. Morava. Some examples of Hopf algebras and Tannakian categories. *Contemp. Math.*, 1993. In Proc. Mahowald Festschrift.

[61] B. Pareigis. A non-commutative non-cocommutative Hopf algebra in nature. *J. Algebra*, 70:356, 1981.

[62] D. Radford. The structure of Hopf algebras with a projection. *J. Alg.*, 92:322–347, 1985.

[63] D. Radford. Solutions of the Yang-Baxter equation: the two-dimensional upper-triangular case. 1991, to appear in *Trans. Amer. Math. Soc.*

[64] N.Yu. Reshetikhin and M.A. Semenov-Tian-Shansky. Quantum R-matrices and factorization problems. *J. Geom. Phys.*, 5:533, 1988.

[65] N.Yu. Reshetikhin and M.A. Semenov-Tian-Shansky. Central extensions of quantum current groups. *Lett. Math. Phys.*, 19:133–142, 1990.

[66] N. Saavedra Rivano. Catégories Tannakiennes. *Springer Lec. Notes in Math*, 265, 1972.

[67] M. Scheunert. Generalized Lie algebras. *J. Math. Phys.*, 20:712–720, 1979.

[68] S. Shnider and S. Sternberg. The cobar resolution and a restricted deformation theory for Drinfeld algebras. 1992, to appear in *J. Algebra*.

[69] M.E. Sweedler. *Hopf Algebras*. Benjamin, 1969.

[70] K.-H. Ulbrich. On Hopf algebras and rigid monoidal categories. *Isr. J. Math*, 72:252–256, 1990.

[71] J.H.C. Whitehead. Combinatorial homotopy, II. *Bull. Amer. Math. Soc.*, 55:453–496, 1949.

[72] D.N. Yetter. Quantum groups and representations of monoidal categories. *Math. Proc. Camb. Phil. Soc.*, 108:261–290, 1990.

Quotient Theory of Hopf Algebras

AKIRA MASUOKA Shimane University, Matsue, Shimane, Japan

Introduction

We work over a field k. There is an anti-equivalence $G \mapsto O(G)$ from the category of affine group schemes to the category of commutative Hopf algebras. The quotients of G correspond to the sub-objects of $O(G)$. More precisely, the quotient affine schemes X of the form $X = H\backslash G$, where $H(\subset G)$ is some closed subgroup scheme, correspond to the right coideal subalgebras $O(X)$ over which $O(G)$ is a faithfully flat module. By abuse of terms, we call the investigation of such sub-objects($\subset O(G)$) "the quotient theory of commutative Hopf algebras". We drop "commutative" to reach the title at the head.

Fix a (non-commutative) Hopf algebra A, and let $B \subset A$ be a right coideal subalgebra, i.e., a subalgebra such that $\Delta(B) \subset B \otimes A$, where $\Delta : A \to A \otimes A$ is the coproduct. Roughly speaking, we classify all B's according to the following conditions:

1) A is faithfully flat as a (left or right) B-module;

2) A is a projective generator as a B-module;

3) A is free as a B-module;

4) A has a normal basis over B, which means that there is a right B-linear and left $\bar{A}$-colinear isomorphism $A \simeq \bar{A} \otimes B$, where $\bar{A}$ is a certain quotient coalgebra of A.

For this purpose the study of the category of (A,B)-Hopf modules is indispensable. In fact the concept of such modules was introduced by Takeuchi [T2] to prove that Condition 1) is satisfied, if A is either commutative or cocommutative and if B is a Hopf subalgebra.

The theory gets more algebraic and comprehensible, and comes to the stage where we may expect its nice contribution to further investigation of Hopf algebras. The purpose of this expository paper is to introduce this interesting subject to as many people as possible. So, among the contents of [Sw] we only assume such minimum knowledge of coalgebras and comodules as in [ibid., Chapter I, II].

0 Preliminaries——Hopf Algebras and Hopf Modules

Unless otherwise stated, everything takes place over a fixed base field k. In particular, unadorned $\otimes$ means $\otimes_k$.

(Co)algebras are assumed to be (co)unitary and (co)associative; (co)modules to be (co)unitary.

Suppose (A, Δ, ε) is a bialgebra, which means A is an algebra and coalgebra such that the coalgebra structures $\Delta : A \to A \otimes A$, $\varepsilon : A \to k$ are algebra maps. We use the usual sigma notation. Thus we write

$$\Delta(a) = \sum a_{(1)} \otimes a_{(2)} \quad (a \in A),$$

and for a right A-comodule (M, ρ)

$$\rho(m) = \sum m_{(0)} \otimes m_{(1)} \quad (m \in M).$$

A linear map $S : A \to A$ is called an <u>antipode</u>, if it is the inverse to the identity map of A under the convolution product, or explicitly

$$\sum S(a_{(1)})a_{(2)} = \varepsilon(a)1 = \sum a_{(1)}S(a_{(2)}) \qquad (a \in A).$$

A bialgebra is called a <u>Hopf algebra</u>, if it has a (unique) antipode.

0.1 LEMMA. For a bialgebra A, the following are equivalent:

a) A is a Hopf algebra;

b) $A \otimes A \to A \otimes A$, $a \otimes a' \mapsto \sum a_{(1)} \otimes a_{(2)} a'$ is an isomorphism;

c) For every right A-comodule M,

$$M \otimes A \to M \otimes A, \quad m \otimes a \mapsto \sum m_{(0)} \otimes m_{(1)} a \tag{0.2}$$

is an isomorphism.

If S is the antipode, the inverse of (0.2) is given by $m\otimes a \mapsto \sum m_{(0)}\otimes S(m_{(1)})a$.

The antipode S of a Hopf algebra A is not necessarily bijective [T1]. But S is bijective in many cases. This is the case, either if A is finite dimensional, if A is commutative, or (as will be seen in § 2) if the coradical of A is cocommutative. The next lemma follows from [DT, Prop.7].

0.3 LEMMA. Let A be a Hopf algebra with antipode S. The following are equivalent:

a) S is bijective;

b) The bialgebra A^{op} got from A by making the product opposite is a Hopf algebra;

c) The bialgebra A^{cop} got from A by making the coproduct opposite is a Hopf algebra.

If these hold true, the composite-inverse of S is the antipode of A^{op} or A^{cop}.

Let A be a bialgebra.

A subalgebra B of A is called a <u>right coideal subalgebra</u>, if it is a right coideal (i.e., $\Delta(B)\subset B\otimes A$). A subbialgebra is in particular a right coideal subalgebra. Dually, a quotient coalgebra of A is called a <u>quotient left A-module coalgebra</u>, if it is a quotient left A-module. <u>Left coideal subalgebras</u> or <u>quotient right A-module coalgebras</u> are defined in an obvious manner. If $B\subset A$ is a right coideal subalgebra, then the quotient space A/AB^+ by the left ideal

AB^+ generated by

$$B^+ := B \cap \operatorname{Ker} \varepsilon$$

is a quotient left A-module coalgebra. Conversely, if $\bar{A}$ is a quotient left A-module coalgebra (with the quotient map $a \mapsto \bar{a}$), then the set

$$^{co\bar{A}}A := \{ a \in A \mid \sum \overline{a_{(1)}} \otimes a_{(2)} = \bar{1} \otimes a \}$$

of the <u>left</u> $\bar{A}$-<u>coinvariants</u> in A is a right coideal subalgebra. When $\bar{A} = A/AB^+$, one has that $B \subset {}^{co\bar{A}}A$, but the equality does not necessarily hold.

Let $B \subset A$ be a right coideal subalgebra. To observe the relationship between A and B, we introduce the categories M_B^A, ${}_B\mathsf{M}^A$. The objects in M_B^A (resp., ${}_B\mathsf{M}^A$) are the right (resp., left) B-modules M with a right A-comodule structure $\rho : M \to M \otimes A$ such that

$$\rho(mb) = \sum m_{(0)}b_{(1)} \otimes m_{(1)}b_{(2)}$$

$$(\text{resp.,} \quad \rho(bm) = \sum b_{(1)}m_{(0)} \otimes b_{(2)}m_{(1)})$$

for $m \in M$, $b \in B$. (An object in M_B^A is called an (A,B)-<u>Hopf module</u> in [T5]. An (A,A)-Hopf module is precisely the A-<u>Hopf module</u> defined in [Sw].) The Morphisms in M_B^A or ${}_B\mathsf{M}^A$ are B-linear and A-colinear maps. These categories are abelian. A and B are both objects in M_B^A and in ${}_B\mathsf{M}^A$ with the obvious structures. We have a natural identification:

$$(0.4) \qquad {}_B\mathsf{M}^A = \mathsf{M}_{B^{op}}^{A^{op}}$$

This (together with (0.3) when A is a Hopf algebra with bijective antipode) permits us to translate a result on either

M_B^A or ${}_B\mathsf{M}^A$ into a result on the other categoty.

1 Faithful Flatness and Projectivity

The quotient theory of Hopf algebras started with the following theorem due to M. Takeuchi.

1.1 THEOREM [T2, 5]. A commutative Hopf algebra is a faithfully flat module, or more strongly a projective generator, over every Hopf subalgebra.

We observe faithful flatness or projectivity of non-commutative Hopf algebras.

Let A be a (non-commutative) Hopf algebra, and $B \subset A$ a right coideal subalgebra. Write $\bar{A} = A/AB^+$, a quotient left A-module coalgebra of A.

We relate M_B^A with the category $\mathsf{M}^{\bar{A}}$ of right $\bar{A}$-comodules. For $M \in \mathsf{M}_B^A$, write $\bar{M} = M/MB^+$, which has an induced right $\bar{A}$-comodule structure. Thus we have a functor

$$\bar{?} : \mathsf{M}_B^A \to \mathsf{M}^{\bar{A}}, \quad M \mapsto \bar{M}.$$

This is left adjoint to the cotensor functor

$$? \,\square_{\bar{A}}\, A : \mathsf{M}^{\bar{A}} \to \mathsf{M}_B^A, \quad V \mapsto V \,\square_{\bar{A}}\, A. \tag{1.2}$$

Here we recall the definition of cotensor products [T3]. Let C be a coalgebra, (V, ρ) a right C-comodule and (W, λ) a left C-comodule. The cotensor product $V \,\square_C\, W$ is is the kernel of the following two maps

$$V \otimes W \underset{\mathrm{id}\otimes\lambda}{\overset{\rho\otimes\mathrm{id}}{\rightrightarrows}} V \otimes C \otimes W.$$

Based on the fact that any comodule is a directed union of finite dimensional subcomodules, the next proposition follows.

1.3 PROPOSITION [T4]. The following are equivalent:

a) W is a coflat (resp., faithfully coflat) left C-comodule (i.e., the functor $? \,\square_C\, W$ is exact (resp., faithfully exact));

b) W is injective (resp., an injective cogenerator) as a left C-comodule.

In (1.2) note that $V \,\square_{\bar{A}}\, A \in \mathbf{M}_B^A$, since the natural left $\bar{A}$-coaction on A commutes with the right B-action and with the right A-coaction. The adjunctions Ξ, Θ determined by $\bar{?}$, $? \,\square_{\bar{A}}\, A$ are given by

$$\Xi_M : M \to \bar{M} \,\square_{\bar{A}}\, A, \quad m \mapsto \sum \overline{m_{(0)}} \otimes m_{(1)},$$

$$\Theta_V : \overline{V \,\square_{\bar{A}}\, A} \to V, \quad \overline{\sum v_i \otimes a_i} \mapsto \sum v_i \varepsilon(a_i),$$

where $M \in \mathbf{M}_B^A$, $V \in \mathbf{M}^{\bar{A}}$.

Combining [T5, Thm.1] with [S1, Thm.4.7], we have:

1.4 THEOREM. Let A be a Hopf algebra, and $B \subset A$ a right coideal subalgebra. Write $\bar{A} = A/AB^+$. Then the following are equivalent:

a) A is faithfully coflat as a left $\bar{A}$-comodule and $B = {}^{co\bar{A}}A$;

b) The functors $\bar{?}$, $? \,\square_{\bar{A}}\, A$ are (mutually quasi-inverse) equivalences;

c) The adjunctions Ξ , Θ are isomorphisms.

If A is faithfully flat as a left B-module, these equivalent conditions hold true.

In case B = A, this yields:

1.5 THEOREM [Sw]. Let A be a Hopf algebra. Then

$$? \otimes A : \begin{matrix}\text{the category of} \\ \text{k-vector spaces}\end{matrix} \to \mathsf{M}_A^A, \quad V \mapsto V \otimes A$$

gives an equivalence.

If the antipode is bijective, we can add something to (1.4).

1.6 THEOREM [MW]. Let A be a Hopf algebra with bijective antipode, and $B \subset A$ a right coideal subalgebra. Then the following are equivalent:

a)-c) the same as in (1.4);

d) A is flat as a left B-module and B is simple as an object in M_B^A;

e) A is faithfully flat as a left B-module;

f) A is a projective generator as a left B-module.

By applying the result to A^{op}, one sees that the right versions of d)-f) are equivalent with each other.

In d), B is simple in M_B^A, if and only if B has no non-trivial such right ideal that is a right coideal of A. By (1.5) this is the case if B is a Hopf subalgebra, since then $B \in \mathcal{M}_B^B$. Taking this into account we have:

1.7 COROLLARY [MW]. Let A be a Hopf algebra, and $B \subset A$ a Hopf subalgebra. Suppose the antipodes of A and B are both bijective. Then the following are equivalent:

a) A is left B-flat;

b) A is left B-faithfully flat;

c) A is a left B-projective generator;

d) Every non-zero object in ${}_B\mathcal{M}^A$ is a left B-projective generator;

a°)-d°) the right versions of a)-d).

Here c°) $\Rightarrow$ d°) follows from:

1.8 THEOREM [D1]. Let A be a Hopf algebra, and $B \subset A$ a Hopf subalgebra. If there is a counitary right B-linear map $A \to B$, then every object in $\mathcal{M}_B^A$ is right B-projective.

1.9 QUESTION. In (1.6), do the equivalent conditions a)-f) imply Condition d) in (1.7)?

1.10 REMARK. We can dualize all of the above results in this section except (1.1), (1.3). (In the dual case, the terms "quotient theory of Hopf algebras" have no abuse. cf. the Introduction.) The dual result of (1.6) is less interesting due to (1.3).

By (1.6) we have:

1.11 THEOREM. Let A be a Hopf algebra with bijective antipode. Then $B \mapsto A/AB^+$, $\bar{A} \mapsto {}^{co\bar{A}}A$ give a 1-1 correspondence between the right coideal subalgebras of A over which A is left faithfully flat and the quotient left A-module coalgebras of A over which A is left faithfully coflat.

The same result is shown in [T5, Thm.4] in the commutative case, which can be restated in terms of group schemes as follows:

1.12 THEOREM [T5]. The dur k-sheaf of left cosets $H\backslash G$ of an affine k-group scheme G by a closed subgroup scheme H is affine, if and only if the affine ring $O(G)$ is faithfully coflat as a left (or right) $O(H)$-comodule.

We observe commutative Hopf algebras further.

1.13 THEOREM [MW]. A commutative Hopf algebra is a flat module over every right coideal subalgebra.

1.14 COROLLARY [MW]. Let A be a commutative Hopf algebra, and $B \subset A$ a right coideal subalgebra. The following are equivalent:

a) B is simple as an object in $\mathcal{M}_B^A$;

b)-d) the same as in (1.7) (with "left" deleted).

If B is a Hopf subalgebra, these hold true.

This follows directly from (1.6) and (1.13). Thus we have a very simple proof of (1.1). By the way, one may wonder:

1.15 QUESTION. Is a non-commutative Hopf algebra A faithfully flat over every Hopf subalgebra?

Yes, if A is finite dimensional (See (4.1)). In the infinite dimensional case this is open, even if the last phrase is replaced by "over every finite dimensional Hopf subalgebra".

2 Hopf Algebras with Cocommutative Coradical

Let C be a coalgebra. C is said to be cocommutative, if $\Sigma\, c_{(1)} \otimes c_{(2)} = \Sigma\, c_{(2)} \otimes c_{(1)}$ for all $c \in C$. The coradical C_0 of C is a (direct) sum of all simple subcoalgebras of C. C is said to be pointed, if C_0 is spanned by the group-likes (i,e., the non-zero elements g such that $\Delta(g) = g \otimes g$). C_0 is cocommutative, if and only if $L \otimes C$ is pointed for some extension L/k of fields.

In this section we see that Hopf algebras with cocommutative coradical (in particular, pointed Hopf algebras) have a nice property concerning coideal subalgebras or quotient module coalgebras.

2.1 EXAMPLE (of pointed Hopf algebras). 1) The commutative pointed Hopf algebras correspond to the solvable group schemes (i.e., the affine group schemes such that every irreducible representation is 1-dimensional) [Wa]. For example the solvable group T_n of upper triangular $n \times n$ matrices is represented by

$$O(T_n) = k[X_{ij} | i \leqslant j]_d,$$

the localization of $k[X_{ij} | i \leqslant j]$ by $d = X_{11}X_{22}\dots X_{nn}$, where the coalgebra structure is given by

$$\Delta(X_{ij}) = \sum_{i \leqslant r \leqslant j} X_{ir} \otimes X_{rj}, \quad \varepsilon(X_{ij}) = \delta_{ij}.$$

2) The quantized enveloping algebra $U_q(\mathfrak{g})$ is pointed. In fact, such a Hopf algebra is pointed that is generated (as an algebra) by group-likes and nearly primitives (i.e., elements u such that $\Delta(u) = g \otimes u + u \otimes h$, where g, h are group-likes).

2.2 PROPOSITION [M1]. A Hopf algebra with cocommutative coradical has bijective antipode.

This follows by combining (0.3) with the following:

2.3 LEMMA [T1]. Let C be a coalgebra, and K an algebra. A linear map $C \to K$ is invertible under the convolution product, if and only if its restriction $C_0 \to K$ is invertible. (Recall the convolution product $*$ is defined by

$$f*f'(c) = \sum f(c_{(1)})f'(c_{(2)}),$$

where $f, f' \in \mathrm{Hom}_k(C, K)$, $c \in C$.)

2.4 LEMMA [M1]. Let A be a Hopf algebra with cocommutative coradical A_0, and $B \subset A$ a right coideal subalgebra. Write $B_0 = B \cap A_0$, a subcoalgebra of A_0. Then the following are equivalent:

a) $S(B_0) = B_0$; b) $S(B_0) \subset B_0$;

c) The monoid $G(L \otimes B)$ of the group-likes contained in $L \otimes B$ forms a goroup, where L/k is some/any extension of fields such that $L \otimes A$ is pointed.

Either if A is finite dimensional or if A is irreducible (i.e., $A_0 = k1$), then any B satisfies these equivalent conditions. A sub-bialgebra $B \subset A$ satisfies these conditions, if and only if it is a Hopf subalgebra.

2.5 THEOREM [M1]. Let A be a Hopf algebra with antipode S. Suppose the coradical A_0 is cocommutative.

1) If $B \subset A$ is a right coideal subalgebra such that $S(B_0) = B_0$, then A is faithfully flat (or a projective generator) as a left and right B-module.

2) If $\bar{A}$ is a quotient left A-module coalgebra of A, then A is faithfully coflat as a left and right $\bar{A}$-comodule.

3) $B \mapsto A/AB^+$, $\bar{A} \mapsto {}^{co\bar{A}}A$ give a 1-1 correspondence between the right coideal subalgebras B such that $S(B_0) = B_0$ and the quotient left A-module coalgebras $\bar{A}$.

In Part 1) a stronger conclusion follows, either if A is pointed or if B is finite dimensional. See (3.5).

From (2.5) or [N, Thm.1] we may expect that quotient module coalgebras have often nice properties rather than coideal subalgebras.

3 Freeness, Normal Basis Property and (Co)cleftness

In this section we observe freeness of Hopf algebras or Hopf modules. First we present two fundamental results.

3.1 PROPOSITION [R]. Let A be a bialgebra, and $B \subset A$ a right coideal subalgebra. Suppose every non-zero object in M_B^A includes such a non-zero sub-object that is B-free (resp., B-projective). (This holds, if every $M \in \mathsf{M}_B^A$ of the form $M = VB$, where $V \subset M$ is some simple A-subcomodule, is B-free (resp., B-projective).) Then every object in M_B^A is B-free (resp., B-projective).

3.2 THEOREM [S3]. Let A be a Hopf algebra, and $B \subset A$ a finite dimensional Hopf subalgebra. Such an infinite dimensional object in M_B^A or ${}_B\mathsf{M}^A$ that is B-projective is B-free.

The next example shows it happens in (3.2) that a finite dimensional object in ${}_B\mathsf{M}^A$ is B-projective but not B-free.

3.3 EXAMPLE [NZ2]. Suppose $\mathrm{ch}\, k \neq 2$. There exists a Hopf algebra A which is universal with respect to the following property: A contains a group like $g (\neq 1)$ and a 3×3 matrix subcoalgebra C with a canonical basis c_{ij} $(1 \leqslant i, j \leqslant 3)$ $(\Delta(c_{ij}) = \sum_r c_{ir} \otimes c_{rj}$, $\varepsilon(c_{ij}) = \delta_{ij})$ such that

$$g^2 = 1, \quad gc_{ij} = \lambda_i \lambda_j c_{ij},$$

where $\lambda_1 = \lambda_2 = 1$, $\lambda_3 = -1$. This A is infinite dimensional.

Hence by (3.2), A is left and right free over the Hopf subalgera $B = k\langle g\rangle$ generated by g, since B is semisimple as an algebra. But $C \in {}_B\mathsf{M}^A$ is not B-free (by counting the dimensions).

In the remainder of this section, we let A be a Hopf algebra, and $B \subset A$ a right coideal subalgebra.

3.4 DEFINITION. A has a <u>normal basis</u> over B, if there is a right B-linear and left $\bar{A}:=A/AB^+$-colinear isomorphism

$$A \simeq \bar{A} \otimes B.$$

In this case such an isomorphism $A \simeq \bar{A} \otimes B$ can be taken so as to be unitary and counitary (See (3.10.2)). Here note that B (resp., $\bar{A}$) has unit 1 and counit $\varepsilon|_B$ (resp., $\bar{1}$, $\varepsilon_{\bar{A}}$), so $\bar{A} \otimes B$ has, too.

The study of Hopf algebras with normal basis was begun by Oberst-Schneider [OS1, 2]. At the present stage we have:

3.5 THEOREM [M1, 4]. Let $A \supset B$ be as above. A has a normal basis over B, if either

i) A is pointed and the monoid $G(B)$ of the group-likes contained in B forms a group,

ii) the coradical A_0 is cocommutative and B is finite dimensional,

iii) A is commutative and B is finite dimensional,

iv) A is commutative, B is a semi-local ring and B is simple as an object in M_B^A, or

v) A is finite dimensional and B is a Hopf subalgebra.

The result in Case v) will be generalized in (4.2). The next proposition follows from (1.4).

3.6 PROPOSITION. If A has a normal basis over B, then every object in $\mathcal{M}_B^A$ is B-free.

Conversely we have:

3.7 THEOREM [MD]. Let $A \supset B$ be as above. Write $\bar{A} = A/AB^+$. Suppose A is faithfully coflat as a left $\bar{A}$-comodule. Then A has a normal basis over B, if either a), b) or c) stated below holds:

a) i) Every simple subcoalgebra of $\bar{A}$ is a matrix coalgebra (i.e., the dual coalgebra of a matrix algebra over k), and ii) Every object in $\mathcal{M}_B^A$ is B-free;

b) i) The quotient algebra $B/\mathrm{Rad}\, B$ by the Jacobson radical is artinian, and

ii) For any finite extension L/k of fields, every object in $\mathcal{M}_{L\otimes B}^{L\otimes A}$ is $L\otimes B$-free;

c) i) B is finite dimensional, and

ii) Every object in $\mathcal{M}_{\bar{k}\otimes B}^{\bar{k}\otimes A}$ is $\bar{k}\otimes B$-free, where $\bar{k}$ is the algebraic closure of k.

Here is a useful characterization of Hopf algebras with normal basis:

3.8 PROPOSITION [MD]. The following are equivalent:

a) A has a normal basis over B;

b) i) A is faithfully coflat as a left $\bar{A}:=A/AB^+$-comodule, and ii) For every simple subcoalgebra $C \subset \bar{A}$, there is a right B-linear and left C-colinear isomorphism $C \,\square_{\bar{A}}\, A \simeq C \otimes B$;

c) There is such a right B-linear map $A \to B$ that is invertible under the convolution product.

3.9 DEFINITION. If Condition c) holds, A is said to be <u>cocleft</u> over B. Dually for an arbitrary quotient left A-module coalgebra $\bar{A}$ of A, A is said to be <u>cleft</u> over $\bar{A}$, if there is an invertible left $\bar{A}$-colinear map $\bar{A} \to A$. (These concepts can be defined in more general contexts. See [MD].)

3.10 PROPOSITION [MD]. Let A be a Hopf algebra.

1) $B \mapsto A/AB^+$, $\bar{A} \mapsto {}^{co\bar{A}}A$ give a 1-1 correspondence between the right coideal subalgebras over which A is cocleft and the quotient left A-module coalgebras over which A is cleft.

2) Suppose $B \leftrightarrow \bar{A}$ in 1). There are natural 1-1 correspondences among the following sets:

a) right B-linear and left $\bar{A}$-colinear isomorphisms $A \simeq \bar{A} \otimes B$ (which is unitary/counitary/unitary and counitary);

b) invertible right B-linear maps $A \to B$ (which is unitary/counitary/unitary and counitary);

c) invertible left $\bar{A}$-colinear maps $\bar{A} \to A$ (which is unitary/counitary/unitary and counitary).

All of these sets are non-empty.

3.11 REMARK. Fix such a unitary and counitary isomorphism

$A \simeq \bar{A} \otimes B$ as in (3.10.2.a). It makes $\bar{A} \otimes B$ into a Hopf algebra with unit $\bar{1} \otimes 1$ and counit $\varepsilon_{\bar{A}} \otimes \varepsilon|_B : \bar{A} \otimes B \to k$.

1)[DT] Suppose $\bar{A}$ is a quotient Hopf algebra. Then the product of $\bar{A} \otimes B$ is described by means of

a (measuring) action $B \otimes \bar{A} \to B$, $b \otimes \bar{a} \mapsto b \leftharpoonup \bar{a}$ and

an invertible linear map $\sigma : \bar{A} \otimes \bar{A} \to B$

(satisfying suitable conditions) as follows:

$$(\overline{a'} \otimes b')(\bar{a} \otimes b) = \Sigma\, \overline{a'_{(1)}}\bar{a}_{(1)} \otimes \sigma(\overline{a'_{(2)}}, a_{(2)})\, b' \leftharpoonup \bar{a}_{(3)}\, b$$

The algebra $\bar{A} \otimes B$ is called a <u>crossed</u> <u>product</u> <u>algebra</u>, and denoted by $\bar{A} \#_\sigma B$ instead.

2) Suppose B is a Hopf subalgebra. Then the coproduct of $\bar{A} \otimes B$ is described by means of

a (comeasuring) coaction $\bar{A} \to B \otimes \bar{A}$, $\bar{a} \mapsto \Sigma\, \bar{a}_B \otimes \bar{a}_{\bar{A}}$ and

an invertible linear map $\pi : \bar{A} \to B \otimes B$, $\bar{a} \mapsto \Sigma\, \bar{a}^{(1)} \otimes \bar{a}^{(2)}$

(satisfying suitable conditions) as follows:

$$\bar{a} \otimes b \mapsto \Sigma\, (\bar{a}_{(1)} \otimes \bar{a}_{(2)B}\bar{a}_{(3)}^{(1)} b_{(1)}) \otimes (\bar{a}_{(2)\bar{A}} \otimes \bar{a}_{(3)}^{(2)} b_{(2)})$$

The coalgebra $\bar{A} \otimes B$ is called a <u>crossed</u> <u>coproduct</u> <u>coalgebra</u>, and denoted by $\bar{A} \overset{\pi}{\#} B$ instead.

We use the cocleftness rather than the normal basis condition to prove the following:

3.12 PROPOSITION [K]. Let A be a Hopf algebra with bijective antipode. If A is cocleft over a right coideal subalgebra B, then A^{op} is cocleft over B^{op}, or equivalently there is a left B-linear and left A/B^+A-colinear isomorphism

$A \simeq B \otimes A/B^{+}A$, so A is left B-free in particular.

4 The Nichols-Zoeller Theorem and its Generalization

The following major result is due to Nichols and Zoeller, which contributes greatly to the theory of finite dimensional Hopf algebras.

4.1 THEOREM [NZ2]. Let A be a finite dimensional Hopf algebra, and $B \subset A$ a Hopf subalgebra. Then every object in $\mathcal{M}_B^A$ or ${}_B\mathcal{M}^A$ is B-free. In particular A is.

One may wonder whether the result is true for a right coideal subalgebra B. We will answer to this question affirmatively in several cases.

4.2 THEOREM [H], [K], [M2]. Let A be a finite dimensional Hopf algebra, and $B \subset A$ a right coideal subalgebra. The following are equivalent:

a) B is a quasi-Frobenius algebra;

b) B is a Frobenius algebra;

c) A has a normal basis over B;

d) Every object in $\mathcal{M}_B^A$ or ${}_B\mathcal{M}^A$ is B-free;

e) A is left (resp., right) B-projective and a right (resp., left) B-generator;

f) There is a left and a right non-zero B-linear map $A \to B$, and at least one of them is identity on B;

g) There is a left and a right non-zero integral in B, and at least one of them generates the right A-comodule B.

Either if B is commutative or if B is a Hopf subalgebra, these equivalent conditions hold true.

Recall that a finite dimensional algebra B is called _quasi-Frobenius_, if B is injective as a left or right B-module. B is called _Frobenius_, if, more strongly, $B \simeq \mathrm{Hom}_k(B, k)$ as a left or right B-module. In g) a _left_ (resp., _right_) _integral_ in B means an element t in B such that

$$bt = \varepsilon(b)t \quad (\text{resp., } tb = \varepsilon(b)t) \quad (b \in B).$$

The last statement in (4.2), in the case where B is a Hopf subalgebra, follows from:

4.3 THEOREM [Sw]. A finite dimensional Hopf algebra is a Frobenius algebra.

Theorem(4.2) is generalized partially as follows:

4.4 THEOREM [K]. Let A be a (possibly infinite dimensional) Hopf algebra with bijective antipode, and $B \subset A$ a finite dimensional right coideal subalgebra. Then the Conditions a), f) and g) in (4.2) are equivalent with each other. If B is commutative, these hold true.

For the proof of (4.2), the isomorphism in (0.2) plays an important role. Let $M \in {}_B\mathcal{M}^A$, where A is an arbitrary Hopf algebra and $B \subset A$ a right coideal subalgebra. There are two ways of defining structure in ${}_B\mathcal{M}^A$ on $M \otimes A$ as follows:

$$b(m\otimes a) = bm\otimes a, \quad m\otimes a \mapsto \sum m_{(0)}\otimes a_{(1)}\otimes m_{(1)}a_{(2)};$$

$$b(m\otimes a) = \sum b_{(1)}m\otimes b_{(2)}a, \quad m\otimes a \mapsto \sum m\otimes a_{(1)}\otimes a_{(2)},$$

where $b\in B$, $m\otimes a\in M \otimes A$. These structures are identified via the isomorphism $M \otimes A \simeq M \otimes A$ in (0.2). The following lemma is one of the consequences of this observation.

4.5 LEMMA [M2]. Let the notation be as above. The ideal $\subset B$ of the annihilator of the left B-module M is a right coideal of A. Therefore B has no non-trivial such ideal that is simultaneously a right coideal of A, if and only if every non-zero object in ${}_B\mathcal{M}^A$ is B-faithful.

To generalize a) $\Rightarrow$ c) in (4.2) in the case where the base commutative ring is semi-local, we had better take the view-point of (co)cleftness rather than of normal basis. Even if the base commutative ring is arbitrary, the Definitions (3.4), (3.9) make sense and the equivalence a) $\Leftrightarrow$ c) in (3.8) holds true.

4.6 THEOREM [MD]. Let R be a commutative semi-local ring. Let A be a Hopf algebra over R, and $\bar{A}$ a quotient left A-module coalgebra of A. Suppose that A is a finitely generated R-module, and that $\bar{A}$ is a (finitely generated) projective R-module and is projective as a left $\bar{A}$-comodule (or equivalently as a right module over the dual algebra $\mathrm{Hom}_R(\bar{A}, R)$). Then A is cleft over $\bar{A}$.

By dualizing this, we have:

4.7 COROLLARY [S2]. Let A be a Hopf algebra over a commutative semi-local ring R, which is a finitely generated projective R-module. Let B be such an R-direct summand of A that is a right coideal subalgebra. Suppose $\mathrm{Hom}_R(B, R)$ is projective as a right B-module (This holds, if B is a Hopf subalgebra). Then A has a normal basis over B.

5 Finite Dimensional Hopf Algebras

Throughout this section, we let A be a finite dimensional Hopf algebra over a field k with (bijective) antipode S.

5.1 DEFINITION. A is said to _have Property_ (Fr), if every right coideal subalgebra of A is a Frobenius algebra (or satisfies the equivalent conditions a), c)-g) in (4.2)).

This condition is equivalent to that every _left_ coideal subalgebra of A is Frobenius, since $B \mapsto S(B)$ gives a 1-1 correspondence between the right coideal subalgebras and the left coideal subalgebras.

5.2 PROPOSITION [M3]. If A has Property (Fr), then every Hopf subalgebra or quotient Hopf algebra of A has, too.

5.3 QUESTION. Does every finite dimensional Hopf algebra have Property (Fr)?

This is still open. But here is a list of Hopf algebras having Property (Fr):

5.4 THEOREM [M2]. A finite dimensional Hopf algebra A has Property (Fr), if either

i) the coradical A_0 is cocommutative,

ii) the quotient algebra $A/\text{Rad } A$ by the Jacobson radical is commutative, or

iii) A is involutory (i.e., $S \circ S = \text{id}$) and the characteristic $\text{ch } k$ is either 0 or $> \dim A$.

The result in Cases i), ii) follows from (2.5). In Case iii) A is semisimple as an algebra, and cosemisimple as a coalgebra (i.e., $A = A_0$). Even the following is open:

5.5 QUESTION. Does every finite dimensional cosemisimple Hopf algebra have Property (Fr)?

To find many more Hopf algebras having Property (Fr), we introduce the concept of extensions of Hopf algebras.

5.6 DEFINITION. A short sequence

$$1 \to K \xrightarrow{i} A \xrightarrow{p} H \to 1$$

of finite dimensional Hopf algebras and Hopf algebra maps is said to be <u>exact</u>, if i is an injection, if p is a surjection, and if the following equivalent conditions a)-d) hold, where we view $K \subset A$ via i:

a) $A/K^{+}A = H$; b) $A/AK^{+} = H$;

c) $K = \{a \in A \mid \sum a_{(1)} \otimes p(a_{(2)}) = a \otimes p(1)\}$;

d) $K = \{a \in A \mid \sum p(a_{(1)}) \otimes a_{(2)} = p(1) \otimes a\} (= {}^{coH}A)$.

Furthermore A is called an <u>extension</u> <u>of</u> H <u>by</u> K, if there exists such a short exact sequence.

If A is an extension of H by K, it follows from (3.5.v), (3.11) that A is isomorphic with the "bicrossed product" Hopf algebra $H \overset{\pi}{\underset{\sigma}{\#}} K$, which is the crossed product $H \underset{\sigma}{\#} K$ as an algebra and the crossed coproduct $H \overset{\pi}{\#} K$ as a coalgebra.

5.7 THEOREM [M3]. Let A be a finite dimensional Hopf algebra which is an extension of H by K, where H, K are finite dimensional Hopf algebras. Suppose K has Property (Fr). Then A has Property (Fr), if either

i) H is irreducible,

ii) H is cocommutative, or

iii) H is cosemisimple and has Property (Fr).

5.8 REMARK. Hoffmann [H] investigates precisely the extensions A of a group Hopf algebra kG_1 by the dual Hopf algebra k^{G_2} of kG_2, where G_1, G_2 are finite groups. For example he shows:

5.9 THEOREM [H]. Let A be as in (5.8). If the characteristic $\operatorname{ch} k$ is either 0 or $> \dim A$, then the number of the right coideal subalgebras of A is finite.

Finally we raise:

5.10 QUESTION. To what extent can the result in (5.9) be generalized?

Added in the revision: In his recent paper "Quotient spaces for Hopf algebras", M. Takeuchi shed light on *normality* to get in particular an improvement of our COROLLARY 1.7.

REFERENCES

[BCM]R.Blattner, M.Cohen and S.Montgomery, Crossed products and inner actions of Hopf algebras, Trans. Amer. Math. Soc. 298(1986), 671-711.

[D1] Y.Doi, On the structure of relative Hopf modules, Comm. Algebra 11(1983), 243-255.

[D2] Y.Doi, Algebras with total integrals, Comm. Algebra 13 (1985), 2137-2159.

[DG] M.Demazure and P.Gabriel, "Groupes algébriques", North-Holland, Amsterdam, 1970.

[DT] Y.Doi and M.Takeuchi, Cleft comodule algebras for a bialgebra, Comm. Algebra 14(1986), 801-817.

[H] Kurt Hoffmann, Coidealunteralgebren in endlich dimensionalen Hopfalgebren, Dissertation, Universität München, 1991.

[K] M.Koppinen, Coideal subalgebras in Hopf algebras: freeness, integrals, smash products, Comm. Algebra 21(1993), 427-444.

[M1] A.Masuoka, On Hopf algebras with cocommutative coradicals, J. Algebra 144(1991), 415-466.

[M2] A.Masuoka, Freeness of Hopf algebras over coideal subalgebras, Comm. Algebra 20(1992), 1353-1373.

[M3] A.Masuoka, Coideal subalgebras in finite Hopf algebras, J. Algebra, to appear.

[M4] A.Masuoka, Faithful flatness, freeness and cleftness of Hopf algebras, Thesis, University of Tsukuba, 1992.

[MD] A.Masuoka and Y.Doi, Generalization of cleft comodule algebras, Comm. Algebra 20(1992), 3703-3721.

[MW] A.Masuoka and D.Wigner, Faithful flatness of Hopf algebras, J. Algebra, to appear.

[N] W.Nichols, Quotients of Hopf algebras, Comm. Algebra 6(1978), 1789-1800.

[NZ1]W.Nichols and M.Zoeller, A Hopf algebra freeness theorem, Amer. J. Math. 111(1989), 381-385.

[NZ2]W.Nichols and M.Zoeller, Freeness of infinite dimensional Hopf algebras over grouplike subalgebras, Comm. Algebra 17(1989), 413-424.

[NZ3]W.Nichols and M.Zoeller, Freeness of infinite dimensional Hopf algebras, Comm. Algebra 20(1992), 1489-1492.

[OS1]U.Oberst and H.-J.Schneider, Über Untergruppen endlicher algebraischer Gruppen, Manuscripta Math. 8(1973), 217-241.

[OS2]U.Oberst and H.-J.Schneider, Untergruppen formeller Gruppen von endlichem Index, J. Algebra 31(1974), 10-44.

[R] D.Radford, Freeness (projectivity) criteria for Hopf algebras over Hopf subalgebras, J. Pure Appl. Algebra 11 (1977), 15-28.

[S1] H.-J.Schneider, Principal homogeneous spaces for arbitrary Hopf algebras, Israel J. Math. 72(1990), 167-195.

[S2] H.-J.Schneider, Normal basis and transitivity of crossed products for Hopf algebras, J. Algebra 152(1992), 289-312.

[S3] H.-J.Schneider, Some remarks on exact sequences of quantum groups, preprint, 1992.

[Sw] M.Sweedler, "Hopf algebras", Benjamin, New York, 1969.

[T1] M.Takeuchi, Free Hopf algebras generated by coalgebras, J. Math. Soc. Japan 23(1971), 561-582.

[T2] M.Takeuchi, A correspondence between Hopf ideals and sub-Hopf algebras, Manuscripta Math. 7(1972), 251-270.

[T3] M.Takeuchi, A note on geometrically reductive groups, J. Fac. Sci. Univ. Tokyo Sect.1 20(1973), 387-396.

[T4] M.Takeuchi, Formal schemes over fields, Comm. Algebra 5(1977), 1483-1528.

[T5] M.Takeuchi, Relative Hopf modules——equivalences and freeness criteria, J. Algebra 60(1979), 452-471.

[Wa] W.Waterhouse, "Introduction to Affine Group Schemes", Graduate Texts in Mathematics Vol.66, Springer-Verlag, New York/Heidelberg/Berlin, 1979.

Cosemisimple Hopf Algebras

WARREN D. NICHOLS Florida State University, Tallahassee, Florida

Introduction.

Cosemisimple Hopf algebras are the Hopf algebras which are most closely related to group algebras. Over an algebraically closed field k, a cosemisimple Hopf algebra H is a group algebra if and only if it is cocommutative; in this case, as a coalgebra, H is a direct sum of copies of k. To get a cosemisimple Hopf algebra H with nontrivial coalgebra structure, take H to be the dual of the group algebra of a finite non-abelian group, over a field of characteristic not dividing the order of the group. In the infinite dimensional case, cosemisimple Hopf algebras can be thought of as generalizing group algebras and completely reducible affine group schemes.

One of the goals of the theory is to show that cosemisimple Hopf algebras resemble group algebras (or algebras dual to groups) as much as possible, for example by developing their representation theory. The representation theory most clearly resembles that of group algebras in the case in which the antipode S of the Hopf algebra H (the mapping which, for group algebras, sends each element of the group to its inverse) satisfies $S^2 = \mathrm{id}_H$. Thus, establishing when this occurs is of considerable interest.

Our goal here is to present an account of some of the major results of the theory of cosemisimple Hopf algebras, as developed by Larson, Radford, and Sweedler, in such a way that all needed background information can be found in Sweedler's book [S]. A major result that we shall show is that if H is a finite dimensional cosemisimple Hopf algebra over a field of characteristic 0, then H is semisimple and $S^2 = \mathrm{id}_H$. Although the theory has enjoyed considerable success, it is still far from being complete, and even in characteristic 0 tantalizing questions remain open. For example, Kaplansky conjectured [K, Appendix 2] that a Hopf algebra H of prime dimension is commutative and cocommutative; we shall show (with the aid of one result not proved here) that in characteristic 0 H must be cosemisimple, but the theory of cosemisimple Hopf algebras is not yet rich enough to decide the matter of commutativity.

0. Cosemisimple coalgebras.

A nonzero coalgebra C is said to be *simple* if it has no nontrivial subcoalgebras. By the Fundamental Theorem on Coalgebras [S, Theorem 2.2.1], such coalgebras are finite dimensional, and thus are the duals of the finite dimensional simple algebras.

A coalgebra C is called *cosemisimple* if it is the direct sum of simple subcoalgebras. An important basic fact [S, Corollary 8.0.6] is that any sum of distinct simple subcoalgebras is direct. This can be thought of as a generalization of the statement that distinct grouplike elements are linearly independent. Thus, a cosemisimple coalgebra is the direct sum of all of its simple subcoalgebras.

A coalgebra C is cosemisimple iff every right C-comodule M is completely reducible [S, Lemma 14.0.1] — that is, iff every subcomodule is a direct summand.

The matrix coalgebra.

A coalgebra C over a field k is called an $n \times n$ *matrix coalgebra* if it has a basis $\{c_{ij}\}_{1 \le i,j \le n}$ with $\Delta(c_{ij}) = \sum\limits_r c_{ir} \otimes c_{rj}$ and $\varepsilon(c_{ij}) = \delta_{ij}$. A basis $\{c_{ij}\}$ for C as above is called a *comatric basis* of C.

Let $\{E_{ij}\}$ be the dual basis of C^* , so that $\langle E_{ij}, c_{rs}\rangle = \delta_{ir}\delta_{js}$. When C^* is made into an algebra using Δ^*, we have $E_{ij}E_{rs} = \delta_{jr}E_{is}$. Thus, C^* is an $n \times n$ matrix algebra, with *matric basis* $\{E_{ij}\}$.

Given a comatric basis $\{c_{ij}\}$ for a matrix coalgebra C, we define a symmetric bilinear form $(\ | \ )$ on C by: $(c_{ij}|c_{rs}) = \delta_{is}\delta_{jr}$. We shall show below that the form does not depend on the choice of comatric basis. We define a linear transformation $\xi : C \to C^*$ by $\langle \xi(c), d\rangle = (c|d)$, all $c, d \in C$. Since $\xi(c_{ij}) = E_{ji}$, ξ is a vector space isomorphism (i.e., our form is nondegenerate). Note that $(c|d) = \mathrm{Tr}(\xi(c)\xi(d))$ for all $c, d \in C$, since this is true for c, d in the comatric basis. We see similarly that $\varepsilon(c) = \sum\limits_i (c|c_{ii}) = \mathrm{Tr}(\xi(c))$, all $c \in C$.

We shall use ξ to identify C with C^*. We use "$\circ$" to denote the algebra structure obtained on C via this identification. In terms of the comatric basis $\{c_{ij}\}$ of C we have $c_{ij} \circ c_{rs} = \delta_{is}c_{rj}$, since c_{ij} corresponds to E_{ji}. The identity of $(C, \circ)$ is $\sum\limits_i c_{ii}$, denoted χ or χ_C. For any $c \in C$, we shall denote the rth power of c in $(C, \circ)$ as $c^{(r)}$.

Proposition 0.1. *Let C be a matrix coalgebra, with form $(\ | \ )$ and multiplication "$\circ$" as above. Then for all $c, d, e \in C$ we have:*

(1) $(c|d) = (d|c)$

(2) $c \circ d = \sum (c_1|d)c_2 = \sum d_1(c|d_2)$

(3) $(c|d) = \varepsilon(c \circ d)$

(4) $(c \circ d|e) = (c|d \circ e) = (d|e \circ c)$

(5) $(c \circ d|e) = \sum (c|e_1)(d|e_2)$.

Proof. Assertions (1) and (2) can readily be verified on a comatric basis. From (2) we have $\varepsilon(c \circ d) = \sum (c_1|d)\varepsilon(c_2) = (c|d)$, yielding (3). Then we obtain $(c \circ d|e) = \varepsilon(c \circ d \circ e) = (c|d \circ e)$. Applying (1) and this result to $(c|d \circ e)$ yields $(d|e \circ c)$, and we have established (4). Assertion (5) records the fact that "$\circ$" corresponds to the algebra structure on C^* obtained from the coproduct on C. $\square$

Proposition 0.2. *Let C be a matrix coalgebra. Then a linear transformation $\varphi : C \to C$ is a coalgebra map iff there exists a $\circ$-invertible $t \in C$ with $\varphi(c) = t^{(-1)} \circ c \circ t$, all $c \in C$. In this case, $\mathrm{Tr}(\varphi) = \varepsilon(t^{(-1)})\varepsilon(t)$.*

Proof. In general, φ is a coalgebra map iff φ^* is an algebra map. Recall that we are identifying C with C^*, so that $(\varphi^*(d)|c) = (d|\varphi(c))$, all $c, d \in C$. By the Noether-Skolem Theorem, φ^* is an algebra map iff there exists a $\circ$-invertible $t \in C$ with $\varphi^*(d) = t \circ d \circ t^{(-1)}$, all $d \in C$. Since $(t \circ d \circ t^{(-1)}|c) = (d|t^{(-1)} \circ c \circ t)$ by Proposition 0.1(4), $\varphi^*(d) = t \circ d \circ t^{(-1)}$ for all d iff $\varphi(c) = t^{(-1)} \circ c \circ t$ for all c.

Now suppose that $\varphi(c) = t^{(-1)} \circ c \circ t$ for all $c \in C$. Then $\varphi(c_{ij}) = t^{(-1)} \circ c_{ij} \circ t = t^{(-1)} \circ \sum_r (c_{ir}|t)c_{rj} = \sum_{rs} (c_{ir}|t)c_{rs}(t^{(-1)}|c_{sj})$, so $\mathrm{Tr}(\varphi) = \sum_{ij} (c_{ii}|t)(t^{(-1)}|c_{jj}) = \varepsilon(t)\varepsilon(t^{(-1)})$. $\square$

Corollary 0.3. *The form $(\ \ |\ \ )$ on a matrix coalgebra C does not depend on the choice of comatric basis.*

Proof. If $\{c'_{ij}\}$ is another comatric basis, then by Proposition 0.2 we have $c'_{ij} = t^{(-1)} \circ c_{ij} \circ t$ for some $t \in C$. So $(c'_{ij}|c'_{rs}) = (t^{(-1)} \circ c_{ij} \circ t|t^{(-1)} \circ c_{rs} \circ t) = (t^{(-1)} \circ c_{ij} \circ t \circ t^{(-1)}|c_{rs} \circ t) = (t^{(-1)} \circ c_{ij}|c_{rs} \circ t) = (t \circ t^{(-1)} \circ c_{ij}|c_{rs}) = (c_{ij}|c_{rs}) = \delta_{is}\delta_{jr}$. Thus we get the same form. $\square$

It is sometimes convenient to use the terminology of matrices/linear transformations for elements of a matrix coalgebra. If C is a matrix coalgebra, we say that the *matrix of $c \in C$ with respect to the comatric basis $\{c_{ij}\}$ of C* is $\underline{c} = ((c|c_{ij}))$. Since $c = \sum (c|c_{ij})c_{ji}$, we have $\xi(c) = \sum (c|c_{ij})E_{ij}$, so $\underline{c}$ is the matrix of $\xi(c)$ with respect to the matric basis $\{E_{ij}\}$ of C^*. Note that Proposition 0.1(5) yields that $\underline{c \circ d} = \underline{c}\ \underline{d}$, all $c, d \in C$, so when t is invertible, the matrix of $t^{(-1)}$ is the inverse of the matrix of t. If we switch to the comatric basis $\{c'_{ij}\}$, where $c'_{ij} = t^{(-1)} \circ c_{ij} \circ t$, then from $(c|c'_{ij}) = (c|t^{(-1)} \circ c_{ij} \circ t) = \sum_{rs} (c_{ir}|t)(c|c_{rs})(t^{(-1)}|c_{sj})$ we obtain that the matrix of c with respect to $\{c'_{ij}\}$ is $\underline{c}_{'} = \underline{t}\ \underline{c}\ \underline{t}^{-1}$. Thus it makes sense to refer to the eigenvalues of $c \in C$, or whether c is diagonalizable, with the understanding that our reference is actually to $\underline{c}$. In particular, when k is algebraically closed we shall say that $\varepsilon(c)$ is the sum of the eigenvalues of c, meaning the sum of the $\sqrt{\dim C}$ roots of the characteristic polynomial of $\underline{c}$.

1. The Fundamental Theorem.

Let H be a Hopf algebra over a field k. Recall that $\lambda \in H^*$ is called a *left integral* in H^* if $p\lambda = \langle p, 1 \rangle \lambda$ for all $p \in H^*$. The fundamental result about cosemisimple Hopf algebras is Sweedler's generalization of Maschke's Theorem:

Theorem 1.1. [S, Theorem 14.0.3] *Let H be a Hopf algebra over a field k. Then: H is cosemisimple iff there exists a left integral λ in H^* with $\langle \lambda, 1 \rangle \neq 0$.*

Proof. We include here a proof of the implication ($\Rightarrow$) only. Suppose H is cosemisimple. Then H is the direct sum of its simple subcoalgebras. The element λ of H^* which is 1 at 1 and is zero on every simple subcoalgebra $C \neq k1$ of H is, by direct computation, a left integral in H^* with $\langle \lambda, 1 \rangle = 1$. In fact, we readily see that λ is a two-sided integral: $\lambda p = \langle p, 1 \rangle \lambda$ for all $p \in H^*$ as well. □

Remark 1.2. It is not hard to show that if $\lambda \in H^*$ is a left integral for the Hopf algebra H over the field k, and K is a field extension of k, then $\lambda \otimes 1$ is a left integral for the K-Hopf algebra $H \otimes K$. (We extend the operations of H by applying $(\ ) \otimes K$, identifying $(H \otimes K) \otimes_K (H \otimes K)$ with $H \otimes H \otimes K$ via $(a \otimes \alpha) \otimes (b \otimes \beta) \to a \otimes b \otimes \alpha\beta$, all $a, b \in H,\ \alpha, \beta \in K$.) Thus, if H is a cosemisimple k-Hopf algebra, $H \otimes K$ is a cosemisimple K-Hopf algebra. This observation allows us to reduce many questions about cosemisimple Hopf algebras to the case k algebraically closed.

Proposition 1.3. *The antipode S of a cosemisimple Hopf algebra H is injective, and the integral $\lambda \in H^*$ satisfies $\lambda \circ S = \lambda$.*

Proof. By Theorem 1.1 and [S, Corollary 5.1.7], S is injective. Since S is a coalgebra anti-endomorphism of H [S, Proposition 4.0.1], S sends a simple subcoalgebra of H other than $k1$ to another such. Using the explicit construction of the integral λ above, and the fact that $S(1) = 1$, we obtain that $\lambda \circ S = \lambda$. □

2. Formulas I.

We first review some notation. Let H be a Hopf algebra over a field k. We always let S denote the antipode of H. For $p \in H^*$, $h \in H$, we

write	say	calculate
$p \rightharpoonup h$	p hit h	$\sum h_1 \langle p, h_2 \rangle$
$h \leftharpoonup p$	h hit by p	$\sum \langle p, h_1 \rangle h_2$
$h \rightharpoonup p$	h hit p	$\langle h \rightharpoonup p, a \rangle = \langle p, ah \rangle$
$p \leftharpoonup h$	p hit by h	$\langle p \leftharpoonup h, a \rangle = \langle p, ha \rangle$
$h \rightharpoondown p$	h twisted-hit p	$p \leftharpoonup S(h)$
$p \leftharpoondown h$	p twisted-hit by h	$S(h) \rightharpoonup p$.

Immediately from the definitions, we have: for $p, q \in H^*$, $a, b \in H$

$$\begin{aligned} \langle pq, a \rangle &= \langle p, q \rightharpoonup a \rangle &= \langle q, a \leftharpoonup p \rangle \\ \langle p, ab \rangle &= \langle p \leftharpoonup a, b \rangle &= \langle b \rightharpoonup p, a \rangle . \end{aligned}$$

Each of the above "arrow operations" yields a module structure in the direction of the arrow. In fact, "$\leftharpoonup$" and "$\rightharpoonup$" make H into an H^*-bimodule, and H^* into an H-bimodule. Our convention is that the arrow operations have lower priority than do operations written as juxtaposition, so that, for example, $pq \rightharpoonup h$ means $(pq) \rightharpoonup h$; however, we sometimes put in unneeded brackets if they seem to make things clearer.

The basic Hopf identity is Lemma 5.1.1 of [S]: for all $p, q \in H^*$, $h \in H$ we have

$$p(q \leftharpoondown h) = \sum ((h_2 \rightharpoonup p)q) \leftharpoondown h_1. \tag{F1}$$

Specializing (F1) to the case $q = \lambda$, a left integral in H^*, we obtain

$$\begin{aligned} p(\lambda \leftharpoondown h) &= \sum \langle h_2 \rightharpoonup p, 1 \rangle \, \lambda \leftharpoondown h_1 \\ &= \sum \lambda \leftharpoondown h_1 \langle p, h_2 \rangle , \end{aligned}$$

so

$$p(\lambda \leftharpoondown h) = \lambda \leftharpoondown (p \rightharpoonup h). \tag{F2}$$

When we evaluate each side on $a \in H$, we obtain the very useful

$$\langle \lambda, (a \leftharpoonup p) S(h) \rangle = \langle \lambda, aS(p \rightharpoonup h) \rangle \text{ for all } p \in H^*, \; a, h \in H. \tag{F3}$$

For later use, we shall need another version of (F2).

For $q \in H^*$, we define $R(q) : H^* \to H^*$ by: $R(q)(p) = pq$, all $p \in H^*$. We consider $H^* \otimes H \subset \mathrm{End}_k(H^*)$ via

$$(q \otimes h)(p) = q \langle p, h \rangle \text{ for } q, p \in H^*,\ h \in H.$$

Then (F2) can be rewritten (check by applying each side to $p \in H^*$) as

$$R(\lambda \leftharpoonup h) = \sum (\lambda \leftharpoonup h_1) \otimes h_2. \tag{F4}$$

For any $f \in \mathrm{End}_k(H)$, we can similarly verify

$$(q \otimes h) \circ f^* = q \otimes f(h), \text{ all } q \in H^*,\ h \in H.$$

Thus from (F4) we obtain [cf. LR2, Proposition 1.1]

$$R(\lambda \leftharpoonup h) \circ f^* = \sum (\lambda \leftharpoonup h_1) \otimes f(h_2) \tag{F5}$$

for all $h \in H$, $f \in \mathrm{End}_k(H)$.

3. Character theory.

Generalities. [cf. L] Let C be a coalgebra, and let V be a finite dimensional right C-comodule with structure map $\rho : V \to V \otimes C$. Let $\{v_1, \ldots, v_d\}$ be a basis of V, and write $\rho(v_j) = \sum_{i=1}^{d} v_i \otimes c_{ij}$, with the c_{ij}'s in C. The element $\chi(V) = \sum_{i=1}^{d} c_{ii}$ of C is called the *character* of V. It does not depend on the choice of basis, because for all $p \in C^*$, $\langle p, \chi(V) \rangle = \mathrm{Tr}(V(p))$, where $V(p) : V \to V$ is the map sending $v \in V$ to $p \rightharpoonup v = \sum v_0 \langle p, v_1 \rangle$. The elements c_{ij} (which may not be linearly independent) satisfy $\Delta c_{ij} = \sum_r c_{ir} \otimes c_{rj}$ and $\varepsilon(c_{ij}) = \delta_{ij}$. Their span D is called the subcoalgebra *associated* with V. If V is simple (i.e., V is nonzero and has no nontrivial subcomodules), then D can be shown to be simple (e.g. by first using [S, Theorem 9.0.3] to obtain $V = \{0_V\} \bigwedge \mathrm{corad}(C)$, and then arguing as in [S, Lemma 14.0.1]), and $\chi(V)$ is called an *irreducible* character. If W is a subcomodule of V, then $\chi(V) = \chi(W) + \chi(V/W)$. Since every finite dimensional right C-comodule contains a simple subcomodule, it follows by induction on dimension that every character is a sum of irreducible characters.

When B is a bialgebra, the product of characters is a character: $\chi(V)\chi(W)$ is the character of $V \otimes W$, where $\rho(v \otimes w) = \sum (v_0 \otimes w_0) \otimes v_1 w_1$. If H is a Hopf algebra with antipode S, then $S(\chi(V))$ is the character of the left H-comodule V^*, made into a right H-comodule via S. (With respect to the dual basis $\{v_1^*, \ldots, v_d^*\}$ of V^*, we have $\rho(v_i^*) = \sum_{j=1}^{d} v_j^* \otimes S(c_{ij})$.)

Proposition 3.1. *Let H be a Hopf algebra, $\lambda \in H^*$ a left integral. Let C, D be subcoalgebras of H with $C \cap D = \{0\}$. Then for all $c \in C$, $d \in D$ we have $\langle \lambda, cS(d) \rangle = 0$.*

Proof. Writing $H = C \oplus D \oplus E$ as a vector space, we see that there exists $p \in H^*$ with $p|_C = \varepsilon|_C$, $p|_D = 0$. Then by (F3) we have $\langle \lambda, cS(d) \rangle = \langle \lambda, (c \leftharpoonup p)S(d) \rangle = \langle \lambda, cS(p \rightharpoonup d) \rangle = \langle \lambda, 0 \rangle = 0$. □

Proposition 3.2. [cf. L, equation (2.5)] *Let H be a Hopf algebra, $\lambda \in H^*$ a left integral. Let $C \subseteq H$ be a matrix coalgebra, with comatric basis $\{c_{ij}\}$. Then:*

$$\langle \lambda, c_{ir}S(c_{sj}) \rangle = 0 \text{ if } i \neq j, \text{ and}$$
$$\langle \lambda, c_{ir}S(c_{si}) \rangle = \langle \lambda, c_{jr}S(c_{sj}) \rangle \text{ all } i, j.$$

Proof. By Proposition 0.1(2) and (F3) we have

$$\langle \lambda, (c_{ir} \circ c_{jv})S(c_{sj}) \rangle = \langle \lambda, c_{ir}S(c_{jv} \circ c_{sj}) \rangle, \text{ so}$$
$$\delta_{iv} \langle \lambda, c_{jr}S(c_{sj}) \rangle = \langle \lambda, c_{ir}S(c_{sv}) \rangle .$$

We obtain the first assertion by setting $v = j$, and the second by setting $v = i$. □

Proposition 3.3. [Orthogonality relations, cf. L, Theorem 2.7] *Let H be a Hopf algebra, $\lambda \in H^*$ a left integral, C, D matrix subcoalgebras of H. Then $\langle \lambda, \chi_C S(\chi_D) \rangle = \delta_{C,D} \langle \lambda, 1 \rangle$.*

Proof. If $C \neq D$, this follows by Proposition 3.1. If $C = D$, let $\{c_{ij}\}$ be a comatric basis. Let us write $\langle \lambda, c_{ir}S(c_{sj}) \rangle = \delta_{ij} \zeta_{rs}$. We have $\langle \lambda, 1 \rangle = \langle \lambda, \varepsilon(c_{11})1 \rangle = \langle \lambda, \sum_r c_{1r}S(c_{r1}) \rangle = \sum_r \zeta_{rr}$. Then we have $\langle \lambda, \chi_C S(\chi_C) \rangle = \sum_{i,j} \langle \lambda, c_{ii}S(c_{jj}) \rangle = \sum_i \zeta_{ii} = \langle \lambda, 1 \rangle$. □

Proposition 3.4 [L, Theorem 3.3] *Let H be a cosemisimple Hopf algebra. Then the antipode S of H is bijective, and for every subcoalgebra C of H we have $S^2(C) = C$.*

Proof. Let $\lambda \in H^*$ be the integral with $\langle \lambda, 1 \rangle = 1$. By Proposition 1.3, S is injective. Since every subcoalgebra of H is a sum of simple subcoalgebras, both of our assertions follow once we establish $S^2(C) = C$ for C simple. When C is simple, $S^2(C)$ is another simple subcoalgebra. If $S^2(C) \neq C$, then $S^2(C) \cap C = \{0\}$, and so by Proposition 3.1 we would have $\langle \lambda, S^2(c)S(d) \rangle = 0$, all $c, d \in C$. Thus we have $0 = \langle \lambda, S^2(c)S(d) \rangle = \langle \lambda, S(dS(c)) \rangle = \langle \lambda \circ S, dS(c) \rangle = \langle \lambda, dS(c) \rangle$, all $c, d \in C$. But then for all $c \in C$ we have $0 = \langle \lambda, \sum c_1 S(c_2) \rangle = \langle \lambda, \varepsilon(c)1 \rangle = \varepsilon(c)$. $(\rightarrow\leftarrow)$ □

It now follows from Proposition 0.2 that if C is a matrix coalgebra in a cosemisimple Hopf algebra, then there exists $t \in C$ so that $S^2(c) = t^{(-1)} \circ c \circ t$, all $c \in C$, where $t^{(-1)}$ denotes the inverse of t in $(C, \circ)$. In particular, $S^2(\chi_C) = \chi_C$.

The next result is a basis-free version of the relationship $\langle \lambda, c_{ir}S(c_{sj}) \rangle = \delta_{ij}\zeta_{rs}$, which incorporates the relationship [L] between the ζ_{rs}'s and the square of the antipode.

Proposition 3.5. *Let C be a matrix coalgebra in a cosemisimple Hopf algebra H, $\lambda \in H^*$ the integral with $\langle \lambda, 1 \rangle = 1$. For $c \in C$ write $S^2(c) = t^{(-1)} \circ c \circ t$. Then for all $c, d \in C$ we have*

$$\langle \lambda, cS(d) \rangle = \varepsilon(t)^{-1}(t|c \circ d), \text{ and}$$
$$\langle \lambda, S(c)d \rangle = \varepsilon(t^{(-1)})^{-1}(t^{(-1)}|c \circ d).$$

Proof. Let $\{c_{ij}\}$ be a comatric basis of C. We have $\langle \lambda, c_{ir}S(c_{sj}) \rangle = \delta_{ij}\zeta_{rs} = (z|c_{ir} \circ c_{sj})$, where $z \in C$ is defined by $(z|c_{sr}) = \zeta_{rs}$. Then $\langle \lambda, cS(d) \rangle = (z|c \circ d)$ for all $c, d \in C$.

Now $(z|c \circ d) = \langle \lambda, cS(d) \rangle = \langle \lambda \circ S, cS(d) \rangle = \langle \lambda, S(cS(d)) \rangle = \langle \lambda, S^2(d)S(c) \rangle = \langle \lambda, (t^{(-1)} \circ d \circ t)S(c) \rangle = (z|t^{(-1)} \circ d \circ t \circ c)$, all $c, d \in C$. For c, d ranging over C, the elements $c \circ d - d \circ c$ span a subspace of C of codimension 1. Thus, the same is true of the elements $(t \circ c) \circ d - d \circ (t \circ c)$, hence of the elements $t^{(-1)} \circ (t \circ c \circ d - d \circ t \circ c) = c \circ d - t^{(-1)} \circ d \circ t \circ c$. We have $(t|t^{(-1)} \circ d \circ t \circ c) = (t \circ t^{(-1)} \circ d|t \circ c) = (d|t \circ c) = (t|c \circ d)$, so $(z|\ )$ and $(t|\ )$ vanish on the same subspace of C of codimension 1 (namely, $\mathrm{span}\{c \circ d - t^{(-1)} \circ d \circ t \circ c : c, d \in C\}$), so $z = \alpha t$, some $\alpha \in k$. Since $\sum_i c_{1i}S(c_{i1}) = \varepsilon(c_{11})1 = 1$, we have $1 = \langle \lambda, 1 \rangle = \langle \lambda, \sum_i c_{1i}S(c_{i1}) \rangle = (z|\sum_i c_{1i} \circ c_{i1}) = (z|\chi) = \varepsilon(z)$, so $\alpha = \varepsilon(t)^{-1}$. This establishes our first assertion. The second can be established similarly, or (better) by reversing the multiplication and the comultiplication of H. □

The above formula includes the important information that $\varepsilon(t) \neq 0$. If $S^2|_C = \mathrm{id}_C$, then we may take $t = \chi_C$, so $\varepsilon(t) = \sqrt{\dim C}\ 1$. Thus when $S^2|_C = \mathrm{id}_C$, $\sqrt{\dim C}$ is not divisible by the characteristic of k [L, Theorem 2.8].

Let H be a cosemisimple Hopf algebra over an algebraically closed field. Write the irreducible characters of H as $1 = \chi_0, \chi_1, \chi_2, \ldots$. Write $S(\chi_i) = \chi_{\bar{i}}$. Since $S^2(\chi_r) = \chi_r$, any r, we have that $r = \bar{s}$ iff $s = \bar{r}$. Let us write $\chi_i\chi_j = \sum_r \alpha_{rj}^{(i)} \chi_r$. Using the orthogonality relations, associativity, and the fact that S is an algebra anti-homomorphism, one can readily derive the relations

$$\alpha_{jk}^{(0)} = \delta_{jk}, \qquad \alpha_{\bar{i}k}^{(j)} = \alpha_{\bar{k}j}^{(i)} = \alpha_{\bar{j}i}^{(k)}, \qquad \alpha_{kj}^{(i)} = \alpha_{\bar{k}\bar{i}}^{(\bar{j})}.$$

4. Formulas II.

We now assume that H is finite dimensional. By [S, Theorem 5.1.3] there is a nonzero left integral $\lambda \in H^*$, and

$$\varphi : H \to H^*$$
$$\varphi(h) = \lambda \leftharpoonup h \qquad \text{all } h \in H$$

is an isomorphism of right H-Hopf modules, which means [cf. R3, Proposition 3] that

$$\varphi(ab) = \varphi(a) \leftharpoonup b \qquad \text{all } a, b \in H, \text{ and}$$
$$\varphi(p \rightharpoonup h) = p\varphi(h) \qquad \text{all } p \in H^*, \quad h \in H.$$

Set $\Lambda = \varphi^{-1}(\varepsilon)$. Then for any $p \in H^*$ we have $p = p\varepsilon = p\varphi(\Lambda) = \varphi(p \rightharpoonup \Lambda)$, so $\varphi^{-1}(p) = p \rightharpoonup \Lambda$. For $h \in H$, $p \in H^*$ we calculate $h = \varphi^{-1}(\varphi(h)) = (\lambda \leftharpoonup h) \rightharpoonup \Lambda$ and $p = \varphi(\varphi^{-1}(p)) = \lambda \leftharpoonup (p \rightharpoonup \Lambda)$. In particular, the cases $h = 1$, $p = \varepsilon$ yield $\lambda \rightharpoonup \Lambda = 1$, $\lambda \leftharpoonup \Lambda = \varepsilon$.

Note that we have $\langle \lambda, \Lambda \rangle = \langle \varepsilon, \lambda \rightharpoonup \Lambda \rangle = \varepsilon(1) = 1$, and $\langle \lambda, S(\Lambda) \rangle = \langle \lambda \leftharpoonup \Lambda, 1 \rangle = \varepsilon(1) = 1$.

Since for $h \in H$ we have $\varphi(\Lambda h) = \varphi(\Lambda) \leftharpoonup h = \varepsilon \leftharpoonup h = \varepsilon(h)\varepsilon = \varphi(\varepsilon(h)\Lambda)$, we have that $\Lambda h = \varepsilon(h)\Lambda$ for all $h \in H$ — i.e., Λ is a right integral in H.

Recall that for all $h \in H$, $f \in \mathrm{End}_k(H)$ we have

$$R(\lambda \leftharpoonup h) \circ f^* = \sum (\lambda \leftharpoonup h_1) \otimes f(h_2). \tag{F5}$$

Since $\lambda \leftharpoonup \Lambda = \varepsilon$, we obtain

$$f^* = \sum (\lambda \leftharpoonup \Lambda_1) \otimes f(\Lambda_2). \tag{F6}$$

Now in general, if for a finite dimensional vector space V we identify $V^* \otimes V$ with $\mathrm{End}_k(V^*)$ via $(p \otimes v)(q) = p\langle q, v \rangle$ for all $p, q \in V^*$ and $v \in V$, we find that $\langle p, v \rangle = \mathrm{Tr}(p \otimes v)$, all $p \in V^*$, $v \in V$. (We just take a basis $v_1, \ldots, v_d$ of V, form the dual basis $v_1^*, \ldots, v_d^*$ of V^*, and calculate that $\mathrm{Tr}(v_i^* \otimes v_j) = \delta_{ij}$.) Thus from (F6) we obtain [cf. R3, Theorem 1]

$$\mathrm{Tr}(f) = \mathrm{Tr}(f^*) = \langle \lambda, \sum f(\Lambda_2) S(\Lambda_1) \rangle, \tag{F7}$$

for all $f \in \mathrm{End}_k(H)$.

For $h \in H$, $p \in H^*$ we define $\ell(h), \ell(p) : H \to H$ by $\ell(h)(a) = ha$, $\ell(p)(a) = p \rightharpoonup a$, all $a \in H$. Noting that for any $a \in H$ we have $\sum a_1 S(p \rightharpoonup a_2) = \sum a_1 S(a_2)\langle p, a_3 \rangle = \sum \varepsilon(a_1)\langle p, a_2 \rangle 1 = \langle p, a \rangle 1$, we compute [cf. R3] using (F7):

$$\begin{aligned}
\mathrm{Tr}(\ell(h) \circ S^2 \circ \ell(p)) &= \langle \lambda, \sum h S^2(p \rightharpoonup \Lambda_2) S(\Lambda_1) \rangle \\
&= \langle \lambda, hS \sum \Lambda_1 S(p \rightharpoonup \Lambda_2) \rangle \\
&= \langle \lambda, h \langle p, \Lambda \rangle \rangle = \langle \lambda, h \rangle \langle p, \Lambda \rangle,
\end{aligned}$$

thus establishing

$$\operatorname{Tr}(\ell(h) \circ S^2 \circ \ell(p)) = \langle \lambda, h \rangle \langle p, \Lambda \rangle \qquad \text{all } h \in H, \quad p \in H^*. \tag{F8}$$

Specializing to $h = 1$, $p = \varepsilon$, we obtain [cf. LR1, equation (3)]

$$\operatorname{Tr}(S^2) = \langle \lambda, 1 \rangle \varepsilon(\Lambda). \tag{F9}$$

Given $f \in \operatorname{End}_k(H)$, define [cf. R3, §3] $x_f \in H$ by

$$\begin{aligned} \langle p, x_f \rangle &= \operatorname{Tr}(f \circ \ell(p)), \quad \text{all } p \in H^* \\ &= \operatorname{Tr}(R(p) \circ f^*) \quad \text{since } \ell(p)^* = R(p). \end{aligned}$$

Thus for $h \in H$ we have $\langle \lambda \leftharpoonup h, x_f \rangle = \operatorname{Tr}(R(\lambda \leftharpoonup h) \circ f^*)$. From (F5),

$$\langle \lambda \leftharpoonup h, x_f \rangle = \langle \lambda, \sum f(h_2) S(h_1) \rangle, \quad \text{all } h \in H.$$

In particular, with $h = 1$ we obtain $\langle \lambda, x_f \rangle = \langle \lambda, f(1) \rangle$. Thus by employing (F8) (with $h = x_f$ and $p = \varepsilon$) and (F9) we obtain that when $f(1) = 1$,

$$\operatorname{Tr}(\ell(x_f) \circ S^2) = \langle \lambda, x_f \rangle \varepsilon(\Lambda) = \langle \lambda, 1 \rangle \varepsilon(\Lambda) = \operatorname{Tr}(S^2).$$

Now suppose that f is a Hopf algebra endomorphism of H. Recalling [S, Corollary 5.1.6] that S is bijective when H is finite dimensional, we compute that for any $h, a \in H$ we have

$$\begin{aligned} \langle \lambda \leftharpoonup h, x_f a \rangle &= \langle aS(h) \rightharpoonup \lambda, x_f \rangle = \langle \lambda \leftharpoonup h S^{-1}(a), x_f \rangle \\ &= \langle \lambda, \sum f(h_2 S^{-1}(a_1)) S(h_1 S^{-1}(a_2)) \rangle \\ &= \langle \lambda, \sum f(h_2) S^{-1}(f(a_1)) a_2 S(h_1) \rangle . \end{aligned}$$

If $\sum S(a_2) f(a_1) = \varepsilon(a)1$, then $\sum S^{-1}(f(a_1)) a_2 = \varepsilon(a)1$, and we obtain $\langle \lambda \leftharpoonup h, x_f a \rangle = \langle \lambda \leftharpoonup h, x_f \varepsilon(a) \rangle$, all $h \in H$, and thus $x_f a = x_f \varepsilon(a)$.

To show that x_f satisfies the above condition, we first note that for any $q \in H^*$ we have $\ell(q) \circ f = f \circ \ell(q \circ f)$: for $h \in H$, we have $(\ell(q) \circ f)(h) = q \rightharpoonup f(h) = \sum f(h_1) \langle q, f(h_2) \rangle = \sum f(h_1) \langle q \circ f, h_2 \rangle = f((q \circ f) \rightharpoonup h) = (f \circ \ell(q \circ f))(h)$. Then for $p, q \in H^*$, we calculate that for $a = x_f$ we have $\langle pq, a \rangle = \operatorname{Tr}(f \circ \ell(pq)) = \operatorname{Tr}(f \circ \ell(p) \circ \ell(q)) = \operatorname{Tr}(\ell(q) \circ f \circ \ell(p)) = \operatorname{Tr}(f \circ \ell(q \circ f) \circ \ell(p)) = \operatorname{Tr}(f \circ \ell((q \circ f)p)) = \langle (q \circ f)p, a \rangle = \langle q \otimes p, \sum f(a_1) \otimes a_2 \rangle = \langle p \otimes q, \sum a_2 \otimes f(a_1) \rangle$. Thus $\sum a_1 \otimes a_2 = \sum a_2 \otimes f(a_1)$, so $\sum S(a_2) f(a_1) = \sum S(a_1) a_2 = \varepsilon(a)1$.

Thus $(x_f)^2 = \varepsilon(x_f) x_f$. That $\varepsilon(x_f) = \operatorname{Tr}(f)$ follows directly from the definition of x_f, so $(x_f)^2 = \operatorname{Tr}(f) x_f$.

If g is a Hopf algebra automorphism of H, then we have $\ell(p \circ g) = g^{-1} \circ \ell(p) \circ g$, all $p \in H^*$, from the above. We calculate $\langle p, g(x_f)\rangle = \langle p \circ g, x_f\rangle = \mathrm{Tr}(f \circ \ell(p \circ g)) = \mathrm{Tr}(f \circ g^{-1} \circ \ell(p) \circ g) = \mathrm{Tr}(g \circ f \circ g^{-1} \circ \ell(p)) = \langle p, x_{g \circ f \circ g^{-1}}\rangle$, showing that $g(x_f) = x_{g \circ f \circ g^{-1}}$. Since f commutes with S, we thus have $S^2(x_f) = x_f$.

Writing $T = \ell(x_f) \circ S^2$, we now have that $T(H) \subseteq x_f H$ and $T|_{x_f H} = \mathrm{Tr}(f) S^2|_{x_f H}$. Thus, $\mathrm{Tr}(T) = \mathrm{Tr}(f)\,\mathrm{Tr}(S^2|_{x_f H})$. Since $\mathrm{Tr}(T) = \mathrm{Tr}(S^2)$ was shown above, this establishes [cf. R3, equation (11)]:

$$\mathrm{Tr}(S^2) = \mathrm{Tr}(f)\,\mathrm{Tr}(S^2|_{x_f H}), \text{ all } f \in \mathrm{End}_{Hopf}(H). \tag{F10}$$

When $f = \mathrm{id}_H$, let us write x_H instead of x_f. Thus (F10) specializes to [cf. LR1, equation (6)]:

$$\mathrm{Tr}(S^2) = (\dim H)\,\mathrm{Tr}(S^2|_{x_H H}). \tag{F11}$$

For $f = S^2$, we calculate, using (F8) with $h = 1$, that for all $p \in H^*$ we have $\langle p, x_{S^2}\rangle = \mathrm{Tr}(S^2 \circ \ell(p)) = \langle \lambda, 1\rangle\langle p, \Lambda\rangle$, showing [cf. R3, equation (5)]

$$x_{S^2} = \langle \lambda, 1\rangle \Lambda.$$

Switching sides and S^4.

Let us observe that $(H^*, \rightharpoonup)$ is a free left H-module with basis $\{\lambda\}$. Since H is finite dimensional, it suffices to show that $h \rightharpoonup \lambda = 0$ implies $h = 0$. We have $h \rightharpoonup \lambda = \lambda \leftharpoonup S^{-1}(h) = \varphi(S^{-1}(h))$, so this is clear. It is also true that $(H^*, \leftharpoonup)$ is a free right H-module with basis $\{\lambda\}$. One way to see this is to consider H^{op}, the Hopf algebra formed by reversing the multiplication of H. (The antipode of H^{op} is S^{-1}). We have that λ is a left integral in $(H^{\mathrm{op}})^*$. For any $h \in H$, $p \in H^*$, $p \leftharpoonup h$ becomes $h \rightharpoonup p$ when we view $p \in (H^{\mathrm{op}})^*$, so our assertion follows from the fact that $((H^{\mathrm{op}})^*, \rightharpoonup)$ is a free left H^{op}-module with basis $\{\lambda\}$.

Thus, for all $h \in H$ there exists a unique $a \in H$ so that $\lambda \leftharpoonup h = a \rightharpoonup \lambda$. We wish to solve for a in terms of h. We have $\varphi^{-1}(a \rightharpoonup \lambda) = (a \rightharpoonup \lambda) \rightharpoonup \Lambda = (\lambda \leftharpoonup S^{-1}(a)) \rightharpoonup \Lambda = S^{-1}(a)$, an identity which we record as [cf. R4, equation (1)]

$$(a \rightharpoonup \lambda) \rightharpoonup \Lambda = S^{-1}(a), \quad \text{all } a \in H. \tag{F12}$$

Thus we have $a = S(\varphi^{-1}(a \rightharpoonup \lambda)) = S(\varphi^{-1}(\lambda \leftharpoonup h))$, so it suffices to calculate $(\lambda \leftharpoonup h) \rightharpoonup \Lambda$.

Note that, for any $b \in H$, $b\Lambda$ is another right integral in H. Since the space of right integrals in H is one-dimensional [S, Corollary 5.1.6], we see that there exists $\gamma \in H^*$ such that $b\Lambda = \langle\gamma, b\rangle\Lambda$, all $b \in H$. It follows readily that $\gamma : H \to k$ is an algebra map, and thus is a grouplike element of H^*. This gives us that "$\leftharpoonup\gamma$" is an algebra automorphism of H. We see that $\Lambda\leftharpoonup\gamma$ is a left integral in H: for $b \in H$ we have $b(\Lambda\leftharpoonup\gamma) = ((b\leftharpoonup\gamma^{-1})\Lambda)\leftharpoonup\gamma = \langle\gamma, b\leftharpoonup\gamma^{-1}\rangle\Lambda\leftharpoonup\gamma = \varepsilon(b)\Lambda\leftharpoonup\gamma$. Moreover, we have $\langle\lambda, \Lambda\leftharpoonup\gamma\rangle = \langle\gamma\lambda, \Lambda\rangle = \langle\lambda, \Lambda\rangle = 1$. Thus, as $\varphi^{-1}(\varepsilon)$ is the unique right integral on which λ has the value 1, $\Lambda\leftharpoonup\gamma$ is the right integral in $H^{\rm op}$ needed to apply the results above. Then (F12) (with a replaced by h) for $H^{\rm op}$, interpreted in H, is $(\lambda\leftharpoonup h)\rightharpoonup(\Lambda\leftharpoonup\gamma) = S(h)$, all $h \in H$. Applying $\leftharpoonup\gamma^{-1}$, we obtain $(\lambda\leftharpoonup h)\rightharpoonup\Lambda = S(h)\leftharpoonup\gamma^{-1} = S(\gamma\rightharpoonup h)$. Thus $\varphi^{-1}(\lambda\leftharpoonup h) = S(\gamma\rightharpoonup h)$, so $a = S^2(\gamma\rightharpoonup\lambda)$. We have established [cf. R4, Theorem 3]

$$\lambda\leftharpoonup h = S^2(\gamma\rightharpoonup h)\rightharpoonup\lambda, \quad \text{all } h \in H. \tag{F13}$$

Surprisingly, there is a second switching formula. It is obtained by interpreting in H (F13) for $H^{\rm op\ cop}$, the Hopf algebra obtained by reversing both the multiplication and the comultiplication of H. (The antipode of $H^{\rm op\ cop}$ is S.) Now Λ is a right integral in H, so $S(\Lambda)$ is a left integral in H, so $S(\Lambda)$ is a right integral in $H^{\rm op\ cop}$. For all $b \in H$ we have $S(\Lambda)b = S(S^{-1}(b)\Lambda) = S(\langle\gamma, S^{-1}(b)\rangle\Lambda) = \langle\gamma^{-1}, b\rangle S(\Lambda)$. Thus, the element of $(H^{\rm op\ cop})^*$ corresponding to γ is γ^{-1}. As above, we find that $\lambda p = \lambda\langle p, g\rangle$, all $p \in H^*$, for some grouplike element g of H, and that $\lambda\leftharpoonup g$ is a right integral in H^*. Thus (F13) for $H^{\rm op\ cop}$, interpreted in H, is $h\rightharpoonup\lambda\leftharpoonup g = \lambda\leftharpoonup g\leftharpoonup S^2(h\leftharpoonup\gamma^{-1})$. Applying $\leftharpoonup g^{-1}$, we obtain [cf. R4, Theorem 3]

$$h\rightharpoonup\lambda = \lambda\leftharpoonup gS^2(h\leftharpoonup\gamma^{-1})g^{-1}. \tag{F14}$$

To simplify the next calculations, let us note that for any grouplike $\beta \in H^*$ we have $S^2(\beta\rightharpoonup h) = \beta\rightharpoonup S^2(h)$ and $S^2(h\leftharpoonup\beta) = S^2(h)\leftharpoonup\beta$ for all $h \in H$, and that for grouplikes $\beta \in H^*$, $b \in H$ we have $\beta\rightharpoonup bhb^{-1} = b(\beta\rightharpoonup h)b^{-1}$ and $bhb^{-1}\leftharpoonup\beta = b(h\leftharpoonup\beta)b^{-1}$ for all $h \in H$.

Combining (F14) and (F13) yields

$$\begin{aligned} h\rightharpoonup\lambda &= \lambda\leftharpoonup gS^2(h\leftharpoonup\gamma^{-1})g^{-1} \\ &= S^2(\gamma\rightharpoonup gS^2(h\leftharpoonup\gamma^{-1})g^{-1})\rightharpoonup\lambda \\ &= (\gamma\rightharpoonup gS^4(h\leftharpoonup\gamma^{-1})g^{-1})\rightharpoonup\lambda, \end{aligned}$$

so $h = \gamma\rightharpoonup gS^4(h\leftharpoonup\gamma^{-1})g^{-1} = \gamma\rightharpoonup g(S^4(h)\leftharpoonup\gamma^{-1})g^{-1} = \gamma\rightharpoonup gS^4(h)g^{-1}\leftharpoonup\gamma^{-1}$, and we obtain Radford's formula [cf. R1, Proposition 6, or R4, equation (8)]:

$$S^4(h) = g^{-1}(\gamma^{-1}\rightharpoonup h\leftharpoonup\gamma)g, \quad \text{all } h \in H. \tag{F15}$$

Since $\gamma \in H^*$ and $g \in H$ are grouplike, $h \to (\gamma^{-1}\rightharpoonup h\leftharpoonup\gamma)$ and $h \to g^{-1}hg$ are commuting algebra automorphisms of H. Thus we have that $S^{4n}(h) = g^{-n}(\gamma^{-n}\rightharpoonup h\leftharpoonup\gamma^n)g^n$ for all positive integers n. The grouplike elements of a Hopf algebra form a finite group, so S has finite order [R1]. In fact, the order of each grouplike divides $\dim H$ [NZ1, NZ2], so the order of S divides $4 \dim H$.

5. Some structure theorems.

In the statements of the theorems in this section, H is always a finite dimensional Hopf algebra over a field k, with antipode denoted S.

Our first result is obtained by combining (F9): $\mathrm{Tr}(S^2) = \langle \lambda, 1 \rangle \varepsilon(\Lambda)$ with Theorem 1.1. The formula gives us that $\mathrm{Tr}(S^2)$ is nonzero iff $\langle \lambda, 1 \rangle$ and $\varepsilon(\Lambda)$ are both nonzero. By Theorem 1.1, $\langle \lambda, 1 \rangle$ nonzero means that H is cosemisimple, and (by duality) $\varepsilon(\Lambda)$ nonzero means that H is semisimple. Thus we have

Theorem 5.1. [LR1, Theorem 1] *H is semisimple and cosemisimple iff* $\mathrm{Tr}(S^2) \neq 0$. □

Using this and (F11): $\mathrm{Tr}(S^2) = (\dim H) \mathrm{Tr}(S^2|_{\mathcal{I}_H H})$, one obtains an affirmative answer to Kaplansky's 7th conjecture on Hopf algebras. If H is semisimple and cosemisimple, then $\mathrm{Tr}(S^2)$ is nonzero, so $(\dim H)1$ is nonzero. Thus we have:

Theorem 5.2. [LR1, Theorem 2] *If H is semisimple and cosemisimple, then the characteristic of k does not divide the dimension of H.* □

Kaplansky's 5th conjecture is that if H is cosemisimple (or semisimple), then $S^2 = \mathrm{id}_H$. Since $\mathrm{Tr}(\mathrm{id}_H) = (\dim H)1$, Theorem 5.1 establishes the converse if k has characteristic 0. If k has characteristic $p > 0$, then the converse is false. If H is the group algebra of a finite group G, then $S^2 = \mathrm{id}_H$, but H is not semisimple if the characteristic of k divides $\dim H$, so in that case the Hopf algebra $H \otimes H^*$ is neither semisimple nor cosemisimple. Thus Theorem 5.1 still gives the best possible converse implication. We state this as

Corollary 5.3. [L, Theorem 4.3] *If $S^2 = \mathrm{id}_H$ and $(\dim H)1 \neq 0$, then H is semisimple and cosemisimple.* □

When k has characteristic 0, Kaplansky's 5th conjecture conjecture has been shown to be true:

Theorem 5.4. [LR2, Theorem 3.3; LR1, Theorem 4] *If k has characteristic* 0, *H is semisimple iff H is cosemisimple iff $S^2 = \mathrm{id}_H$.*

Proof. By Corollary 5.3 (and duality), it suffices to show that if H is cosemisimple, then H is semisimple and $S^2 = \mathrm{id}_H$.

By Remark 1.2, we may assume that k is algebraically closed. Let C be a simple subcoalgebra of H. Since k is algebraically closed, C is a matrix coalgebra. As noted after Proposition 3.4, there exists $t \in C$ such that $S^2(c) = t^{(-1)} \circ c \circ t$, all $c \in C$. Let r denote the order of S^2 (which we have seen is finite). Then for all $c \in C$ we have $c = S^{2r}(c) = t^{(-r)} \circ c \circ t^{(r)}$, so $t^{(r)}$ lies in the center of $(C, \circ)$. Say $t^{(r)} = \alpha\chi$, some $\alpha \in k$. Since t is invertible, $\alpha \neq 0$. Replacing t by $\alpha^{-\frac{1}{r}}t$, we may assume $t^{(r)} = \chi$. Thus t is diagonalizable, and the eigenvalues of t are all roots of unity. We have by Proposition 0.2 that $\mathrm{Tr}(S^2|_C) = \varepsilon(t^{(-1)})\varepsilon(t)$. Identifying the algebraic closure of $\mathbb{Q}$ in k with a subfield of $\mathbb{C}$, we find that the eigenvalues of $t^{(-1)}$ are the conjugates of those of t, and thus that $\mathrm{Tr}(S^2|_C)$ is a non-negative real number. By Proposition 3.5, $\varepsilon(t) \neq 0$, so $\mathrm{Tr}(S^2|_C)$ must in fact be positive. Then $\mathrm{Tr}(S^2)$ is a sum of positive numbers, so is nonzero. By Theorem 5.1, H is semisimple.

We saw in the proof of Theorem 1.1 that if H is cosemisimple, then there is a nonzero two-sided integral in H^*. Since H is semisimple, there is also a nonzero two-sided integral in H. Thus in (F15) we have $\gamma = \varepsilon$, $g = 1$, so $S^4 = \mathrm{id}_H$. Thus, the eigenvalues of S^2 are all ± 1. So now we have $0 < \mathrm{Tr}(S^2) \leq \dim H$. Since $\mathrm{Tr}(S^2) = (\dim H)\,\mathrm{Tr}(S^2|_{\mathcal{I}_H H})$, and $\mathrm{Tr}(S^2|_{\mathcal{I}_H H})$ is an integer, we must have $\mathrm{Tr}(S^2) = \dim H$. Thus every eigenvalue of S^2 must be 1. Since S^2 is diagonalizable, we have $S^2 = \mathrm{id}_H$. □

Similar techniques yield partial results in positive characteristic, if one makes the additional assumption that $S^4 = \mathrm{id}_H$. This assumption is particularly potent with regard to Hopf algebras of odd dimension.

Theorem 5.5. [LR3, Theorem 2.2] *Suppose $S^4 = \mathrm{id}_H$ and $\dim H$ is odd. If the characteristic of k is* 0 *or* 2 *or $p > \dim H$, then H is semisimple and cosemisimple.*

Proof. We have that $\mathrm{Tr}(S^2)$ is the sum of the $\dim H$ eigenvalues of S^2, each of which is ± 1. Our assumptions yield that Theorem 5.1 applies. □

Note that something more than $S^4 = \mathrm{id}_H$ is needed to assure that H is semisimple and cosemisimple, even in characteristic 0: there is a Hopf algebra of dimension 4 with $S^4 = \mathrm{id}_H$ which is neither semisimple nor cosemisimple. We have seen above that in characteristic 0, H cosemisimple implies $S^2 = \mathrm{id}_H$. In positive characteristic, we have the following:

Theorem 5.6. *Suppose $S^4 = \mathrm{id}_H$, H is cosemisimple, and k has characteristic $p > 0$.*

(i) [LR2, Theorem 2.7] *There is a nonzero two-sided integral in H.*

(ii) [LR2, Proposition 3.4] *If $p > \dim H$, then H is semisimple.*

(iii) [cf. LR1, Theorem 3] *If $p > (\dim H)^2$, then $S^2 = \mathrm{id}_H$.*

Proof.

(i) Recall that $x_{S^2} = \langle \lambda, 1 \rangle \Lambda$ is a nonzero right integral in H. Since the antipode of H^{op} is S^{-1}, and the meaning of $\ell(p)$ is unchanged in H^{op}, the corresponding nonzero right integral in H^{op} is $x_{S^{-2}}$. Since $S^2 = S^{-2}$, we conclude that x_{S^2} is a nonzero two-sided integral in H.

(ii) We now assume $p > \dim H$, and proceed as in Theorem 5.4. Since $(S^2)^2 = \mathrm{id}_H$, we may assume that the element t of C satisfies $t \circ t = \chi_C$. Then $t^{(-1)} = t$. Since the eigenvalues of t are ± 1, we have $\varepsilon(t) = n_C 1$, where n_C is an integer such that $0 < |n_C| \leq \sqrt{\dim(C)}$. Thus $\mathrm{Tr}(S^2|_C) = \varepsilon(t^{(-1)})\varepsilon(t) = n_C^2 1$, and $\mathrm{Tr}(S^2) = \sum_C n_C^2 1 = n_H 1$, where $n_H = \sum_C n_C^2$. We have $1 \leq n_H \leq \dim H$, so $\mathrm{Tr}(S^2) \neq 0$, and our result follows from Theorem 5.1.

(iii) Assuming $p > (\dim H)^2$, we now consider the formula $\mathrm{Tr}(S^2) = (\dim H)\,\mathrm{Tr}(S^2|_{x_H H})$. By (ii), $0 \neq \mathrm{Tr}(S^2) = n_H 1$, where n_H is an integer between 1 and $\dim H$. Thus $x_H \neq 0$. Calculating $\mathrm{Tr}(S^2|_{x_H H})$ as a sum of eigenvalues (all of which are ± 1) and recalling that $S^2(x_H) = x_H$ (showing that 1 is an eigenvalue), we write $\mathrm{Tr}(S^2|_{x_H H}) = m1$, where m is a nonzero integer satisfying $-(\dim H) + 1 \leq m \leq \dim H$. Thus we have $-(\dim H)^2 + \dim H \leq (\dim H)m \leq (\dim H)^2$, and so

$$-p < -(\dim H)^2 + \dim H - n_H \leq (\dim H)m - n_H \leq (\dim H)^2 - n_H < p.$$

Since $(\dim H)m = \mathrm{Tr}(S^2) = n_H 1$, $(\dim H)m - n_H$ is divisible by p, so we must have $n_H = (\dim H)m$, forcing $n_H = \dim H$. Since S^2 is diagonalizable, $S^2 = \mathrm{id}_H$ follows. □

Kaplansky's 8th conjecture is that if the dimension of H is prime, then H is commutative and cocommutative. (This is trivial when $\dim H = 2$.) If H is either commutative or cocommutative, then $S^2 = \mathrm{id}_H$ [S, Proposition 4.0.1], so, as long as $\dim H \neq \mathrm{char}\ k$, a Hopf algebra H of prime dimension which is commutative or commutative is semisimple and cosemisimple by Corollary 5.3. Thus the next result is a theorem in the direction of establishing the conjecture.

Theorem 5.7. [cf. LR3, Theorem 2.3] *Suppose that the dimension of H is a prime larger than 2. If the characteristic of k is 0 or 2 or $p > \dim H$, then H is semisimple and cosemisimple.*

Proof. If H or H^* is spanned by grouplikes, then $S^2 = \mathrm{id}_H$ and Corollary 5.3 applies. Otherwise, H and H^* have no nontrivial grouplikes $[NZ1, NZ2]$, so $S^4 = \mathrm{id}_H$ by (F15). Now Theorem 5.5 applies. □

The results above tell us nothing in positive characteristic unless we already know that $S^4 = \mathrm{id}_H$. However, further progress can be made if we restrict the class of cosemisimple Hopf algebras under consideration:

Theorem 5.8. [cf. R2, Corollary 2] *Suppose H is cosemisimple, k has characteristic $p > 0$, and p does not divide the order of S^2 (as will be the case if $2(\dim H)1 \neq 0$). If H is generated as an algebra by its subcoalgebras of dimension at most 8, then $S^2 = \mathrm{id}_H$.*

Theorem 5.9. [cf. E, Theorem 4.3.11] *Suppose H is cosemisimple, k has characteristic $p > 0$, and $p > (\dim H)^2$. If every simple subcoalgebra of H has dimension at most 15, then $S^2 = \mathrm{id}_H$.*

The proofs of these last two results are not all of the same flavour as those of the earlier ones in this section. They are much more complicated, and much longer — perhaps indicating that a lot more work needs to be done before the characteristic p case is very well understood.

References

[E] M. Eberwein, *Cosemisimple Hopf Algebras*, Ph.D. Dissertation, Florida State University (1992).

[K] I. Kaplansky, *Bialgebras*, University of Chicago Lecture Notes in Mathematics, Chicago (1975).

[L] R.G. Larson, Characters of Hopf algebras, *J. Algebra* **17** (1971), 352-368.

[LR1] R.G. Larson and D.E. Radford, Semisimple cosemisimple Hopf algebras, *Amer. J.*

Math. **109** (1987), 187-195.

[LR2] ———, Finite dimensional cosemisimple Hopf algebras in characteristic 0 are semisimple, *J. Algebra* **117** (1988), 267-289.

[LR3] ———, Semisimple Hopf algebras, *J. Algebra*, to appear.

[NZ1] W.D. Nichols and M.B. Zoeller, Finite-dimensional Hopf algebras are free over grouplike subalgebras, *J. Pure Appl. Algebra* **56** (1989), 51-57.

[NZ2] ———, A Hopf algebra freeness theorem, *Amer. J. Math.* **111** (1989), 381-385.

[R1] D.E. Radford, The order of the antipode of a finite-dimensional Hopf algebra is finite, *Amer. J. Math.* **98** (1976), 333-355.

[R2] ———, On the antipode of a cosemisimple Hopf algebra, *J. Algebra* **88** (1984), 68-88.

[R3] ———, The group of automorphisms of a semisimple Hopf algebra over a field of characteristic 0 is finite, *Amer. J. Math.* **112** (1990), 331-357.

[R4] ———, The trace function and Hopf algebras, *J. Algebra*, to appear.

[S] M.E. Sweedler, *Hopf Algebras*, W.A. Benjamin, New York (1969).

Endormorphism Bialgebras of Diagrams and of Noncommutative Algebras and Spaces

BODO PAREIGIS University of Munich, Munich, Germany

Bialgebras and Hopf algebras have a very complicated structure. It is not easy to construct explicit examples of such and check all the necessary properties. This gets even more complicated if we have to verify that something like a comodule algebra over a bialgebra is given.

Bialgebras and comodule algebras, however, arise in a very natural way in non-commutative geometry and in representation theory. We want to study some general principles on how to construct such bialgebras and comodule algebras.

The leading idea throughout this paper is the notion of an action as can been seen most clearly in the example of vector spaces. Given a vector space V we can associate with it its endomorphism *algebra* $\mathrm{End}(V)$ that, in turn, defines an action $\mathrm{End}(V) \times V \longrightarrow V$. There is also the general linear *group* $\mathrm{GL}(V)$ that defines an action $\mathrm{GL}(V) \times V \longrightarrow V$. In the case of the endomorphism algebra we are in the pleasant situation that $\mathrm{End}(V)$ is a vector space itself so that we can write the action also as $\mathrm{End}(V) \otimes V \longrightarrow V$. The action of $\mathrm{GL}(V)$ on V can also be described using the tensor product by expanding the group $\mathrm{GL}(V)$ to the group algebra $K(\mathrm{GL}(V))$ to obtain $K(\mathrm{GL}(V)) \otimes V \longrightarrow V$.

We are going to find analogues of $\mathrm{End}(V)$ or $K(\mathrm{GL}(V))$ acting on non-commutative geometric spaces or on certain diagrams. This will lead to bialgebras, Hopf algebras, and comodule algebras.

There are two well-known procedures to obtain bialgebras from endomorphisms of certain objects. In the first section we will construct endomorphism spaces in the category of non-commutative spaces. These endomorphism spaces are described through bialgebras.

In the second section we find (co-)endomorphism coalgebras of certain diagrams of vector spaces, graded vector spaces, differential graded vector spaces, or others. Under additional conditions they again will turn out to be bialgebras.

The objects constructed in the first section will primarily be algebras, whereas in the second section the objects coend(ω) will have the natural structure of a coalgebra. Nevertheless we will show in the third section that the constructions of bialgebras from non-commutative spaces and of bialgebras from diagrams of vector spaces, remote as they may seem, are closely related, in fact that the case of an endomorphism space of a non-commutative space is a special case of a coendomorphism bialgebra of a certain diagram. Some other constructions of endomorphism spaces from the literature will also be subsumed under the more general construction of coendomorphism bialgebras of diagrams. We also will find such bialgebras coacting on Lie algebras.

This indicates that a suitable setting of non-commutative geometry might be obtained by considering (monoidal) diagrams of vector spaces (which can be considered as partially defined algebras) as a generalization of affine non-commutative spaces. So the problem of finding non-commutative (non-affine) schemes might be resolved in this direction.

In the last section we show that similar results hold for Hopf algebras acting on non-commutative spaces resp. on diagrams. The universal Hopf algebra coacting on an algebra is usually obtained as the Hopf envelope of the universal bialgebra acting on this algebra. We show that this Hopf algebra can also be obtained as the (co-)endomorphism bialgebra of a specific diagram constructed from the given algebra.

Throughout this paper K shall denote an arbitrary field and $\otimes$ stands for $\otimes_K$.

1 Endomorphisms of Non-Commutative Geometric Spaces

In this section we discuss some simple background on non-commutative spaces and their endomorphisms.

1.1 Affine non-commutative spaces

In algebraic geometry an affine algebraic space is given as a subset of K^n consisting of all points satisfying certain polynomial equalities. A typical example is the unit circle in $\mathbf{R}^2$ given by

$$\mathrm{Circ}(\mathbf{R}) = \{(x, y) \in \mathbf{R}^2 |\ x^2 + y^2 = 1\}.$$

Actually one is interested in circles with arbitrary radius r over any commutative $\mathbf{R}$-algebra B. They are defined in a similar way by

$$\mathrm{Circ}(B) = \{(x, y) \in B^2 |\ x^2 + y^2 - r^2 = 0\}.$$

If for example r^2 is -1 instead of 1, there are no points with coefficients in the field of real numbers but lots of points with coefficients in the field of complex numbers.

Furthermore this defines a functor $\mathrm{Circ} : \mathbf{R}\text{-}\mathrm{Alg}_c \longrightarrow \mathrm{Sets}$ from the category of all commutative $\mathbf{R}$-algebras to the category of sets, since an algebra homomorphism $f : B \longrightarrow B'$ induces a map of the corresponding unit circles $\mathrm{Circ}(f) : \mathrm{Circ}(B) \longrightarrow \mathrm{Circ}(B')$. The coordinates for the points which we consider are taken from algebra B, the *coordinate domain*, and Circ(B) is the set of all points of the space, that is the given manifold or geometric space. One has this set for all choices of coordinate domains B.

In general a functor $\mathcal{X} : K\text{-Alg}_c \longrightarrow \text{Sets}$ with

$$\mathcal{X}(B) := \{(b_1, \ldots, b_n) \in B^n \mid p_1(b_1, \ldots, b_n) = \ldots = p_r(b_1, \ldots, b_n) = 0\} \tag{1}$$

is called an *affine algebraic space*. This functor is represented by the algebra

$$A = K[x_1, \ldots, x_n]/(p_1(x_1, \ldots, x_n), \ldots, p_r(x_1, \ldots, x_n)),$$

hence $\mathcal{X}(B) = K\text{-Alg}_c(A, B)$. So for the circle we have $\text{Circ}(B) \cong \mathbf{R}\text{-Alg}_c(\mathbf{R}[x,y]/(x^2+y^2-r^2), B)$.

The Yoneda Lemma shows that this representing algebra A is uniquely determined (up to isomorphism) by the functor $\mathcal{X}$. It is usually considered as the "function algebra" of the geometric space under consideration. Indeed there is a map $A \times \mathcal{X}(K) \longrightarrow K$, $(a, f) \mapsto f(a)$, where $f \in \mathcal{X}(K) \cong K\text{-Alg}_c(A, K)$.

All of this has nothing to do with the commutativity of the algebras under consideration. Hence one can use non-commutative K-algebras A and B as well. Certain questions in physics in fact require such algebras. Instead of representing algebras $A = K[x_1, \ldots, x_n]/(p_1(x_1, \ldots, x_n), \ldots, p_r(x_1, \ldots, x_n))$ we take now representing algebras $A = K\langle x_1, \ldots, x_n\rangle/(p_1(x_1, \ldots, x_n), \ldots, p_r(x_1, \ldots, x_n))$ where $K\langle x_1, \ldots, x_n\rangle$ is the algebra of polynomials in non-commuting variables or, equivalently, the tensor algebra of the finite-dimensional vector space with basis $x_1, \ldots, x_n$.

DEFINITION 1.1 (B-points of quantum spaces) *A functor* $\mathcal{X} = K\text{-Alg}(A, -)$ *is called an* affine non-commutative space *or a* quantum space *and the elements of* $\mathcal{X}(B) = K\text{-Alg}(A, B)$ *the* B-points *of* $\mathcal{X}$.

As in (1) the B-points of a quantum space are elements of B^n, so that $\mathcal{X}(B)$ is a subset of B^n.

Well known examples of quantum spaces are the *quantum planes* with function algebras [5]

$$\mathcal{O}(A_q^{2|0}) = K\langle x, y\rangle/(xy - q^{-1}yx), \qquad q \in K \setminus \{0\}$$

and [6]

$$\mathcal{O}(A_0^{2|0}) = K\langle x, y\rangle/(xy - yx + y^2).$$

As in (commutative) algebraic geometry one can consider the algebra $A = \mathcal{O}(\mathcal{X}) = K\langle x_1, \ldots, x_n\rangle/(p_1(x_1, \ldots, x_n), \ldots, p_r(x_1, \ldots, x_n))$ (which represents the quantum space $\mathcal{X}$) as the *function algebra* of $\mathcal{X}$ consisting of the (natural) functions from $\mathcal{X}(B)$ to the coordinate algebra $\mathcal{U}(B)$ where $\mathcal{U} : K\text{-Alg} \longrightarrow \text{Sets}$ is the underlying functor. So the elements of A are certain functions from $\mathcal{X}(B)$ to B. We denote the set of all natural transformations or functions from $\mathcal{X}$ to $\mathcal{U}$ by $\mathcal{M}\text{ap}(\mathcal{X}, \mathcal{U})$

LEMMA 1.2 (the function algebra of a space) *Let* $\mathcal{X} : K\text{-Alg} \longrightarrow \text{Sets}$ *be a representable functor with representing algebra A. Then there is an isomorphism* $A \cong \mathcal{M}ap(\mathcal{X}, \mathcal{U})$ *inducing a natural transformation* $A \times \mathcal{X}(B) \longrightarrow B$.

Proof: The underlying functor $\mathcal{U} : K\text{-Alg} \longrightarrow \text{Sets}$ is represented by the algebra $K[x] = K\langle x\rangle$. So by the Yoneda Lemma we get

$$\mathcal{M}\text{ap}(\mathcal{X}, \mathcal{U}) \cong \mathcal{M}\text{ap}(K\text{-Alg}(A, -), K\text{-Alg}(K[x], -)) \cong K\text{-Alg}(K[x], A) \cong A.$$

□

COROLLARY 1.3 (the universal property of the function algebra) *Let $\mathcal{X}$ and A be as in Lemma 1.2. If $\varphi : C \times \mathcal{X}(B) \longrightarrow B$ is a natural transformation then there exists a unique map $\tilde{\varphi} : C \longrightarrow A$ such that the following diagram commutes*

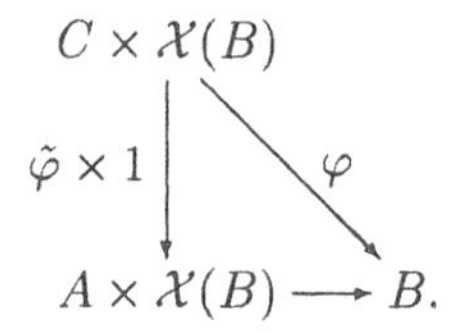

Proof: This follows from the Yoneda lemma. □

If $\mathcal{X}$ and $\mathcal{Y}$ are quantum spaces, then we will call a natural transformation $f : \mathcal{X} \longrightarrow \mathcal{Y}$ simply a *map of quantum spaces*. So the quantum spaces form a category $\mathcal{Q}$ which is antiequivalent to the category of finitely generated K-algebras K-Alg. The set of maps from $\mathcal{X}$ to $\mathcal{Y}$ will be denoted by $\mathcal{M}\text{ap}(\mathcal{X}, \mathcal{Y})$.

If $\mathcal{X}$ is a quantum space, $A = \mathcal{O}(\mathcal{X})$ its function algebra, and $I \subseteq A$ a two-sided ideal of A, then $A \longrightarrow A/I$ is an epimorphism which induces a monomorphism $\iota : \mathcal{X}_I \longrightarrow \mathcal{X}$ for the associated quantum spaces. In particular the B-points $\mathcal{X}_I(B) = K\text{-Alg}(A/I, B)$ can be identified with a certain subset of B-points in $\mathcal{X}(B)$. Conversely every subspace $\mathcal{Y} \subseteq \mathcal{X}$ (i. e. $\mathcal{Y}(B) \subseteq \mathcal{X}(B)$ functorially for all B) is induced by an epimorphism $\mathcal{O}(\mathcal{X}) \longrightarrow \mathcal{O}(\mathcal{Y})$. (Observe, however, that not every epimorphism in the category K-Alg is surjective, e. g. $K[x] \longrightarrow K(x)$ is an epimorphism but not a surjection.)

1.2 The commutative part of a quantum space

The quantum plane $A_q^{2|0}$ defines a functor from the category of algebras to the category of sets. We call its restriction to commutative algebras the commutative part $(A_q^{2|0})_{\text{comm}}$ of the quantum plane. In general the restriction of a quantum space $\mathcal{X}$ to commutative algebras is called the *commutative part* of the quantum space and is denoted by $\mathcal{X}_{\text{comm}}$. The commutative part of a quantum space represented by an algebra A is always an affine algebraic (commutative) scheme, since it is represented by the algebra $A/[A, A]$, where $[A, A]$ denotes the two-sided ideal generated by all commutators $[a, b] = ab - ba$ for $a, b \in A$. In particular, the commutative part $\mathcal{X}_{\text{comm}}$ of a quantum space $\mathcal{X}$ is indeed a subspace of $\mathcal{X}$.

For a commutative algebra B the spaces $\mathcal{X}_{\text{comm}}$ and $\mathcal{X}$ have the same B-points: $\mathcal{X}_{\text{comm}}(B) = \mathcal{X}(B)$.

For the quantum plane $A_q^{2|0}$ and commutative algebras B the set of B-points consists exactly of the two coordinate axes in B^2 since $(A_q^{2|0})_{\text{comm}}$ is represented by $K[x, y]/(xy - qyx) \cong K[x, y]/(xy)$ for $q \neq 1$.

1.3 Commuting points

In algebraic geometry any two points $(b_1, \ldots, b_m)$ and $(b'_1, \ldots, b'_n)$ with coefficients in the same coordinate algebra B have the property that their coordinates mutually commute under the multiplication $b_i b'_j = b'_j b_i$ of B since B is commutative. This does not hold any longer for non-commutative algebras B and arbitrary quantum spaces $\mathcal{X}$ and $\mathcal{Y}$.

DEFINITION 1.4 (commuting points) *If* $A = \mathcal{O}(\mathcal{X})$ *and* $A' = \mathcal{O}(\mathcal{Y})$ *and if* $p : A \longrightarrow B \in \mathcal{X}(B)$ *and* $p' : A' \longrightarrow B \in \mathcal{Y}(B)$ *are two points with coordinates in* B*, we say that they are* commuting points *if for all* $a \in A, a' \in A'$ *we have* $p(a)p'(a') = p'(a')p(a)$*, i. e. the images of the algebra homomorphisms* p *and* p' *commute elementwise.*

Obviously it is sufficient to require this just for a set of algebra generators a_i of A and a'_j of A'. In particular if A is of the form $A = K\langle x_1, \ldots, x_m\rangle/I$ and A' is of the form $A' = K\langle x_1, \ldots, x_n\rangle/J$ then the B-points are given by $(b_1, \ldots, b_m)$ resp. $(b'_1, \ldots, b'_n)$ with coordinates in B, and the two points commute iff $b_i b'_j = b'_j b_i$ for all i and j.

The set of commuting points

$$(\mathcal{X} \perp \mathcal{Y})(B) = \mathcal{X}(B) \perp \mathcal{Y}(B) := \{(p, q) \in \mathcal{X}(B) \times \mathcal{Y}(B) |\ p \text{ and } q \text{ commute}\}$$

is a subfunctor of $\mathcal{X} \times \mathcal{Y}$ since commutativity of two elements in B is preserved by algebra homomorphisms $f : B \longrightarrow B'$. We call it the *orthogonal product* of the quantum spaces $\mathcal{X}$ and $\mathcal{Y}$.

LEMMA 1.5 (the orthogonal product) *If* $\mathcal{X}$ *and* $\mathcal{Y}$ *are quantum spaces with function algebras* $A = \mathcal{O}(\mathcal{X})$ *and* $A' = \mathcal{O}(\mathcal{Y})$ *then the orthogonal product* $\mathcal{X} \perp \mathcal{Y}$ *is a quantum space with (representing) function algebra* $\mathcal{O}(\mathcal{X} \perp \mathcal{Y}) = A \otimes A' = \mathcal{O}(\mathcal{X}) \otimes \mathcal{O}(\mathcal{Y})$.

Proof: Let $(f, g) \in (\mathcal{X} \perp \mathcal{Y})(B)$ be a pair of commuting points. Then there is a unique homomorphism $h : A \otimes A' \longrightarrow B$ such that

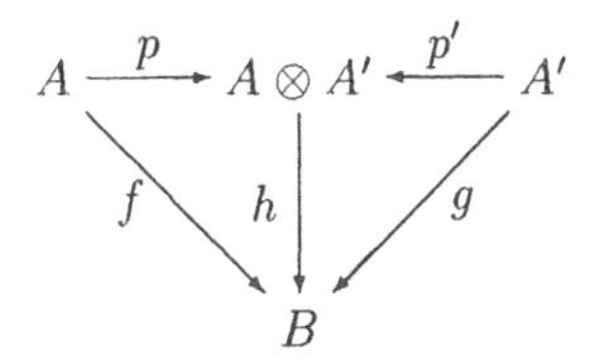

commutes; in fact the map is given by $h(a \otimes a') = f(a)g(a')$. Conversely every algebra homomorphism $h : A \otimes A' \longrightarrow B$ defines a pair of commuting points by the above diagram. The algebra homomorphism $p : A \longrightarrow A \otimes A'$ is defined by $p(a) = a \otimes 1$, and p' is defined similarly. □

This lemma shows that the set of commuting points

$$((b_1, \ldots, b_m), (b'_1, \ldots, b'_n)) = (b_1, \ldots, b_m, b'_1, \ldots, b'_n)$$

with $b_i b'_j - b'_j b_i = 0$ forms again a quantum space. It is now easy to show that the category $\mathcal{Q}$ of quantum spaces is a monoidal category with the orthogonal product $\mathcal{X} \perp \mathcal{Y}$ (in the sense of [3]). The associativity of the orthogonal product arises from the associativity of the tensor product of the function algebras.

The preceding lemma sheds some light on the reason, why we have restricted our considerations to commuting points. There is a general credo that the function algebra of a "non-commutative" space should be graded and have polynomial growth that is some kind of a Poincaré-Birkhoff-Witt theorem should hold. But the free product of algebras (which would correspond to the product of the quantum spaces) grows exponentially (with the degree). Some kind of commutation relation among the elements of the function algebra is required and this is given by letting the elements of the two function algebras A and B in the orthogonal product of the quantum spaces commute. This is done by the tensor product.

1.4 The endomorphisms of a quantum space

In the category of quantum spaces we want to find an analogue of the endomorphism algebra End(V) of a vector space V and of its action on V. Actually we consider a somewhat more general situation of an action $\mathcal{H}(\mathcal{X},\mathcal{Y})\perp\mathcal{X} \longrightarrow \mathcal{Y}$ which resembles the action $\mathrm{Hom}(V,W) \otimes V \longrightarrow W$ for vector spaces. The tensor product $\otimes$ of vector spaces will be replaced by the orthogonal product of quantum spaces $\perp$.

DEFINITION 1.6 (the hom of quantum spaces) *Let $\mathcal{X}$ and $\mathcal{Y}$ be a quantum spaces. A (universal) quantum space $\mathcal{H}(\mathcal{X},\mathcal{Y})$ together with a map $\rho : \mathcal{H}(\mathcal{X},\mathcal{Y})\perp\mathcal{X} \longrightarrow \mathcal{Y}$, such that for every quantum space $\mathcal{Z}$ and every map $\alpha : \mathcal{Z}\perp\mathcal{X} \longrightarrow \mathcal{Y}$ there is a unique map $\beta : \mathcal{Z} \longrightarrow \mathcal{H}(\mathcal{X},\mathcal{Y})$ such that the diagram*

$$\begin{array}{ccc} \mathcal{Z}\perp\mathcal{X} & & \\ {\scriptstyle \beta\perp 1}\downarrow & \searrow{\scriptstyle \alpha} & \\ \mathcal{H}(\mathcal{X},\mathcal{Y})\perp\mathcal{X} & \xrightarrow[\rho]{} & \mathcal{Y} \end{array}$$

commutes, is called a homomorphism space. $\mathcal{E}_{\mathcal{X}} := \mathcal{H}(\mathcal{X},\mathcal{X})$ *is called an* endomorphism space.

Apart from the map $\mathcal{H}(\mathcal{X},\mathcal{Y})\perp\mathcal{X} \longrightarrow \mathcal{Y}$, which we will regard as a multiplication of $\mathcal{H}(\mathcal{X},\mathcal{Y})$ on $\mathcal{X}$ (with values in $\mathcal{Y}$), this construction leads to further multiplications.

LEMMA 1.7 (the multiplication of homs) *If there exist homomorphism spaces $\mathcal{H}(\mathcal{X},\mathcal{Y})$, $\mathcal{H}(\mathcal{Y},\mathcal{Z})$, and $\mathcal{H}(\mathcal{X},\mathcal{Z})$ for the quantum spaces $\mathcal{X}$, $\mathcal{Y}$, and $\mathcal{Z}$, then there is a multiplication $m : \mathcal{H}(\mathcal{Y},\mathcal{Z})\perp\mathcal{H}(\mathcal{X},\mathcal{Y}) \longrightarrow \mathcal{H}(\mathcal{X},\mathcal{Z})$ with respect to the orthogonal product structure in $\mathcal{Q}$. Furthermore this product is associative and unitary if the necessary homomorphism spaces exist.*

Proof: Consider the diagram

$$\begin{array}{ccccc} (\mathcal{H}(\mathcal{Y},\mathcal{Z})\perp\mathcal{H}(\mathcal{X},\mathcal{Y}))\perp\mathcal{X} & \stackrel{\cong}{\to} \mathcal{H}(\mathcal{Y},\mathcal{Z})\perp(\mathcal{H}(\mathcal{X},\mathcal{Y})\perp\mathcal{X}) & \xrightarrow{1\perp\rho} & \mathcal{H}(\mathcal{Y},\mathcal{Z})\perp\mathcal{Y} \\ {\scriptstyle m\perp 1}\downarrow & & & \downarrow{\scriptstyle \rho} \\ \mathcal{H}(\mathcal{X},\mathcal{Z})\perp\mathcal{X} & & \xrightarrow{\rho} & \mathcal{Z} \end{array}$$

which induces a unique homomorphism $m : \mathcal{H}(\mathcal{Y},\mathcal{Z})\perp\mathcal{H}(\mathcal{X},\mathcal{Y}) \longrightarrow \mathcal{H}(\mathcal{X},\mathcal{Z})$, the multiplication. □

COROLLARY 1.8 (the endomorphism space is a monoid) *The endomorphism space $\mathcal{E}_{\mathcal{X}}$ of a quantum space $\mathcal{X}$ is a monoid (an algebra) in the category $\mathcal{Q}$ w. r. t. the orthogonal product. The space $\mathcal{H}(\mathcal{X},\mathcal{Y})$ is a left $\mathcal{E}_{\mathcal{Y}}$-space and a right $\mathcal{E}_{\mathcal{X}}$-space (an $\mathcal{E}_{\mathcal{Y}},\mathcal{E}_{\mathcal{X}}$-bimodule) in $\mathcal{Q}$.*

Proof: The multiplication is given in Lemma 1.7. To get the unit we consider the one point quantum space $\mathcal{J}(B) = \{u_B\}$ with function algebra K and the diagram

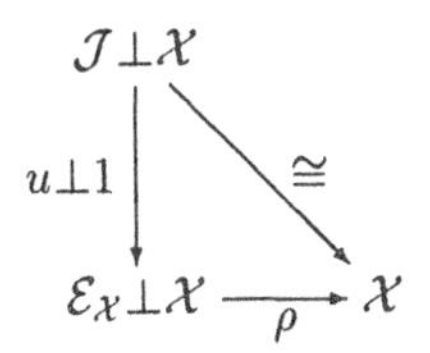

which induces a unique homomorphism $u : \mathcal{J} \longrightarrow \mathcal{E}_{\mathcal{X}}$, the unit for $\mathcal{E}_{\mathcal{X}}$.
Standard arguments show now that $(\mathcal{E}_{\mathcal{X}}, m, u)$ forms a monoid in the category of quantum spaces and that the spaces $\mathcal{H}(\mathcal{X}, \mathcal{Y})$ are $\mathcal{E}_{\mathcal{Y}}$- resp. $\mathcal{E}_{\mathcal{X}}$-spaces. □

We will call a *quantum monoid* a space $\mathcal{E}$ together with $\mathcal{E} \perp \mathcal{E} \longrightarrow \mathcal{E}$ in $\mathcal{Q}$ which is a monoid w.r.t. to the orthogonal product. If $\mathcal{E}$ acts on a quantum space $\mathcal{X}$ by $\mathcal{E} \perp \mathcal{X} \longrightarrow \mathcal{X}$ such that this action is associative and unitary then we call $\mathcal{X}$ an *$\mathcal{E}$-space.*

PROPOSITION 1.9 (the universal properties of the endomorphism space) *The endomorphism space $\mathcal{E}_{\mathcal{X}}$ and the map $\rho : \mathcal{E}_{\mathcal{X}} \perp \mathcal{X} \longrightarrow \mathcal{X}$ have the following universal properties:*
a) *For every quantum space $\mathcal{Z}$ and map $\alpha : \mathcal{Z} \perp \mathcal{X} \longrightarrow \mathcal{X}$ there is a unique map $\beta : \mathcal{Z} \longrightarrow \mathcal{E}_{\mathcal{X}}$ such that the diagram*

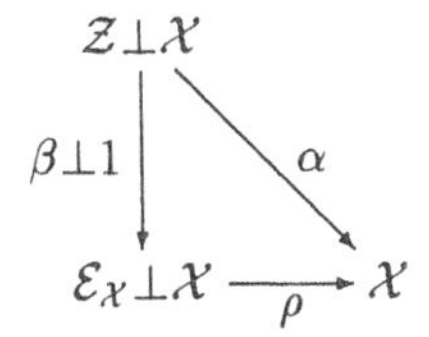

commutes.
b) *For every quantum monoid $\mathcal{M}$ and map $\alpha : \mathcal{M} \perp \mathcal{X} \longrightarrow \mathcal{X}$ which makes $\mathcal{X}$ an $\mathcal{M}$-space there is a unique monoid map $\beta : \mathcal{M} \longrightarrow \mathcal{E}_{\mathcal{X}}$ such that the diagram*

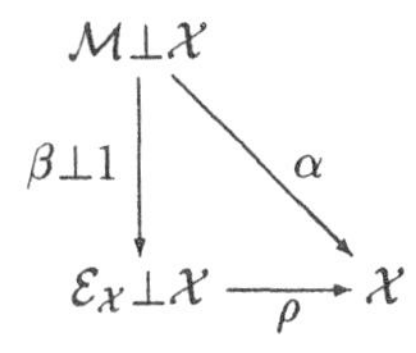

commutes.

Proof: a) is the definition of $\mathcal{E}_{\mathcal{X}}$.
b) We consider the diagram

$$\begin{array}{ccccc} \mathcal{M} \perp \mathcal{M} \perp \mathcal{X} & \overset{\bar{m} \perp 1}{\underset{1 \perp \alpha}{\rightrightarrows}} & \mathcal{M} \perp \mathcal{X} & \overset{\alpha}{\searrow} & \\ \downarrow \beta \perp \beta \perp 1 & & \downarrow \beta \perp 1 & & \mathcal{X} \\ \mathcal{E}_{\mathcal{X}} \perp \mathcal{E}_{\mathcal{X}} \perp \mathcal{X} & \overset{m \perp 1}{\underset{1 \perp \rho}{\rightrightarrows}} & \mathcal{E}_{\mathcal{X}} \perp \mathcal{X} & \overset{\rho}{\nearrow} & \end{array}$$

The right triangle commutes by definition of β. The lower square commutes since $(\beta \perp 1)(1 \perp \alpha) = \beta \perp \alpha = \beta \perp \rho(\beta \perp 1) = (1 \perp \rho)(\beta \perp \beta \perp 1)$. The horizontal pairs are equalized since $\mathcal{X}$ is an $\mathcal{M}$-space and an $\mathcal{E}_{\mathcal{X}}$-space. So we get $\rho(\beta \perp 1)(\bar{m} \perp 1) = \rho(m \perp 1)(\beta \perp \beta \perp 1)$. By the universal property a) of $\mathcal{E}_{\mathcal{X}}$, the map β is compatible with the multiplication. Similarly one shows that it is compatible with the unit. □

It is not so clear which homomorphism spaces or endomorphisms spaces exist. Tambara [15] has given a construction for homomorphism spaces $\mathcal{H}(\mathcal{X},\mathcal{Y})$ where $\mathcal{X}$ is represented by a finite dimensional algebra. In 3.2 we shall give another proof of his theorem. There are other examples of such coendomorphism bialgebras e. g. for "quadratic algebras" as considered in [5]. We will reconstruct them in 3.3.

1.5 Bialgebras and comodule algebras

With quantum homomorphism spaces and endomorphism spaces we have already obtained bialgebras and comodule algebras.

Let $\mathcal{X}$ resp. $\mathcal{Y}$ be quantum spaces with function algebras $A = \mathcal{O}(\mathcal{X})$ resp. $B = \mathcal{O}(\mathcal{Y})$. Assume that $\mathcal{E}_{\mathcal{X}}$ and $\mathcal{E}_{\mathcal{Y}}$ exist and have function algebras $E_A = \mathcal{O}(\mathcal{E}_{\mathcal{X}})$ resp. $E_B = \mathcal{O}(\mathcal{E}_{\mathcal{Y}})$. Then the operations $\mathcal{E}_{\mathcal{X}} \perp \mathcal{E}_{\mathcal{X}} \longrightarrow \mathcal{E}_{\mathcal{X}}$, $\mathcal{E}_{\mathcal{X}} \perp \mathcal{X} \longrightarrow \mathcal{X}$, $\mathcal{E}_{\mathcal{Y}} \perp \mathcal{H}(\mathcal{X},\mathcal{Y}) \longrightarrow \mathcal{H}(\mathcal{X},\mathcal{Y})$, and $\mathcal{H}(\mathcal{X},\mathcal{Y}) \perp \mathcal{E}_{\mathcal{X}} \longrightarrow \mathcal{H}(\mathcal{X},\mathcal{Y})$ (if the quantum space $\mathcal{H}(\mathcal{X},\mathcal{Y})$ exists) lead to algebra homomorphisms

$$\begin{aligned} E_A &\longrightarrow E_A \otimes E_A, \\ A &\longrightarrow E_A \otimes A, \\ \mathcal{O}(\mathcal{H}(\mathcal{X},\mathcal{Y})) &\longrightarrow E_B \otimes \mathcal{O}(\mathcal{H}(\mathcal{X},\mathcal{Y})), \\ \mathcal{O}(\mathcal{H}(\mathcal{X},\mathcal{Y})) &\longrightarrow \mathcal{O}(\mathcal{H}(\mathcal{X},\mathcal{Y})) \otimes E_A. \end{aligned}$$

In particular E_A and E_B are bialgebras, and A and $\mathcal{O}(\mathcal{H}(\mathcal{X},\mathcal{Y}))$ are comodule algebras over E_A, and B and $\mathcal{O}(\mathcal{H}(\mathcal{X},\mathcal{Y}))$ are comodule algebras over E_B. Furthermore the map $\mathcal{H}(\mathcal{X},\mathcal{Y}) \perp \mathcal{X} \longrightarrow \mathcal{Y}$ induces an algebra homomorphism $B \longrightarrow \mathcal{O}(\mathcal{H}(\mathcal{X},\mathcal{Y})) \otimes A$. Write

$$a(A,B) := \mathcal{O}(\mathcal{H}(\mathcal{X},\mathcal{Y})).$$

Then we have

$$\mathcal{M}\text{ap}(\mathcal{Z},\mathcal{H}(\mathcal{X},\mathcal{Y})) \cong \mathcal{M}\text{ap}(\mathcal{Z} \perp \mathcal{X},\mathcal{Y})$$

and

$$K\text{-Alg}(a(A,B),C) \cong K\text{-Alg}(B, C \otimes A).$$

and a universal algebra homomorphism $B \longrightarrow a(A,B) \otimes A$. Here we have used the notation of Tambara [15] for the universal algebra.

From Definition 1.6 and Lemma 1.7 we get the following

COROLLARY 1.10 (the coendomorphism bialgebra of an algebra) *Let A be a (non-commutative) algebra. Let E_A be an algebra and $\delta : A \longrightarrow E_A \otimes A$ be an algebra homomorphism, such that for every algebra B and algebra homomorphism $\alpha : A \longrightarrow B \otimes A$ there is a unique algebra homomorphism $\beta : E_A \longrightarrow B$ such that*

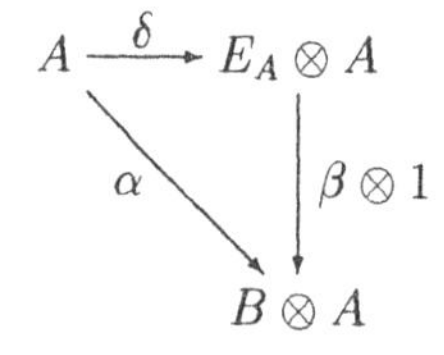

commutes. Then E_A represents the endomorphism space of $\mathcal{X} = K\text{-Alg}(A, \text{-})$, E_A is a bialgebra, and A is an E_A-comodule algebra.

Since the category of algebras is dual to the category of quantum space, we call E_A the *coendomorphism bialgebra* of A and its elements *coendomorphisms.* From Proposition 1.9 we obtain immediately

COROLLARY 1.11 (the universal properties of the coendomorphism bialgebra) *The coendomorphism bialgebra E_A of an algebra A and the map $\delta : A \longrightarrow E_A \otimes A$ have the following universal properties:*
a) *For every algebra B and algebra homomorphism $\alpha : A \longrightarrow B \otimes A$ there is a unique algebra homomorphism $\beta : E_A \longrightarrow B$ such that the diagram*

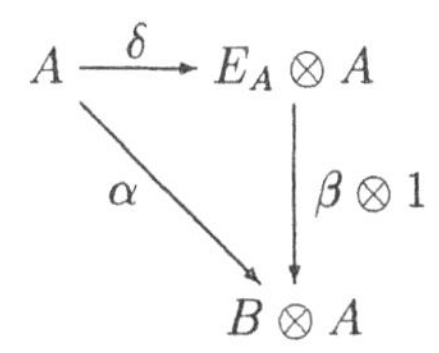

commutes.
b) *For every bialgebra B and algebra homomorphism $\alpha : A \longrightarrow B \otimes A$ making A into a comodule algebra there is a unique bialgebra homomorphism $\beta : E_A \longrightarrow B$ such that the diagram*

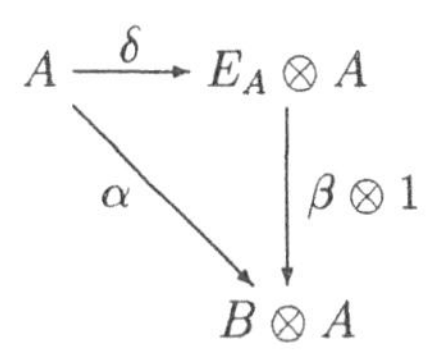

commutes. □

2 Coendomorphism Bialgebras of Diagrams

We have seen that, similar to commutative algebraic geometry, non-commutative spaces represented by non-commutative algebras induce non-commutative endomorphism spaces which are represented by bialgebras.

Now we move to a seemingly unrelated subject which is studied in representation theory and Tannaka duality. The principal question here is whether a group, a monoid, or an algebra are completely determined, if all their representations or modules are "known". We certainly have to specify what the term "known" should mean in this context. Certain reasons, mainly the fundamental theorem on the structure of coalgebras and comodules, make it easier to consider comodules rather than modules as representations.

The purpose of this section is to show that for each diagram of finite dimensional vector spaces there is an associated coalgebra which behaves like the dual of an endomorphism algebra. In particular we associate with the trivial diagram that consists of just one finite dimensional vector space V a coalgebra C that is the dual of the endomorphism algebra of this vector space: $C \cong \mathrm{End}(V)^* \cong V \otimes V^*$. If the diagram of vector spaces has additional properties then the associated coalgebra will have additional properties. This construction renders bialgebras and comodule algebras associated with certain diagrams.

If we start with a coalgebra C then we can construct the diagram of all its finite dimensional comodules and comodule homomorphisms, and also the diagram or category of all comodules. Then the coalgebra C can be recovered from the underlying functor $\omega : \mathcal{C}\text{omod-}C \longrightarrow \mathcal{A}$ as the coalgebra associated with this diagram.

Each bialgebra B induces a tensor product in the category of comodules $\mathcal{C}\text{omod-}B$ over B. This bialgebra can also be recovered from the underlying functor as the bialgebra associated with the given diagram. Additional properties of B induce additional features of $\mathcal{C}\text{omod-}B$ and conversely.

2.1 The base category

All the structures considered in this paper are built on underlying vector spaces over a given field. Certain generalizations of our constructions are straightforward whence we will use a general category $\mathcal{A}$ instead of the category of vector spaces.

We assume that $\mathcal{A}$ is an abelian category and a monoidal category with an associative unitary tensor product $\otimes : \mathcal{A} \times \mathcal{A} \longrightarrow \mathcal{A}$ as in [8]. We also assume that the category $\mathcal{A}$ is cocomplete ([7] and [13] p. 23) and that colimits commute with tensor products. Finally we assume that the monoidal category $\mathcal{A}$ is *quasisymmetric* in the sense of [13] or *braided* that is there is a bifunctorial isomorphism $\sigma : X \otimes Y \longrightarrow Y \otimes X$ such that the diagram

$$\begin{array}{ccccc}
X \otimes (Y \otimes Z) & \xrightarrow{\sigma} & (Y \otimes Z) \otimes X & \xleftarrow{\alpha} & Y \otimes (Z \otimes X) \\
\Big\downarrow \alpha & & & & \Big\uparrow 1 \otimes \sigma \\
(X \otimes Y) \otimes Z & \xrightarrow{\sigma \otimes 1} & (Y \otimes X) \otimes Z & \xleftarrow{\alpha} & Y \otimes (X \otimes Z)
\end{array}$$

and the corresponding diagram for σ^{-1} commute. σ is called a *symmetry* if in addition $\sigma_{X,Y}^{-1} = \sigma_{Y,X}$ holds.

We give a few interesting examples of such categories.

1) The category $\mathcal{V}\text{ec}$ of all vector spaces over K with the usual tensor product is a symmetric monoidal category and satisfies the conditions for $\mathcal{A}$.

2) The category $\mathcal{C}\text{omod-}H$ of comodules over a coquasitriangular or braided bialgebra H [13] with the diagonal action of H on the tensor product of comodules over K and with colimits in $\mathcal{V}\text{ec}$ and the canonical H-comodule structure on these is a quasisymmetric monoidal category and satisfies the conditions for $\mathcal{A}$ since tensor products commute with colimits.

3) The category of $\mathbf{N}$-graded vector spaces is isomorphic to the category $\mathcal{C}\text{omod-}H$, with $H = K[\mathbf{N}]$ the monoid algebra over the monoid of integers. Hence the category of graded vector spaces (with the usual graded tensor product) is a symmetric monoidal category which satisfies the conditions for $\mathcal{A}$.

4) The category of chain complexes of vector spaces $K\text{-}\mathcal{C}\text{omp}$ is isomorphic to the category $\mathcal{C}\text{omod-}H$ of comodules over the Hopf algebra $H = K\langle x, y, y^{-1}\rangle/(xy + yx, x^2)$ with $\Delta(x) = x \otimes 1 + y^{-1} \otimes x$ and $\Delta(y) = y \otimes y$ [10, 11]. H is a cotriangular Hopf algebra ([10] p. 373) and $K\text{-}\mathcal{C}\text{omp}$ is a symmetric monoidal category with the total tensor product of complexes which satisfies the conditions for $\mathcal{A}$.

5) Super-symmetric spaces are defined as the symmetric monoidal category $\mathcal{C}omod\text{-}H$ of comodules over the Hopf algebra $H = K[\mathbf{Z}/2\mathbf{Z}]$ which is cotriangular [1].

We are going to assume throughout that the category $\mathcal{A}$ is symmetric, so that we can use the full strength of the coherence theorems for monoidal categories. Thus most of the time we will delete all associativity, unity, and symmetry morphisms assuming that our categories are strict symmetric monoidal categories. Most of the given results hold also over quasisymmetric monoidal categories, cf.[4], but it gets quite technical if one wants to check all the details.

We say that an object X of a category $\mathcal{A}$ has a *dual* (X^*,ev) where X^* is an object and ev: $X^* \otimes X \longrightarrow I$ is a morphism in $\mathcal{A}$, if there is a morphism db: $I \longrightarrow X \otimes X^*$ such that

$$(X \xrightarrow{\mathrm{db}\otimes 1} X \otimes X^* \otimes X \xrightarrow{1\otimes \mathrm{ev}} X) = 1_X ,$$
$$(X^* \xrightarrow{1\otimes \mathrm{db}} X^* \otimes X \otimes X^* \xrightarrow{\mathrm{ev}\otimes 1} X^*) = 1_{X^*} .$$

The category $\mathcal{A}$ is *rigid* if every object of $\mathcal{A}$ has a dual.

The full subcategory of objects in $\mathcal{A}$ having duals in the sense of [12] is denoted by $\mathcal{A}_0$. The category $\mathcal{A}_0$ then is a rigid symmetric monoidal category. Observe that a vector space has a dual in this sense iff it is finite dimensional.

2.2 Cohomomorphisms of diagrams

In this subsection the category $\mathcal{A}$ does not have to be symmetric or quasisymmetric.

We consider diagrams (commutative or not) in $\mathcal{A}$. They are given by the objects at the vertices and the morphisms along the edges. The vertices and the (directed) edges or arrows alone define the shape of the diagram, e. g. a triangle or a square. This shape can be made into a category of its own right [3, 7], a *diagram scheme*, and the concrete diagram can then be considered as a functor, sending the vertices to the objects at the vertices and the arrows to the morphisms of the diagram. So the diagram scheme for commutative triangles

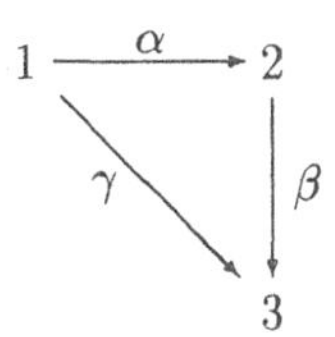

has a total of three objects $\{1, 2, 3\}$ and 6 morphisms $\{\alpha, \beta, \gamma, \mathrm{id}_1, \mathrm{id}_2, \mathrm{id}_3\}$, identities included.

Let $\omega : \mathcal{D} \longrightarrow \mathcal{A}$ be a diagram in $\mathcal{A}$. The category $\mathcal{D}$ and thus the diagram ω is always assumed to be small. We call the diagram *finite* if the functor $\omega : \mathcal{D} \longrightarrow \mathcal{A}$ factors through $\mathcal{A}_0$ that is if all objects of the diagram have duals.

THEOREM 2.1 (the existence of cohom of diagrams) *Let $(\mathcal{D}, \omega)$ and $(\mathcal{D}, \omega')$ be two diagrams in $\mathcal{A}$ over the same diagram scheme and let $(\mathcal{D}, \omega')$ be finite. Then the set of all natural transformations $\mathcal{M}or_f(\omega, M \otimes \omega')$ is representable as a functor in M.*

Proof: This has been proved in various special case in [2], [10], [16], and [12]. We follow [12] 2.1.9 and give the explicit construction here since this construction will be used later on. As remarked in [12] 2.1.7 and 2.1.9 there are isomorphisms

$$\mathcal{M}\mathrm{or}_f(\omega, M \otimes \omega') \quad = \int_{X\in\mathcal{D}} \mathrm{Hom}(\omega(X), M \otimes \omega'(X)) \qquad ([3]\ \mathrm{IX.5(2)})$$

$$= \int_{X \in \mathcal{D}} \mathrm{Hom}(\omega(X) \otimes \omega'(X)^*, M) \qquad [\omega'(X) \text{ must have a dual!}]$$

$$= \mathrm{Hom}(\int^{X \in \mathcal{D}} \omega(X) \otimes \omega'(X)^*, M)$$

where the integral means the end resp. coend of the given bifunctors in the sense of [3]. Hence the representing object is

$$\int^{X \in \mathcal{D}} \omega(X) \otimes \omega'(X)^*$$

$$= \mathrm{diffcoker}\Big(\coprod_{f \in \mathrm{Mor}(\mathcal{D})} \omega(\mathrm{dom}(f)) \otimes \omega'(\mathrm{cod}(f))^* \overset{F_C}{\underset{F_D}{\rightrightarrows}} \coprod_{X \in \mathrm{Ob}(\mathcal{D})} \omega(X) \otimes \omega'(X)^* \Big),$$

whence it can be written as a quotient of a certain coproduct. The two morphisms F_C and F_D are composed of morphisms of the form $1 \otimes \omega'(f)^* : \omega(\mathrm{dom}(f)) \otimes \omega'(\mathrm{cod}(f))^* \longrightarrow \omega(\mathrm{dom}(f)) \otimes \omega'(\mathrm{dom}(f))^*$ respectively $\omega(f) \otimes 1 : \omega(\mathrm{dom}(f)) \otimes \omega'(\mathrm{cod}(f))^* \longrightarrow \omega(\mathrm{cod}(f)) \otimes \omega'(\mathrm{cod}(f))^*$. □

DEFINITION 2.2 (cohomomorphisms) *The representing object of* $\mathcal{M}or_f(\omega, M \otimes \omega')$ *is written as*

$$\mathrm{cohom}(\omega', \omega) := \mathrm{diffcoker}\Big(\coprod_{f \in \mathrm{Mor}(\mathcal{D})} \omega(\mathrm{dom}(f)) \otimes \omega'(\mathrm{cod}(f))^* \overset{F_C}{\underset{F_D}{\rightrightarrows}} \coprod_{X \in \mathrm{Ob}(\mathcal{D})} \omega(X) \otimes \omega'(X)^* \Big).$$

The elements of cohom(ω', ω) *are called* cohomomorphisms. *If* $\omega = \omega'$ *then the representing object of* $\mathcal{M}or_f(\omega, M \otimes \omega)$ *is written as* coend(ω), *the set of* coendomorphisms *of* ω.

It is essential for us to describe very explicitly the equivalence relation for the difference cokernel. Let $f : X \longrightarrow Y$ be a morphism in $\mathcal{D}$. It induces homomorphisms $\omega(f) : \omega(X) \longrightarrow \omega(Y)$ and $\omega'(f)^* : \omega'(Y)^* \longrightarrow \omega'(X)^*$. We thus obtain the two components of the maps F_C and F_D

$$\begin{array}{ccc} & & \omega(X) \otimes \omega'(X)^* \\ & \overset{1 \otimes \omega'(f)^*}{\nearrow} & \\ \omega(X) \otimes \omega'(Y)^* & & \\ & \underset{\omega(f) \otimes 1}{\searrow} & \\ & & \omega(Y) \otimes \omega'(Y)^* \end{array}$$

which map elements $x \otimes \eta \in \omega(X) \otimes \omega'(Y)^*$ to two equivalent elements $x \otimes \omega'(f)^*(\eta) \sim \omega(f)(x) \otimes \eta$. So cohom$(\omega', \omega)$ is the direct sum of the objects $\omega(X) \otimes \omega'(X)^*$ for all $X \in \mathcal{D}$ modulo this equivalence relation.

Since a representable functor defines a universal arrow we get

COROLLARY 2.3 (the universal property of cohoms) *Let* $(\mathcal{D}, \omega)$ *and* $(\mathcal{D}, \omega')$ *be two diagrams in* $\mathcal{A}$ *and let* $(\mathcal{D}, \omega')$ *be finite. Then there is a natural transformation* $\delta : \omega \longrightarrow \mathrm{cohom}(\omega', \omega) \otimes \omega'$, *such that for each object* $M \in \mathcal{A}$ *and each natural transformation* $\varphi : \omega \longrightarrow M \otimes \omega'$ *there is a unique morphism* $\tilde{\varphi} : \mathrm{cohom}(\omega', \omega) \longrightarrow M$ *such that the diagram*

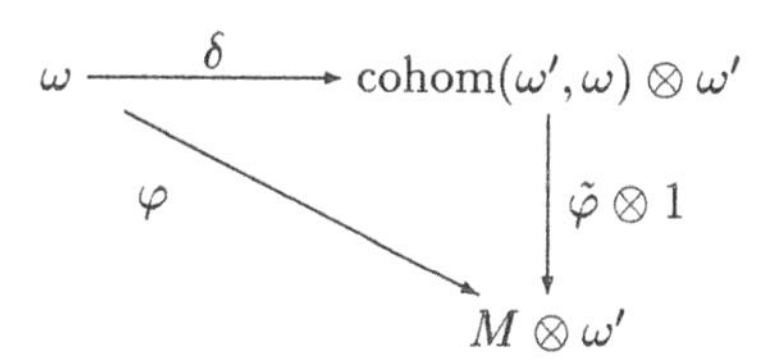

commutes.

By [12] 2.1.12 the coendomorphism set coend(ω) is a coalgebra. The comultiplication arises from the commutative diagram

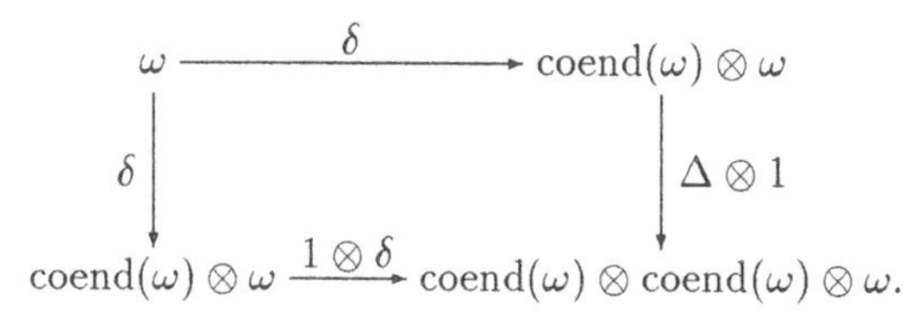

Somewhat more generally one shows

PROPOSITION 2.4 (the comultiplication of cohoms) *Let* $(\mathcal{D},\omega), (\mathcal{D},\omega')$, *and* $(\mathcal{D},\omega'')$ *be diagrams in* $\mathcal{A}$ *and let* $(\mathcal{D},\omega')$ *and* $(\mathcal{D},\omega'')$ *be finite. Then there is a comultiplication*

$$\Delta : \mathrm{cohom}(\omega'',\omega) \longrightarrow \mathrm{cohom}(\omega',\omega) \otimes \mathrm{cohom}(\omega'',\omega')$$

which is coassociative and counitary.

Proof: By 2.3 the homomorphism Δ is induced by the following commutative diagram

$$\begin{array}{ccc} \omega & \xrightarrow{\quad\delta\quad} & \mathrm{cohom}(\omega'',\omega)\otimes\omega'' \\ \delta\downarrow & & \downarrow\Delta\otimes 1 \\ \mathrm{cohom}(\omega',\omega)\otimes\omega' & \xrightarrow{1\otimes\delta} & \mathrm{cohom}(\omega',\omega)\otimes\mathrm{cohom}(\omega'',\omega')\otimes\omega''. \end{array}$$

A simple calculation shows that this comultiplication is coassociative and counitary. □

COROLLARY 2.5 (coend(ω) is a coalgebra) *Let* $(\mathcal{D},\omega)$ *and* $(\mathcal{D},\omega')$ *be finite diagrams in* $\mathcal{A}$. *Then* coend(ω) *and* coend(ω') *are coalgebras, all objects* $\omega(X)$ *resp.* $\omega'(X)$ *are comodules, and* cohom(ω',ω) *is a right* coend(ω')*-comodule and a left* coend(ω)*-comodule.*

Proof: A consequence of the preceding proposition is that coend(ω) and coend(ω') are coalgebras. The comodule structure on $\omega(X)$ comes from $\delta_X : \omega(X) \longrightarrow \mathrm{coend}(\omega)\otimes\omega(X)$. The other comodule structures are clear. □

What we have found here is in essence the universal coalgebra $C = \mathrm{coend}(\omega)$ such that all vector spaces and all homomorphisms of the diagram are comodules over C resp. comodule homomorphisms. In fact an easy consequence of the last corollary is

COROLLARY 2.6 (the universal coalgebra for a diagram of comodules) *Let $(\mathcal{D}, \omega)$ be a finite diagram in $\mathcal{A}$. Then all objects $\omega(X)$ are comodules over* $\operatorname{coend}(\omega)$ *and all morphisms $\omega(f)$ are comodule homomorphisms. If D is another coalgebra and all $\omega(X)$ are comodules over D and all $\omega(f)$ are comodule homomorphisms, then there is a unique coalgebra homomorphism* $\varphi : \operatorname{coend}(\omega) \longrightarrow D$ *such that the diagrams*

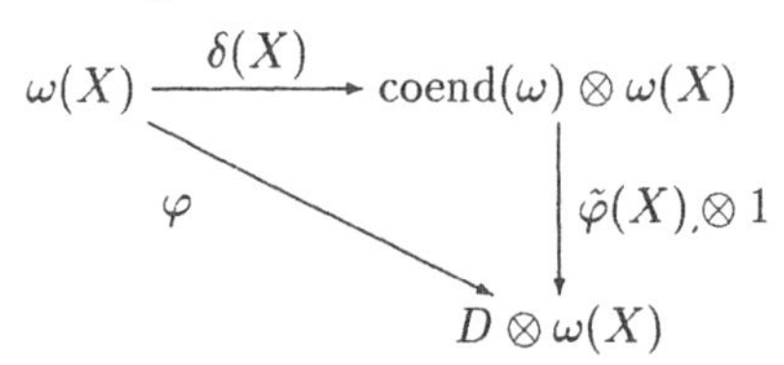

commute.

Proof: $\varphi(X) : \omega(X) \longrightarrow D \otimes \omega(X)$ is in fact a natural transformation. Then the existence and uniqueness of a homomorphism of vector spaces $\tilde{\varphi} : \operatorname{coend}(\omega) \longrightarrow D$ is obvious. The fact that this is a homomorphism of coalgebras follows from the universal property of $C = \operatorname{coend}(\omega)$ by

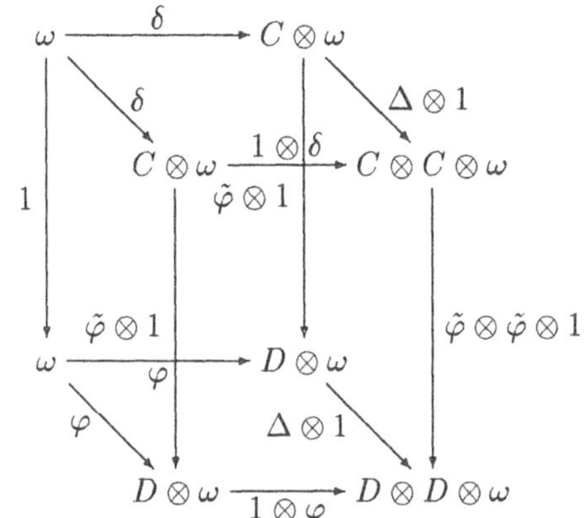

where the right side of the cube commutes by the universal property. Similarly one proves that $\tilde{\varphi}$ preserves the counit. □

To describe the coalgebra structure of $\operatorname{coend}(\omega)$ we denote the image of $x \otimes \xi \in \omega(X) \otimes \omega(X)^*$ in $\operatorname{coend}(\omega)$ by $\overline{x \otimes \xi}$. Let furthermore $\sum x_i \otimes \xi_i$ denote the dual basis in $\omega(X) \otimes \omega(X)^*$. Then $\delta : \omega(X) \longrightarrow \operatorname{coend}(\omega) \otimes \omega(X)$ is induced by the map $\omega(X) \otimes \omega(X)^* \longrightarrow \operatorname{coend}(\omega)$ and is given by

$$\delta(x \otimes \xi) = \sum \overline{x \otimes \xi_i} \otimes x_i.$$

The construction of $\Delta : \operatorname{coend}(\omega) \longrightarrow \operatorname{coend}(\omega) \otimes \operatorname{coend}(\omega)$ furnishes

$$\Delta(\overline{x \otimes \xi}) = \sum \overline{x \otimes \xi_i} \otimes \overline{x_i \otimes \xi}.$$

COROLLARY 2.7 (diagram restrictions induce morphisms for the cohoms) *Let $\mathcal{D}$ and $\mathcal{D}'$ be diagram schemes and let $\omega : \mathcal{D}' \longrightarrow \mathcal{A}$ and $\omega' : \mathcal{D}' \longrightarrow \mathcal{A}$ be finite diagrams. Let $\mathcal{F} : \mathcal{D} \longrightarrow \mathcal{D}'$ be a functor. Then $\mathcal{F}$ induces a homomorphism $\tilde{\varphi}$:* $\operatorname{cohom}(\omega'\mathcal{F}, \omega\mathcal{F}) \longrightarrow \operatorname{cohom}(\omega', \omega)$ *that is compatible with the comultiplication on cohoms as described in 2.4. In particular $\mathcal{F}$ induces a coalgebra homomorphism* $\operatorname{coend}(\omega\mathcal{F}) \longrightarrow \operatorname{coend}(\omega)$.

Proof: We obtain a morphism $\tilde{\varphi} : \mathrm{cohom}(\omega''\mathcal{F}, \omega'\mathcal{F}) \longrightarrow \mathrm{cohom}(\omega'', \omega')$ from

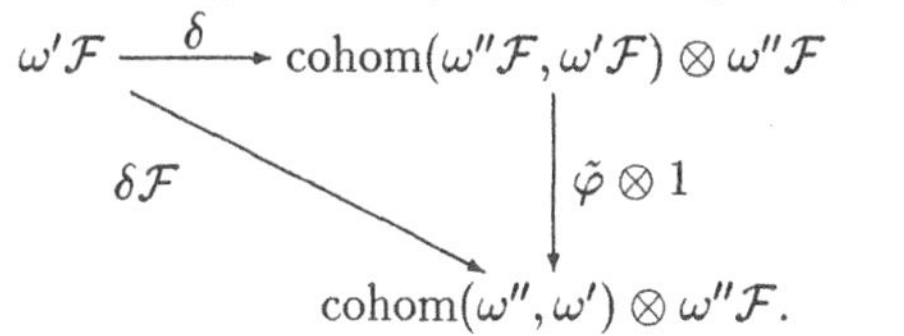

that induces a commutative diagram

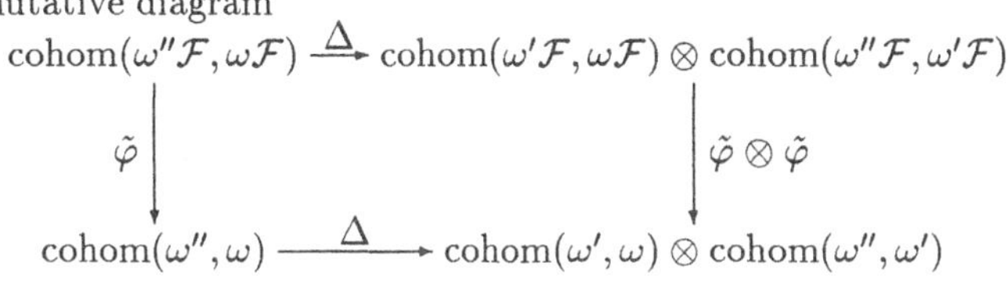

□

PROPOSITION 2.8 (isomorphic cohoms) *Let $\mathcal{D}$ and $\mathcal{D}'$ be diagram schemes and let $\omega : \mathcal{D}' \longrightarrow \mathcal{A}$ and $\omega' : \mathcal{D}' \longrightarrow \mathcal{A}$ be diagrams such that $(\mathcal{D}', \omega')$ is finite. Let $\mathcal{F} : \mathcal{D} \longrightarrow \mathcal{D}'$ be a functor which is bijective an the objects and surjective on the morphism sets. Then the map $\tilde{\varphi} : \mathrm{cohom}(\omega'\mathcal{F}, \omega\mathcal{F}) \longrightarrow \mathrm{cohom}(\omega', \omega)$ is an isomorphism.*

Proof: Since by definition the objects $\mathrm{cohom}(\omega'\mathcal{F}, \omega\mathcal{F})$ and $\mathrm{cohom}(\omega', \omega)$ represent the functors $\mathcal{M}\mathrm{or}_f(\omega\mathcal{F}, M \otimes \omega'\mathcal{F})$ resp. $\mathcal{M}\mathrm{or}_f(\omega, M \otimes \omega')$ it suffices to show that these functors are isomorphic. But every natural transformation $\varphi : \omega\mathcal{F} \longrightarrow M \otimes \omega'\mathcal{F}$ makes the diagrams

$$\begin{array}{ccc} \omega\mathcal{F}(X) & \xrightarrow{\varphi(X)} & M \otimes \omega'\mathcal{F}(X) \\ {\scriptstyle \omega\mathcal{F}(f)}\downarrow & & \downarrow{\scriptstyle M \otimes \omega'\mathcal{F}(f)} \\ \omega\mathcal{F}(Y) & \xrightarrow{\varphi(Y)} & M \otimes \omega'\mathcal{F}(Y) \end{array}$$

commute for all f in $\mathcal{D}$. Similarly every natural transformation $\psi : \omega \longrightarrow \omega'$ makes the diagrams

$$\begin{array}{ccc} \omega(X') & \xrightarrow{\psi(X')} & M \otimes \omega'(X') \\ {\scriptstyle \omega(f')}\downarrow & & \downarrow{\scriptstyle M \otimes \omega'(X')} \\ \omega(Y') & \xrightarrow{\psi(Y')} & M \otimes \omega'(Y') \end{array}$$

commute for all f' in $\mathcal{D}'$. Since $\mathcal{F}$ is bijective on the objects and surjective on the morphisms, we can identify the natural transformations φ and ψ. □

2.3 Monoidal diagrams and bialgebras

If we consider additional structures like tensor products on the diagrams, we get additional structure for the objects $\mathrm{cohom}(\omega', \omega)$. In particular we find examples of bialgebras and comodule algebras. For this we have to assume now that the category $\mathcal{A}$ is (quasi)symmetric. We first recall some definitions about monoidal categories.

DEFINITION 2.9 (monoidal diagrams) *Let $(\mathcal{D},\omega)$ be a diagram in $\mathcal{A}$. Assume that $\mathcal{D}$ is a monoidal category and that ω is a monoidal functor. Then we call the diagram $(\mathcal{D},\omega)$ a* monoidal diagram.

DEFINITION 2.10 (monoidal transformations) *Let $(\mathcal{D},\omega)$ and $(\mathcal{D},\omega')$ be monoidal diagrams in $\mathcal{A}$. Let $C \in \mathcal{A}$ be an algebra. A natural transformation $\varphi : \omega \longrightarrow C\otimes\omega'$ is called* monoidal, *if the diagrams*

$$\begin{array}{ccc} \omega(X)\otimes\omega(Y) & \xrightarrow{\varphi(X)\otimes\varphi(Y)} & C\otimes C\otimes\omega'(X)\otimes\omega'(Y) \\ \downarrow{\scriptstyle\rho} & & \downarrow{\scriptstyle m\otimes\rho} \\ \omega(X\otimes Y) & \xrightarrow{\varphi(X\otimes Y)} & C\otimes\omega'(X\otimes Y) \end{array}$$

and

$$\begin{array}{ccc} K & \xrightarrow{\cong} & K\otimes K \\ \downarrow & & \downarrow \\ \omega(I) & \xrightarrow{\varphi(I)} & C\otimes\omega'(I) \end{array}$$

commute.

Let $\omega : \mathcal{D} \longrightarrow \mathcal{A}$ and $\omega' : \mathcal{D}' \longrightarrow \mathcal{A}$ be diagrams in $\mathcal{A}$. We define the tensor product of these diagrams by $(\mathcal{D},\omega)\otimes(\mathcal{D}',\omega') = (\mathcal{D}\times\mathcal{D}',\omega\otimes\omega')$ where $(\omega\otimes\omega')(X,Y) = \omega(X)\otimes\omega'(Y)$. The tensor product of two diagrams can be viewed as the diagram consisting of all the tensor products of all the objects of the first diagram and all the objects of the second diagram; similarly for the morphisms of the diagrams.

PROPOSITION 2.11 (the tensor product of cohoms) *Let $(\mathcal{D}_1,\omega_1)$, $(\mathcal{D}_2,\omega_2)$, $(\mathcal{D}_1,\omega_1')$, and $(\mathcal{D}_2,\omega_2')$ be diagrams in $\mathcal{A}$ and let $(\mathcal{D}_1,\omega_1')$ and $(\mathcal{D}_2,\omega_2')$ be finite. Then*

$$\mathrm{cohom}(\omega_1'\otimes\omega_2',\omega_1\otimes\omega_2) \cong \mathrm{cohom}(\omega_1',\omega_1)\otimes\mathrm{cohom}(\omega_2',\omega_2).$$

Proof: Similar to the proof in [12] 2. 3. 6 we get

$$\begin{aligned} \mathrm{cohom}(\omega_1',\omega_1) \otimes\ & \mathrm{cohom}(\omega_2',\omega_2) \\ & \cong \Big(\int^{X\in\mathcal{D}_1} \omega_1(X)\otimes\omega_1'(X)^*\Big)\otimes\Big(\int^{Y\in\mathcal{D}_2} \omega_2(Y)\otimes\omega_2'(Y)^*\Big) \\ & \cong \int^{X\in\mathcal{D}_1}\int^{Y\in\mathcal{D}_2} (\omega_1(X)\otimes\omega_1'(X)^*\otimes\omega_2(Y)\otimes\omega_2'(Y)^*) \\ & \cong \int^{(X,Y)\in\mathcal{D}_1\times\mathcal{D}_2} (\omega_1(X)\otimes\omega_1'(X)^*\otimes\omega_2(Y)\otimes\omega_2'(Y)^*) \\ & \cong \int^{(X,Y)\in\mathcal{D}_1\times\mathcal{D}_2} (\omega_1\otimes\omega_2)(X,Y)\otimes(\omega_1'\otimes\omega_2')(X,Y)^* \\ & \cong \mathrm{cohom}(\omega_1'\otimes\omega_2',\omega_1\otimes\omega_2). \end{aligned}$$

This uses the fact that colimits commute with tensor products in $\mathcal{A}$. □

If we look at the representation of coend(ω) in Definition 2.2 then this isomorphism identifies an element $\overline{x\otimes\xi}\otimes\overline{y\otimes\eta} \in \mathrm{cohom}(\omega_1',\omega_1)\otimes\mathrm{cohom}(\omega_2',\omega_2)$ with representative element

$$(x\otimes\xi)\otimes(y\otimes\eta) \in (\omega_1(X)\otimes\omega_1'(X)^*)\otimes(\omega_2(Y)\otimes\omega_2'(Y)^*)$$

with the element $\overline{(x\otimes y)\otimes(\xi\otimes\eta)} \in \mathrm{cohom}(\omega_1'\otimes\omega_2',\omega_1\otimes\omega_2)$.

COROLLARY 2.12 (the universal property of the tensor product of cohoms) *Under the assumptions of Proposition 2.11 there is a natural transformation*

$$\delta : \omega_1 \otimes \omega_2 \longrightarrow \mathrm{cohom}(\omega_1', \omega_1) \otimes \mathrm{cohom}(\omega_2', \omega_2) \otimes \omega_1' \otimes \omega_2',$$

such that for each object $M \in \mathcal{A}$ and each natural transformation $\varphi : \omega_1 \otimes \omega_2 \longrightarrow M \otimes \omega_1' \otimes \omega_2'$ there is a unique morphism $\tilde{\varphi} : \mathrm{cohom}(\omega_1', \omega_1) \otimes \mathrm{cohom}(\omega_2', \omega_2) \longrightarrow M$ such that the diagram

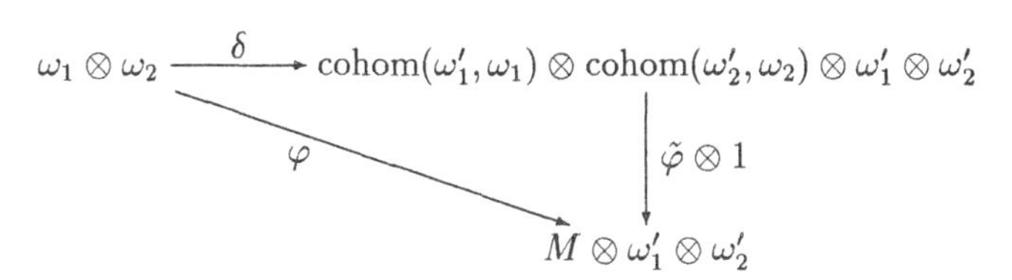

commutes.

THEOREM 2.13 (cohom is an algebra) *Let $(\mathcal{D}, \omega)$ and $(\mathcal{D}, \omega')$ be two monoidal diagrams in $\mathcal{A}$ and let $(\mathcal{D}, \omega')$ be finite. Then $\mathrm{cohom}(\omega', \omega)$ is an algebra in $\mathcal{A}$ and $\delta : \omega \longrightarrow \mathrm{cohom}(\omega', \omega) \otimes \omega'$ is monoidal.*

Proof: The multiplication on $\mathrm{cohom}(\omega', \omega)$ results from the commutative diagram

$$\begin{array}{ccc} \omega(X) \otimes \omega(Y) & \xrightarrow{\delta \otimes \delta} & \mathrm{cohom}(\omega', \omega) \otimes \mathrm{cohom}(\omega', \omega) \otimes \omega'(X) \otimes \omega'(Y) \\ \downarrow & & \downarrow \\ \omega(X \otimes Y) & \xrightarrow{\delta} & \mathrm{cohom}(\omega', \omega) \otimes \omega'(X \otimes Y). \end{array}$$

and from an application of Proposition 2.11.

Let $\mathcal{D}_0 = (\{I\}, \{\mathrm{id}\})$ together with $\omega_0 : \mathcal{D}_0 \longrightarrow \mathcal{A}$ and $\omega_0(I) = K$ be the monoidal "unit object" diagram of the monoidal category $\mathrm{Diag}(\mathcal{A})$. Then $(K \longrightarrow K \otimes K) = (\omega_0 \longrightarrow \mathrm{coend}(\omega_0) \otimes \omega_0)$ is the universal arrow. The unit on $\mathrm{cohom}(\omega', \omega)$ is given by the commutative diagram

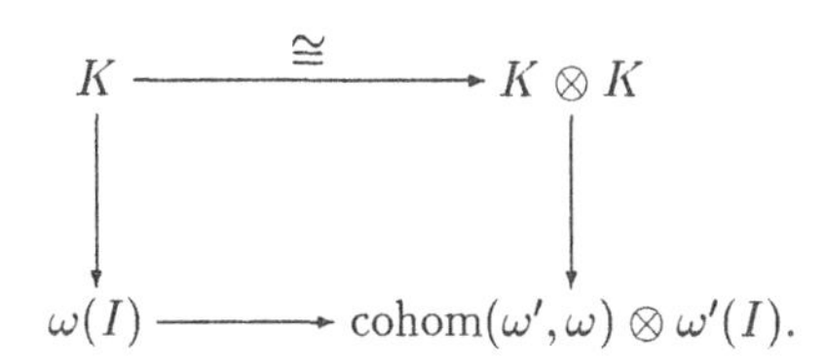

It is easy to verify the algebra laws and the additional claims of the theorem. □

PROPOSITION 2.14 (the product in the algebra cohom) *The product in $\mathrm{cohom}(\omega', \omega)$ is given by*

$$(\overline{x \otimes \xi}) \cdot (\overline{y \otimes \eta}) = \overline{x \otimes y \otimes \xi \otimes \eta}.$$

Proof: We apply the diagram defining the multiplication of $\mathrm{cohom}(\omega', \omega)$ to an element $\overline{x \otimes y}$ and obtain $\sum (\overline{x \otimes \xi_i}) \cdot (\overline{y \otimes \eta_j}) \otimes x_i \otimes y_j = \sum \overline{x \otimes y \otimes \xi_i \otimes \eta_j} \otimes x_i \otimes y_j$, where for simplicity of notation $\omega(X) \otimes \omega(Y) = \omega(X \otimes Y)$ and $\omega'(X) \otimes \omega'(Y) = \omega'(X \otimes Y)$ are identified and where

$\sum x_i \otimes \xi_i \in \omega'(X) \otimes \omega'(X)^*$ and $\sum y_j \otimes \eta_j \in \omega'(Y) \otimes \omega'(Y)^*$ are dual bases. This implies the proposition. □

One easily proves the following corollaries

COROLLARY 2.15 (Δ is an algebra morphism) *Let the assumptions of Proposition2.4 be satisfied and let all diagrams $(\mathcal{D},\omega)$, $(\mathcal{D},\omega')$, and $(\mathcal{D},\omega'')$ be monoidal. Then the morphism*

$$\Delta : \mathrm{cohom}(\omega'',\omega) \longrightarrow \mathrm{cohom}(\omega',\omega) \otimes \mathrm{cohom}(\omega'',\omega')$$

in 2.4 is an algebra homomorphism. □

COROLLARY 2.16 ($\mathrm{coend}(\omega)$ is a bialgebra) *Let $(\mathcal{D},\omega)$ and $(\mathcal{D},\omega')$ be two monoidal diagrams in $\mathcal{A}$ and let $(\mathcal{D},\omega')$ be finite. Then* $\mathrm{coend}(\omega')$ *and, if $(\mathcal{D},\omega)$ is finite,* $\mathrm{coend}(\omega)$ *are bialgebras and* $\mathrm{cohom}(\omega',\omega)$ *is a right* $\mathrm{coend}(\omega')$*-comodule algebra and a left* $\mathrm{coend}(\omega)$*-comodule algebra.* □

COROLLARY 2.17 (monoidal diagram morphisms induce algebra morphisms for the cohoms) *Let $\mathcal{D}$ and $\mathcal{D}'$ be monoidal diagram schemes and let $\omega : \mathcal{D}' \longrightarrow \mathcal{A}$ and $\omega' : \mathcal{D}' \longrightarrow \mathcal{A}$ be monoidal diagrams. Let $\mathcal{F} : \mathcal{D} \longrightarrow \mathcal{D}'$ be a monoidal functor. Then $\mathcal{F}$ induces algebra homomorphism $f :$* $\mathrm{cohom}(\omega'\mathcal{F},\omega\mathcal{F}) \longrightarrow \mathrm{cohom}(\omega',\omega)$ *which is compatible with the comultiplication on cohoms as described in 2.4. Furthermore $\mathcal{F}$ induces a bialgebra homomorphism* $\mathrm{coend}(\omega'\mathcal{F}) \longrightarrow \mathrm{coend}(\omega')$. □

After having established the structure of an algebra on $\mathrm{cohom}(\omega',\omega)$, we are now interested in algebra homomorphisms from $\mathrm{cohom}(\omega',\omega)$ to other algebras.

COROLLARY 2.18 (the universal property of the algebra cohom) *Let $(\mathcal{D},\omega)$ and $(\mathcal{D},\omega')$ be monoidal diagrams in $\mathcal{A}$ and let $(\mathcal{D},\omega')$ be finite. Then there is a natural monoidal transformation $\delta : \omega \longrightarrow$* $\mathrm{cohom}(\omega',\omega) \otimes \omega'$*, such that for each algebra $C \in \mathcal{A}$ and each natural monoidal transformation $\varphi : \omega \longrightarrow C \otimes \omega'$ there is a unique algebra morphism $f :$* $\mathrm{cohom}(\omega',\omega) \longrightarrow C$ *such that the diagram*

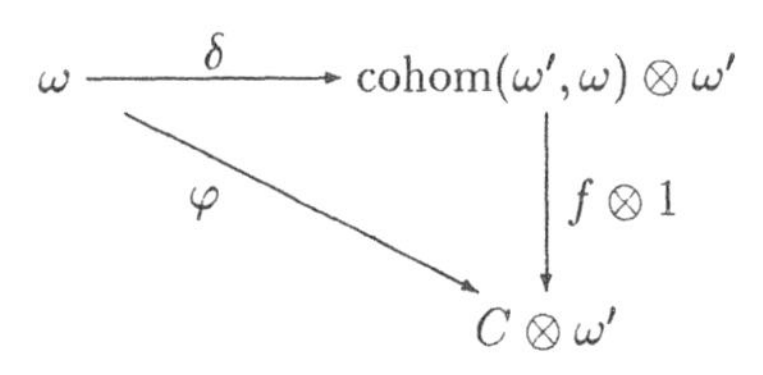

commutes.

Proof: The multiplication of $\mathrm{cohom}(\omega',\omega)$ was given in the proof of Theorem 2.13. Hence we get the following commutative diagram

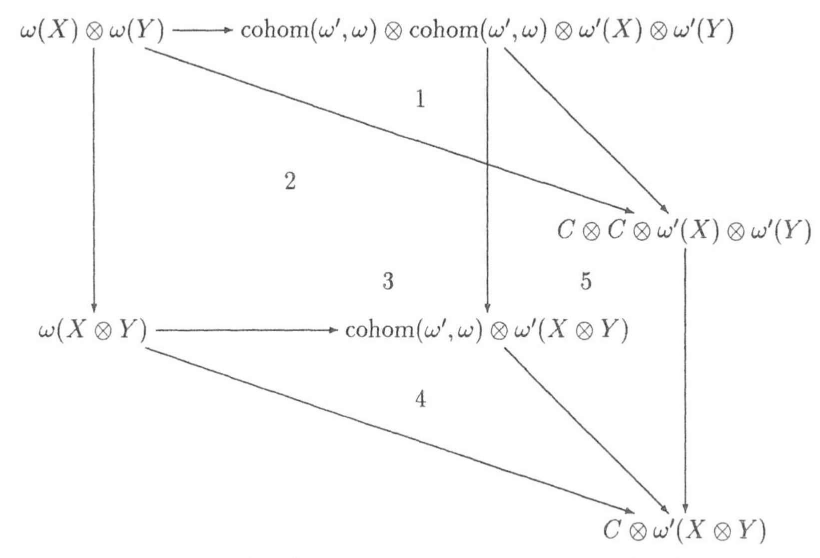

where triangle 1 commutes by Corollary 2.12, square 2 commutes by the definition of the multiplication in Theorem 2.13, square 3 commutes since φ is a natural monoidal transformation, and triangle 4 commutes by the universal property of $\mathrm{cohom}(\omega', \omega)$. Thus by the universal property from Corollary 2.12 the square 5 also commutes so that the universal map $\mathrm{cohom}(\omega', \omega) \longrightarrow C$ is compatible with the multiplication.

We leave it to the reader to check that this map is also compatible with the unit, whence it is an algebra homomorphism. □

We denote by $\mathcal{M}\mathrm{or}_f^{\otimes}(\omega, C \otimes \omega')$ the set of natural monoidal transformations $\varphi : \omega \longrightarrow C \otimes \omega'$. It is easy to see that this is a functor in algebras C. Thus we have proved

COROLLARY 2.19 (cohom as a representing object) *For diagrams $(\mathcal{D}, \omega)$ and $(\mathcal{D}, \omega')$, with $(\mathcal{D}, \omega')$ finite, there is a natural isomorphism*

$$\mathcal{M}or_f(\omega, M \otimes \omega') \cong \mathrm{Hom}(\mathrm{cohom}(\omega', \omega), M).$$

If the diagrams are monoidal then there is a natural isomorphism

$$\mathcal{M}or_f^{\otimes}(\omega, C \otimes \omega') \cong K\text{-}\mathrm{Alg}(\mathrm{cohom}(\omega', \omega), C).$$

We end this section with an interesting observation about a particularly large "diagram" namely the underlying functor $\omega : \mathcal{C}omod\text{-}C \longrightarrow \mathcal{A}$. It says that if all "representations" of a coalgebra resp. bialgebra are known then the coalgebra resp. the bialgebra can be recovered.

THEOREM 2.20 (Tannaka duality) *Let C be a coalgebra. Then $C \cong \mathrm{coend}(\omega)$ as coalgebras where $\omega : \mathcal{C}omod\text{-}C \longrightarrow \mathcal{A}$ is the underlying functor. If C is a bialgebra and ω monoidal, then the coalgebras are isomorphic as bialgebras.*

For a proof we refer to [9] Corollary 6. 4. and [10] Theorem 15 and Corollary 16. □

3 Coendomorphism Bialgebras of Quantum Spaces

In this section we will establish the connection between the bialgebras arising as coendomorphism bialgebras of quantum spaces and those arising as coendomorphism bialgebras of diagrams. It will turn out that the construction via diagrams is far more general. So by Theorem 2.1 and Corollary 2.16 we have established the general existence of these bialgebras. And we can apply this to give an explicit description of coendomorphism bialgebras of finite dimensional algebras, of (finite) quadratic algebras, of families of quadratic algebras, and of finite dimensional Lie algebras.

3.1 Generators and relations of the algebra of cohomomorphisms

We will construct monoidal diagrams generated by a given (finite) family of objects, of morphisms, and of commutativity conditions. Then we will calculate the associated coendomorphism bialgebras. Since in a monoidal category there are two compositions, the tensor product and the composition of morphisms, we are constructing a free (partially defined) algebra from the objects, morphisms and relations.

Let $X_1, \ldots, X_n$ be a given set of objects. Then there is a free monoidal category $\mathcal{C}[X_1, \ldots, X_n]$ generated by the $X_1, \ldots, X_n$ constructed in an analogous way as the free monoidal category on one generating object in [3].

Let $X_1, \ldots, X_n$ be a given set of objects and let $\varphi_1, \ldots, \varphi_m$ be *additional* morphisms between the objects of the free monoidal category $\mathcal{C}[X_1, \ldots, X_n]$ generated by $X_1, \ldots, X_n$. Then there is a free monoidal category $\mathcal{C}[X_1, \ldots, X_n; \varphi_1, \ldots, \varphi_m]$ generated by $X_1, \ldots, X_n$; $\varphi_1, \ldots, \varphi_m$.

If the objects $X_1, \ldots, X_n$ and morphisms $\varphi_1, \ldots, \varphi_m$ are taken in $\mathcal{A}$, then they induce a unique monoidal functor $\omega : \mathcal{C}[X_1, \ldots, X_n; \varphi_1, \ldots, \varphi_m] \longrightarrow \mathcal{A}$.

We indicate how the various free monoidal categories can be obtained. $\mathcal{C}[X_1, \ldots, X_n]$ is generated as follows. The set of objects is given by

(O_1) $X_1, \ldots, X_n$ are objects,

(O_2) I is an object,

(O_3) if Y_1 and Y_2 are objects then $Y_1 \otimes Y_2$ is an object (actually this object should be written as $(Y_1 \otimes Y_2)$ to avoid problems with the explicit associativity conditions),

(O_4) these are all objects.

The set of morphisms is given by

(M_1) for each object there is an identity morphism,

(M_2) for each object Y there are morphisms $\lambda : I \otimes Y \longrightarrow Y$, $\lambda^- : Y \longrightarrow I \otimes Y$, $\rho : Y \otimes I \longrightarrow Y$, and $\rho^- : Y \longrightarrow Y \otimes I$,

(M_3) for any three objects Y_1, Y_2, Y_3 there are morphisms $\alpha : Y_1 \otimes (Y_2 \otimes Y_3) \longrightarrow (Y_1 \otimes Y_2) \otimes Y_3$ and $\alpha^- : (Y_1 \otimes Y_2) \otimes Y_3 \longrightarrow Y_1 \otimes (Y_2 \otimes Y_3)$,

(M_4) if $f : Y_1 \longrightarrow Y_2$ and $g : Y_3 \longrightarrow Y_4$ are morphisms then $f \otimes g : Y_1 \otimes Y_3 \longrightarrow Y_2 \otimes Y_4$ is a morphism,

(M_5) if $f: Y_1 \longrightarrow Y_2$ and $g: Y_2 \longrightarrow Y_3$ are morphisms then $gf: Y_1 \longrightarrow Y_3$ is a morphism,

(M_7) these are all morphisms.

The morphisms are subject to the following congruence conditions with respect to the composition and the tensor product

(R_1) the associativity and unitary conditions of composition of morphisms,

(R_2) the associativity and unitary coherence condition for monoidal categories,

(R_3) the conditions that λ and λ^-, ρ and ρ^-, α and α^- are inverses of each other.

The free monoidal category $\mathcal{C}[X_1, \ldots, X_n; \varphi_1, \ldots, \varphi_m]$ for given objects $X_1, \ldots, X_n$ and morphisms $\varphi_1, \ldots, \varphi_m$ (where the $\varphi_1, \ldots, \varphi_m$ are additional new morphisms between objects of $\mathcal{C}[X_1, \ldots, X_n]$) is obtained by adding the following to the list of conditions for generating the set of morphisms

(M_6) $\varphi_1, \ldots, \varphi_m$ are morphisms.

If there are additional commutativity relations $r_1, \ldots, r_k$ for the morphisms expressed by the φ_i, they can be added to the defining congruence relations to define the free monoidal category $\mathcal{C}[X_1, \ldots, X_n; \varphi_1, \ldots, \varphi_m; r_1, \ldots, r_k]$ by

(R_4) $r_1, \ldots, r_k$ are in the congruence relation.

THEOREM 3.1 (the invariance of the coendomorphism bialgebra) *Let $X_1, \ldots, X_n$ be objects in $\mathcal{A}_0$ and let $\varphi_1, \ldots, \varphi_m$ be morphisms in $\mathcal{A}_0$ between tensor products of the objects $X_1, \ldots, X_n, I$. Let $r_1, \ldots, r_k$ be relations between the given morphisms in $\mathcal{A}_0$. Define $\mathcal{D} := \mathcal{C}[X_1, \ldots, X_n; \varphi_1, \ldots, \varphi_m]$ and $\mathcal{D}' := \mathcal{C}[X_1, \ldots, X_n; \varphi_1, \ldots, \varphi_m; r_1, \ldots, r_k]$ and let $\omega : \mathcal{D} \longrightarrow \mathcal{A}$ and $\omega' : \mathcal{D}' \longrightarrow \mathcal{A}$ be the corresponding underlying functors. Then* $\mathrm{coend}(\omega) \cong \mathrm{coend}(\omega')$ *as bialgebras.*

Proof: follows immediately from Proposition 2.8 applied to the functor $\mathcal{F} : \mathcal{D} \longrightarrow \mathcal{D}'$ that is the identity on the objects and that sends morphisms to their congruence classes so that $\omega = \omega'\mathcal{F}$. □

LEMMA 3.2 (the generating set for cohom) *Let $\mathcal{D} = \mathcal{C}[X_1, \ldots, X_n; \varphi_1, \ldots, \varphi_m]$ be the freely generated monoidal category generated by the objects $X_1, \ldots, X_n$ and the morphisms $\varphi_1, \ldots, \varphi_m$. Let $(\mathcal{D}, \omega)$ and $(\mathcal{D}, \omega')$ be monoidal diagrams and let $(\mathcal{D}, \omega')$ be finite. Then* $\mathrm{cohom}(\omega', \omega)$ *is generated as an algebra by the vector spaces $\omega(X_i) \otimes \omega'(X_i)^*$, $i = 1, \ldots, n$.*

Proof: The multiplication in $\mathrm{cohom}(\omega', \omega)$ is given by taking tensor products of the representatives as in Proposition 2.14. □

For objects $X_1, \ldots, X_n$ and morphisms $\varphi_1, \ldots, \varphi_m$ generating a free monoidal category we now get a complete explicit description of the algebra $\mathrm{cohom}(\omega', \omega)$ in terms of generators and relations. This result resembles that of [12] Lemma 2.1.16 and will be central for our further studies.

THEOREM 3.3 (representation of the algebra cohom)
Let $\mathcal{D} = \mathcal{C}[X_1, \ldots, X_n; \varphi_1, \ldots, \varphi_m]$ be the freely generated monoidal category generated by the objects $X_1, \ldots, X_n$ and the morphisms $\varphi_1, \ldots, \varphi_m$. Let $(\mathcal{D}, \omega)$ and $(\mathcal{D}, \omega')$ be monoidal diagrams and let $(\mathcal{D}, \omega')$ be finite. Then

$$\mathrm{cohom}(\omega', \omega) \cong T(\bigoplus_1^n \omega(X_i) \otimes \omega'(X_i)^*)/I$$

where $T(\bigoplus_1^n \omega(X_i) \otimes \omega'(X_i)^*)$ is the (free) tensor algebra generated by the spaces $\omega(X_i) \otimes \omega'(X_i)^*$ and where I is the two-sided ideal generated by the differences of the images of the $\varphi_1, \ldots, \varphi_m$ under the maps

$$\omega(\mathrm{dom}(\varphi_i)) \otimes \omega'(\mathrm{cod}(\varphi_i))^* \overset{F_C}{\underset{F_D}{\rightrightarrows}} \coprod \omega(X) \otimes \omega'(X)^* \cong T(\bigoplus_1^n \omega(X_i) \otimes \omega'(X_i)^*).$$

Proof: The tensor algebra contains all the spaces of the form $\omega(X) \otimes \omega'(X)$, $X \in \mathcal{D}$, since the objects X are iterated tensor products of the generating objects X_i and since the multiplication in $\mathrm{cohom}(\omega', \omega)$ as described in Proposition 2.14 identifies the images of $(\omega(X) \otimes \omega'(X)^*) \cdot ((\omega(Y) \otimes \omega'(Y)^*)$ with $(\omega(X \otimes Y) \otimes \omega'(X \otimes Y)^*)$.

Assume that $(f : X \longrightarrow Y) \in \mathcal{D}$ is a morphism and that for all $x \otimes \eta \in \omega(X) \otimes \omega'(Y)^*$ the elements $x \otimes \omega'(f)^*(\eta) - \omega(f)(x) \otimes \eta$ are in I. Let $Z \in \mathcal{D}$ and apply $f \otimes Z$ to $x \otimes z \otimes \eta \otimes \zeta \in \omega(X \otimes Z) \otimes \omega'(X \otimes Z)^* \cong \omega(X) \otimes \omega'(Y)^* \otimes \omega(Z) \otimes \omega'(Z)^*$. Then we have

$$\begin{aligned}
&x \otimes z \otimes \omega'(f \otimes Z)^*(\eta \otimes \zeta) - \omega(f \otimes Z)(x \otimes z) \otimes \eta \otimes \zeta = \\
&\quad (x \otimes \omega'(f)^*(\eta)) \cdot (z \otimes \zeta) - (\omega(f)(x) \otimes \eta) \cdot (z \otimes \zeta) = \\
&\quad (x \otimes \omega'(f)^*(\eta) - \omega(f(x)) \otimes \eta) \cdot (z \otimes \zeta) \in I.
\end{aligned}$$

Assume now that $(f : X \longrightarrow Y), (g : Y \longrightarrow Z) \in \mathcal{D}$ are morphisms and that for all $x \otimes \eta \in \omega(X) \otimes \omega'(Y)^*$ and all $y \otimes \zeta \in \omega(Y) \otimes \omega'(Z)^*$ the elements $x \otimes \omega'(f)^*(\eta) - \omega(f)(x) \otimes \eta$ and $y \otimes \omega'(f)^*(\zeta) - \omega(f)(y) \otimes \zeta$ are in I. Then we have

$$\begin{aligned}
&x \otimes \omega'(gf)^*(\zeta) - \omega(gf)(x) \otimes \zeta = \\
&\quad \{x \otimes \omega'(f)^*[\omega'(g)^*(\zeta)] - \omega(f)(x) \otimes [\omega'(g)^*(\zeta)]\} + \\
&\quad \{[\omega(f)(x)] \otimes \omega'(g)^*(\zeta) - \omega(g)[\omega(f)(x)] \otimes \zeta\} \in I.
\end{aligned}$$

□

3.2 Finite quantum spaces

Let $\mathcal{D} = \mathcal{C}[X; m, u]$ be the free monoidal category generated by one object X, a multiplication $m : X \otimes X \longrightarrow X$ and a unit $u : I \longrightarrow X$. The objects of $\mathcal{D}$ are n-fold tensor products $X^{\otimes n}$ of X with itself. (Because of coherence theorems we can assume without of loss of generality that we have strict monoidal categories, i.e. that the associativity morphisms and the left and right unit morphisms are the identities.) Observe that we do not require that m is associative or that u acts as a unit, since by Theorem 3.1 any such relations are irrelevant for the construction of the coendomorphism bialgebras of diagrams over $\mathcal{D}$.

Let A be in K-Alg. Let $\omega_A : \mathcal{D} \longrightarrow \mathcal{A}$ be defined by sending X to the object $A \in \mathcal{A}$, and the multiplication and the unit in $\mathcal{D}$ to the multiplication resp. the unit of the algebra

A in $\mathcal{A}$. Then ω_A is a monoidal functor. If A is finite dimensional, then the diagram $(\mathcal{D}, \omega_A)$ is finite.

For a finite dimensional algebra $A \in K$-Alg we can construct the coendomorphism bialgebra E_A of A as in subsection 1.5. We also can consider the corresponding diagram $(\mathcal{D}, \omega_A)$ in $\mathcal{A}$ and construct its coendomorphism bialgebra $\mathrm{coend}(\omega_A)$.

THEOREM 3.4 (cohomomorphisms of non-commutative spaces and of diagrams) *Let $\mathcal{X}$ and $\mathcal{Y}$ be quantum spaces with function algebras $A = \mathcal{O}(\mathcal{X})$ and $B = \mathcal{O}(\mathcal{Y})$. Let A be finite dimensional. Then $\mathcal{H}(\mathcal{X}, \mathcal{Y})$ exists with $\mathcal{O}(\mathcal{H}(\mathcal{X}, \mathcal{Y})) = a(A, B) \cong$* $\mathrm{cohom}(\omega_A, \omega_B)$.

Proof: Given a quantum space $\mathcal{Z}$ and a map of quantum spaces $\mathcal{Z} \perp \mathcal{X} \longrightarrow \mathcal{Y}$. Let $C = \mathcal{O}(\mathcal{Z})$. Then the map induces an algebra homomorphism $f : B \longrightarrow C \otimes A$. We construct the associated diagrams $(\mathcal{D}, \omega_A)$ and $(\mathcal{D}, \omega_B)$.

We will show now that there is a bijection between the algebra homomorphisms $f : B \longrightarrow C \otimes A$ and the monoidal natural transformations $\varphi : \omega_B \longrightarrow C \otimes \omega_A$. Given f we get φ by

$$\varphi(X^{\otimes n}) : \omega_B(X^{\otimes n}) = B^{\otimes n} \longrightarrow C^{\otimes n} \otimes A^{\otimes n} \overset{m_n \otimes 1}{\longrightarrow} C \otimes A^{\otimes n} = C \otimes \omega_A(X^{\otimes n}),$$

where $m_n : C^{\otimes n} \longrightarrow C$ is the n-fold multiplication. This is a natural transformation since the diagrams

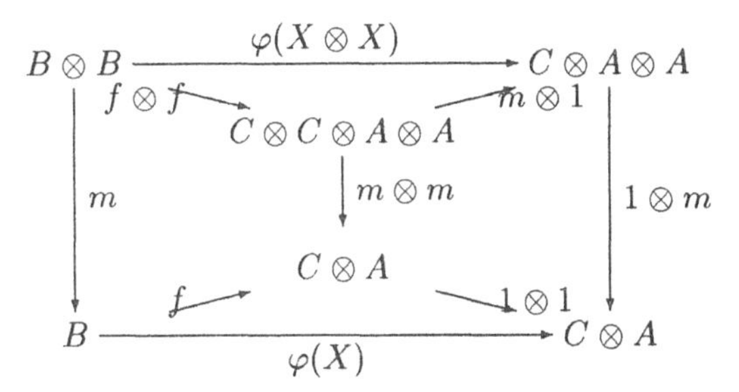

and

$$\begin{array}{ccc} K & \xrightarrow{\varphi(I)} & C \otimes K \\ \downarrow u & & \downarrow 1 \otimes u \\ B & \xrightarrow[\varphi(X)]{} & C \otimes A \end{array}$$

commute. Furthermore the diagrams

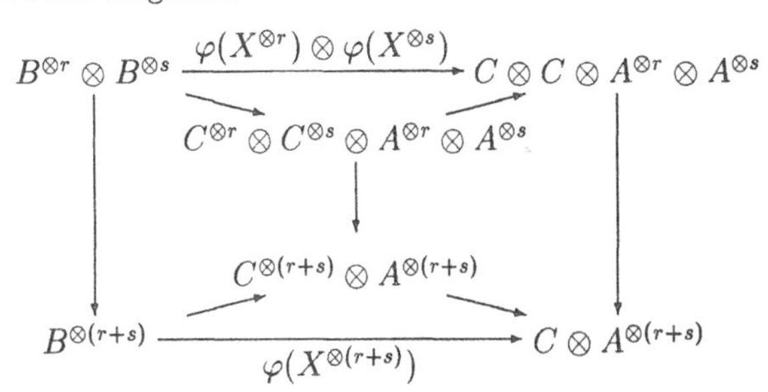

commute so that $\varphi : \omega_B \longrightarrow C \otimes \omega_A$ is a natural monoidal transformation.

Conversely given a natural transformation $\varphi : \omega_B \longrightarrow C \otimes \omega_A$ we get a morphism $f = \varphi(X) : B \longrightarrow C \otimes A$. The diagrams

$$\begin{array}{ccc} B \otimes B & \xrightarrow{f \otimes f} & C \otimes C \otimes A \otimes A \\ \downarrow = & & \downarrow m \otimes 1 \\ B \otimes B & \xrightarrow{\varphi(X \otimes X)} & C \otimes A \otimes A \\ \downarrow m & & \downarrow 1 \otimes m \\ B & \xrightarrow{f} & C \otimes A \end{array}$$

and

$$\begin{array}{ccc} K & \xrightarrow{\cong} & K \otimes K \\ \downarrow = & & \downarrow \\ K & \longrightarrow & C \otimes K \\ \downarrow u & & \downarrow 1 \otimes u \\ B & \xrightarrow{f} & C \otimes A \end{array}$$

commute. Hence $f : B \longrightarrow C \otimes A$ is an algebra homomorphism. This defines a natural isomorphism $K\text{-Alg}(B, C \otimes A) \cong \mathcal{M}\mathrm{or}^{\otimes}_f(\omega_B, C \otimes \omega_A)$. If A is finite dimensional, then the left side is represented by $a(A, B)$ and the right side by $\mathrm{cohom}(\omega_A, \omega_B)$ (2.19). □

COROLLARY 3.5 (isomorphic coendomorphism bialgebras) *There is a unique isomorphism* $E_A \cong \mathrm{coend}(\omega_A)$ *of bialgebras such that the diagram*

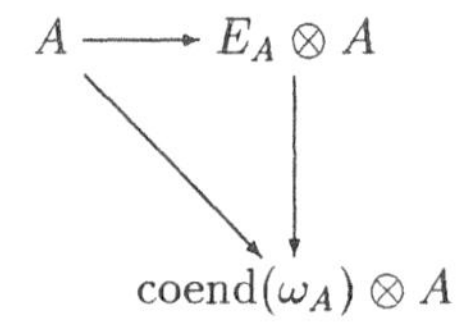

commutes.

Proof: If the coendomorphism bialgebra E_A exists, then it satisfies the universal property given in Corollary 1.10. □

COROLLARY 3.6 (Tambara [15] Thm.1.1) *Let* A, B *be algebras and let* A *be finite dimensional. Then*

$$a(A, B) \cong T(B \otimes A^*)/(xy \otimes \zeta - \sum x \otimes y \otimes \zeta_{(1)} \otimes \zeta_{(2)}, \zeta(1) - 1_A \otimes \zeta | x, y \in B; \zeta \in A^*).$$

Proof: This is an immediate consequence of Theorem 3.3 and Theorem 3.4. In fact the map $m : X \otimes X \longrightarrow X$ induces by the construction in Theorem 2.1 the relations $xy \otimes \zeta \sim \sum x \otimes y \otimes \zeta_{(1)} \otimes \zeta_{(2)}$ for all $x \otimes y \otimes \zeta \in B \otimes B \otimes A^*$. Similarly the map $u : I \longrightarrow X$ induces $\zeta(1) - 1_A \otimes \zeta$ for all $1 \otimes \zeta \in K \otimes A^*$. □

3.3 Quadratic quantum spaces

We now consider quadratic algebras in the sense of Manin [5]. They are $\mathbf{N}$-graded algebras A with $A_0 = K$, A generated by A_1 with (homogeneous) relations generated by $R \subseteq A_1 \otimes A_1$. The quadratic algebra corresponding to (A_1, R) is $T(A_1)/(R)$ where $T(A_1)$ is the tensor algebra generated by A_1. The admissible algebra homomorphisms are those generated by homomorphisms of vector spaces $f : A_1 \longrightarrow B_1$ such that $(f \otimes f)(R_A) \subseteq R_B$. Thus we obtain the category $\mathcal{QA}$ of quadratic algebras. The dual $\mathcal{QQ}$ of this category is the category of quadratic quantum spaces. We will simply denote quadratic algebras by (A, R) where we assume $R \subseteq A \otimes A$.

We consider the free monoidal category $\mathcal{D} = \mathcal{C}[X, Y; \iota]$ where $\iota : Y \longrightarrow X \otimes X$ in $\mathcal{D}$. Then each quadratic algebra (A, R) induces a monoidal functor $\omega_{(A,R)} : \mathcal{D} \longrightarrow \mathcal{A}$ with $\omega(X) = A$, $\omega(Y) = R$, and $\omega(\iota : Y \longrightarrow X \otimes X) = \iota : R \longrightarrow A \otimes A$.

For any two quadratic algebras (A, R) and (B, S), where A and R are finite dimensional, we can construct the universal algebra $\mathrm{cohom}(\omega_{(A,R)}, \omega_{(B,S)})$ satisfying Corollary 2.18. We show that this is the same algebra as the quantum homomorphism space $\mathrm{hom}((A, R), (B, S))$ constructed by Manin [5] 4.4. that has the universal property given in [5] 4. Theorem 5. In particular it is again a quadratic algebra.

THEOREM 3.7 (cohom of quadratic algebras) *Let (A, R) and (B, S) be quadratic algebras with (A, R) finite. Then*

$$\mathrm{cohom}(\omega_{(A,R)}, \omega_{(B,S)}) \cong (B \otimes A^*, S \otimes R^{\perp}),$$

where $R^{\perp}$ is the annihilator of R in $(A \otimes A)^* = A^* \otimes A^*$.

Proof: By Theorem 3.3 the algebra $\mathrm{cohom}(\omega_{(A,R)}, \omega_{(B,S)})$ is generated by the vector spaces $B \otimes A^*$ and $S \otimes R^*$. It satisfies the relations generated by the morphism $\iota : Y \longrightarrow X \otimes X$, which induces relations through the diagram

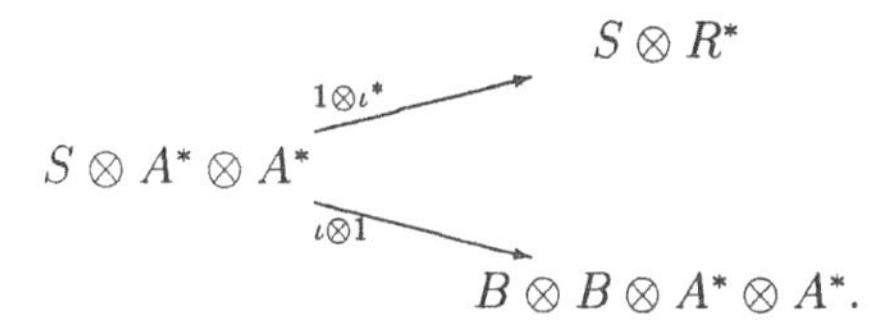

Given an element $s \otimes \alpha \otimes \alpha' \in S \otimes A^* \otimes A^*$ we get equivalent elements $s \otimes (\alpha \otimes \alpha')|_R \sim s \otimes \alpha \otimes \alpha'$. Since the map $1 \otimes \iota^*$ is surjective, every element in $S \otimes R^*$ is equivalent to an element in $B \otimes B \otimes A^* \otimes A^*$ so that we can dispose of the generating set $S \otimes R^*$ altogether. Furthermore elements of the form $s \otimes \alpha \otimes \alpha' \in B \otimes B \otimes A^* \otimes A^*$ are equivalent to zero if $s \in S$ and $\alpha \otimes \alpha'$ induces the zero map on R that is if it is in $R^{\perp}$, so that the set of relations is induced by $S \otimes R^{\perp}$. $\square$

Given an algebra C in $\mathcal{A}$, we call an algebra homomorphism $f : (A, R) \longrightarrow C \otimes (B, S)$ *quadratic*, if it satisfies $f(A) \subseteq C \otimes B$ and $(m_C \otimes B \otimes B)(f \otimes f)(R) \subseteq C \otimes S$ that is if

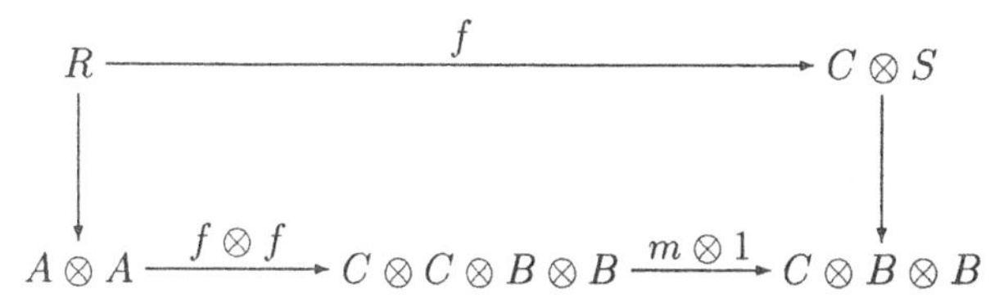

commutes. The set of all quadratic homomorphisms from (A,R) to $C \otimes (B,S)$ is denoted by $K\text{-Alg}^q((A,R), C \otimes (B,S))$. Then one proves as in Theorem 3.4 that $K\text{-Alg}^q((A,R), C \otimes (B,S)) \cong \mathcal{M}\mathrm{or}_f^{\otimes}(\omega_{(A,R)}, C \otimes \omega_{(B,S)})$. Hence we get the following universal property which is different from the one in Manin [5] 4. Theorem 5.

THEOREM 3.8 (universal property of cohom for quadratic algebras) *Let (A,R) and (B,S) be quadratic algebras and let (A,R) be finite. Then there is a quadratic algebra homomorphism $\delta : (A,R) \longrightarrow \mathrm{cohom}(\omega_{(A,R)}, \omega_{(B,S)}) \otimes (B,S)$ such that for every algebra C and every quadratic algebra homomorphism $\varphi : (A,R) \longrightarrow C \otimes (B,S)$ there is a unique algebra homomorphism $\tilde{\varphi} : \mathrm{cohom}(\omega_{(A,R)}, \omega_{(B,S)}) \longrightarrow C$ such that the diagram*

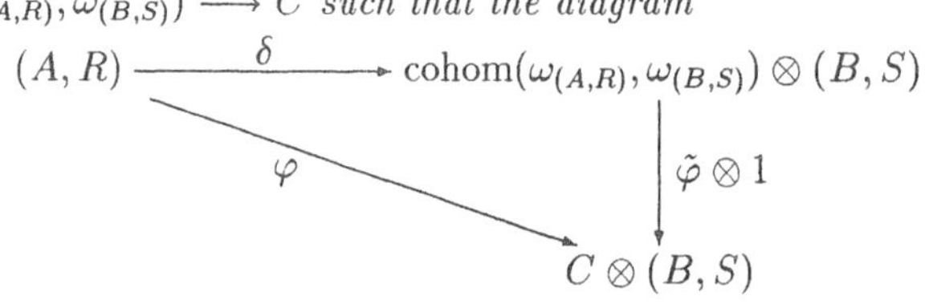

commutes. □

3.4 Complete quadratic quantum spaces

The most interesting diagram for constructing comodule algebras over bialgebras is defined over the free monoidal category $\mathcal{D} = \mathcal{C}[X,\rho]$ with $\rho : X \otimes X \longrightarrow X \otimes X$. If $\omega : \mathcal{D} \longrightarrow \mathcal{A}$ is a finite diagram over this diagram scheme with $\omega(X) = V$ and $\omega(\rho) = f : V \otimes V \longrightarrow V \otimes V$, then we can define a quadratic algebra $A_{(V,f)} := T(V)/(\mathrm{Im}(f))$. Let us furthermore define $B = B_{(V,f)} := \mathrm{coend}(\omega)$. We say that B is a *bialgebra with R-matrix.*

LEMMA 3.9 (spaces for a bialgebra with R-matrix) *$A_{(V,f)}$ is a $B_{(V,f)}$-comodule algebra.*

Proof: Certainly all vector spaces $V^{\otimes n}$ are $B_{(V,f)}$-comodules and f is a comodule homomorphism. Thus $\mathrm{Im}(f) \subseteq V \otimes V$ is a subcomodule. But then one checks easily that $A_{(V,f)} := T(V)/(\mathrm{Im}(f))$ is a comodule as well and that it is in fact a comodule algebra over $B_{(V,f)}$. □

Now the algebra $A_{(V,f)}$ can become very small, in fact degenerate, namely if $f : V \otimes V \longrightarrow V \otimes V$ is bijective. Then $A = K \oplus V$ where the multiplication on V is the zero map. This happens in "most" cases, since for f to be bijective it suffices that $\det(f) \neq 0$. But the next proposition shows that even in the degenerate case we still have room to move.

PROPOSITION 3.10 (change of the R-matrix) *Let $(V, f : V \otimes V \longrightarrow V \otimes V)$ with V finite dimensional be given. Then for every $\lambda \in K$ we have $B_{(V,f-\lambda\cdot\mathrm{id})} = B_{(V,f)}$.*

Proof: By Theorem 3.3 we know that $B_{(V,f)} = T(V \otimes V^*)/I$. We look at the relations. The ideal I is generated by elements of the form

$$x \otimes y \otimes f^*(\xi \otimes \eta) - f(x \otimes y) \otimes \xi \otimes \eta$$

with $x, y, \in V, \xi, \eta \in V^*$. But then the same ideal is also generated by elements of the form

$$x \otimes y \otimes (f^* - \lambda \cdot \mathrm{id}^*)(\xi \otimes \eta) - (f - \lambda \cdot \mathrm{id})(x \otimes y) \otimes \xi \otimes \eta,$$

since the λ-terms simply cancel. Thus $B_{(V,f-\lambda\cdot\mathrm{id})} = B_{(V,f)}$. □

COROLLARY 3.11 (the spectrum of quantum spaces for an R-matrix) *Let $(V, f : V \otimes V \longrightarrow V \otimes V)$ with V finite dimensional be given. Then for every $\lambda \in K$ the algebra $A_{(V,f-\lambda\cdot\mathrm{id})}$ is a $B_{(V,f)}$-comodule algebra. It is non-degenerate if and only of λ is an eigenvalue of f.*

Proof: Instead of changing f in the definition of $B_{(V,f)}$ by a multiple of the identity, we can as well change it in the definition of $A_{(V,f)}$.

So for $B_{(V,f)}$ we have obtained a one-parameter family $A_{(V,f-\lambda\cdot\mathrm{id})}$ of comodule algebras. Since V is finite dimensional and f has only finitely many eigenvalues, all but finitely many of these comodule algebras are degenerate by Lemma 3.9. □

EXAMPLE 3.12 (two parameter quantum matrices) Let us take $V = Kx \oplus Ky$ two dimensional and $f : V \otimes V \longrightarrow V \otimes V$ given by the matrix (with $q, p \neq 0$)

$$R = \begin{pmatrix} 1 & 0 & 0 & 0 \\ 0 & 0 & q^{-1} & 0 \\ 0 & p^{-1} & 1 - q^{-1}p^{-1} & 0 \\ 0 & 0 & 0 & 1 \end{pmatrix}.$$

The bialgebra generated by this matrix is generated by the elements $a = x \otimes \xi$, $b = x \otimes \eta$, $c = y \otimes \xi$, $d = y \otimes \eta$, where ξ, η is the dual basis to x, y. The relations are

$$ac = q^{-1}ca,\ bd = q^{-1}db,\ ad - q^{-1}cb = da - qbc,\ ab = p^{-1}ba,\ cd = p^{-1}dc,\ ad - p^{-1}bc = da - pcb.$$

From this follows $qbc = pcb$. This is the two parameter version of a quantum matrix bialgebra constructed in [6] Chap. 4, 4.10. The matrix R has two eigenvalues $\lambda_1 = 1$ (of multiplicity three) and $\lambda_2 = -q^{-1}p^{-1}$ (of multiplicity one) which lead to algebras

$$A_1 = K\langle x, y\rangle/(xy - q^{-1}yx)$$

and

$$A_2 = K\langle x, y\rangle/(x^2,\ y^2,\ xy + pyx).$$

These are the quantum plane with parameter q and the dual quantum plane with parameter p.

EXAMPLE 3.13 (two further quantum 2×2-matrices) In we replace the generating matrix in previous example by

$$R = \begin{pmatrix} 1 & 0 & 0 & 0 \\ 0 & 0 & q^{-1} & 0 \\ 0 & 1 & 1 - q^{-1} & 0 \\ 0 & -1 & q^{-1} & 1 \end{pmatrix},$$

then the relations for the corresponding bialgebra are given by

$$ac = q^{-1}ca,\ bd = q^{-1}db,\ ad - q^{-1}cb = da - qbc,$$
$$ba = ab + b^2,\ cb + cd + d^2 = da - db + dc,\ ad - b(c-d) = da - (c+d)b.$$

The bialgebra coacts on the algebras

$$A_1 = K\langle x, y\rangle/(xy - q^{-1}yx)$$

and

$$A_2 = K\langle x, y\rangle/(x^2,\ xy + yx,\ xy - y^2).$$

Finally if we take

$$R = \begin{pmatrix} 1 & 0 & 0 & 0 \\ 0 & 0 & 1 & -1 \\ 0 & 1 & 0 & 1 \\ 0 & -1 & 1 & 0 \end{pmatrix},$$

then the bialgebra satisfies the relations

$$ca = ac + c^2,\ bc + bd + d^2 = da + db - dc,\ ad - c(b-d) = da - (b+d)c,$$
$$ba = ab + b^2,\ cb + cd + d^2 = da - db + dc,\ ad - b(c-d) = da - (c+d)b.$$

The nondegenerate algebras it coacts on are

$$A_1 = K\langle x, y\rangle/(yx - xy - y^2)$$

and

$$A_2 = K\langle x, y\rangle/(x^2,\ xy + yx,\ xy - y^2).$$

The endomorphism $f : V \otimes V \longrightarrow V \otimes V$ may have more than two eigenvalues thus inducing more than two non-degenerate comodule algebras. In general the following holds

THEOREM 3.14 (realization of algebras as spectrum of quantum spaces) *Let V be a finite dimensional vector space and let (V, R_i), $i = 1, \ldots, n$ be quadratic algebras. Let $\lambda_1, \ldots, \lambda_n \in K$ be pairwise distinct. Assume that there are subspaces $W_i \subset V_i \subseteq V \otimes V$ such that $V \otimes V = \bigoplus_{i=1}^n V_i$ and $R_i = \bigoplus_{j=1, j\neq i}^k V_j \oplus W_i$. Then there is a homomorphism $f : V \otimes V \longrightarrow V \otimes V$ such that the non-degenerate algebras for B_f are precisely the $(V, R_i) = (V, \mathrm{Im}(f - \lambda_i \cdot \mathrm{id}) = A_{(V, f - \lambda_i \cdot \mathrm{id})}$ for all $i = 1, \ldots, n$.*

Proof: We form the Jordan matrix

$$R = \begin{pmatrix} J_1 & & \\ & \ddots & \\ & & J_n \end{pmatrix}$$

with

$$J_i = \begin{pmatrix} \lambda_i & 1 & & & \\ & \ddots & 1 & & \\ & & \ddots & \ddots & \\ & & & \ddots & 0 \\ & & & & \lambda_i \end{pmatrix}$$

having $\dim(V_i)$ entries λ_i and $\dim(W_i)$ entries 1. Then the induced homomorphism $f : V \otimes V \longrightarrow V \otimes V$ with respect to a suitable basis through the W_i and V_i satisfies $R_i = \mathrm{Im}(f - \lambda_i \cdot \mathrm{id})$. □

3.5 Lie algebras

Similar to our considerations about finite quantum spaces, let $\mathcal{D} = \mathcal{C}[X; m]$ be a free monoidal category on an object X with a multiplication $m : X \otimes X \longrightarrow X$. Then every finite dimensional Lie algebra $\mathbf{g}$ induces a diagram $(\mathcal{D}, \omega)$ with $\omega(X) = \mathbf{g}$ and $\omega(m)$ the Lie bracket. Essentially the same arguments as in Lemma 3.9 show that the bialgebra coend(ω) makes $\mathbf{g}$ a comodule Lie algebra (the Lie multiplication is a comodule homomorphism) and its universal enveloping algebra a comodule bialgebra.

Again coend(ω) has a universal property with respect to its coaction on $\mathbf{g}$. To show this let $\mathbf{g}$ and $\mathbf{g}'$ be Lie algebras. We say that a linear map $f : \mathbf{g} \longrightarrow C \otimes \mathbf{g}'$ is *multiplicative* if the diagram

$$\begin{array}{ccc} \mathbf{g} \otimes \mathbf{g} & \xrightarrow{f \otimes f} & C \otimes C \otimes \mathbf{g}' \otimes \mathbf{g}' \\ \downarrow [\, , \,] & & \downarrow m \otimes [\, , \,] \\ \mathbf{g} & \xrightarrow{f} & C \otimes \mathbf{g}' \end{array}$$

commutes. Then the set of all multiplicative maps Mult$(\mathbf{g}, C \otimes \mathbf{g}')$ is certainly a functor in C.

THEOREM 3.15 (the universal bialgebra coacting on a Lie algebra) *Let* $\mathbf{g}$ *and* $\mathbf{g}'$ *be Lie algebras and let* $\mathbf{g}'$ *be finite dimensional. Then* Mult$(\mathbf{g}, C \otimes \mathbf{g}')$ *is a representable functor with representing object* cohom$(\omega_{\mathbf{g}}, \omega_{\mathbf{g}'})$.
In particular the multiplicative map $\mathbf{g}' \longrightarrow$ cohom$(\omega_{\mathbf{g}}, \omega_{\mathbf{g}'}) \otimes \mathbf{g}$ *is universal.*

Proof: It is analogous to the proof of Theorem 3.4. In particular we get an isomorphism Mult$(\mathbf{g}, C \otimes \mathbf{g}') \cong \mathcal{M}\text{or}_f^{\otimes}(\omega_{\mathbf{g}}, C \otimes \omega_{\mathbf{g}'})$. □

We compute now a concrete example of a universal bialgebra coacting on a Lie algebra.

EXAMPLE 3.16 (the universal bialgebra coacting on a three-dimensional Lie algebra of upper triangular matrices) Let $\mathbf{g} = K\{x, y, z\}$ be the three dimensional Lie algebra with basis x, y, z and Lie bracket $[x, y] = 0$ and $[x, z] = [z, y] = z$. Then $\mathbf{g}^*$ is a Lie coalgebra with dual basis ξ, η, ζ and cobracket $\Delta(\xi) = \Delta(\eta) = 0$ and $\Delta(\zeta) = (\xi - \eta) \otimes \zeta - \zeta \otimes (\xi - \eta)$. By Theorem 3.3 the universal bialgebra coend(ω) for the diagram induced by $\mathbf{g}$ is generated by the elements of the matrix

$$M = \begin{pmatrix} a & b & c \\ d & e & f \\ g & h & i \end{pmatrix} = \begin{pmatrix} x \otimes \xi & x \otimes \eta & x \otimes \zeta \\ y \otimes \xi & y \otimes \eta & y \otimes \zeta \\ z \otimes \xi & z \otimes \eta & z \otimes \zeta \end{pmatrix}.$$

The comultiplication is described in the text after Corollary 2.6 and is given by $\Delta(M) = M \otimes M$. With some straightforward computations one gets the relations as

$$(a-b)c = c(a-b), \ (d-e)f = f(d-e), \ (a-b)f = c(d-e), \ f(a-b) = (d-e)c,$$
$$(a-b)i = i(a-b) = 0, \ (d-e)i = i(d-e) = 0, \ g = 0, \ h = 0.$$

EXAMPLE 3.17 (the universal bialgebra coacting on $sl(2)$)
Let $\mathbf{g} = K\{x, y, z\} = sl(2)$ be the three dimensional Lie algebra with basis x, y, z and Lie

bracket $[x,y]=z$, $[z,x]=x$ and $[y,z]=y$. Then $\mathbf{g}^*$ is a Lie coalgebra with dual basis ξ,η,ζ and cobracket $\Delta(\xi)=\zeta\otimes\xi-\xi\otimes\zeta$, $\Delta(\eta)=\eta\otimes\zeta-\zeta\otimes\eta$, and $\Delta(\zeta)=\xi\otimes\eta-\eta\otimes\xi$. The universal bialgebra coend(ω) for the diagram induced by $\mathbf{g}$ again is generated by the elements of the matrix

$$M=\begin{pmatrix} a & b & c \\ d & e & f \\ g & h & i \end{pmatrix}=\begin{pmatrix} x\otimes\xi & x\otimes\eta & x\otimes\zeta \\ y\otimes\xi & y\otimes\eta & y\otimes\zeta \\ z\otimes\xi & z\otimes\eta & z\otimes\zeta \end{pmatrix}.$$

The comultiplication is given by $\Delta(M)=M\otimes M$. Here one gets the relations as

$$\begin{gathered} ab=ba,\ ac=ca,\ bc=cb,\ de=ed,\ df=fd,\ ef=fe,\ gh=hg,\ gi=ig, \\ a=ai-cg=ia-gc,\ b=ch-bi=hc-ib,\ c=bg-ah=gb-ha, \\ d=fg-di=gf-id,\ e=ei-fh=ie-hf,\ f=dh-eg=hd-ge, \\ g=cd-af=dc-fa,\ h=bf-ce=fb-ec,\ i=ae-db=ea-bd. \end{gathered}$$

4 Automorphisms and Hopf Algebras

In this last short section we want to extend our techniques of using diagrams for determining bialgebras to Hopf algebras.

Hopf algebras arise as function rings of affine algebraic groups which act as group of automorphisms on algebraic varieties. The problem in non-commutative geometry is that the definition of an automorphism group is not that clear. So one defines the *coautomorphism Hopf algebra* of a space X to be the Hopf envelope of the coendomorphism bialgebra of X. The construction of the Hopf envelope H of a bialgebra B was given in a paper of Takeuchi [14].

Hopf algebras also arise as coendomorphism algebras of rigid diagrams [17]. So we first study some properties of rigid monoidal categories and then show that coautomorphism Hopf algebras can also be obtained from diagrams. One of the main theorems in this context is

THEOREM 4.1 (coendomorphism Hopf algebras of rigid diagrams) *(a) Let H be a Hopf algebra. Then the category of right H-comodules that are finite dimensional as vector spaces, is rigid, and the underlying functor $\omega:\mathcal{C}omod\text{-}H\longrightarrow\mathcal{V}ec$ is monoidal and preserves dual objects (up to isomorphism).*
(b) Let $(\mathcal{D},\omega)$ be a finite monoidal diagram and let $\mathcal{D}$ be rigid. Then the coendomorphism bialgebra coend(ω) *is a Hopf algebra.*

Proof: [12] and [16]. □

LEMMA 4.2 (the rigidization of a monoidal category) *Let $\mathcal{D}$ be a small monoidal category. Then there exists a unique (left-)rigidization (up to isomorphism), i.e. a (left-)rigid small monoidal category $\mathcal{D}^*$ and a monoidal functor $\iota:\mathcal{D}\longrightarrow\mathcal{D}^*$ such that for every (left-) rigid small monoidal category $\mathcal{E}$ and monoidal functor $\tau:\mathcal{D}\longrightarrow\mathcal{E}$ there is a unique monoidal functor $\rho:\mathcal{D}^*\longrightarrow\mathcal{E}$ such that*

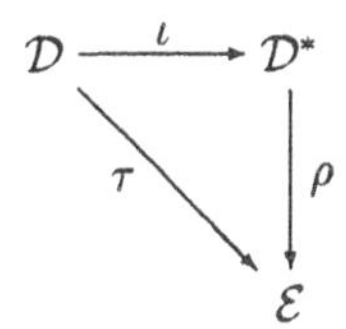

commutes.

Proof: The construction follows essentially the same way as the construction of a free monoidal category over a given (finite) set of objects and morphisms in section 3.1. □

COROLLARY 4.3 (the rigidization of a diagram) *Each finite monoidal diagram $\omega : \mathcal{D} \longrightarrow \mathcal{A}$ has a unique rigidization $\omega^* : \mathcal{D}^* \longrightarrow \mathcal{A}$ such that $\omega^*\iota = \omega$.*

Proof: A finite monoidal diagram is by definition a monoidal diagram in the rigid category $\mathcal{A}_0$. □

PROPOSITION 4.4 (extending monoidal transformations to the rigidization) *Given a finite monoidal diagram $(\mathcal{D}, \omega)$ and its (left-)rigidization $(\mathcal{D}^*, \omega^*)$. Given a (left-)Hopf algebra K and a monoidal natural transformation $f : \omega \longrightarrow K \otimes \omega$. Then there is a unique extension $g : \omega^* \longrightarrow K \otimes \omega^*$, a monoidal natural transformation, such that $(g\iota : \omega^*\iota \longrightarrow K \otimes \omega^*\iota) = (f : \omega \longrightarrow K \otimes \omega)$.*

Proof: For $X \in \mathcal{D}$ let $X^* \in \mathcal{D}^*$ be its dual. Let $M := \omega(X)$, $M^* := \omega^*(X^*)$ and ev: $M^* \otimes M \longrightarrow K$ be the evaluation. M is a K-comodule. We define a comodule structure on M^*. Since M is finite-dimensional (or more generally has a left dual in $\mathcal{A}$) there is a canonical isomorphism $M^* \otimes K \cong \mathrm{Hom}(M, K)$. Then $\delta(m^*) \in M^* \otimes K$ is defined by

$$\sum m^*_{(0)}(m) \otimes m^*_{(1)} := \sum m^*(m_{(0)}) \otimes S(m_{(1)}).$$

It is tedious but straightforward to check that this is a right comodule structure on M^*. We check the coassociativity:

$$\begin{aligned}
\sum m^*_{(0)(0)}(m) \otimes S(m^*_{(0)(1)}) \otimes S(m^*_{(1)}) &= \sum m^*_{(0)}(m_{(0)}) \otimes m_{(1)} \otimes S(m^*_{(1)}) \\
&= \sum m^*(m_{(0)(0)}) \otimes m_{(1)} \otimes m_{(0)(1)} \\
&= \sum m^*(m_{(0)}) \otimes \tau\Delta(m_{(1)}) \\
&= \sum m^*_{(0)}(m) \otimes \tau\Delta(S(m^*_{(1)})) \\
&= \sum m^*_{(0)}(m) \otimes S(m^*_{(1)}) \otimes S(m^*_{(2)}).
\end{aligned}$$

Then the evaluation ev: $M^* \otimes M \longrightarrow k$ satisfies

$$\sum m^*_{(0)}(m_{(0)}) \otimes m^*_{(1)}m_{(1)} = \sum m^*_{(0)}(m_{(0)}) \otimes S(m_{(1)})m_{(2)} = m^*(m) \otimes 1_K$$

hence it is a comodule homomorphism.

We need a unique K-comodule structure on M^* such that ev becomes a comodule homomorphism. Let M^* have such a comodule structure with $\delta(m^*) = \sum m^*_{(0)} \otimes m^*_{(1)}$ and let $m \in M$ with $\delta(m) = \sum m_{(0)} \otimes m_{(1)}$ then we get

$$\sum m^*_{(0)}(m_{(0)}) \otimes m^*_{(1)}m_{(1)} = m^*(m) \otimes 1_K$$

from the fact that ev is a comodule homomorphism. Hence we get

$$\begin{aligned}\sum m^*_{(0)}(m) \otimes m^*_{(1)} &= \sum m^*_{(0)}(m_{(0)}) \otimes m^*_{(1)}\varepsilon(m_{(1)}) \\ &= \sum m^*_{(0)}(m_{(0)}) \otimes m^*_{(1)}m_{(1)}S(m_{(2)}) \\ &= \sum m^*(m_{(0)}) \otimes S(m_{(1)}).\end{aligned}$$

But this is precisely the induced comodule-structure on M^* by the antipode S of K given above.

For iterated duals and tensor products of objects the K-comodule structure arises from iterating the process given above resp. from using the multiplication of K to give the tensor product a comodule structure. □

THEOREM 4.5 (the Hopf envelope of a coendomorphism bialgebra) *Given a monoidal diagram* $(\mathcal{D},\omega)$ *and its rigidization* $(\mathcal{D}^*,\omega^*)$. *Let* $B := \mathrm{coend}(\omega)$, *a bialgebra, and* $H := \mathrm{coend}(\omega^*)$, *a Hopf algebra. Then there is a bialgebra homomorphism* $\sigma : B \longrightarrow H$ *such that for every Hopf algebra* K *and every bialgebra homomorphism* $f : B \longrightarrow K$ *there is a unique bialgebra homomorphism* $g : H \longrightarrow K$ *such that*

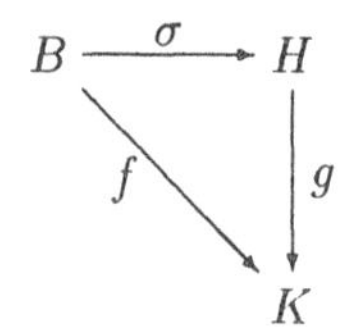

commutes.

Proof: The monoidal natural transformation $\omega^* \longrightarrow H \otimes \omega^*$ induces a unique homomorphism $\sigma : B \longrightarrow H$ such that

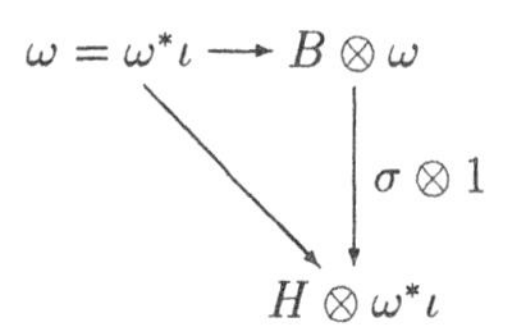

The homomorphism $f : B \longrightarrow K$ induces a monoidal natural transformation $(f \otimes 1)\delta : \omega \longrightarrow B \otimes \omega \longrightarrow K \otimes \omega$, which may be extended to $\omega^* \longrightarrow K \otimes \omega^*$. Hence there is a unique $g : H \longrightarrow K$ such that

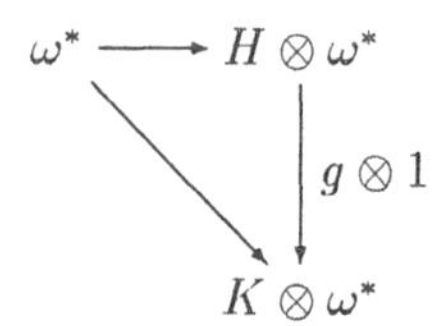

commutes. Then the following diagrams commute

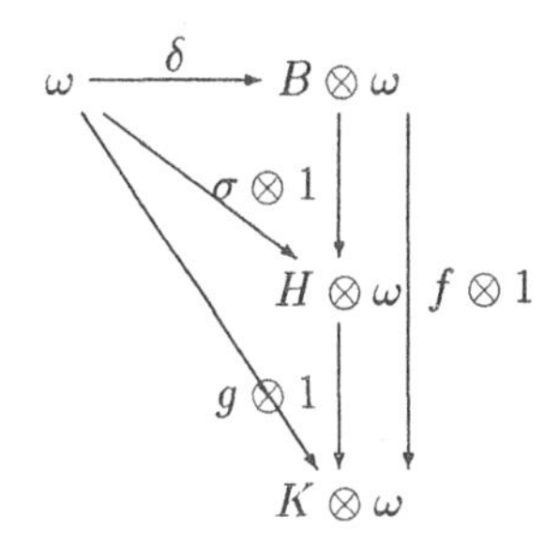

and

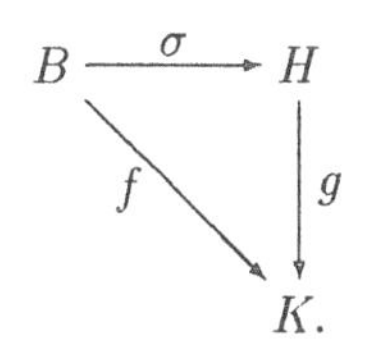

□

We close with a few observations and applications of this result.

Every bialgebra B is the coendomorphism bialgebra of the diagram of all finite-dimensional B-comodules by [10]. So the construction with diagrams given above gives the Hopf envelope of an arbitrary bialgebra.

If a finite-dimensional non-commutative algebra A is given then the rigidization of the finite monoidal diagram $\omega_A : \mathcal{D} \longrightarrow \mathcal{A}$ with $\mathcal{D} := \mathcal{C}[X; m, u]$ generated by A as in 3.2 has a coendomorphism bialgebra H which is the Hopf envelope of the coendomorphism bialgebra of A. A similar remark holds for quadratic algebras.

In the category of quantum spaces the endomorphism quantum space $\mathcal{E}$ of a quantum space $\mathcal{X}$ can be restricted to the "automorphism" quantum space $\mathcal{A}$ by using the homomorphism from the representing bialgebra of $\mathcal{E}$ to its Hopf envelope. Thus $\mathcal{A}$ acts also on $\mathcal{X}$. The existence and the construction in the relevant cases can be obtained from associated rigid monoidal diagrams as in the above theorem.

References

[1] Miriam Cohen, Sara Westreich: *From Supersymmetry to Quantum Commutativity.* Preprint 1991.

[2] P. Deligne, J. S. Milne: *Tannakian Categories. Hodge Cycles, Motives and Shimura Varieties.* Springer LN Math 900, 1982, 101-228.

[3] Saunders Mac Lane: *Categories for the Working Mathematician.* Springer-Verlag New York, Heidelberg, Berlin, 1971.

[4] Shahn Majid: *Rank of Quantum Groups and Braided Groups in Dual Form.* DAMTP/90-44, Euler Inst. Programme on Quantum Groups, Leningrad, 1990.

[5] Yuri I. Manin: *Quantum Groups and Non-Commutative Geometry.* Les publications CRM, Université de Montréal, 1988.

[6] Yuri I. Manin: *Topics in Noncommutative Geometry.* Princeton University Press, Princeton, N. J. , 1991.

[7] Bodo Pareigis: *Categories and Functors.* Academic Press 1970.

[8] Bodo Pareigis: *Non-additive Ring and Module Theory I. General Theory of Monoids.* Publ. Math. (Debrecen) 24, 1977, 189-204.

[9] Bodo Pareigis: *Non-additive Ring and Module Theory III. Morita equivalences.* Publ. Math. (Debrecen) 25, 1978, 177-186.

[10] Bodo Pareigis: *A Non-Commutative Non-Cocommutative Hopf Algebra in "Nature".* J. Alg. 70, 1981, 356-374.

[11] Bodo Pareigis: *Four Lectures on Hopf Algebras.* Centre de Recerca Matemàtica Institut d'Estudis Catalan, No 6, Octubre 1984, 47 pp.

[12] Peter Schauenburg: *Tannaka Duality for Arbitrary Hopf Algebras.* Algebra Berichte No 66, Verlag Reinhard Fischer, München, 1992, 57 pp.

[13] Peter Schauenburg: *On Coquasitriangular Hopf Algebras and the Quantum Yang-Baxter Equation.* Algebra Berichte No 67, Verlag Reinhard Fischer, München, 1992, 78 pp.

[14] M. Takeuchi: Free Hopf Algebras generated by Coalgebras. J. Math. Soc. Japan 23, No.4, 1971.

[15] D. Tambara: *The coendomorphism bialgebra of an algebra.* J. Fac. Sci. Univ. Tokyo 37, 1990, 425-456.

[16] Karl-H. Ulbrich: *Determining Bialgebras from k-Valued Monoidal Functors.* preprint 1989, 8pp.

[17] Karl-H. Ulbrich: *Tannakian Categories for Non-Commutative Hopf Algebras.* Israel J. Math. 72, 1990.

The (Almost) Right Connes Spectrum

D. S. PASSMAN University of Wisconsin, Madison, Wisconsin

§1. INTRODUCTION

Let H be a finite-dimensional semisimple Hopf algebra over a field K and assume that K is a splitting field for H. If A is an H-module algebra, then the action of H on A determines a collection of irreducible H-modules, the Connes spectrum, and this in turn yields information on the structure of the smash product $A\#H$. While a number of papers have already appeared on this subject, the precise definition of the Connes spectrum still seems to be in flux. Certainly, the correct definition should satisfy the following three criteria:

1. The Connes spectrum determines when $A\#H$ is prime.
2. It is relatively easy to compute.
3. It affords information that is understandable and useful in general.

Furthermore, one would like the definition to be elegant and aesthetically pleasing. In this note, we discuss a simplified one-sided definition and indicate how close it comes to meeting the above criteria. Since it is probably not the correct answer, we hedge our bets and call it the *almost right Connes spectrum*.

The Connes spectrum was introduced by A. Connes in his work [Co] on von Neumann algebras and was used in the classification of factors of type III. Later, D. Olesen and G. Pedersen [OlPd] extended these ideas to locally compact abelian groups acting as *-preserving continuous automorphisms on C*-algebras. About five years ago, S. Montgomery and the author defined an algebraic analog of the Connes spectrum in our study [MP] of smash products $A\#K[G]^*$ with G a finite group and with A a G-graded ring. Notice that if G is abelian and acts as automorphisms on A, then the skew group ring $AG = A\#K[G]$ is isomorphic to $A\#K[\hat{G}]^*$ and thus the above applies. On the other hand, the first skew group ring results for nonabelian groups G turned out to be of a negative nature. Indeed, it was shown in [Ch1] and [P] that the primeness of AG could not be determined in general by just studying the $K[G]$-module structure of various G-stable subalgebras of A. Then, it was observed by J. Osterburg and C. Peligrad [OPe] that necessary and sufficient conditions for the simplicity of AG require that we consider the $d_\pi \times d_\pi$ matrix rings over suitable subalgebras of A. Here $d_\pi = \deg \pi$ and π runs through all the irreducible representations of $K[G]$. The latter argument was simplified and extended to handle the primeness of AG in [OP1].

The three papers we are most concerned with are due to J. Osterburg, D. Quinn and the author in various combinations. To start with, a Connes spectrum for finite-dimensional Hopf algebras was introduced in [OPQ]. Later, [OP2] computed a number of interesting examples, and [OQ] made important progress towards understanding the information encoded in the spectrum. It is, in fact, the latter paper which motivates us to rethink the appropriate definitions, since the best results of [OQ] require that H be cocommutative. It would certainly be nice if that assumption could be dropped. We remark that C. Chen takes an entirely different and quite efficient approach to the spectrum in [Ch1] and [Ch2].

Finally, the author would like to thank Caoyu Chen, James Osterburg and Declan Quinn for their helpful comments and suggestions.

§2. SMASH PRODUCTS

The Hopf algebra Connes spectrum is based, in retrospect, on several simple ideas which we briefly outline in this and the next section. Usually, there are certain subalgebras, called hereditary subalgebras, which play a crucial role in this definition. However, to simplify the exposition, we have chosen to finesse these objects by relating the primeness and simplicity of the smash product to the ring of invariants rather than to the base ring. This makes the work much more transparent and, in addition, it allows us to replace the strongly semiprime hypothesis with some weaker, more natural assumptions. A convenient reference for many of the quoted results is [OPQ].

Let H be a finite-dimensional Hopf algebra over a field K and let A be an H-module algebra (with 1). Then the smash product $A\#H$ is an algebra extension of A, containing H and having the additive structure of $A \otimes_K H$. To avoid later confusion, we write the elements of $A\#H$ without tensors so that, if $a \in A$ and $h \in H$, then $a \otimes h$ corresponds to the element $ah \in A\#H$. Thus $A\#H$ is a free left A-module of rank equal to $\dim_K H$ with free A-basis given by any K-basis of H. Furthermore, the multiplication in $A\#H$ extends that of A and H and is determined by the formula

$$ha = \sum_{(h)} (h_1 \cdot a) h_2$$

where $h \cdot a$ denotes the image of a under the given left action of h and where $\Delta h = \sum_{(h)} h_1 \otimes h_2$ is the comultiplication of h in the sigma notation of R. Heyneman and M. Sweedler.

In the case of infinite-dimensional Hopf algebras, properties of the smash product are not right-left symmetric in general. However, when H is finite dimensional, as we are assuming throughout, then symmetry does in fact occur. To start with, we know that the antipode $S: H \to H$ is an algebra antiautomorphism,

$$ah = \sum_{(h)} h_2 \big(S^{-1}(h_1) \cdot a\big)$$

for all $a \in A$ and $h \in H$, and $A\#H$ is a free right A-module with free A-basis given by any K-basis of H. Furthermore, one can use the smash product to show that if X is an H-stable subset of A, then its right and left annihilators in A are also H-stable.

Suppose now that H is semisimple and let e denote its principal idempotent, so that e is both a right and left integral for H. Indeed, if $\epsilon: H \to K$ is the counit of H, then $he = eh = \epsilon(h)e$ for all $h \in H$ and $\epsilon(e) = 1$. Now if V is any left H-module, then the map $\mathrm{Tr}: V \to V$ given by $v \mapsto ev$ is a K-linear projection of V onto the subspace of invariants

$$V^H = \{\, v \in V \mid hv = \epsilon(h)v \text{ for all } h \in H\,\}.$$

Of course, if V is an H-module algebra, then V^H is a subalgebra of V.

For the remainder of this paper, we assume that H is a finite-dimensional semisimple Hopf algebra over K and that K is a splitting field for H. Thus, H is a finite direct sum of its simple two-sided ideals and each such simple component is isomorphic to a full matrix ring over K. Since finite-dimensional semisimple Hopf algebras are separable and since we are concerned with the irreducible modules for H, the splitting assumption is quite reasonable. We will need a few elementary Galois-theoretic facts. Part (iii) below is a classical observation, while (i) and (ii) come from [BM].

LEMMA 2.1. *Let $A\#H$ be semiprime.*

i. *If I is a nonzero H-stable right or left ideal of A, then $I^H = I \cap A^H \neq 0$.*
ii. *A^H is a semiprime algebra. Furthermore, if $A\#H$ is prime, then so is A^H.*
iii. *If I is a nonzero right or left ideal of $A\#H$ and if I is closed under both right and left multiplication by H, then $I \cap \mathbb{C}_{A\#H}(H) \neq 0$.*

Proof. We suppose, when relevant, that I is a right ideal.

(i) Since $0 \neq I$ is H-stable, it follows easily that eI is a nonzero right ideal of the semiprime ring $A\#H$. Thus $(eI)^2 \neq 0$ and hence $eIe = e\,\mathrm{Tr}(I) \neq 0$. In particular, we conclude that $0 \neq \mathrm{Tr}(I) = I \cap A^H$.

(ii) This follows since $A^H \cong eA^H = e(A\#H)e$.

(iii) Since the centrally primitive idempotents of H sum to 1, there exists such an idempotent f with $fI \neq 0$. Then fI is a nonzero right ideal of the semiprime ring $A\#H$, so $(fI)^2 \neq 0$ and hence $fIf \neq 0$. Now set $T = fH + fIf \subseteq f(A\#H)f$. Then T is a K-algebra with identity f, fIf is a nonzero two-sided ideal of T, and $fH = \mathrm{M}_n(K)$ is a full matrix ring over K by the splitting assumption. The latter implies that $T = \mathrm{M}_n(T')$ where T' is the centralizer of fH in T and therefore, since $0 \neq fIf \lhd T$, it follows that $fIf \cap T' \neq 0$. But T' centralizes both fH and $(1-f)H$, so T' centralizes H and $0 \neq fIf \cap T' \subseteq I \cap \mathbb{C}_{A\#H}(H)$ as required. □

Recall that the left adjoint action of H on $A\#H$ is defined by

$$h \cdot \alpha = \sum_{(h)} h_1 \alpha S(h_2)$$

for all $h \in H$ and $\alpha \in A\#H$. It is easy to see that this extends the original action of H on A and that, in this way, $A\#H$ becomes an H-module algebra. Furthermore, $\mathbb{C}_{A\#H}(H) = (A\#H)^H$ and therefore (iii) above asserts that $I^H = I \cap (A\#H)^H \neq 0$.

The following is proved in [BCF, §3]; it is also a simplified version of the work in [OPQ, §4].

PROPOSITION 2.2. *If $A\#H$ is semiprime, then the following are equivalent.*

i. *$A\#H$ is prime.*
ii. *eA has trivial right annihilator in $A\#H$ and A^H is prime.*
iii. *eA has trivial right annihilator in $(A\#H)^H$ and A^H is prime.*

Proof. Since AeA is a nonzero two-sided ideal of $A\#H$, it follows that $\mathrm{r.ann}\, eA = \mathrm{r.ann}\, AeA$ is a two-sided ideal of $A\#H$. In particular, (ii) and (iii) above are equivalent by Lemma 2.1(iii). Furthermore, (i) implies (ii) by Lemma 2.1(ii). Thus all that remains to be proved is that (ii) implies (i). To this end, assume that (ii) holds and let I be a nonzero two-sided ideal of $A\#H$. The goal is to show that $\mathrm{r.ann}\, I = 0$.

Now it is easy to see that $eI = eR$ for some H-stable right ideal R of A and, since $\mathrm{r.ann}\, eA = 0$, we have $eI = eAI \neq 0$. Thus $R \neq 0$ and, by Lemma 2.1(i), $J = R^H$ is a nonzero right ideal of A^H. Set $B = \mathrm{r.ann}_A\, J$. Then B is an H-stable right ideal of A and $B^H = 0$ since A^H is prime. Thus $B = 0$, by Lemma 2.1(i) again, and J is right regular in A. Indeed, J is right regular in $A\#H$ since $A\#H$ is a free left A-module. Finally, since $R \supseteq JA$ and e commutes with J, it follows that $I \supseteq eI \supseteq J(eA)$. But J and eA are both right regular in $A\#H$, and therefore the same is true of I. □

We remark that eA is a right ideal of $A\#H$ and its structure as a right $A\#H$-module is well known. Indeed, the map $a \mapsto ea$ yields a right A-module isomorphism $A \cong eA$. Furthermore, the formula $eah = e(S^{-1}(h)\cdot a)$ shows that right multiplication by $h \in H$ on eA corresponds to the original left action of $S^{-1}(h)$ on A.

The next rather standard result is an analog of the above for simple rings.

PROPOSITION 2.3. *The following are equivalent.*

i. *$A\#H$ is simple.*
ii. *$1 \in AeA$ and A^H is simple.*
iii. *$1 \in (AeA)^H$ and A^H is simple.*

Proof. Parts (ii) and (iii) are clearly equivalent. Furthermore, since $(A\#H)e(A\#H) = AeA$ is a nonzero two-sided ideal of $A\#H$, it is obvious that both (i) and (ii) imply that $1 \in AeA$. But then $A\#H$ is Morita equivalent to $e(A\#H)e = eA^H \cong A^H$, and therefore $A\#H$ is simple if and only if A^H has the same property. □

As we will see, the Connes spectrum gives the third conditions of the previous two propositions an interpretation in terms of the irreducible representations of H.

§3. THE CONNES SPECTRUM

We continue with the preceding notation and assumptions. Furthermore, we let $A\otimes H$ denote the tensor product of A and H as K-algebras and we define the two maps $\theta, \theta' \colon A\#H \to A\otimes H$ by

$$\theta \colon ax \mapsto a\otimes S^{-1}(x)$$
$$\theta' \colon ax \mapsto a\otimes x$$

for all $a \in A$ and $x \in H$. These are, of course, well-defined K-vector space isomorphisms since $A\#H$ has the additive structure of the tensor product and since $S^{-1}\colon H \to H$ exists. Moreover, we have

LEMMA 3.1. *Let $\alpha \in A\#H$ and $\beta \in (A\#H)^H$. Then $\theta(\alpha\beta) = \theta(\alpha)\theta(\beta)$. In particular, the restriction of θ to $(A\#H)^H$ is an algebra monomorphism.*

Proof. Write $\alpha = \sum_x a_x x$ and $\beta = \sum_y b_y y$ with $x, y \in H$ and $a_x, b_y \in A$. Since $\beta \in (A\#H)^H = \mathbb{C}_{A\#H}(H)$, we have

$$\alpha\beta = \sum_x a_x x\beta = \sum_x a_x \beta x = \sum_{x,y} a_x b_y yx.$$

Thus

$$\begin{aligned}\theta(\alpha\beta) &= \sum_{x,y} a_x b_y \otimes S^{-1}(x)S^{-1}(y) \\ &= \Big(\sum_x a_x \otimes S^{-1}(x)\Big)\Big(\sum_y b_y \otimes S^{-1}(y)\Big) = \theta(\alpha)\theta(\beta)\end{aligned}$$

since S^{-1} is an algebra antiautomorphism. □

Following [OPQ] we define the left, right and middle terms of A by $A^\ell = \theta'(eA)$, $A^{\mathrm{r}} = \theta(eA)$ and $A^{\mathrm{m}} = \theta((A\#H)^H)$, respectively. In view of the preceding lemma, these are all K-subspaces of $A \otimes H$ with A^{m} algebra-isomorphic to $(A\#H)^H$ and with A^{r} a right A^{m}-module. Furthermore, it can be shown that A^ℓ is a left A^{m}-module. The left term A^ℓ comes into play because of

LEMMA 3.2. *For all $a, b \in A$, we have*

$$\theta(\mathrm{Tr}(aeb)) = \theta'(ea)\theta(eb).$$

Proof. First observe that $eb = \sum_{(e)}(e_1 \cdot b)e_2$ and then note that

$$hae = \sum_{(h)}(h_1 \cdot a)h_2 e = \sum_{(h)}(h_1\epsilon(h_2) \cdot a)e = (h \cdot a)e$$

for all $h \in H$.

Now e occurs in both aeb and in the definition of Tr. Therefore, to avoid confusion, we write f for e in the formula for Tr. In particular, we conclude from the above that

$$\begin{aligned}\mathrm{Tr}(aeb) &= \sum_{(f)} f_1(aeb)S(f_2) = \sum_{(f)}(f_1 \cdot a)ebS(f_2) \\ &= \sum_{(f),(e)} (f_1 \cdot a)(e_1 \cdot b)e_2 S(f_2).\end{aligned}$$

Finally, since S^{-1} is an algebra antiautomorphism, the definitions of θ and θ' yield

$$\begin{aligned}\theta(\mathrm{Tr}(aeb)) &= \sum_{(f),(e)} (f_1 \cdot a)(e_1 \cdot b) \otimes f_2 S^{-1}(e_2) \\ &= \Big(\sum_{(f)}(f_1 \cdot a) \otimes f_2\Big)\Big(\sum_{(e)}(e_1 \cdot b) \otimes S^{-1}(e_2)\Big) \\ &= \theta'(fa)\theta(eb).\end{aligned}$$

Since $f = e$, the lemma is proved. □

In particular, since $(AeA)^H = \mathrm{Tr}(AeA)$, we see that $\theta\big((AeA)^H\big) = \theta'(eA)\theta(eA) = A^\ell A^\mathrm{r}$ is a two-sided ideal of $\theta\big((A\#H)^H\big) = A^\mathrm{m}$. We now introduce local versions of all these terms. To start with, recall that H is a semisimple K-algebra and that K is a splitting field for H. In particular, if $\mathrm{Irr}(H) = \hat{H}$ denotes the set of irreducible representations of H, then H has centrally primitive idempotents e_π, one for each $\pi \in \hat{H}$, and

$$H = \oplus\sum_{\pi\in\hat{H}} e_\pi H = \oplus\sum_{\pi\in\hat{H}} \mathrm{M}_{d_\pi}(K)$$

where $d_\pi = \deg\pi$. Furthermore, in the algebra $A \otimes H$, the central idempotents $1 \otimes e_\pi$ yield an orthogonal decomposition of $1 = 1 \otimes 1$ and thus

$$A \otimes H = \oplus\sum_{\pi\in\hat{H}} (1 \otimes e_\pi)(A \otimes H) = \oplus\sum_{\pi\in\hat{H}} \mathrm{M}_{d_\pi}(A).$$

Notice that multiplication by $1 \otimes e_\pi$ affords the algebra epimorphism

$$1 \otimes \pi\colon A \otimes H \to \mathrm{M}_{d_\pi}(A)$$

from $A \otimes H$ to its π-component and we set

$$\begin{aligned} A^\ell_\pi &= (1 \otimes e_\pi)A^\ell = (1 \otimes \pi)(A^\ell) \\ A^\mathrm{r}_\pi &= (1 \otimes e_\pi)A^\mathrm{r} = (1 \otimes \pi)(A^\mathrm{r}) \\ A^\mathrm{m}_\pi &= (1 \otimes e_\pi)A^\mathrm{m} = (1 \otimes \pi)(A^\mathrm{m}) \end{aligned}$$

for all $\pi \in \hat{H}$.

LEMMA 3.3. *With respect to the preceding decomposition of $A \otimes H$, we have*

$$A^\ell = \oplus\sum_{\pi\in\hat{H}} A^\ell_\pi \qquad A^\mathrm{r} = \oplus\sum_{\pi\in\hat{H}} A^\mathrm{r}_\pi \qquad A^\mathrm{m} = \oplus\sum_{\pi\in\hat{H}} A^\mathrm{m}_\pi.$$

In particular, each A^m_π is a subalgebra of $\mathrm{M}_{d_\pi}(A)$, A^r_π is a right A^m_π-submodule of $\mathrm{M}_{d_\pi}(A)$, and $A^\ell_\pi A^\mathrm{r}_\pi \triangleleft A^\mathrm{m}_\pi$.

Proof. It suffices to show that each of A^ℓ, A^r and A^m is closed under multiplication by the various $1\otimes e_\pi$. For A^r and A^m, this follows immediately from Lemma 3.1 since each of eA and $(A\#H)^H$ is closed under right multiplication by $S(e_\pi) \in H^H = \mathbb{Z}(H)$ and since $\theta\big(S(e_\pi)\big) = 1 \otimes e_\pi$. Finally, the formula $\theta'(\alpha h) = \theta'(\alpha)(1 \otimes h)$, for all $\alpha \in A\#H$ and $h \in H$, yields the result for $A^\ell = \theta'(eA)$. □

It is now a simple matter to define the Connes spectrum. In fact, there are two versions which we denote here by CS′ and CS″, the latter being an algebraic analog of the Kishimoto spectrum [K]. Specifically, we let

$$\begin{aligned} \mathrm{CS}'(A,H) &= \{\, \pi \in \hat{H} \mid A^\mathrm{m}_\pi \cap \mathrm{r.ann}\, A^\mathrm{r}_\pi = 0 \,\} \\ \mathrm{CS}''(A,H) &= \{\, \pi \in \hat{H} \mid 1 \in A^\ell_\pi A^\mathrm{r}_\pi \,\}. \end{aligned}$$

The motivation for these choices is apparent in the following.

THEOREM 3.4. *Assume that $A\#H$ is semiprime.*

i. *$A\#H$ is prime if and only if A^H is prime and $\mathrm{CS}'(A,H) = \hat{H}$.*

ii. *$A\#H$ is simple if and only if A^H is simple and $\mathrm{CS}''(A,H) = \hat{H}$.*

Proof. (i) According to Proposition 2.2, $A\#H$ is prime if and only if A^H is prime and $(A\#H)^H \cap \mathrm{r.ann}\, eA = 0$. Moreover, by Lemma 3.1, the latter fact is equivalent to $A^{\mathrm{m}} \cap \mathrm{r.ann}\, A^{\mathrm{r}} = 0$. Thus Lemma 3.3 yields the result.

(ii) By Proposition 2.3, $A\#H$ is simple if and only if A^H is simple and $1 \in (AeA)^H$. This time, Lemma 3.2 shows that the latter fact is equivalent to $1 \in A^{\ell}A^{\mathrm{r}}$ and again Lemma 3.3 yields the result. □

While this is the basic idea, an additional amount of work must still be done. Specifically, it is necessary to obtain an internal characterization of the subsets A^{ℓ}_{π}, A^{r}_{π} and A^{m}_{π} of $\mathrm{M}_{d_\pi}(A)$. This is a routine task which unfortunately leads to somewhat unpleasant formulas. We begin with the left and right terms.

First, define the $\cdot$ action of H on $A \otimes H$ by $h \cdot (a \otimes x) = (h \cdot a) \otimes x$ for all $h, x \in H$ and $a \in A$. It is easy to see that this is the tensor product of the original $\cdot$ action on A with the trivial action on H and, in this way, $A \otimes H$ becomes an H-module algebra. Since $1 \otimes H \subseteq (A \otimes H)^H$, it follows that each $\mathrm{M}_{d_\pi}(A) = (1 \otimes e_\pi)(A \otimes H)$ is also an H-module algebra by restriction. In fact, the action is merely given by $h \cdot [a_{i,j}] = [h \cdot a_{i,j}]$ for all $h \in H$ and matrices $[a_{i,j}] \in \mathrm{M}_{d_\pi}(A)$.

Next, observe that $h(ax) = \sum_{(h)} (h_1 \cdot a) h_2 x$ and hence

$$\theta(h(ax)) = \sum_{(h)} (h_1 \cdot a) \otimes S^{-1}(x) S^{-1}(h_2) = \sum_{(h)} (h_1 \cdot \theta(ax)) \left(1 \otimes S^{-1}(h_2)\right).$$

Thus

$$\theta(h\alpha) = \sum_{(h)} (h_1 \cdot \theta(\alpha)) \left(1 \otimes S^{-1}(h_2)\right)$$

for all $h \in H$ and $\alpha \in A\#H$, and similarly we have

$$\theta'(h\alpha) = \sum_{(h)} (1 \otimes h_2)(h_1 \cdot \theta'(\alpha)).$$

In particular, since $A^{\ell} = \theta'(eA)$, $A^{\mathrm{r}} = \theta(eA)$ and

$$eA = \{\, \alpha \in A\#H \mid \epsilon(h)\alpha = h\alpha, \quad \forall\, h \in H \,\},$$

it follows that

$$A^{\ell} = \{\, \beta \in A \otimes H \mid \epsilon(h)\beta = \sum_{(h)} (1 \otimes h_2)(h_1 \cdot \beta), \quad \forall\, h \in H \,\}$$

$$A^{\mathrm{r}} = \{\, \beta \in A \otimes H \mid \epsilon(h)\beta = \sum_{(h)} (h_1 \cdot \beta)(1 \otimes S^{-1}(h_2)), \quad \forall\, h \in H \,\}.$$

Finally, since the map $1 \otimes \pi$ commutes with the $\cdot$ action of H, we obtain

LEMMA 3.5. *For all* $\pi \in \hat{H}$,

$$A_\pi^\ell = \{\, X \in \mathrm{M}_{d_\pi}(A) \mid \epsilon(h)X = \sum_{(h)} \pi(h_2)(h_1 \cdot X), \quad \forall\, h \in H \,\}$$

$$A_\pi^{\mathrm{r}} = \{\, X \in \mathrm{M}_{d_\pi}(A) \mid \epsilon(h)X = \sum_{(h)} (h_1 \cdot X)\pi(S^{-1}(h_2)), \quad \forall\, h \in H \,\}.$$

For the middle term, we define the $*$ action of H on $A \otimes H$ to be the tensor product of the original $\cdot$ action on A with the contragredient left adjoint action of H on H which is given by

$$h{:}\, x = \sum_{(h)} h_2 x S^{-1}(h_1)$$

for all $h, x \in H$. In this case, $A \otimes H$ is a left H-module under $*$, but unfortunately not an H-module algebra in general. Since

$$\begin{aligned} h * (a \otimes x) &= \sum_{(h)} (h_1 \cdot a) \otimes (h_2 : x) = \sum_{(h)} (h_1 \cdot a) \otimes \big(h_3 x S^{-1}(h_2)\big) \\ &= \sum_{(h)} (1 \otimes h_3)\big(h_1 \cdot (a \otimes x)\big)\big(1 \otimes S^{-1}(h_2)\big), \end{aligned}$$

we have

$$h * \beta = \sum_{(h)} (1 \otimes h_3)(h_1 \cdot \beta)\big(1 \otimes S^{-1}(h_2)\big)$$

for all $h \in H$ and $\beta \in A \otimes H$. Furthermore, if $ax \in A\#H$ then

$$h \cdot (ax) = \sum_{(h)} h_1(ax)S(h_2) = \sum_{(h)} (h_1 \cdot a) h_2 x S(h_3).$$

Therefore

$$\begin{aligned} \theta(h \cdot (ax)) &= \sum_{(h)} (h_1 \cdot a) \otimes h_3 S^{-1}(x) S^{-1}(h_2) \\ &= \sum_{(h)} (1 \otimes h_3)\big(h_1 \cdot \theta(\alpha)\big)\big(1 \otimes S^{-1}(h_2)\big) = h * \theta(ax) \end{aligned}$$

and thus $\theta(h \cdot \alpha) = h * \theta(\alpha)$ for all $h \in H$, $\alpha \in A\#H$. Since $A^{\mathrm{m}} = \theta\big((A\#H)^H\big)$ and

$$(A\#H)^H = \{\, \alpha \in A\#H \mid \epsilon(h)\alpha = h \cdot \alpha, \quad \forall\, h \in H \,\}$$

it therefore follows that

$$A^{\mathrm{m}} = \{\, \beta \in A \otimes H \mid \epsilon(h)\beta = h * \beta, \quad \forall\, h \in H \,\}$$

and hence we obtain

LEMMA 3.6. *For all $h \in \hat{H}$,*

$$A_\pi^{\mathrm{m}} = \{ X \in \mathrm{M}_{d_\pi}(A) \mid \epsilon(h)X = \sum_{(h)} \pi(h_3)(h_1 \cdot X)\pi(S^{-1}(h_2)), \quad \forall h \in H \}.$$

While the formulas in the preceding two lemmas are indeed unpleasant, they do have rather nice interpretations in some special cases of interest.

§4. SPECIFIC COMPUTATIONS

The first few computations merely require that we plug elements into the formulas. For example

LEMMA 4.1. *If $H = K[G]$ is a group algebra, then*

$$\begin{aligned} A_\pi^{\ell} &= \{ X \in \mathrm{M}_{d_\pi}(A) \mid g \cdot X = \pi(g^{-1})X, \quad \forall g \in G \} \\ A_\pi^{\mathrm{r}} &= \{ X \in \mathrm{M}_{d_\pi}(A) \mid g \cdot X = X\pi(g), \quad \forall g \in G \} \\ A_\pi^{\mathrm{m}} &= \{ X \in \mathrm{M}_{d_\pi}(A) \mid g \cdot X = \pi(g^{-1})X\pi(g), \quad \forall g \in G \}. \end{aligned}$$

Furthermore, it is easy to see that the formulas for A_π^{ℓ}, A_π^{r} and A_π^{m} in Lemmas 3.5 and 3.6 need only be checked on a generating subcoalgebra of H. This explains why we only consider elements of L in the following.

LEMMA 4.2. *If $H = u(L)$ is a restricted enveloping algebra, then*

$$\begin{aligned} A_\pi^{\ell} &= \{ X \in \mathrm{M}_{d_\pi}(A) \mid l \cdot X = -\pi(l)X, \quad \forall l \in L \} \\ A_\pi^{\mathrm{r}} &= \{ X \in \mathrm{M}_{d_\pi}(A) \mid l \cdot X = X\pi(l), \quad \forall l \in L \} \\ A_\pi^{\mathrm{m}} &= \{ X \in \mathrm{M}_{d_\pi}(A) \mid l \cdot X = X\pi(l) - \pi(l)X, \quad \forall l \in L \}. \end{aligned}$$

If H is commutative, then the contragredient left adjoint action of H on H is trivial and hence $*$ and $\cdot$ agree on $A \otimes H$. In addition, all irreducible representations of H are linear and hence $\mathrm{M}_{d_\pi}(A) = A$ for all $\pi \in \hat{H}$. With this identification, it then follows easily that $A_\pi^{\mathrm{m}} = A^H$. Now let $H = K[G]^*$ be the dual of a group algebra so that $A = \oplus\sum_g A_g$ is a G-graded ring and $\hat{H} = G$. Then we have

LEMMA 4.3. *If $H = K[G]^*$, then*

$$A_g^{\ell} = A_{g^{-1}} \qquad A_g^{\mathrm{r}} = A_g \qquad A_g^{\mathrm{m}} = A_1.$$

In particular,

$$\begin{aligned} \mathrm{CS}'(A, H) &= \{ g \in G \mid A_g \text{ is a faithful right } A_1\text{-module} \} \\ \mathrm{CS}''(A, H) &= \{ g \in G \mid A_{g^{-1}} A_g = A_1 \}. \end{aligned}$$

Probably the most interesting computation so far is the inner example of [OP2] which we now briefly describe. To start with, let H be given and let A be a simple two-sided ideal of H. Then A is an H-module algebra via the restriction of the adjoint action of H on H and we let $\phi: H \to A$ denote the irreducible representation of H

corresponding to A. Since H is semisimple, we expect its antipode S to be involutory, that is to satisfy $S^2 = 1$. This has of course been verified for fields of characteristic 0 in [LR]. In any case, since H is semisimple with K a splitting field, we do know at least, by an earlier result of [L], that S^2 is an inner automorphism of H. Say u is a unit of H and $S^2(h) = u^{-1}S(h)u$ for all $h \in H$.

Now let $\pi \in \hat{H}$ and set $B = \mathrm{M}_{d_\pi}(K) = \pi(H)$. Notice that $\mathrm{M}_{d_\pi}(A) = A \otimes B$ and define the maps $D, E\colon H \to A \otimes B$ by

$$D(h) = \sum_{(h)} \phi(h_2) \otimes \pi(uh_1u^{-1})$$
$$E(h) = \phi(h) \otimes 1.$$

Since D is the composition of a number of algebra homomorphisms, we see that D and E are both algebra homomorphisms. The first key step is to prove

LEMMA 4.4. *If H is inner on A, then using the above notation we have*

$$A_\pi^\ell = \{\, X \in A \otimes B \mid D(h)X = XE(h), \quad \forall\, h \in H \,\}$$
$$A_\pi^{\mathrm{r}} = \{\, X \in A \otimes B \mid XD(h) = E(h)X, \quad \forall\, h \in H \,\}$$
$$A_\pi^{\mathrm{m}} = \{\, X \in A \otimes B \mid D(h)X = XD(h), \quad \forall\, h \in H \,\}.$$

With this in hand, we need just one more observation. Let V be the irreducible left H-module associated with the representation ϕ and let $W = W(\pi)$ be the irreducible left H-module associated with π. Since K is a splitting field for H, it follows that $A = \mathrm{End}_K(V) = \mathrm{M}_{d_\phi}(K)$ and $B = \mathrm{End}_K(W) = \mathrm{M}_{d_\pi}(K)$. Thus

$$A \otimes B = \mathrm{End}_K(V) \otimes \mathrm{End}_K(W) = \mathrm{End}_K(V \otimes W)$$

with appropriate identification. In particular, any algebra homomorphism from H to $A \otimes B = \mathrm{End}_K(V \otimes W)$ defines a left H-module structure on $V \otimes W$. There are two such homomorphisms of interest here, namely D and E, and we denote the corresponding H-modules by $(V \otimes W)_D$ and $(V \otimes W)_E$, respectively. In view of the definitions of D and E, it is easy to see that $(V \otimes W)_D \cong W \otimes V$, where the latter is the usual tensor module corresponding to the representation $\pi \otimes \phi$. Furthermore, $(V \otimes W)_E \cong d_\pi V$, the direct sum of $d_\pi = \dim_K W = \deg \pi$ copies of V. Since the formulas defining the sets A_π^ℓ, A_π^{r} and A_π^{m} in Lemma 4.4 can be interpreted as intertwining relations, we obtain

THEOREM 4.5. *Let V be an irreducible H-module with corresponding representation ϕ, and let $A = \mathrm{End}_K(V)$ become an H-module algebra by restricting the adjoint action of H on H. If $W = W(\pi)$ is the irreducible H-module associated with $\pi \in \hat{H}$, then we have*

$$A_\pi^\ell = \mathrm{Hom}_H(d_\pi V, W \otimes V) \qquad A_\pi^{\mathrm{r}} = \mathrm{Hom}_H(W \otimes V, d_\pi V)$$
$$A_\pi^{\mathrm{m}} = \mathrm{Hom}_H(W \otimes V, W \otimes V) = \mathrm{End}_H(W \otimes V),$$

with the H-homomorphisms acting on the left. In particular, $\pi \in \mathrm{CS}'(A, H)$ or $\mathrm{CS}''(A, H)$ if and only if $W \otimes V \cong d_\pi V$.

If $H = K[G]$ is a group algebra, let χ_π denote the group character associated with the representation π. Then the condition $W \otimes V \cong d_\pi V$ of the previous theorem is equivalent to $\chi_\pi \chi_\phi = \chi_\pi(1)\chi_\phi$ and this in turn means that χ_ϕ vanishes off

$$\ker \chi_\pi = \{\, g \in G \mid \chi_\pi(g) = \chi_\pi(1) \,\} \triangleleft G.$$

Thus we can easily compute a number of examples in this case. For instance, if $G/\mathbb{Z}(G)$ is a simple group, then we must have $\mathrm{CS}'(A, H) = \{\, \epsilon \,\}$. On the other hand, for appropriate choices of G and ϕ, we can have $\mathrm{CS}'(A, H)$ contain irreducible representations of arbitrarily large degree.

In terms of the Connes spectrum, the study of outer actions of H on A is far less interesting. To start with, it is not even clear what to use for the definition of outer. One possibility might be that linearly independent elements of H give rise to independent operators on A. To be precise, let $\{\, x_1, x_2, \ldots, x_n \,\}$ be a K-basis for H, let $c_1, c_2, \ldots, c_n$ be elements of A which are not all zero, and define the trace form $t\colon A \to A$ by $t(a) = \sum_{i=1}^{n} (x_i \cdot a)c_i$ for all $a \in A$. Then we say that H is trace outer on A if and only if $t(A) \neq 0$ for all such t. Of course, this is closely related to the right $A\#H$-module structure of eA and indeed it just asserts that $A\#H$ acts faithfully on the module. But if $A\#H$ acts faithfully, then certainly $(A\#H)^H$ also acts faithfully, so Lemmas 3.1 and 3.3 imply that $\mathrm{CS}'(A, H) = \hat{H}$.

To conclude this section, let us return to the general situation and consider the irreducible representation $\epsilon \in \hat{H}$. Then $\deg \epsilon = 1$, so $\mathrm{M}_{d_\epsilon}(A) = A$ and, with this understanding it is easy to see that

LEMMA 4.6. *For any H-module algebra A,*

$$A_\epsilon^\ell = A_\epsilon^{\mathrm{r}} = A_\epsilon^{\mathrm{m}} = A^H.$$

In particular, ϵ is contained in both $\mathrm{CS}'(A, H)$ and $\mathrm{CS}''(A, H)$.

This leads to the rather natural question of determining what $\mathrm{CS}'(A, H) = \{\, \epsilon \,\}$ says about the smash product $A\#H$. For example, it was suggested by J. Bergen and D. Haile that this condition might imply that $A\#H$ is algebra-isomorphic to $A \otimes H$. However, this turns out not to be the case and, indeed, [OP2, Example 3.7] constructs an appropriate counterexample as follows. First, observe that the group $G = \mathrm{SL}_2(5)$ has an irreducible representation of degree 2 over any algebraically closed field K of characteristic 0 and, by using this, we can embed G in $\mathrm{GL}_2(K)$. Then G acts by conjugation on the matrix ring $\mathrm{M}_2(K)$ and, since $\mathbb{Z}(G)$ acts trivially, we obtain an action of $G/\mathbb{Z}(G) = \mathrm{PSL}_2(5) \cong \mathrm{Alt}_5$. In this way, $A = \mathrm{M}_2(K)$ becomes an H-module algebra for $H = K[\mathrm{Alt}_5]$ and it can be shown, using the Hopf algebra epimorphism $K[G] \to H$ and Theorem 4.5 applied to $K[G]$, that $\mathrm{CS}'(A, H) = \{\, \epsilon \,\}$. Finally, since G is a nontrivial central cover of Alt_5, it follows that $A\#H$ has no algebra direct summand isomorphic to $\mathrm{M}_2(K)$, and therefore $A\#H$ is not isomorphic to $A \otimes H$.

§5. THE COCOMMUTATIVE CASE

As we indicated earlier, the real problem here is to understand the information encoded in the Connes spectrum. To get an idea of what one could hope for, let us first assume that H is commutative. In this situation, all irreducible H-modules are linear, $\mathrm{M}_{d_\pi}(A) = A$ and, as we observed earlier, $A_\pi^{\mathrm{m}} = A^H$ for all $\pi \in \hat{H}$. Furthermore, $\hat{H}$ forms a group under tensor product with ϵ playing the role of the identity element. The following is motivated by the work of [OQ].

PROPOSITION 5.1. *If H is commutative, then A^{ℓ}_{π} and A^{r}_{π} are H-stable subspaces of A. Furthermore, for any $\sigma, \tau \in \hat{H}$, we have*

$$A^{\ell}_{\tau}A^{\ell}_{\sigma} \subseteq A^{\ell}_{\sigma\otimes\tau} \qquad A^{\mathrm{r}}_{\sigma}A^{\mathrm{r}}_{\tau} \subseteq A^{\mathrm{r}}_{\sigma\otimes\tau}.$$

In particular, if we assume that $A\#H$ is semiprime, then $\mathrm{CS}'(A,H)$ and $\mathrm{CS}''(A,H)$ are both subgroups of $\hat{H}$.

Proof. First observe that $\pi(H) = K$ is central in A. Now let $X \in A^{\ell}_{\pi}$ and let $h, k \in H$. Then $\sum_{(h)} \pi(h_2)(h_1 \cdot X) = \epsilon(h)X$ and, since H is commutative, we have

$$\begin{aligned}\sum_{(h)} \pi(h_2)(h_1 \cdot (k \cdot X)) &= \sum_{(h)} \pi(h_2)(k \cdot (h_1 \cdot X)) \\ &= k \cdot \sum_{(h)} \pi(h_2)(h_1 \cdot X) \\ &= k \cdot \epsilon(h)X = \epsilon(h)(k \cdot X).\end{aligned}$$

Thus $k \cdot X$ is contained in A^{ℓ}_{π}, and A^{ℓ}_{π} is indeed H-stable. The argument for the right term A^{r}_{π} is similar.

Now let $\sigma, \tau \in \hat{H}$ and let $X \in A^{\ell}_{\tau}$, $Y \in A^{\ell}_{\sigma}$. Then

$$\sum_{(h)} \sigma(h_2)(h_1 \cdot X) = \epsilon(h)X \qquad \sum_{(h)} \tau(h_2)(h_1 \cdot Y) = \epsilon(h)Y$$

for all $h \in H$. Thus, for any $h \in H$, we have

$$\begin{aligned}\sum_{(h)} (\sigma \otimes \tau)(h_2)(h_1 \cdot YX) &= \sum_{(h)} \sigma(h_3)\tau(h_4)(h_1 \cdot Y)(h_2 \cdot X) \\ &= \sum_{(h)} \tau(h_4)(h_1 \cdot Y)\big[\sigma(h_3)(h_2 \cdot X)\big] \\ &= \sum_{(h)} \tau(h_3)(h_1 \cdot Y)\epsilon(h_2)X \\ &= \sum_{(h)} \big[\tau(h_2)(h_1 \cdot Y)\big]X = \epsilon(h)YX\end{aligned}$$

and we conclude from Lemma 3.5 that $YX \in A^{\ell}_{\sigma\otimes\tau}$. Of course, the argument for the right terms is similar.

Finally, suppose $A\#H$ is semiprime. Since A^{r}_{π} is H-stable, we know that its right annihilator in A is also H-stable. Hence, by Lemma 2.1(i), A^{r}_{π} has trivial right annihilator in $A^{\mathrm{m}}_{\pi} = A^{H}$ if and only if it has trivial right annihilator in A. In other words, $\pi \in \mathrm{CS}'(A,H)$ if and only if $\mathrm{r.ann}\, A^{\mathrm{r}}_{\pi} = 0$. It therefore follows from the inclusion $A^{\mathrm{r}}_{\sigma}A^{\mathrm{r}}_{\tau} \subseteq A^{\mathrm{r}}_{\sigma\otimes\tau}$ that if $\sigma, \tau \in \mathrm{CS}'(A,H)$, then $\mathrm{r.ann}\, A^{\mathrm{r}}_{\sigma\otimes\tau} = 0$ and we conclude that $\sigma \otimes \tau \in CS'(A,H)$, as required.

On the other hand, suppose $\sigma, \tau \in \mathrm{CS}''(A,H)$, so that $1 \in A^\ell_\sigma A^{\mathrm{r}}_\sigma$ and $1 \in A^\ell_\tau A^{\mathrm{r}}_\tau$. Then the inclusions $A^\ell_\tau A^\ell_\sigma \subseteq A^\ell_{\sigma\otimes\tau}$ and $A^{\mathrm{r}}_\sigma A^{\mathrm{r}}_\tau \subseteq A^{\mathrm{r}}_{\sigma\otimes\tau}$ imply that

$$1 \in A^\ell_\tau A^{\mathrm{r}}_\tau \subseteq A^\ell_\tau (A^\ell_\sigma A^{\mathrm{r}}_\sigma) A^{\mathrm{r}}_\tau \subseteq A^\ell_{\sigma\otimes\tau} A^{\mathrm{r}}_{\sigma\otimes\tau}.$$

Thus $\sigma \otimes \tau \in \mathrm{CS}''(A,H)$ and the proposition is proved. □

Note that a semisimple commutative Hopf algebra H over a splitting field K is necessarily isomorphic to the dual of a group algebra. Thus the preceding argument can be somewhat simplified by using Lemma 4.3. Nevertheless, the longer proof is more interesting because of its relationship to the work in Lemma 5.4.

If H is not commutative, then a tensor product of irreducible H-modules need not be irreducible in general. It will, however, be a finite direct sum of its irreducible constituents. Thus an appropriate analog of the above result might assert that if $\sigma, \tau \in \mathrm{CS}'(A,H)$ or $\mathrm{CS}''(A,H)$ and if π is an irreducible constituent of $\sigma \otimes \tau$, then $\pi \in \mathrm{CS}'(A,H)$ or $\mathrm{CS}''(A,H)$, respectively. Indeed, it follows easily from Theorem 4.5 that this holds in the inner case, and it is certainly true for $\mathrm{CS}'(A,H) = \hat{H}$ when H is trace outer on A. More interesting is the cocommutative result and argument of [OQ]. We start with some observations from [OP2].

Suppose A is an H-module algebra and that H is cocommutative. If $\pi \in \hat{H}$, then the $*$ action of H on $A \otimes H$ determines an action on $\mathrm{M}_{d_\pi}(A)$ given by

$$h * X = \sum_{(h)} \pi(h_3)(h_1 \cdot X)\pi(S^{-1}(h_2))$$

for all $h \in H$ and $X \in \mathrm{M}_{d_\pi}(A)$. Since H is cocommutative, it is easy to see that $*$ is a measuring and therefore $\mathrm{M}_{d_\pi}(A)$ is an H-module algebra under this operation. To avoid confusion, we will usually write $*H$ for H whenever the action of H is given as above. In particular, by Lemma 3.6, we have $A^{\mathrm{m}}_\pi = \mathrm{M}_{d_\pi}(A)^{*H}$, and [OP2, Lemmas 5.4 and 5.5] yield

LEMMA 5.2. *If H is cocommutative, then*

$$A^\ell_\pi = \{\, X \in \mathrm{M}_{d_\pi}(A) \mid \epsilon(h)X = \sum_{(h)} (h_1 * X)\pi(h_2), \quad \forall\, h \in H \,\}$$

$$A^{\mathrm{r}}_\pi = \{\, X \in \mathrm{M}_{d_\pi}(A) \mid \epsilon(h)X = \sum_{(h)} \pi(S^{-1}(h_2))(h_1 * X), \quad \forall\, h \in H \,\}.$$

Furthermore, $\mathrm{l.ann}\, A^\ell_\pi$ *and* $\mathrm{r.ann}\, A^{\mathrm{r}}_\pi$ *are $*H$-stable left and right ideals of* $\mathrm{M}_{d_\pi}(A)$*, respectively.*

If $A\#H$ is semiprime, it is reasonable to expect Lemma 2.1 to come into play at this point. But a direct application of that lemma requires $\mathrm{M}_{d_\pi}(A)\#(*H)$ to be semiprime. This latter fact can be proved, but in a somewhat roundabout manner. To start with, the nature of the $\cdot$ action of H on $\mathrm{M}_{d_\pi}(A)$ implies that $\mathrm{M}_{d_\pi}(A)\#H \cong \mathrm{M}_{d_\pi}(A\#H)$ and therefore the smash product is semiprime. It then follows that $\mathrm{M}_{d_\pi}(A)$ is an H-semiprime ring, or in other words that $\mathrm{M}_{d_\pi}(A)$ has no nonzero H-stable nilpotent two-sided ideal. But, by [OP2, Lemma 5.3(iii)], the $*H$-stable ideals of $\mathrm{M}_{d_\pi}(A)$ are precisely the same as the H-stable ones. Thus $\mathrm{M}_{d_\pi}(A)$ is also $*H$-semiprime and, since H is cocommutative and semisimple, [Chi, Theorem 2] yields the result. We can now apply Lemma 2.1(i) to the $*$ action and, since $A^{\mathrm{m}}_\pi = \mathrm{M}_{d_\pi}(A)^{*H}$, we obtain

LEMMA 5.3. *Let $\pi \in \hat{H}$. If $A\#H$ is semiprime, then $\pi \in \mathrm{CS}'(A, H)$ if and only if A^{r}_{π} has trivial right annihilator in $\mathrm{M}_{d_\pi}(A)$.*

With this in hand, we can now move on to the argument of [OQ]. First, observe that the characterizations of A^{ℓ}_{π}, A^{r}_{π} and A^{m}_{π} in Lemmas 3.5 and 3.6 make sense even if π is not irreducible. Thus we can use the formulas in those lemmas to define the sets A^{ℓ}_{π}, A^{r}_{π} and A^{m}_{π} for any finite-dimensional representation π of H. Next, observe that the multiplication map $A \otimes A \to A$ extends to algebra homomorphisms $\wedge\colon \mathrm{M}_r(A) \otimes \mathrm{M}_s(A) \to \mathrm{M}_{rs}(A)$ for all positive integers r, s. To be precise, if $X = a \otimes x \in A \otimes \mathrm{M}_r(K)$ and $Y = b \otimes y \in A \otimes M_s(K)$, then

$$\begin{aligned} X \wedge Y = \wedge(X \otimes Y) &= ab \otimes x \otimes y \\ &\in A \otimes \mathrm{M}_r(K) \otimes \mathrm{M}_s(K) \cong A \otimes \mathrm{M}_{rs}(K). \end{aligned}$$

The key fact here is

LEMMA 5.4. *Let H be cocommutative. If σ and τ are finite-dimensional representations of H, then*

$$A^{\ell}_{\sigma} \wedge A^{\ell}_{\tau} \subseteq A^{\ell}_{\sigma\otimes\tau} \qquad A^{\mathrm{r}}_{\sigma} \wedge A^{\mathrm{r}}_{\tau} \subseteq A^{\mathrm{r}}_{\sigma\otimes\tau}$$
$$A^{\mathrm{m}}_{\sigma} \wedge A^{\mathrm{m}}_{\tau} \subseteq A^{\mathrm{m}}_{\sigma\otimes\tau}.$$

Proof. If π is any finite-dimensional representation of H, then $\mathrm{M}_{d_\pi}(A) = A \otimes \mathrm{M}_{d_\pi}(K)$ can be given a left H-module structure by defining

$$h(a \otimes x) = \sum_{(h)} (h_1 \cdot a) \otimes \pi(h_2)x$$

for all $h \in H$, $a \in A$ and $x \in \mathrm{M}_{d_\pi}(K)$. This is, of course, just the tensor module determined by the $\cdot$ action of H on A and by left multiplication by the elements of $\pi(H)$ on $\mathrm{M}_{d_\pi}(K)$. Furthermore, we know that A^{ℓ}_{π} is the subspace of H-invariants of this module.

Now let σ and τ be given. Then $\mathrm{M}_{d_\sigma}(A)$ and $\mathrm{M}_{d_\tau}(A)$ are left H-modules in the above manner and $\mathrm{M}_{d_\sigma}(A) \otimes \mathrm{M}_{d_\tau}(A)$ becomes a left H-module via tensor product. Of course, $d_{\sigma\otimes\tau} = d_\sigma d_\tau$ and therefore $\mathrm{M}_{d_\sigma d_\tau}(A)$ is also a left H-module using the representation $\sigma \otimes \tau$. We claim that $\wedge\colon \mathrm{M}_{d_\sigma}(A) \otimes \mathrm{M}_{d_\tau}(A) \to \mathrm{M}_{d_\sigma d_\tau}(A)$ is an H-module homomorphism for these particular structures.

To see this, let $X = a \otimes x \in A \otimes \mathrm{M}_{d_\sigma}(K)$, let $Y = b \otimes y \in A \otimes \mathrm{M}_{d_\tau}(K)$ and take any $h \in H$. Then

$$\begin{aligned} h(X \otimes Y) &= \sum_{(h)} h_1 X \otimes h_2 Y \\ &= \sum_{(h)} (h_1 \cdot a) \otimes \sigma(h_2)x \otimes (h_3 \cdot b) \otimes \tau(h_4)y \\ &= \sum_{(h)} (h_1 \cdot a) \otimes \sigma(h_3)x \otimes (h_2 \cdot b) \otimes \tau(h_4)y \end{aligned}$$

by cocommutativity. Thus the image of this element under the wedge map is

$$\begin{aligned}\wedge(h(X\otimes Y)) &= \sum_{(h)}(h_1\cdot a)(h_2\cdot b)\otimes\sigma(h_3)x\otimes\tau(h_4)y\\ &= \sum_{(h)}(h_1\cdot ab)\otimes\big[(\sigma\otimes\tau)(h_2)(x\otimes y)\big]\\ &= h(ab\otimes x\otimes y) = h(X\wedge Y)\end{aligned}$$

since the module structure on $\mathrm{M}_{d_\sigma d_\tau}(A)\cong A\otimes\mathrm{M}_{d_\sigma}(K)\otimes\mathrm{M}_{d_\tau}(K)$ is determined by the representation $\sigma\otimes\tau$.

Finally, since A^ℓ_π is the set of H-invariants for the appropriate module action on $\mathrm{M}_{d_\pi}(A)$, it follows first that $A^\ell_\sigma\otimes A^\ell_\tau$ is contained in the set of H-invariants of the tensor module $\mathrm{M}_{d_\sigma}(A)\otimes\mathrm{M}_{d_\tau}(A)$ and then that the image of the latter under $\wedge$ is contained in $A^\ell_{\sigma\otimes\tau}$. In particular, we have $A^\ell_\sigma\wedge A^\ell_\tau\subseteq A^\ell_{\sigma\otimes\tau}$, as required. The argument for the right and middle terms is similar. □

Notice that, if σ and τ are linear representations, then the corresponding matrix rings are all equal to A and $\wedge$ reduces to ordinary multiplication. In this case, we obtain $A^\ell_\sigma A^\ell_\tau\subseteq A^\ell_{\sigma\otimes\tau}$, a formula at variance to that of Proposition 5.1. However, here H is cocommutative, so $\sigma\otimes\tau=\tau\otimes\sigma$ and the formulas do indeed agree.

It is convenient to make one more observation. Let $\pi\in\hat{H}$ and recall that the $\cdot$ action of H on $\mathrm{M}_{d_\pi}(A)$ behaves in an entry-wise fashion, that is $h\cdot[a_{i,j}]=[h\cdot a_{i,j}]$ for all $h\in H$ and matrices $[a_{i,j}]\in\mathrm{M}_{d_\pi}(A)$. Furthermore, the defining relations for A^ℓ_π, as given in Lemma 3.5, only involve this action and left multiplication by elements of $\pi(H)$. It therefore follows that $X\in A^\ell_\pi$ if and only if each column of X satisfies these same defining conditions. In other words, A^ℓ_π is naturally a direct sum of d_π copies its column space $\mathrm{col}(A^\ell_\pi)$. In the same way, A^{r}_π is naturally a direct sum of d_π copies of its row space $\mathrm{row}(A^{\mathrm{r}}_\pi)$. The following is essentially [OQ, Theorem 4].

THEOREM 5.5. *Let A be an H-module algebra and assume that $A\#H$ is semiprime and that H is cocommutative. If $\sigma,\tau\in\mathrm{CS}'(A,H)$ or $\mathrm{CS}''(A,H)$ and if π is an irreducible constituent of $\sigma\otimes\tau$, then $\pi\in\mathrm{CS}'(A,H)$ or $\mathrm{CS}''(A,H)$, respectively.*

Proof. Suppose that $\sigma,\tau\in\mathrm{CS}'(A,H)$ so that, by Lemma 5.3, A^{r}_σ is right regular in $\mathrm{M}_{d_\sigma}(A)$ and A^{r}_τ is right regular in $\mathrm{M}_{d_\tau}(A)$. Notice that $A\otimes\mathrm{M}_{d_\sigma}(K)$ embeds in $A\otimes\mathrm{M}_{d_\sigma}(K)\otimes\mathrm{M}_{d_\tau}(K)$ via the map $f\colon a\otimes x\mapsto a\otimes x\otimes 1$ and that $A\otimes\mathrm{M}_{d_\tau}(K)$ embeds via $g\colon b\otimes y\mapsto b\otimes 1\otimes y$. Furthermore, it is easy to see that these maps preserve the right regularity of sets. Thus $f(A^{\mathrm{r}}_\sigma)$ and $g(A^{\mathrm{r}}_\tau)$ are both right regular in $\mathrm{M}_{d_\sigma d_\tau}(A)$ and hence, so is $A^{\mathrm{r}}_\sigma\wedge A^{\mathrm{r}}_\tau=f(A^{\mathrm{r}}_\sigma)g(A^{\mathrm{r}}_\tau)$. By Lemma 5.4, the same is also true of $A^{\mathrm{r}}_{\sigma\otimes\tau}$.

Similarly, if $\sigma,\tau\in CS''(A,H)$, then $1\in A^\ell_\sigma A^{\mathrm{r}}_\sigma$ and $1\in A^\ell_\tau A^{\mathrm{r}}_\tau$. Thus, since $\wedge$ is an algebra homomorphism, we have

$$\begin{aligned}1&\in(A^\ell_\sigma A^{\mathrm{r}}_\sigma)\wedge(A^\ell_\tau A^{\mathrm{r}}_\tau)\\ &=(A^\ell_\sigma\wedge A^\ell_\tau)(A^{\mathrm{r}}_\sigma\wedge A^{\mathrm{r}}_\tau)\subseteq A^\ell_{\sigma\otimes\tau}A^{\mathrm{r}}_{\sigma\otimes\tau}\end{aligned}$$

by Lemma 5.4.

Finally, $\sigma\otimes\tau=\pi+\pi'$ for some representation π' and, by means of a similarity, we can write the matrix ring $\mathrm{M}_{d_\sigma d_\tau}(K)$ in such a way that $(\sigma\otimes\tau)(h)=\mathrm{diag}(\pi(h),\pi'(h))$

for all $h \in H$. Corresponding to this decomposition, we can partition the matrices in $M_{d_\sigma d_\tau}(A)$ as $\left[\frac{Z}{Z'}\right]$ where Z contains the first d_π rows or as $[Z \mid Z']$ where Z contains the first d_π columns. With this notation, we see that the defining relations for $A^\ell_{\sigma\otimes\tau}$ imply that if $\left[\frac{Z}{Z'}\right] \in A^\ell_{\sigma\otimes\tau}$, then each column of Z is contained in $\mathrm{col}(A^\ell_\pi)$. Similarly, if $[Z \mid Z'] \in A^{\mathrm{r}}_{\sigma\otimes\tau}$, then each row of Z is contained in $\mathrm{row}(A^{\mathrm{r}}_\pi)$.

It now follows from the latter information that if $W \in M_{d_\pi}(A)$ annihilates A^{r}_π on the right, then $\mathrm{diag}(W, 0) \in M_{d_\sigma d_\tau}(A)$ annihilates $A^{\mathrm{r}}_{\sigma\otimes\tau}$ on the right. In particular, if $\sigma, \tau \in \mathrm{CS}'(A, H)$, then $W = 0$ and $\pi \in \mathrm{CS}'(A, H)$. It also follows fairly easily from the partition information and the nature of A^ℓ_π and A^{r}_π that the upper left $d_\pi \times d_\pi$ corners of matrices in $A^\ell_{\sigma\otimes\tau}A^{\mathrm{r}}_{\sigma\otimes\tau}$ are all contained in $A^\ell_\pi A^{\mathrm{r}}_\pi$. If particular, if $\sigma, \tau \in \mathrm{CS}''(A, S)$, then $1 \in A^\ell_\pi A^{\mathrm{r}}_\pi$ and hence $\pi \in \mathrm{CS}''(A, H)$. □

As a consequence of this and [PQ, Corollary 9] we have

COROLLARY 5.6. *Let A be an H-module algebra with $A\#H$ semiprime and with H cocommutative. Then there exist Hopf ideals I', I'' of H such that $\mathrm{CS}'(A, H)$ and $\mathrm{CS}''(A, H)$ are precisely the sets of those irreducible representations of H with kernels containing I' and I'', respectively. In particular, if π is in either spectrum, then so is its contragredient $\pi \circ S$.*

Thus in the cocommutative situation, the Connes spectrum CS' corresponds to a certain Hopf ideal of H and hence to a certain Hopf algebra homomorphic image of H. One wonders if this property is true in general. One also wonders what information the Hopf ideal might contribute to our understanding of the action of H on A. Paper [OQ] then goes on to take a closer look at the case of group algebras $K[G]$ since the Hopf ideals here are just the kernels of the natural maps $K[G] \to K[G/N]$ with $N \triangleleft G$. In particular, it investigates the relationship between the Connes spectrum and the action of G/N on the fixed ring A^N.

REFERENCES

[BCF] J. Bergen, M. Cohen and D. Fischman, *Irreducible actions and faithful actions of Hopf algebras*, Israel J. Math. **72** (1990), 5–18.

[BM] J. Bergen and S. Montgomery, *Smash products and outer derivations*, Israel J. Math. **53** (1986), 321–345.

[Ch1] C. Chen, *The Connes spectrum of certain finite non-abelian groups*, Comm. Algebra **18** (1990), 1505–1515.

[Ch2] ______, *The Connes spectrum of finite dimensional Hopf algebras*, Chin. Annals Math. **12** (1991), ser. B, 520–524.

[Chi] W. Chin, *Crossed products of semisimple cocommutative Hopf algebras*, Proc. AMS **116** (1992), 321–327.

[Co] A. Connes, *Une classification des facteurs de type III*, Ann. Sci. École Norm. Sup. **6** (1973), 133–252.

[K] A. Kishimoto, *Simple crossed products of C*-algebras by locally compact abelian groups*, Yokohama Math. J. **28** (1980), 69–85.

[L] R. G. Larson, *Characters of Hopf algebras*, J. Algebra **17** (1971), 352–368.

[LR] R. G. Larson and D. E. Radford, *Semisimple cosemisimple Hopf algebras*, Amer. J. Math. **109** (1987), 187–195.

[MP] S. Montgomery and D. S. Passman, *Algebraic analogs of the Connes spectrum*, J. Algebra **115** (1988), 92–124.

[OlPd] D. Olesen and G. K. Pedersen, *Applications of the Connes spectrum to C^*-dynamical systems*, J. Funct. Anal. **30** (1976), 179–197.

[OP1] J. Osterburg and D. S. Passman, *What makes a skew group ring prime?*, Contemporary Math. **124** (1992), 165–177.

[OP2] ———, *Computing the Connes spectrum of a Hopf algebra*, Israel J. Math (to appear).

[OPQ] J. Osterburg, D. S. Passman and D. Quinn, *A Connes spectrum for Hopf algebras*, Contemporary Math. **130** (1992), 311–334.

[OPe] J. Osterburg and C. Peligrad, *A strong Connes spectrum for finite group actions of simple rings*, J. Algebra **142** (1991), 424–434.

[OQ] J. Osterburg and D. Quinn, *Cocommutative Hopf algebra actions and the Connes spectrum*, J. Algebra (to appear).

[P] D. S. Passman, *Skew group rings and free rings*, J. Algebra **131** (1990), 502–512.

[PQ] D. S. Passman and D. Quinn, *Burnside's theorem for Hopf algebras*, Proc. AMS.

On Kauffman's Knot Invariants Arising from Finite-Dimensional Hopf Algebras

DAVID E. RADFORD University of Illinois—Chicago, Chicago, Illinois

Hopf algebras arise in variety of settings in the theory of quantum groups and in this context have interesting connections with representation theory, topology and physics. Perhaps one of the most intriguing areas to explore is the relationship between finite-dimensional Hopf algebras and invariants of knots, links and 3-manifolds. It appears that the deepest structural aspects of finite-dimensional Hopf algebras are closely related to the topology of these structures.

The purpose of this paper is to discuss the knot invariants described by Kauffman [6] which arise from finite-dimensional ribbon Hopf algebras (H, R, v) over a field k and to discuss the algebra involved in their construction and computation. We also construct an extensive family of finite-dimensional ribbon Hopf algebras for generating invariants. Their evaluation will be the subject of future research. Some computations, in specific cases, have been made in [9, 13]. Our family includes the familiar $U_q(sl_2)'$ when q is a root of unity.

Kauffman's invariants studied in this paper are functions $\mathrm{Tr}_\chi : \mathcal{K} \longrightarrow k$ from the set of knots $\mathcal{K}$ to the field k, where $\chi : H \longrightarrow k$ is a certain functional

*Research partially supported by NSF Grants DMS-9106222 and DMS-9308106. The author wishes to express appreciation to the organizers of the conference on Hopf algebras held at DePaul University August 3–10, 1992 for the opportunity to present the rudimentary ideas which eventually led to this paper.

which behaves like a trace. For a knot K, an element $G^d w$ of H is computed according to a diagrammatic representation of K in the plane, and the value of $\mathrm{Tr}_\chi(K)$ is given by $\mathrm{Tr}_\chi(K) = \chi(G^d w)$. These trace-like functionals χ form a subspace of H^* which is related to a subalgebra of the center of H in a very natural way when H is unimodular. The Hopf algebras in the family we construct are unimodular. They seem to be interesting enough in their own right.

When H is unimodular, there is a trace-like functional χ as described above which is also a (generalized) right integral. Under very general conditions, which is the case when the quasitriangular Hopf algebra (H, R) is factorizable, χ can be used to construct a 3-manifold invariant. Hennings was the first to do so [5]. His work generalizes ideas found in [25]. In Kauffman's paper [6] Hennings' ideas are simplified and the invariant is reformulated.

The direct connection between the generalized right integrals of a finite-dimensional ribbon Hopf algebra and 3-manifold invariants is a very interesting relationship to explore. We will not do so here, but will at least point out in this paper when the invariants are defined. It turns out that they are defined for a very large subfamily of the examples we construct.

There is a rather extensive literature concerning the theory of invariants of knots, links and 3-manifolds. The reader is directed to [5, 6, 7, 25, 26] for a rudimentary introduction to the subject, indication of current directions and as a source of further references. Certain Hopf algebras $\mathcal{U}_q(\mathcal{G})$ have been shown to be related to interesting invariants. Many recent papers are framed in the language of category theory. The setting of [5, 6, 13] is formally different and relatively simple. The reader is encouraged to contrast, at some point, the approaches to invariant theory found in [5, 6, 13] with approaches taken by other authors. How the various invariants are related is a very interesting question.

The paper is organized as follows. In Section 1 we discuss briefly some aspects of the theory of finite-dimensional Hopf algebras which play an important role in the sequel. In particular we describe the connection between grouplike elements and integrals. We will assume that the reader is familiar with the general theory of finite-dimensional Hopf algebras [27].

In Section 2 we review the notions of finite-dimensional quasitriangular, ribbon and factorizable Hopf algebras and explore connections between them. The Drinfel'd double is central to our discussion. We introduce the notion of quasitriangular envelope and show that the Drinfel'd double of a finite-dimensional Hopf algebra has a universal description in this context. We find it convenient to characterize ribbon Hopf algebras in terms of grouplike elements and describe two different ways of doing so. We show that factorizable Hopf algebras are necessarily unimodular and minimal quasitriangular.

In Section 3 we describe Kauffman's knot invariants which arise from a finite-dimensional ribbon Hopf algebra and illustrate how they are computed through some simple examples. In Section 4 we consider the algebra of knot invariants and address what it means for an invariant to be "new" in relation to given ones from both algebraic and topological points of view.

In Section 5.1 we construct a family of finite-dimensional pointed unimodular ribbon Hopf algebras for the purpose of computing knot and 3-manifold invariants. We show that these Hopf algebras are quotients of the Drinfel'd double. We determine the Hopf algebras in this family for which 3-manifold invariants mentioned above are defined, and the ones which are factorizable. Our results apply to $U_q(sl_2)'$ in particular. Although the family we construct is simple to describe, its combinatorial features seem complex enough to allow one to hope that interesting invariants will emerge.

Throughout k is a field, not necessarily of characteristic 0.

It is a pleasure to acknowledge the many conversations with Louis Kauffman concerning the topology and algebra of knots, links and 3-manifolds. At various points during the development of this paper, David Krebes and Shlomo Gelaki had comments to make which proved very useful. The author wishes to express gratitude to the referee for bringing other references to his attention [13, 23] which are relevant to this paper.

1 Preliminaries

We will focus on the more important background material from the theory of finite-dimensional Hopf algebras needed for this paper, principally on the relationship between grouplike elements and integrals, cocommutative elements and the Drinfel'd double. The reader is referred to Sweedler's book [27] as a general reference. We will need a few technical results on certain sums of roots of unity, and we will discuss how the Diamond Lemma for rings is used for finitely generated k-algebras.

1.1 Finite-Dimensional Hopf algebras and the Drinfel'd Double

Throughout this section, unless otherwise stated, H is a finite-dimensional Hopf algebra with antipode s over the field k. Regard H as an H-bimodule under left and right multiplication. The transpose actions of H on H^*, which are described by

$$h \cdot p(a) = p(ah) \qquad \text{and} \qquad p \cdot h(a) = p(ha)$$

for h, $a \in H$ and $p \in H^*$, give H^* the structure of an H-bimodule. In the same manner the algebra H^* affords $H^{**} = H$ an H^*-bimodule structure. The module actions of H^* on H are given by

$$p \rightharpoonup h = \sum h_{(1)}p(h_{(2)}) \qquad \text{and} \qquad h \leftharpoonup p = \sum p(h_{(1)})h_{(2)}$$

for all $p \in H^*$ and $h \in H$. We use the Heyneman–Sweedler notation $\Delta(h) = \sum h_{(1)} \otimes h_{(2)}$ for comultiplication.

"Twisting" the multiplication and comultiplication in H gives rise to bialgebras H^{op} and H^{cop} respectively. As a coalgebra $H^{op} = H$, and multiplication in H^{op} is defined by the rule $a \cdot b = ba$ for a, $b \in H$. As an algebra $H^{cop} = H$, and comultiplication for H^{cop} is defined by $\Delta^{cop}(h) = \sum h_{(2)} \otimes h_{(1)}$ for $h \in H$, where $\Delta(h) = \sum h_{(1)} \otimes h_{(2)}$. The antipode s of H is an algebra and a coalgebra anti-endomorphism, that is $s : H \longrightarrow H^{op\,cop}$ is a bialgebra map, by [27, Proposition 4.0.1]. Since H is finite-dimensional, s is bijective [27, Corollary 5.1.6]. Thus H^{op} and H^{cop} are Hopf algebras with antipode s^{-1}. Suppose that H' is also a Hopf algebra over k. Then a map $f : H \longrightarrow H'$ of bialgebras is a map of Hopf algebras by [27, Lemma 4.0.4]. Thus we can use the terms bialgebra map and Hopf algebra map interchangeably since the bialgebras involved are Hopf algebras.

The set of grouplike elements $G(H)$ of H is linearly independent by [27, Proposition 3.2.1]. Since H is a Hopf algebra $G(H)$ is a group under multiplication. Observe that $G(H^*) = \mathrm{Alg}_k(H, k)$. There is an intimate connection between grouplike elements and integrals.

Recall that a left (respectively right) integral for H is an element $\Lambda \in H$ which satisfies $h\Lambda = \epsilon(h)\Lambda$ (respectively $\Lambda h = \epsilon(h)\Lambda$) for all $h \in H$. The space of left integrals for H is a one-dimensional ideal of H by [27, Corollary 5.1.6]. Let Λ be a non-zero left integral for H. Then there is a unique

$\alpha \in \mathrm{Alg}_k(H, k) = G(H^*)$ determined by $\Lambda h = \alpha(h)\Lambda$ for all $h \in H$, and α does not depend on the choice of Λ. We refer to α as the distinguished grouplike element of H^*. Likewise the space of right integrals for H^* is a one-dimensional ideal of H^*. Let λ be a non-zero right integral for H^*. Then there is a unique $g \in \mathrm{Alg}_k(H^*, k) = G(H)$ such that $p\lambda = p(g)\lambda$ for all $p \in H^*$, and g does not depend on the particular choice of λ. We refer to g as the distinguished grouplike element of H.

The distinguished grouplike elements play a very important role in the theory of finite-dimensional Hopf algebras. We will use the equations

$$\lambda \circ s = \lambda \cdot g \qquad \text{and} \qquad \lambda(ab) = \lambda(s^2(b \leftharpoonup \alpha)a) \qquad \text{for all} \quad a, b \in H \quad (1)$$

from [22, Section 2] to study cocommutative elements of H^*.

By [27, Theorem 5.1.8] the algebra structure of H is semisimple if and only if $\epsilon(\Lambda) \neq 0$. In this case the ideal of left integrals for H and the ideal of right integrals for H are the same. These ideals are usually different. The Hopf algebra H is said to be unimodular if left integrals and right integrals for H are the same, or equivalently if $\alpha = \epsilon$. Consequently H^* is unimodular if and only if $g = 1$.

There is a very important fact about integrals which we use repeatedly, namely that $(H, \leftharpoonup)$ a free right H^*-module with basis Λ, and likewise $(H^*, \cdot)$ is a free right H-module with basis λ. See [27, Section 5]. In particular every functional $p \in H^*$ can be represented as $p = \lambda \cdot a$ for a unique $a \in H$.

We will call a functional $p \in H^*$ a *generalized right integral* if p belongs to a one-dimensional right ideal of H^*. One-dimensional right (or left) ideals of a Hopf algebra are one-dimensional ideals. The assignment $a \mapsto k(\lambda \cdot a)$ determines a one-one correspondence between $G(H)$ and the set of one-dimensional ideals of H^*. See [20] for details.

Now suppose that $p \in H^*$ is a non-zero generalized right integral. Then there are unique $a, b \in H$ such that $pq = q(a)p$ for all $q \in H^*$, or equivalently

$$\sum p(h_{(1)})h_{(2)} = p(h)a \qquad \text{for all} \quad h \in H, \qquad (2)$$

and $qp = q(b)p$ for all $q \in H^*$, or equivalently

$$\sum h_{(1)}p(h_{(2)}) = p(h)b \qquad \text{for} \quad h \in H. \qquad (3)$$

In particular (2) characterizes right integrals for H^* when $a = 1$ and (3) characterizes left integrals for H^* when $b = 1$.

The cocommutative elements of H^* will play an important role in the study of knot invariants arising from H. Recall that $h \in H$ is said to be cocommutative if $\Delta^{cop}(h) = \Delta(h)$, which translates to $\sum h_{(2)} \otimes h_{(1)} = \sum h_{(1)} \otimes h_{(2)}$. Since H is a Hopf algebra, the set of cocommutative elements $\mathrm{Cocom}(H)$ of H is a subalgebra of H invariant under the antipode s of H. Observe that $\mathrm{Cocom}(H^*)$ consists of all functionals $p \in H^*$ which satisfy $p(ab) = p(ba)$ for all $a, b \in H$. Thus the span $\mathrm{Char}(H)$ of the characters of left H-modules is a subalgebra of $\mathrm{Cocom}(H^*)$. Of importance is the subspace $\mathrm{Cocom}_s(H^*)$ of $\mathrm{Cocom}(H^*)$ consisting of those cocommutative elements χ of H^* such that $\chi \circ s = \chi$.

We will give a very brief description of the underlying Hopf algebra structure $D(H)$ of the Drinfel'd double of H. The reader is refered to [3, 11, 16] for details, a perspective on the role which the double plays in the theory of quantum groups and for other references. We follow the conventions of [16] in this paper in describing the double. More will be said about the double as the paper develops.

As a coalgebra $D(H) = H^{*cop} \otimes H$. Multiplication in the double can be described in several ways. The following are very convenient:

$$(p \otimes a)(q \otimes b) = \sum p(a_{(1)} \cdot q \cdot s^{-1}(a_{(3)})) \otimes a_{(2)}b \tag{4}$$

and

$$(p \otimes a)(q \otimes b) = \sum pq_{(2)} \otimes (S^{-1}(q_{(1)}) \rightharpoonup a \leftharpoonup q_{(3)})b \tag{5}$$

for all $p, q \in H^*$ and $a, b \in H$, where $S = s^*$ is the antipode of H^*. These equations can be found in [16, Section 3]. The antipode s of $D(H)$ is given by $\mathsf{s}(p \otimes a) = (\epsilon \otimes s(a))(S^{-1}(p) \otimes 1)$ for $p \in H^*$ and $a \in H$.

Observe that $(p \otimes 1)(q \otimes b) = pq \otimes b$ and $(p \otimes a)(\epsilon \otimes b) = p \otimes ab$. As a consequence, the one-one maps $\imath_H : H \longrightarrow D(H)$ and $\imath_{H^*} : H^{*cop} \longrightarrow D(H)$ defined by $\imath_H(h) = \epsilon \otimes h$ and $\imath_{H^*}(p) = p \otimes 1$ respectively for $h \in H$ and $p \in H^*$ are Hopf algebra maps.

The group algebra $k[G]$ of the cyclic group $G = (a)$ of order n will play a role in many aspects of this paper. Suppose that $n > 1$ and k has a primitive n^{th} root of unity ω. The idempotent generators of the minimal ideals of $H = k[G]$ form a very natural and useful basis for computing in H and related Hopf algebras. For $\ell \in Z$ set $e_\ell = \sum_{i=0}^{n-1} \frac{\omega^{-i\ell}}{n} a^i$. The reader can easily verify that

$$1 = e_0 + \cdots + e_{n-1}, \qquad \text{and} \quad e_i e_j = \delta_{i,j} e_i, \tag{6}$$

$$a^i e_j = \omega^{ij} e_j, \tag{7}$$

$$\Delta(e_\ell) = \sum_{p=0}^{n-1} e_p \otimes e_{\ell-p} \qquad \text{and} \qquad \epsilon(e_i) = \delta_{0,i} \tag{8}$$

for all $0 \leq i, j, \ell < n$. Notice that $e_0, \ldots, e_{n-1}$ generate the minimal ideals of $k[G]$ and are idempotents by (6).

Suppose that $A : H \longrightarrow k$ is the algebra homomorphism determined by $A(a) = \omega$. Then A generates the group $G(H^*)$, and there is an isomorphism of Hopf algebras $H \simeq H^*$ determined by $a \mapsto A$. Observe that $\{\epsilon, A, \ldots, A^{n-1}\}$ and $\{e_0, e_1, \ldots, e_{n-1}\}$ are dual bases for each other since $A^\ell(e_m) = \delta_{\ell,m}$ for all $0 \leq \ell, m < n$. The multiplicative function $\Theta : k \longrightarrow k[G]$ defined by

$$\Theta(\beta) = \sum_{\ell=0}^{n-1} \beta^{\ell^2} e_\ell$$

for $\beta \in k$ will play an interesting role for us in this paper. Using (8) it is easy to see that $\Theta(\omega^m)$ is a grouplike element of $k[G]$ if and only if $\omega^{2m} = 1$.

The Hopf algebras which arise in this paper are for the most part pointed. Recall that a coalgebra over k is pointed if its simple subcoalgebras are one-dimensional. We denote the nth term of the coradical filtration of a coalgebra C over k by C_n. Whether or not a bialgebra is pointed depends on whether or not a generating coalgebra is pointed. If V is a subspace of an algebra A over k, we let (V) denote the subalgebra of A generated by V.

Lemma 1 *Suppose that H is a bialgebra over the field k and let C be a subcoalgebra of H. Then:*

a) $(C)_0 \subseteq (C_0)$.

b) *If $H = (C)$ then $H_0 \subseteq (C_0)$.*

c) *Suppose that C is pointed and $H = (C)$. Then H is pointed and $H_0 = k[(G(C))]$.*

PROOF: Various forms of the lemma appear in the literature; see [12, 15] for example. The proof is very simple.

Let $K = C+C^2+C^3+\cdots$. For $n \geq 0$ set $K_{(n)} = \sum_{0\leq i_1+\cdots+i_m\leq n} C_{i_1}\cdots C_{i_m}$. Then $K_{(0)} \subseteq K_{(1)} \subseteq K_{(2)} \subseteq \ldots \subseteq \cup_{i=0}^\infty K_{(i)} = K$ and $\Delta(K_{(n)}) \subseteq \sum_{i+j=n} K_{(i)}\otimes$

$K_{(j)}$ for $n \geq 0$. Thus $K_0 \subseteq K_{(0)} = C_0 + C_0C_0 + C_0C_0C_0 + \cdots$ by [27, Proposition 11.1.1]. Note that the subalgebra of H which C generates is $k1+K$. Since the coradical of $k1+K$ is $k1+K_0$ by part a) of [27, Proposition 8.0.3], the proof follows.

1.2 Some Calculations Involving Roots of Unity and the Diamond Lemma

We will need to know that certain sums of roots of unity are not zero for the proof of Proposition 12. The result we need is not very deep, but has interesting implications for the construction of 3-manifold invariants.

Lemma 2 *Suppose that $n > 1$ and the field k has a primitive n^{th} root of unity ω. Then:*

a) $\sum_{\ell=0}^{n-1} \omega^{\ell(\ell+2m)} = \omega^{-m^2}(\sum_{\ell=0}^{n-1} \omega^{\ell^2})$ *for all $m \in Z$.*

b) $\sum_{\ell=0}^{n-1} \omega^{\ell(\ell+2m+1)} = \omega^{-m(m+1)}(\sum_{\ell=0}^{n-1} \omega^{\ell(\ell+1)})$ *for all $m \in Z$.*

c) *Either $\sum_{\ell=0}^{n-1} \omega^{\ell^2} \neq 0$ or $\sum_{\ell=0}^{n-1} \omega^{\ell(\ell+1)} \neq 0$.*

d) *If n is odd, then $\sum_{\ell=0}^{n-1} \omega^{\ell(\ell+m)} \neq 0$ for all $m \in Z$.*

e) *If $n = 2m$ and m is odd, then $\sum_{\ell=0}^{n-1} \omega^{\ell^2} = 0 \neq \sum_{\ell=0}^{n-1} \omega^{\ell(\ell+1)}$.*

f) *If $n = 2m$ and m is even, then $\sum_{\ell=0}^{n-1} \omega^{\ell^2} \neq 0 = \sum_{\ell=0}^{n-1} \omega^{\ell(\ell+1)}$.*

PROOF: Parts a) and b) follow from the observations $\ell(\ell+2m) = (\ell+m)^2 - m^2$ and $\ell(\ell+2m+1) = (\ell+m)(\ell+m+1) - m^2 - m$ respectively. Suppose that $\sum_{\ell=0}^{n-1} \omega^{\ell^2} = 0 = \sum_{\ell=0}^{n-1} \omega^{\ell(\ell+1)}$. Then by parts a) and b) we have $\sum_{\ell=0}^{n-1} \omega^{\ell(\ell+i)} = 0$ for all $i \in Z$. But then $0 = \sum_{i=0}^{n-1}\sum_{\ell=0}^{n-1} \omega^{\ell(\ell+i)} = \sum_{\ell=0}^{n-1}\sum_{i=0}^{n-1} \omega^{\ell^2}\omega^{\ell i} = n1_k$. But the characteristic of k does not divide n since ω is a primitive n^{th} root of unity. Thus part c) follows. If $n = 2m+1$, then $\omega^{\ell(\ell+1)} = \omega^{\ell(\ell-2m)}$. Thus part d) follows from parts a) – c).

Now suppose that $n = 2m$. Then $\omega^m = -1$. Therefore $\sum_{\ell=0}^{n-1} \omega^{\ell^2} = \sum_{\ell=0}^{m-1}(1+\omega^{m^2})\omega^{\ell^2}$, which is 0 when m is odd, and $\sum_{\ell=0}^{n-1} \omega^{\ell(\ell+1)} = \sum_{\ell=0}^{m-1}(1+\omega^{m(m+1)})\omega^{\ell(\ell+1)}$, which is 0 when m is even. Thus parts e) and f) follow from part c), and the proof is complete.

We will need a workable description of the expansion of $(a+x)^n$, where $xa = \omega ax$ for some non-zero $\omega \in k$, for some results found in Section 5. Our description involves ω-binomial coefficients. Their use, and the use of related symbols, is very common in work related to quantum groups.

Set $(0)_\omega = 0$, and for $m > 0$ set $(m)_\omega = 1 + \omega + \cdots + \omega^{m-1}$. For $m > 0$ observe that $(m)_\omega = m$ when $\omega = 1$, and

$$(m)_\omega = \frac{1-\omega^m}{1-\omega}$$

when $\omega \neq 1$. Define $(0)_\omega! = 1$, and for $m > 0$ set $(m)_\omega! = (m)_\omega(m-1)_\omega \cdots (1)_\omega$. The binomial symbol $\binom{m}{\ell}_\omega$ is defined as follows for $m \geq 0$ and $\ell \in Z$:

$$\binom{m}{0}_\omega = 1 = \binom{m}{m}_\omega \qquad \text{for} \quad m \geq 0,$$

$$\binom{m}{\ell}_\omega = \frac{(m)_\omega!}{(\ell)_\omega!(m-\ell)_\omega!} \qquad \text{for} \quad m > \ell > 0$$

whenever $(m-1)_\omega! \neq 0$, and $\binom{m}{\ell}_\omega = 0$ otherwise. By induction it follows that

$$\binom{m+1}{\ell}_\omega = \omega^\ell \binom{m}{\ell}_\omega + \binom{m}{\ell-1}_\omega$$

whenever $m = 0$, or the three conditions $m > 0$, $m+1 \geq \ell \geq 0$ and $(m-1)_\omega! \neq 0$ hold. The last equation is the basis for an inductive proof of part a) of the following proposition.

Proposition 1 *Suppose that A is an algebra over the field k and $\omega \in k$ is a non-zero scalar. Let $a, x \in A$ satisfy $xa = \omega ax$. Then:*

a) $(a+x)^m = \sum_{\ell=0}^{m} \binom{m}{\ell}_\omega a^{m-\ell}x^\ell$ *if* $m = 0$, *or* $m > 0$ *and* $(m-1)_\omega! \neq 0$.

b) *If ω is a primitive n^{th} root of unity, then* $(a+x)^n = a^n + x^n$.

We will need to determine linear bases for certain algebras which are described in terms of generators and relations in the last section of this paper. Possibly the result which has widest application for this purpose is the Diamond Lemma for rings [1]. We are interested in its application to finitely generated algebras over the field k.

Suppose that R is a commutative ring. Let $R\{x_1, \ldots, x_n\}$ be the free ring in non-commuting indeterminants $x_1, \ldots, x_n$ over R, in other words the tensor algebra of the free R-module with basis $\{x_1, \ldots, x_n\}$. We order monomials in the x_i's lexicographically, reading left to right, where we set $x_1 < x_2 < \cdots < x_n$. A *relation* is a pair (W, σ_W), where W is a monomial and σ_W is a R-linear combination of monomials less than W. The substitution rule $W \longleftarrow \sigma_W$ applied to an expression replaces W by σ_W. Types of ambiguities which may arise when substitution rules are applied to a monomial are:

- *inclusion ambiguities:* $W = AW'B$, where A and B are monomials, (W, σ_W) and $(W', \sigma_{W'})$ are relations;
- *overlap ambiguities:* $W = AB$ and $W' = BC$, where A, B, C are monomials not equal to 1, (W, σ_W) and $(W', \sigma_{W'})$ are relations.

An inclusion ambiguity $W = AW'B$ is said to be resolvable if σ_W and $A\sigma_{W'}B$ reduce to the same expression on application of substitution rules. Likewise an overlap ambiguity $W = AB$ and $W' = BC$ is said to be resolvable if $\sigma_W C$ and $A\sigma_{W'}$ reduce to the same expression on application of substitution rules. Let I be the ideal of $R\{x_1, \ldots, x_n\}$ generated by the differences $W - \sigma_W$. By the Diamond Lemma, if all ambiguities are resolvable, then the quotient $A = R\{x_1, \ldots, x_n\}/I$ is a free R-module with basis of cosets represented by monomials which do not contain one of the "words" W belonging to a relation.

2 Quasitriangular, ribbon and factorizable Hopf algebras

In this section we recall the definitions of finite-dimensional quasitriangular, ribbon and factorizable Hopf algebras, describe some of their basic properties and explore connections between them. Of particular interest to us is the Drinfel'd double.

In [5] Hennings shows how to construct 3-manifold invariants using certain finite-dimensional ribbon Hopf algebras, notably those which are unimodular and factorizable. Kauffman shows to how use finite-dimensional ribbon Hopf algebras to construct knot invariants and reworks Hennings' 3-manifold invariant construction in [6].

Both ribbon Hopf algebras and factorizable Hopf algebras belong to the class of quasitriangular Hopf algebras.

2.1 Quasitriangular Hopf Algebras

Let H be a finite-dimensional Hopf algebra over a field k, and let $R = \sum R^{(1)} \otimes R^{(2)} \in H \otimes H$. Define a linear map $f_R : H^* \longrightarrow H$ by $f_R(p) = \sum p(R^{(1)})R^{(2)}$ for $p \in H^*$. The pair (H, R) is said to be a quasitriangular Hopf algebra in the category of finite-dimensional vector spaces over k if the following axioms hold ($r = R$):

(QT.1) $\sum \Delta(R^{(1)}) \otimes R^{(2)} = \sum R^{(1)} \otimes r^{(1)} \otimes R^{(2)}r^{(2)}$,

(QT.2) $\sum \epsilon(R^{(1)})R^{(2)} = 1$,

(QT.3) $\sum R^{(1)} \otimes \Delta^{cop}(R^{(2)}) = \sum R^{(1)}r^{(1)} \otimes R^{(2)} \otimes r^{(2)}$,

(QT.4) $\sum R^{(1)}\epsilon(R^{(2)}) = 1$ and

(QT.5) $(\Delta^{cop}(h))R = R(\Delta(h))$ for all $h \in H$;

or equivalently if $f_R : H^* \longrightarrow H^{cop}$ is a bialgebra map and (QT.5) is satisfied. Observe that (QT.5) is equivalent to

(QT.5') $\sum(p_{(1)} \rightharpoonup h)f_R(p_{(2)}) = \sum f_R(p_{(1)})(h \leftharpoonup p_{(2)})$

for all $p \in H^*$ and $h \in H$. This equation is obtained by applying $p \otimes I$ to both sides of (QT.5).

Identify the vector spaces $H \otimes H$ and $\mathrm{Hom}_k(H^*, H)$ by $(h \otimes a)(p) = p(h)a$ for $h, a \in H$ and $p \in H^*$. Then $R \mapsto f_R$ determines a one-one correspondence between the set of quasitriangular structures (H, R) on H and the set of bialgebra maps $F : H^* \longrightarrow H^{cop}$ which satisfy

$$\sum(p_{(1)} \rightharpoonup h)F(p_{(2)}) = \sum F(p_{(1)})(h \leftharpoonup p_{(2)}) \tag{9}$$

for all $p \in H^*$ and $h \in H$. Thus there is a very natural formulation of quasitriangular Hopf algebra in the finite-dimensional case in terms of bialgebra maps.

This definition of quasitriangular Hopf algebra given by (QT.1)–(QT.5) is a minor reformulation, in the context of finite-dimensional vector spaces, of the definition Drinfel'd gives [3]. In particular R is invertable. We shall refer to a pair (H, R) which satisfies the first five axioms above simply as a quasitriangular Hopf algebra. See [17] for a fuller discussion and references, notably to Majid's work on the double.

We define a morphism $f : (H, R) \longrightarrow (H', R')$ of quasitriangular Hopf algebras over k to be a map of Hopf algebras $f : H \longrightarrow H'$ such that $R' = (f \otimes f)(R)$. The class of quasitriangular Hopf algebras over k together with morphisms under composition form a category. Suppose that (H, R) and (H', R') are quasitriangular Hopf algebras over k. Then $(H \otimes H', R'')$ is a quasitriangular Hopf algebra, where $H \otimes H'$ is the tensor product of Hopf algebras over k and $R'' = \sum(R^{(1)} \otimes R'^{(1)}) \otimes (R^{(2)} \otimes R'^{(2)})$.

Suppose that (H, R) is a finite-dimensional quasitriangular Hopf algebra over k. Suppose that $f : H \longrightarrow H'$ is a surjective Hopf algebra map, and set $R' = (f \otimes f)(R)$. Then (H', R') is quasitriangular and $f : (H, R) \longrightarrow (H', R')$ is a morphism. If H is commutative then H must be cocommutative by (QT.5). Also notice that $(H^{op}, \widetilde{R})$ and $(H^{cop}, \widetilde{R})$ are quasitriangular, where $\widetilde{R} = \sum R^{(2)} \otimes R^{(1)}$. Consequently $(H^{op\ cop}, R)$ is quasitriangular. With the usual identification of vector spaces H and H^{**}, the maps $f_{\widetilde{R}}$ and f_R are related by the equation $f_{\widetilde{R}} = f_R^*$. If $R^{-1} = \widetilde{R}$ then (H, R) is said to be triangular.

Let H_R be the sub-Hopf algebra of H generated by $\operatorname{Im} f_R$ and $\operatorname{Im} f_{\widetilde{R}}$. If B is a sub-Hopf algebra of H such that $R \in B \otimes B$, then $H_R \subseteq B$. We say that (H, R) is a minimal quasitriangular Hopf algebra if $H = H_R$. Minimal quasitriangular Hopf algebras are introduced and studied in [16].

Fundamental properties of finite-dimensional quasitriangular Hopf algebras (H, R) are discussed by Drinfel'd in [2]. We will need to use several of them in this paper.

$$R^{-1} = (s \otimes I)(R) = (I \otimes s^{-1})(R), \tag{10}$$

and consequently

$$R = (s \otimes s)(R), \tag{11}$$

where s is the antipode of H. Now set $u = \sum s(R^{(2)})R^{(1)}$. Then u is invertable and

$$u^{-1} = \sum R^{(2)} s^2(R^{(1)}), \tag{12}$$

$$\epsilon(u) = 1, \tag{13}$$

$$\Delta(u) = (u \otimes u)(\tilde{R}R)^{-1} = (\tilde{R}R)^{-1}(u \otimes u) \tag{14}$$

and

$$s^2(h) = uhu^{-1} \tag{15}$$

for all $h \in H$.

Since $s^2(h) = h$ for all $h \in G(H)$, it follows by (15) that u commutes with the grouplike elements of H. Note that (11) implies $s(u) = u$ when H is commutative. By (14) it follows that u is a grouplike element of H if and only if (H, R) is triangular.

A finite-dimensional Hopf algebra H over k always admits a quasitriangular structure in the cocommutative case, namely $(H, 1 \otimes 1)$. For H to admit a quasitriangular structure s^2 must be inner by (15). The square of the antipode of a finite-dimensional Hopf algebra is usually not inner [18].

By the virtue of the Drinfel'd double construction every finite-dimensional Hopf algebra H over k can be embedded into a finite-dimensional quasitriangular Hopf algebra over k. Let $\mathcal{C} \in H \otimes H^*$ correspond to the identity map under the identification of $H \otimes H^*$ and $\mathrm{End}_k(H)$ determined by $(h \otimes h^*)(a) = h^*(a)h$ for $h, a \in H$ and $h^* \in H^*$. Notice that $\mathcal{C} = \sum_{i=1}^n h_i \otimes \overline{h_i}$, where $\{h_1, \ldots, h_n\}$ is any linear basis for H and $\{\overline{h_1}, \ldots, \overline{h_n}\}$ is the corresponding dual basis for H^*. The underlying Hopf algebra $D(H)$ of the Drinfel'd double of H admits a quasitriangular structure $(D(H), \mathcal{R})$, where

$$\mathcal{R} = \epsilon \otimes \mathcal{C} \otimes 1.$$

Observe that $(D(H), \mathcal{R})$ is a minimal quasitriangular Hopf algebra, and that $\mathrm{Im} f_{\widetilde{\mathcal{R}}} = \imath_H(H)$. The triple $(D(H), \mathcal{R}, \imath_H)$ satisfies a universal mapping property which we now describe. We call a triple $(A, R, \imath)$ a *quasitriangular envelope for* H if (A, R) is a quasitriangular Hopf algebra over k and $\imath : H \longrightarrow A$ is a one-one Hopf algebra map such that $\mathrm{Im} f_{\widetilde{R}} = \imath(H)$. Every finite-dimensional quasitriangular Hopf algebra (A, R) over k is the quasitriangular envelope for some Hopf algebra over k as $\mathrm{Im} f_{\widetilde{R}} = H$ is a sub-Hopf algebra of A [16, Proposition 2].

Theorem 1 *Suppose that H is a finite-dimensional Hopf algebra over the field k. Then:*

a) *$(D(H), \mathcal{R}, \imath_H)$ is a quasitriangular envelope for H.*

b) *If $(A, R, \imath)$ is a quasitriangular envelope for H, there exists a unique morphism of quasitriangular Hopf algebras $F : (D(H), \mathcal{R}) \longrightarrow (A, R)$ such that $F \circ \imath_H = \imath$. Furthermore F is onto if (A, R) is minimal.*

PROOF: We need only establish part b). First we show uniqueness. Suppose that $F : (D(H), \mathcal{R}) \longrightarrow (A, R)$ is a morphism of quasitriangular Hopf algrebras such that $F \circ \imath_H = \imath$, and write $\mathcal{C} = \sum_{i=1}^{n} h_i \otimes \overline{h_i}$ as above. Since $R = (F \otimes F)(\mathcal{R})$ and $F \circ \imath_H = \imath$, we compute that

$$\begin{aligned} R &= \sum_{i=1}^{n} F(\epsilon \otimes h_i) \otimes F(\overline{h_i} \otimes 1) \\ &= \sum_{i=1}^{n} F \circ \imath_H(h_i) \otimes F \circ \imath_{H^*}(\overline{h_i}) \\ &= \sum_{i=1}^{n} \imath(h_i) \otimes F \circ \imath_{H^*}(\overline{h_i}). \end{aligned}$$

Therefore $f_R = F \circ \imath_{H^*} \circ \imath^*$. Since $\imath : H \longrightarrow A$ is one-one, the map $\imath^* : A^* \longrightarrow H^*$ is onto. Since F is an algebra map and $D(H) = \imath_{H^*}(H^*)\imath_H(H)$, it follows that F must be unique.

To show existence, we first use part a) of [16, Theorem 2] to conclude that there is a morphism of quasitriangular Hopf algebras $F' : (D(\imath(H)), \mathcal{R}) \longrightarrow (A, R)$ such that $F' \circ \imath_{\imath(H)} = I|_{\imath(H)}$. The proof of [16, Proposition 5] yields an isomorphism of quasitriangular Hopf algebras $F'' : (D(H), \mathcal{R}) \longrightarrow (D(\imath(H)), \mathcal{R})$ such that $F'' \circ \imath_H = \imath_{\imath(H)} \circ \imath$. $F = F' \circ F''$ is the desired morphism. The fact that F is onto when (A, R) is minimal quasitriangular follows by part c) of [16, Theorem 2]. This concludes the proof.

The quasitriangular structures admitted by the group algebra $H = k[G]$ of a finite abelian group G over k are interesting in their own right. Let e be the exponent of G and suppose that k contains a primitive e^{th} root of unity ω. In this case G and the character group of G are isomorphic; thus there is a Hopf algebra isomorphism $H^* \simeq H$. Since H is commutative and cocommutative, the quasitriangular structures on H can be identified with

$$\mathrm{Hom}_{Hopf}(H^*, H) = \mathrm{Hom}_{Hopf}(H, H) = \mathrm{Hom}_{Group}(G, G).$$

We examine two cases in detail.

Let n be a fixed positive integer and $G = (a)$ be the cyclic group of order n. We identify H^* and H by the Hopf algebra isomorphism determined by $A \mapsto a$, where $A : H \longrightarrow k$ is the algebra map which satisfies $A(a) = \omega$. It is a straightforward exercise to see that $H = k[G]$ admits n quasitriangular structures which are given by

$$R_m = \sum_{0 \leq i,j < n} \frac{\omega^{-ij}}{n} a^i \otimes a^{mj}$$

for $0 \leq m < n$. The quasitriangular Hopf algebra (H, R_1) is considered in [17]. Observe that $u_m = \sum s(R_m^{(2)})R_m^{(1)} = \Theta(\omega^{-m})$.

Now let $\mathcal{G} = G \times G$ be the direct product of cyclic groups of order n, and make the identifications $b = (1, a)$ and $a = (a, 1)$. Then $\mathcal{H} = k[\mathcal{G}]$ admits n^4 quasitriangular structures given by

$$R_{pqrs} = \sum_{0 \leq i,\, j,\, k,\, \ell < n} \frac{\omega^{-(ij+k\ell)}}{n^2} a^i b^k \otimes a^{pj+r\ell} b^{qj+s\ell}$$

for $0 \leq p, q, r, s < n$. It is easy to see that $D(H) \simeq \mathcal{H}$ $(A^i \otimes a^j \mapsto b^i a^j)$ is an isomorphism of Hopf algebras. Under this identification

$$\mathcal{R} = R_{0010} = \sum_{0 \leq i,j < n} \frac{\omega^{-ij}}{n} b^i \otimes a^j.$$

Sweedler's 4-dimensional Hopf algebra $T_{2,\omega}$ over k, when the characteristic of k is not 2, admits quasitriangular structures, the set of which is parameterized by k. As an algebra $T_{2,\omega}$ is generated over k by symbols a and x which satisfy the relations

$$a^2 = 1, \quad x^2 = 0 \quad \text{and} \quad xa = -ax.$$

The coalgebra structure of $T_{2,\omega}$ is determined by

$$\Delta(a) = a \otimes a \quad \text{and} \quad \Delta(x) = x \otimes a + 1 \otimes x.$$

Consequently the antipode s of $T_{2,\omega}$ satisfies

$$s(a) = a, \quad s(x) = ax \quad \text{and} \quad s(ax) = -x.$$

For each $\alpha \in k$

$$R_\alpha = \frac{1}{2}(1 \otimes 1 + 1 \otimes a + a \otimes 1 - a \otimes a) + \frac{\alpha}{2}(x \otimes x + x \otimes ax + ax \otimes ax - ax \otimes x)$$

affords $T_{2,\omega}$ the structure of a quasitriangular Hopf algebra $(T_{2,\omega}, R_\alpha)$. If $(T_{2,\omega}, R)$ is quasitriangular, then $R = R_\alpha$ for some $\alpha \in k$. For details concerning these assertions the reader is referred to [16, Section 2].

Let $\alpha \in k$ be fixed. Then $u = \sum s(R_\alpha^{(2)})R_\alpha^{(1)} = a$. Therefore $(T_{2,\omega}, R_\alpha)$ is triangular by (14). Notice that $(T_{2,\omega}, R_\alpha)$ is a minimal quasitriangular Hopf algebra except when $\alpha = 0$. We also determine the quasitriangular structures on $T_{2,\omega}$ at the end of Section 5.1.

2.2 Ribbon Hopf Algebras

Ribbon Hopf algebras are quasitriangular Hopf algebras with a designated element having very special properties in connection with the topology of knots, links and 3-manifolds. A finite-dimensional ribbon Hopf algebra over k is a triple (H, R, v), where (H, R) is a finite-dimensional quasitriangular Hopf algebra over k and $v \in H$ satisfies the following:

(R.0) v is in the center of H,

(R.1) $v^2 = us(u)$,

(R.2) $s(v) = v$,

(R.3) $\epsilon(v) = 1$ and

(R.4) $\Delta(v) = (v \otimes v)(\tilde{R}R)^{-1} = (\tilde{R}R)^{-1}(v \otimes v)$.

Ribbon Hopf algebras were introduced and studied by Reshetikhin and Turaev in [26]. The element v is refered to as a special element or ribbon element in the literature. We shall refer to v as a ribbon element for (H, R). If v satisfies (R.1)–(R.4) then we say that v is a quasi-ribbon element for (H, R) [8]. By (R.1) and the fact that u is invertable it follows that v is invertable.

Suppose that (H, R, v) and (H', R', v') are ribbon Hopf algebras over k. We define a morphism $f : (H, R, v) \longrightarrow (H', R', v')$ of ribbon Hopf algebras to be a morphism $f : (H, R) \longrightarrow (H', R')$ of quasitriangular Hopf algebras such that $f(v) = v'$. The class of ribbon Hopf algebras over k with morphisms under composition form a category. Observe that $(H \otimes H', R'', v \otimes v')$ is a ribbon Hopf algebra, where $(H \otimes H', R'')$ is the tensor product of the quasitriangular Hopf algebras (H, R) and (H', R').

Suppose that (H, R) is quasitriangular. Then (H, R, u) is a ribbon Hopf algebra if H is commutative. In particular the example $(k[G], R_m)$ of Section 2.1 has the structure of a ribbon Hopf algebra $(k[G], R_m, \Theta(\omega^{-m}))$. If (H, R, v) is a ribbon Hopf algebra, observe that v is a grouplike element of H if and only if (H, R) is triangular.

The notion of ribbon Hopf algebra can be formulated very simply in terms of grouplike elements in several different ways, and in the unimodular case in terms of cocommutative generalized right integrals [5, 6]. Suppose that (H, R, v) is a ribbon Hopf algebra. Write $v = au$, where $a = vu^{-1} = u^{-1}v$. Then by (R.4) and (14) it follows that a is a grouplike element of H. We will determine the grouplike elements $a \in G(H)$ such that au is a ribbon element for (H, R). For the connection between ribbon elements and generalized right integrals we will need:

Lemma 3 *Suppose that H is a finite-dimensional Hopf algebra with antipode s over the field k. Let λ be a non-zero right integral for H^*. Suppose that $a \in H$ is invertable and let $\chi = \lambda \cdot a$. Then the following are equivalent:*

a) *χ is cocommutative.*

b) *H is unimodular and $s^2(h) = aha^{-1}$ for all $h \in H$.*

PROOF: Let α be the distinguished grouplike element of H. Then χ is cocommutative if and only if $s^2(h \leftharpoonup \alpha)a = ah$ for all $h \in H$ by [22, Proposition 6]. Since a is invertable, this condition is equivalent to $s^2(h \leftharpoonup \alpha) = aha^{-1}$ for all $h \in H$. Applying ϵ to both sides of this equation yields $\alpha = \epsilon$, which is to say that H is unimodular.

As a consequence of the lemma, the existence of a non-zero cocommutative generalized right integral for H^* implies that H is unimodular.

Suppose that (H, R) is a finite-dimensional quasitriangular Hopf algebra over k, and let g and α be the distinguished grouplike elements of H and H^* respectively. Observe that $g_\alpha = \sum R^{(1)}\alpha(R^{(2)}) \in G(H)$ by virtue of (QT.1) and (QT.2). Let

$$\hbar = g^{-1}g_\alpha. \tag{16}$$

We have the important relationship $us(u) = u^2\hbar$, or $u^{-1}s(u) = \hbar$, which is discussed in [2, 8]. Observe that $\hbar = g^{-1}$ when H is unimodular.

For an algebra A over k we let $Z(A)$ denote the center of A.

Proposition 2 *Suppose that (H, R) is a finite-dimensional quasitriangular Hopf algebra with antipode s over the field k. Let $u = \sum s(R^{(2)})R^{(1)}$ and $\hbar$ be as in (16). Then:*

a) *Every ribbon element v for (H, R) has the form $v = au$ for some $a \in G(H)$.*

b) *Suppose $a \in G(H)$. Then the following are equivalent:*

   i) *au is a ribbon element for (H, R).*
   ii) *$a^2 = \hbar$ and $s^2(h) = a^{-1}ha$ for all $h \in H$.*
   iii) *$G = a^{-1}$ satisfies $s(u) = G^{-2}u$ and $G^{-1}u \in Z(H)$.*

Suppose further that H is unimodular and λ is a non-zero right integral for H^*. Then i)–iii) *are equivalent to*

   iv) *$\chi = \lambda \cdot a^{-1}$ is cocommutative and $\chi \circ s = \chi$.*

PROOF: The equivalence of i) and ii) is part b) of [8, Theorem 1]. The equivalence of i) and iii) is noted in [6]. Suppose that H is unimodular, and let $\chi = \lambda \cdot a^{-1}$. Then χ is cocommutative if and only if $s^2(h) = a^{-1}ha$ for all $h \in H$ by Lemma 3. Let g be the distinguished grouplike element of H. Then $\lambda \circ s = \lambda \cdot g$ by (1). Let $h \in H$. Since $\chi \circ s(h) = \lambda(a^{-1}s(h)) = \lambda \circ s(ha)$, we have that $\chi \circ s(h) = \lambda(gha) = \lambda(agh)$ by (1) again. Thus $\chi \circ s = \chi$ if and only if $\lambda \cdot (ag) = \lambda \cdot a^{-1}$, or equivalently $ag = a^{-1}$. This last equation holds if and only if $a^2 = g^{-1} = \hbar$. Thus ii) and iv) are equivalent in the unimodular case, and the proof is complete.

Let $a \in G(H)$. Apropos of the proposition, we note that $v = au$ is a quasi-ribbon element for (H, R) if and only if $a^2 = \hbar$, or equivalently $s(u) = G^{-2}u$, where $G = a^{-1}$. The first assertion is part a) of [8, Theorem 1]. Note that iv) makes no mention of the quasitriangular structure on H.

From an algebraic point of view, a finite-dimensional ribbon Hopf algebra is most easily described as a triple (H, R, G), where (H, R) is a finite-dimensional quasitriangular Hopf algebra and $G \in G(H)$ satisfies condition iii) of the preceding proposition.

Suppose that the characteristic of k is not 2. Since 1 and a are the grouplike elements of Sweedler's 4-dimensional example $T_{2,\omega}$, it follows by the previous proposition that $(T_{2,\omega}, R_\alpha)$ has a unique ribbon element 1 for all $\alpha \in k$.

When the Drinfel'd double of a finite-dimensional Hopf algebra H over k has a ribbon element is related in a very interesting way to the formula for the fourth power of the antipode s of H

$$s^4(h) = g(\alpha \rightharpoonup h \leftharpoonup \alpha^{-1})g^{-1} \quad \text{for all} \quad h \in H$$

which can be found in [8]. This equation is basically that of [20, Proposition 6]. The Hopf algebra $D(H)$ is unimodular by part a) of [16, Theorem 4], and the distinguished grouplike element of $D(H)$ is $\alpha \otimes g$ by part a) of [16, Corollary 7]. Proposition 2 and basic algebraic properties of the Drinfel'd double are all that is needed for a proof of the following:

Theorem 2 [8, Theorem 3b)] *Suppose that H is a finite-dimensional Hopf algebra with antipode s over the field k. Let g and α be the distinguished grouplike elements of H and H^* respectively. Then $(D(H), \mathcal{R})$ has a ribbon element if and only if there are $\ell \in G(H)$ and $\beta \in G(H^*)$ such that $\ell^2 = g, \beta^2 = \alpha$ and*

$$s^2(h) = \ell(\beta \rightharpoonup h \leftharpoonup \beta^{-1})\ell^{-1}$$

for all $h \in H$.

The existence of grouplike elements which are square roots of g and α is a necessary and sufficient condition for $(D(H), \mathcal{R})$ to have a quasi-ribbon element by part a) of [8, Theorem 3].

Since Sweedler's 4-dimensional example $T_{2,\omega}$ is not unimodular (as we will show in Section 5.1), it follows that the distinguished grouplike element of $T_{2,\omega}$ is $g = a$. Therefore $(D(T_{2,\omega}), \mathcal{R})$ does not have a quasi-ribbon element, hence does not have a ribbon element, by Theorem 2. In particular quasitriangular Hopf algebras need not possess ribbon elements. In [26] Reshetikhin and Turaev show how to embed any finite-dimensional quasitriangular Hopf algebra into a finite-dimensional ribbon Hopf algebra.

2.3 Factorizable Hopf algebras

Factorizable quasitriangular Hopf algebras are introduced and studied in [24]. They are important in Hennings' investigation of 3-manifold invariants [5].

Suppose that (H, R) is a finite-dimensional quasitriangular Hopf algebra over the field k. Then (H, R) is said to be factorizable if $f_{\widetilde{R}R} : H^* \longrightarrow H$ is a linear isomorphism. Thus finite-dimensional triangular Hopf algebras are never factorizable unless they are one-dimensional. In particular the ribbon Hopf algebras $(T_{2,\omega}, R_\alpha, 1)$ are not factorizable. Observe that the tensor product of two factorizable quasitriangular Hopf algebras over k is factorizable with the tensor product quasitriangular structure.

Proposition 3 *Suppose that (H, R) is a finite-dimensional quasitriangular Hopf algebra over the field k. Then:*

a) $f_{\widetilde{R}R} = f_{\widetilde{R}} * f_R$ *in the convolution algebra* $\mathrm{Hom}_k(H^*, H)$.

b) $\mathrm{Im}\, f_{\widetilde{R}R} \subseteq H_R$. *Thus factorizable implies minimal quasitriangular.*

c) *Let α be the distinguished grouplike element of H^*. Then $f_{\widetilde{R}R}(\alpha) = 1 = f_{\widetilde{R}R}(\epsilon)$. Thus factorizable implies unimodular.*

d) *Suppose that $x \in H$ satisfies $\Delta(x) = (x \otimes x)(\widetilde{R}R)$. Then*

$$x \leftharpoonup (p \cdot x) = x(f_{\widetilde{R}R}(x \cdot p))$$

for all $p \in H^*$.

e) *(H, R) is factorizable if and only* $\mathrm{Rank}\, \widetilde{R}R = \mathrm{Dim}\, H$.

PROOF: Part a) follows by the calculation $(R = r)$

$$\begin{aligned} f_{\widetilde{R}R}(p) &= \sum p(R^{(2)}r^{(1)})R^{(1)}r^{(2)} \\ &= \sum p_{(1)}(R^{(2)})p_{(2)}(r^{(1)})R^{(1)}r^{(2)} \\ &= f_{\widetilde{R}} * f_R(p) \end{aligned}$$

for $p \in H^*$. By part a) we have that $\mathrm{Im} f_{\widetilde{R}R} \subseteq (\mathrm{Im}\, f_{\widetilde{R}})(\mathrm{Im}\, f_R) \subseteq H_R$, and thus part b) follows from part a).

The proof of part c) involves comparison of $(H^{cop}, \widetilde{R})$ and (H, R). By part a) we have $f_{\widetilde{R}R}(\eta) = f_{\widetilde{R}}(\eta) f_R(\eta)$ for all $\eta \in G(H^*)$. Therefore $f_{\widetilde{R}R}(\epsilon) = 1$ by (QT.2) and (QT.4), and $f_{\widetilde{R}R}(\alpha) = g_\alpha \widetilde{g_\alpha}$, where $\widetilde{g_\alpha} = \sum \alpha(R^{(1)})R^{(2)}$. We will show that $\widetilde{g_\alpha} = g_\alpha^{-1}$.

Recall that $(H^{cop}, \widetilde{R})$ is a quasitriangular Hopf algebra with antipode s^{-1}. Let g be the distinguished grouplike element of H. Then g^{-1} and α are the distinguished grouplike elements of H^{cop} and H^{cop*} respectively. Let $\hbar$ be defined as in (16) for H and $\widetilde{\hbar}$ be its counterpart for H^{cop}. Then $g^{-1}g_\alpha = \hbar$ and $g\widetilde{g_\alpha} = \widetilde{\hbar}$. Therefore to show that $\widetilde{g_\alpha} = g_\alpha^{-1}$ it is sufficient to show that $\widetilde{\hbar} = \hbar^{-1}$.

Let $u = \sum s(R^{(2)})R^{(1)}$ and $\widetilde{u} = \sum s^{-1}(\widetilde{R}^{(2)})\widetilde{R}^{(1)} = \sum s^{-1}(R^{(1)})R^{(2)}$. Then $u^{-1}s(u) = \hbar$ and $\widetilde{u}^{-1}s^{-1}(\widetilde{u}) = \widetilde{\hbar}$. Note that $\widetilde{u} = s^{-1}(u)$. Now $s^2(u) = u$ and $s(u)$ and u^{-1} commute by (15). Thus we calculate

$$\begin{aligned} \widetilde{\hbar} &= (s^{-1}(u))^{-1}s^{-2}(u) \\ &= s^{-1}(u^{-1})u \\ &= s^{-1}(s(u)u^{-1}) \\ &= s^{-1}(\hbar). \end{aligned}$$

But $\hbar \in G(H)$. Therefore $\widetilde{\hbar} = \hbar^{-1}$, and the proof of part c) is complete.

We now show part d). If $x \in H$ and $\Delta(x) = (x \otimes x)(\widetilde{R}R)$ then

$$\begin{aligned} x \leftharpoonup p &= (p \otimes I)(\Delta(x)) \\ &= x((p \cdot x \otimes I)(\widetilde{R}R)) \\ &= x(f_{\widetilde{R}R}(p \cdot x)) \end{aligned}$$

for all $p \in H^*$, and part d) is established. Part e) is clear.

As a consequence of part d) of the previous proposition:

Corollary 1 *Suppose that (H, R) is a finite-dimensional quasitriangular Hopf algebra over the field k. Suppose further that $x \in H$ is invertable and satisfies $\Delta(x) = (x \otimes x)(\widetilde{R}R)$. Then the following are equivalent:*

a) *(H, R) is factorizable.*

b) Rank $\Delta(x) = \operatorname{Dim} H$.

c) *$x \leftharpoonup H^* = H$, or equivalently H is a free right H^*-module with generator x.*

Hennings' and Kauffman's description of 3-manifold invariants [5] start with a finite-dimensional ribbon Hopf algebra (H, R, v) such that $\lambda(v) \neq 0 \neq$

$\lambda(v^{-1})$, where λ is a non-zero right integral for H^*. Hennings noted that $\lambda(v^{-1}) \neq 0$ (and further that $\lambda(v) \neq 0$) when (H, R) is factorizable. The same calculation yields more generally:

Corollary 2 *Suppose that (H, R) is a finite-dimensional factorizable quasi-triangular Hopf algebra over the field k. Assume that $x \in H$ is invertable and satisfies $\Delta(x) = (x \otimes x)(\tilde{R}R)$. If μ is a non-zero generalized right integral for H^*, then $\mu(x) \neq 0$.*

PROOF: Since μ generates a one-dimensional right ideal of H^*, there is an $a \in H$ such that $\mu p = p(a)\mu$ for all $p \in H^*$. From our discussion of (2) it follows that this condition is equivalent to

$$h \leftharpoonup \mu = \sum \mu(h_{(1)})h_{(2)} = \mu(h)a$$

for all $h \in H$. Since part c) of Corollary 1 holds, it follows that $0 \neq x \leftharpoonup \mu = \mu(x)a$. Therefore $\mu(x) \neq 0$.

Let H be a finite-dimensional Hopf algebra with antipode s over the field k. Hennings notes [5] that the Drinfel'd double $(D(H), \mathcal{R})$ is shown to be factorizable in [24]. Factorizability of the double in the finite-dimensional case can be seen by Corollary 1. For let $\{h_1, \ldots, h_n\}$ be a linear basis for H and $\{\overline{h_1}, \ldots, \overline{h_n}\}$ the corresponding dual basis for H^*. Then $\mathcal{R} = \sum_{i=1}^{n}(\epsilon \otimes h_i) \otimes (\overline{h_i} \otimes 1)$, from which the calculation $\mathsf{u} = \sum_{i=1}^{n} \mathsf{s}(\overline{h_i} \otimes 1)(\epsilon \otimes h_i) = \sum_{i=1}^{n}(S^{-1}(\overline{h_i}) \otimes 1)(\epsilon \otimes h_i)$ gives

$$\mathsf{u} = \sum_{i=1}^{n}(\overline{h_i} \circ s^{-1}) \otimes h_i. \tag{17}$$

Likewise $\mathsf{u}^{-1} = \sum_{i=1}^{n} \overline{h_i} \otimes s^2(h_i)$. We regard $D(H) = H^* \otimes H = \mathrm{End}_k(H)$ as vector spaces. Identifying $D(H)^*$ and $H^{op} \otimes H^*$ as algebras, the vector spaces $H \otimes H^*$ and $\mathrm{End}_k(H)^*$ by $(a \otimes q)(f) = q(f(a))$ for $a \in H$, $q \in H^*$ and $f \in \mathrm{End}_k(H)$, we have for $a, h \in H$ and $q, p \in H^*$ that

$$\begin{aligned}(a \otimes q)(\mathsf{u}^{-1} \leftharpoonup (h \otimes p)) &= ((h \otimes p) * (a \otimes q))(\mathsf{u}^{-1}) \\ &= (ah \otimes pq)(\mathsf{u}^{-1}) \\ &= pq(s^2(ah)) \\ &= (a \otimes q)(r(p) \circ r(s^2(h)) \circ s^2),\end{aligned}$$

and therefore $u^{-1} \leftharpoonup (h \otimes p) = r(p) \circ r(s^2(h)) \circ s^2$. The details are straightforward and left to the reader. By [22, Proposition 12] the map $\mathrm{End}_k(H) \longrightarrow \mathrm{End}_k(H)$ defined by $h \otimes p \mapsto r(p) \circ r(h)$ is bijective. Therefore $u^{-1} \leftharpoonup \mathcal{H}^* = \mathcal{H}$, where $\mathcal{H} = D(H)$, and condition c) of Corollary 1 is established.

Suppose that the characteristic of k is not 2 and let $T_{2,\omega}$ be Sweedler's 4-dimensional example. Since $(D(T_{2,\omega}), \mathcal{R})$ is not a ribbon Hopf algebra, factorizable Hopf algebras do not necessarily possess a ribbon element.

Suppose that $n > 1$ and k has a primitive n^{th} root of unity ω. Let $G = (a)$ be the cyclic group of order n. We will show that the example $(k[G], R_m)$ of Section 2.1 is factorizable if and only if n is odd and $(m, n) = 1$. Recall that $u = \sum s(R_m^{(2)}) R_m^{(1)} = \Theta(\omega^{-m})$. Therefore $u^{-1} = \Theta(\omega^m)$. Since $\Delta(e_\ell) = \sum_{i+j=\ell} e_i \otimes e_j$, we have that $\Delta(\Theta(\omega^m)) = \sum_{i=0}^{n-1} f_i \otimes e_i$, where $f_i = \sum_{j=0}^{n-1} \omega^{m(i+j)^2} e_j$. By Corollary 1 it follows that $(k[G], R_m)$ is factorizable if and only if the set $\{f_0, \ldots, f_{n-1}\}$ is linearly independent. This is the case if and only if the $n \times n$ matrix $\mathcal{A} = (a_{ij})_{0 \leq i,j < n}$ is invertable, where $a_{ij} = \omega^{m(i+j)^2}$.

Consider the $n \times n$ matrix $\mathcal{B} = (b_{ij})_{0 \leq i,j < n}$, where $b_{ij} = \omega^{2mij}$, and let σ be any permutation of $\{0, \ldots, n-1\}$. Since $a_{0\,\sigma 0} \cdots a_{n-1\,\sigma(n-1)} = \omega^{2m(0^2+\cdots+(n-1)^2)} b_{0\,\sigma 0} \cdots b_{n-1\,\sigma(n-1)}$, it follows that $\mathrm{Det}\,\mathcal{A} \neq 0$ if and only if $\mathrm{Det}\,\mathcal{B} \neq 0$. But $\mathrm{Det}\,\mathcal{B} \neq 0$ if and only if ω^{2m} is a primitive n^{th} root of unity, which is the case if and only if n is odd and $(m, n) = 1$.

3 Construction of the Knot Invariants

We describe the construction of Kauffman's knot invariants in this section. The reader is referred to [6] for a detailed discussion of these and similar invariants of links and a reformulation of Hennings' 3-manifold invariant. The reader is encouraged to study Hennings' work [5] as well.

We start with a quadruple (H, R, G, χ), where (H, R) is a finite-dimensional quasitriangular Hopf algebra with antipode s over the field k, $G \in G(H)$ which satisfies $s(u) = G^{-2}u$ and $G^{-1}u \in Z(H)$, where $u = \sum s(R^{(2)})R^{(1)}$, and $\chi \in \mathrm{Cocom}_s(H^*)$. Notice that (H, R, v) is a ribbon Hopf algebra, where $v = G^{-1}u$, by Proposition 2. For a knot K an element $w \in H$ is computed by permuting tensorands in an expression of the form

$$s^{i_1}(R_1^{(1)}) \otimes s^{j_1}(R_1^{(2)}) \otimes \cdots \otimes s^{i_n}(R_n^{(1)}) \otimes s^{j_n}(R_n^{(2)}),$$

where $R_1 = R_2 = \ldots = R_n = R$, according to the knot presentation and then multiplying. Kauffman defines a function $\mathrm{Tr}_\chi : \mathcal{K} \longrightarrow k$ from the set of all knots $\mathcal{K}$ to k by

$$\mathrm{Tr}_\chi(K) = \chi(G^d w),$$

where d is the Whitney degree of the knot (orientation to be explained). The function Tr_χ is a *regular isotopy* invariant of knots. The reader is directed to Kauffman's book on knots for the definition and discussion of regular isotopy [7].

Suppose that (H', R', G', χ') also satisfies the conditions stated above. Then $(H \otimes H', R'', G \otimes G', \chi \otimes \chi')$ does as well, where $(H \otimes H', R'')$ is the tensor product of the quasitriangular Hopf algebras (H, R) and (H', R'). Based on the description of Tr_χ we can conclude that

$$\mathrm{Tr}_{\chi \otimes \chi'}(K) = \mathrm{Tr}_\chi(K)\mathrm{Tr}_{\chi'}(K) \tag{18}$$

for all knots K.

In application it is very natural to assume that H is unimodular. We will first describe the subspaces $\mathrm{Cocom}(H^*)$ and $\mathrm{Cocom}_s(H^*)$ of H^* in terms of the center of H and one of its subalgebras. Then we will describe how to compute the expressions $G^d w$ for a knot K and give some elementary examples.

3.1 The Structure of $\mathrm{Cocom}_s(H^*)$

Let H be a finite-dimensional Hopf algebra over k and suppose that λ is a non-zero right integral for H^*. Then any $p \in H^*$ can be written $p = \lambda \cdot a$ for a unique $a \in H$. This representation of functionals provides a natural vehicle for connecting $\mathrm{Cocom}_s(H^*)$ with $Z(H)$. In the unimodular case:

Proposition 4 *Suppose that (H, R) is a finite-dimensional quasitriangular Hopf algebra with antipode s over the field k, and let $u = \sum s(R^{(2)})R^{(1)}$. Suppose further that H is unimodular, λ is a non-zero right integral for H^* and g is the distinguished grouplike element of H. Let $\chi \in H^*$ and write $\chi = \lambda \cdot a$ for some $a \in H$. Then:*

a) $\chi \in \mathrm{Cocom}(H^*)$ *if and only if* $a = zu$ *for some* $z \in Z(H)$.

b) $\chi \circ s = \chi$ *if and only if* $s(a) = ag^{-1}$.

c) $\chi \in \mathrm{Cocom}_s(H^*)$ *if and only if* $a = zu$ *for some* $z \in Z(H)$ *which satisfies* $s(z) = z$.

PROOF: We first show part a). By [22, Proposition 6] it follows that $\chi \in \mathrm{Cocom}\,(H^*)$ if and only if $s^2(h)a = ah$ for all $h \in H$. Since $s^2(h) = uhu^{-1}$ for all $h \in H$, part a) now follows. To show part b) we use the equations $\lambda(ab) = \lambda(s^2(b)a)$ and $\lambda \circ s = \lambda \cdot g$, which come from (1). These imply that

$$\begin{aligned}\chi \circ s(h) &= \lambda(as(h)) \\ &= \lambda \circ s(hs^{-1}(a)) \\ &= \lambda(ghs^{-1}(a)) \\ &= \lambda(s(a)gh)\end{aligned}$$

for all $h \in H$. Therefore $\chi \circ s = \chi$ if and only if $\lambda \cdot (s(a)g) = \lambda \cdot a$, or $s(a)g = a$, and part b) is established.

To show part c) we first note that $\hbar = g^{-1}$ since H is unimodular. On the other hand $u^{-1}s(u) = \hbar$. Therefore $s(u) = ug^{-1}$. By parts a) and b) we have that $\chi \in \mathrm{Cocom}_s(H^*)$ if and only if $a = zu$ for some $z \in Z(H)$ and $s(a) = ag^{-1}$. Suppose that $a = zu$. Then $s(a) = ag^{-1}$ if and only if $s(u)s(z) = zug^{-1} = zs(u)$. Thus part c) follows.

Let (H, R) be a finite-dimensional quasitriangular Hopf algebra over k and suppose that H is unimodular. Let $Z_s(H)$ be the set of all $z \in Z(H)$ such that $s(z) = z$. Then $Z_s(H)$ is a subalgebra of $Z(H)$. By virtue of the lemma the subspaces $\mathrm{Cocom}\,(H^*)$ and $\mathrm{Cocom}_s(H^*)$ of H^* correspond to the subalgebras $Z(H)$ and $Z_s(H)$ of H. In particular

$$\mathrm{Dim}\,\mathrm{Cocom}\,(H^*) = \mathrm{Dim}\,Z(H)$$

and

$$\mathrm{Dim}\,\mathrm{Cocom}_s(H^*) = \mathrm{Dim}\,Z_s(H).$$

Also see Drinfel'd's paper [2].

The restriction $\chi \circ s = \chi$ poses no real problem when the characteristic of k is not 2. For suppose that the characteristic of k is not 2 and let $\chi \in \mathrm{Cocom}\,(H^*)$. Since s^2 is inner and χ is cocommutative it follows that $\chi \circ s^2 = \chi$. Thus $\chi_s = \frac{1}{2}(\chi + \chi \circ s) \in \mathrm{Cocom}_s(H^*)$. Observe that the map $\mathrm{Cocom}\,(H^*) \longrightarrow \mathrm{Cocom}_s(H^*)$ given by $\chi \mapsto \chi_s$ is a linear projection.

Now suppose that (H, R, v) is a ribbon Hopf algebra and λ is a non-zero right integral for H^*. Set $G = v^{-1}u$. Then the generalized right integral $\mu_G = \lambda \cdot G$ is cocommutative and satisfies $\mu_G \circ s = \mu_G$ by Proposition 2. The quadruple (H, R, G, μ_G) satisfies the conditions above, and therefore Tr_{μ_G} is a regular isotopy invariant of knots.

Hennings and Kauffman use μ_G (Hennings implicitly) to describe a 3-manifold invariant when $\lambda(v) \neq 0 \neq \lambda(v^{-1})$. We note that the equations

$$u(u^{-1} \leftharpoonup \mu_G) = \lambda(v^{-1})v \qquad \text{and} \qquad s(u^{-1})(s(u) \leftharpoonup \mu_G) = \lambda(v)v^{-1}$$

needed for invariance under the Kirby moves are essentially very simple consequences of the definition of λ. Recall that λ satisfies (2) with $a = 1$. Thus for $a \in G(H)$ and $h \in H$ we have

$$a(h \leftharpoonup (\lambda \cdot a)) = \sum \lambda(ah_{(1)})ah_{(2)} = \sum \lambda((ah)_{(1)})(ah)_{(2)} = \lambda(ah)1,$$

and consequently

$$h \leftharpoonup \lambda \cdot a = \lambda(ah)a^{-1}.$$

Therefore $u^{-1} \leftharpoonup \mu_G = \lambda(v^{-1})G^{-1}$ and $s(u) \leftharpoonup \mu_G = \lambda(v)G^{-1}$ since $u^{-1} = G^{-1}v^{-1}$ and $s(u) = G^{-1}v$.

Suppose that H is not semisimple. Then $\mu_G \notin \mathrm{Char}(H)$. To see this, let Λ be a non-zero left integral for H. Then $\mu_G(\Lambda) = \lambda(\Lambda) \neq 0$ by part e) of [22, Proposition 1]. But if χ is the character of a finite-dimensional left H-module, $\chi(\Lambda) = 0$ since $\Lambda^2 = \epsilon(\Lambda)\Lambda = 0$. We remark that μ_G produces a 3-manifold invariant essentially different from the Reshetikhin-Turaev invariant [9]. This is shown independently by Ohtsuki in [13]. Theirs is a (normalized) linear combination of irreducible characters.

3.2 Kauffman's Knot Invariants which Arise from a Finite-Dimensional Ribbon Hopf Algebra

Knots can be represented topologically by figures such as

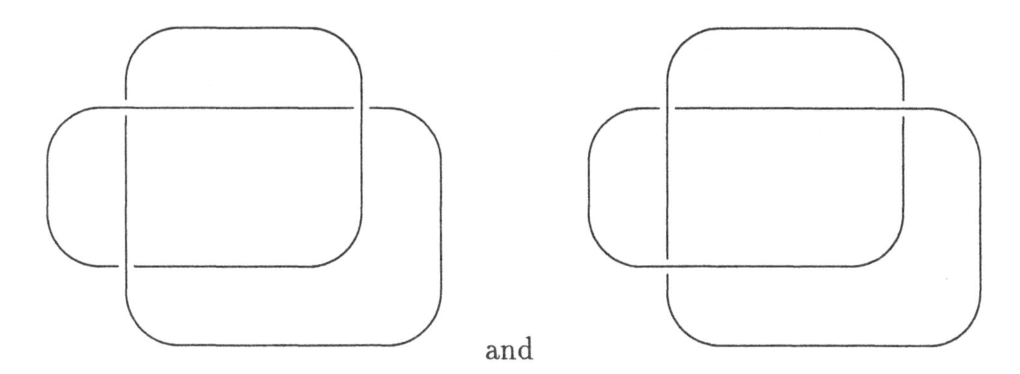

These particular ones are derived from

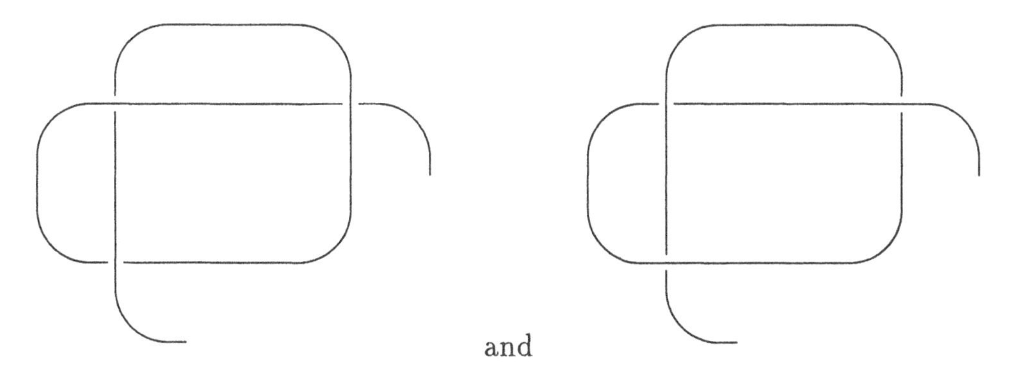

respectively by joining endpoints. The first two figures represent the trefoil knot and its mirror image (figure with crossings reversed). Manipulating figures representing these knots to give the more formal topological equivalents

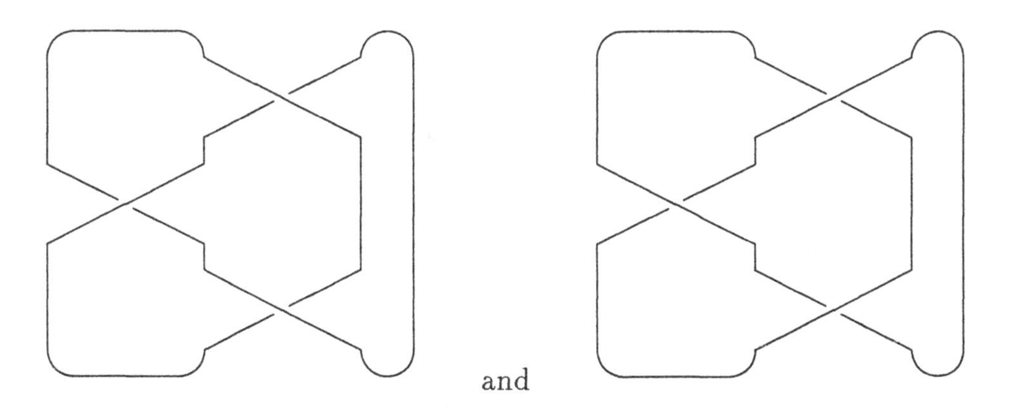

suggests how to describe knot diagrams in terms of "vertical" line segments, arcs for designating local maxima and minima, and crossings.

We will first give an informal explanation of how to compute the knot invariants of [6] and then we will consider this computation from a more formal algebraic point of view.

Let (H, R, G, χ) be a quadruple as described in Section 3.1. Think of a knot diagram K (or knot) as a closed loop of wire around which beads can be moved. We will require the diagram to have a vertical line segment which is not part of a crossing. Number the crossings and place a pair of labeled beads at each crossing according to the crossing type as indicated below:

The formal expression $e_i \otimes f_i$ represents $R_i = R$. Now choose a point on a vertical line segment in the diagram which is not part of a crossing. Moving *upward* from that point, slide the beads around the figure until they are positioned together at the chosen point. As a bead moves through an arc its label will change according to the following rules:

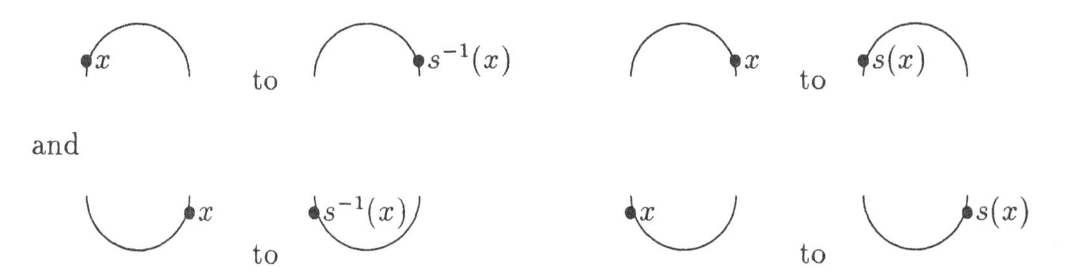

In other words, movement through the arc in a clockwise direction introduces s^{-1} and movement through the arc in a counterclockwise direction introduces s.

The final juxtaposition of labelled beads at the chosen point gives rise to a product w in H (down to up is interpreted as multiplication left to right). The symbol $s^j(e_i) \otimes s^\ell(f_i)$ represents $\sum s^j(R_i^{(1)}) \otimes s^\ell(R_i^{(2)})$. Let d be the Whitney degree of the diagram computed by traversing the diagram starting at the chosen point in the upward direction. Then $\mathrm{Tr}_\chi(K) = \chi(G^d w)$. We illustrate with some simple examples. The chosen point is designated by $\times$.

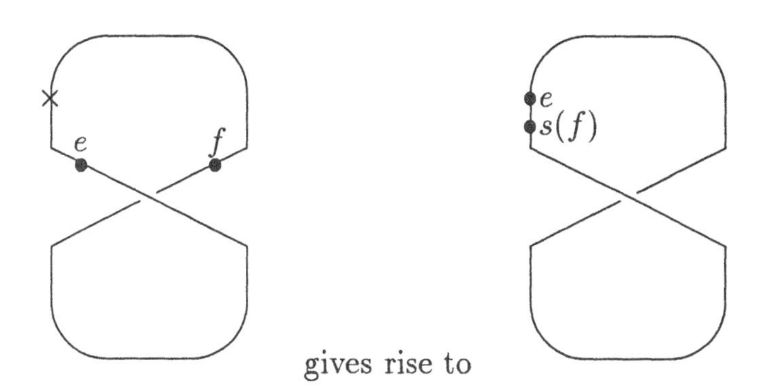

$$s(f)e \qquad \text{represents} \quad w = \sum s(R^{(2)})R^{(1)} = u. \qquad d = 0.$$

Thus $\mathrm{Tr}_\chi(K) = \chi(G^0 u) = \chi(u)$. Recomputing using a different starting position can give a different element of H.

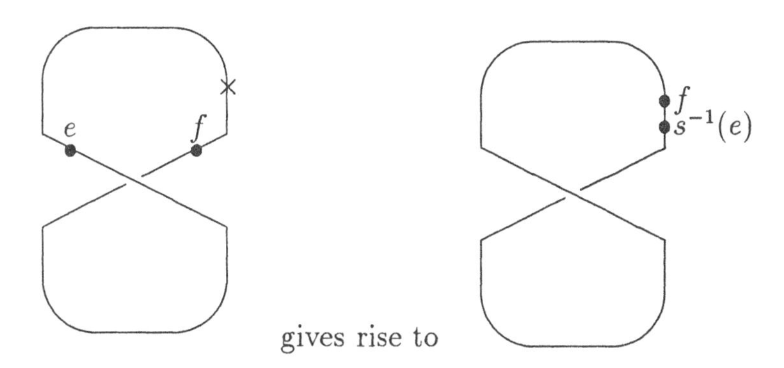

$$s^{-1}(e)f \qquad \text{represents} \quad w = \sum s^{-1}(R^{(1)})R^{(2)} = s^{-1}(u). \qquad d = 0.$$

Thus $\mathrm{Tr}_\chi(K) = \chi(G^0 s^{-1}(u)) = \chi(s^{-1}(u))$. The property $\chi = \chi \circ s$ guarantees that $\chi(s^{-1}(u)) = \chi(u)$.

The trefoil knot and its mirror image are relatively simple knots. Even so notice that the sums arising in the computation of Kauffman's invariant are rather involved. We use (11) to simplify the formal algebraic expressions.

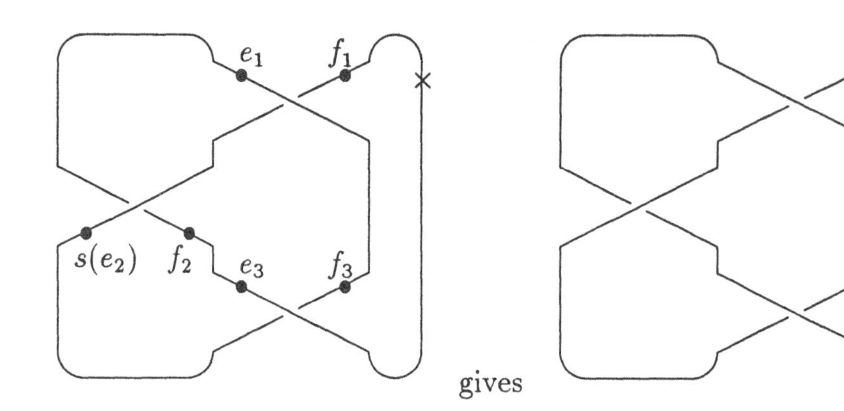

$$s(f_1)s^3(e_2)s(f_3)e_1f_2e_3 \quad \text{represents}$$

$$w = \sum s(R_1^{(2)})s^3(R_2^{(1)})s(R_3^{(2)})R_1^{(1)}R_2^{(2)}R_3^{(1)}. \qquad d = -2.$$

Thus $\mathrm{Tr}_\chi(K) = \sum \chi(G^{-2}s(R_1^{(2)})s^3(R_2^{(1)})s(R_3^{(2)})R_1^{(1)}R_2^{(2)}R_3^{(1)})$.

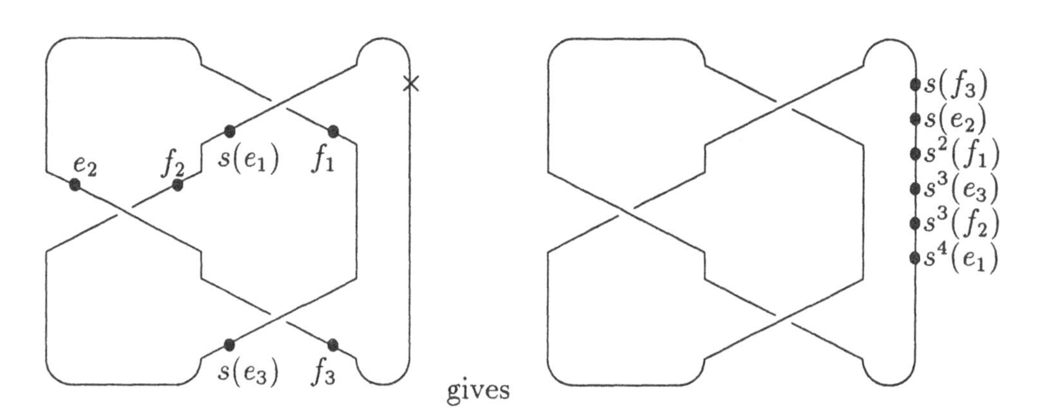

$$s^2(e_1)s^2(f_2)s^2(e_3)f_1e_2f_3 \quad \text{represents}$$

$$w = \sum s^2(R_1^{(1)})s^2(R_2^{(2)})s^2(R_3^{(1)})R_1^{(2)}R_2^{(1)}R_3^{(2)}. \qquad d = -2.$$

Thus $\mathrm{Tr}_\chi(K) = \sum \chi(G^{-2}s^2(R_1^{(1)})s^2(R_2^{(2)})s^2(R_3^{(1)})R_1^{(2)}R_2^{(1)}R_3^{(2)})$.

The "unknot"

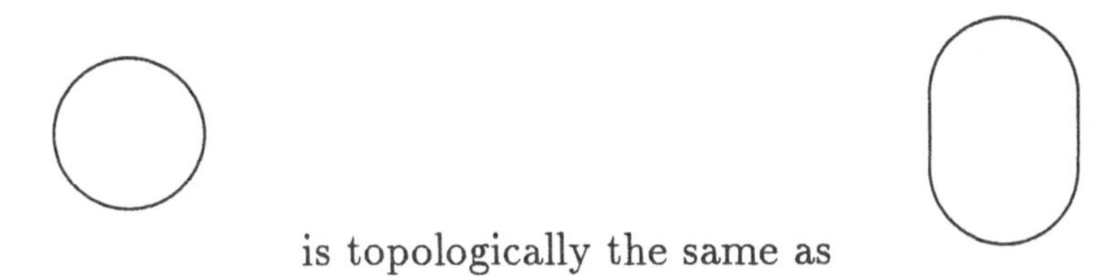

is topologically the same as

The requirement that a knot diagram K have a vertical line which is not in a crossing is a justifiable matter of convenience. For the diagram on the right $w = 1$, and $d = \pm 1$ depending on which point is chosen to begin the transversal.

To describe a rigorous algorithm for computing $\mathrm{Tr}_\chi(K)$, we will define a knot (or more precisely a knot diagram) in combinatorial terms. By *vertical line segment* we will mean a closed straight line segment in the plane of positive length which does not lie on a horizontal line. By an *arc* we will mean a semicircular arc in the plane with endpoints which lie on a horizontal line. Arcs are of two types, here depicted by the following symbols:

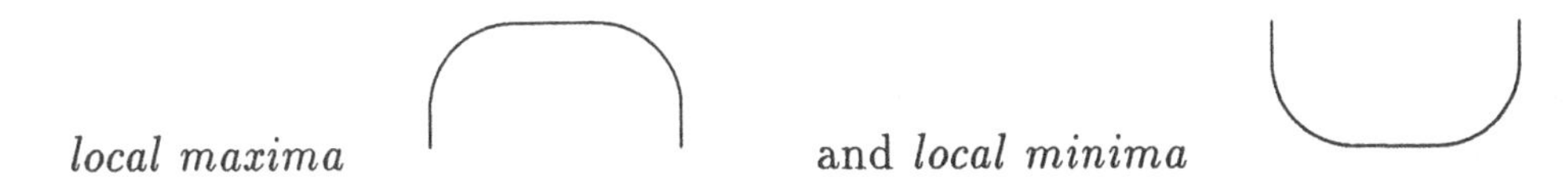

local maxima and *local minima*

By *crossing* we will mean an ordered pair of vertical line segments (x, y) with finite slopes which have opposite signs and $x \cap y = \{p\}$, where p is not an endpoint of either x or y. We adopt the convention that x crosses over y. Thus there are two possibilities for a crossing (x, y), depending on whether or not x has negative or positive slope:

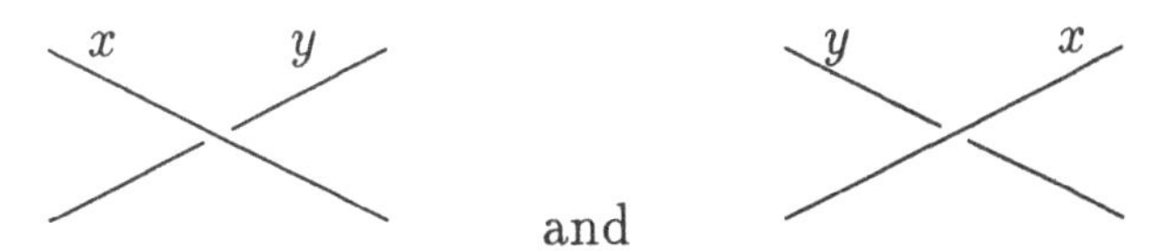

For a vertical line segment or an arc x we let $x' = x \backslash \{\text{endpoints}\}$. If $\mathcal{S}$ is a set of vertical line segments and/or arcs, we let $\mathcal{G}(\mathcal{S})$ denote the multiline graph whose points are the endpoints of the members of $\mathcal{S}$ and whose lines between points p and q are those objects of $\mathcal{S}$ which have p and q as endpoints. If $\mathcal{L}$ is a set of vertical line segments, we call a set of crossings $\mathcal{C} \subseteq \mathcal{L} \times \mathcal{L}$ a *crossing set* if $(x, y) \in \mathcal{C}$ implies that $(y, x) \notin \mathcal{C}$, and whenever $(x, y), (w, z) \in \mathcal{C}$ and are different, $(x \cup y) \cap (w \cup z) = \emptyset$. We will define a knot as a triple

$K = (\mathcal{L}, \mathcal{A}, \mathcal{C})$, where $\mathcal{L}$ and $\mathcal{A}$ are finite sets of vertical line segments and arcs respectively, $\mathcal{C} \subseteq \mathcal{L} \times \mathcal{L}$ is a crossing set such that:

i) $\mathcal{G}(\mathcal{L} \cup \mathcal{A})$ is a cyclic graph,

ii) If $x, y \in \mathcal{L} \cup \mathcal{A}$ are different then the highest and lowest points of $x \cup y$ are not common endpoints,

iii) If $x, y \in \mathcal{L} \cup \mathcal{A}$ are different then $x' \cup y'$ has two connected components, unless $(x, y) \in \mathcal{C}$ or $(y, x) \in \mathcal{C}$, and

iv) Some line in $\mathcal{L}$ does not belong to a crossing in $\mathcal{C}$.

Let $K = (\mathcal{L}, \mathcal{A}, \mathcal{C})$ be a knot, let $\mathcal{P}$ be the set of points of $\mathcal{G} = \mathcal{G}(\mathcal{S})$, and let $x_1 \in \mathcal{L}$ not be a line of a crossing in $\mathcal{C}$. Let p_1, p_2 be the endpoints of x_1, where p_2 is located higher in the plane. Since the graph $\mathcal{G}$ is a cycle, there is a unique arrangement of the points and lines of $\mathcal{G}$

$$p_1 \xrightarrow{x_1} p_2 \xrightarrow{x_2} p_3 \xrightarrow{x_3} \cdots \xrightarrow{x_{n-1}} p_n \xrightarrow{x_n} p_1,$$

where $p \xrightarrow{x} q$ denotes the unique line of $\mathcal{G}$ (a vertical line segment or an arc) joining p and q.

Let $\mathcal{K} = k[\mathsf{s}, \mathsf{s}^{-1}]$ be the group algebra of the free abelian group with generator s over k, and let $\mathcal{H} = \mathcal{K} \otimes H$. We regard $\mathcal{H}$ as a free left $\mathcal{K}$-module and write $ah = a \otimes h$ for $a \in \mathcal{K}$ and $h \in H$. To calculate $G^d w$ we will first construct a product $\mathsf{h} = \mathsf{h}_1 \cdots \mathsf{h}_n$ in the n-fold tensor product $\mathcal{H}^{(n)} = \mathcal{H} \otimes \cdots \otimes \mathcal{H}$. The factor h_i corresponds to x_i.

1) If x_i is a vertical line segment which is not the first line in a crossing, then let $\mathsf{h}_i = 1 \otimes \cdots \otimes 1$ be the unity of $\mathcal{H}^{(n)}$.

2) If x_i is an arc, then set $\mathsf{h}_i = 1 \otimes \cdots \otimes \mathsf{s}^{\ell} \otimes \cdots \otimes 1$ (1's except in the ith position) according to:

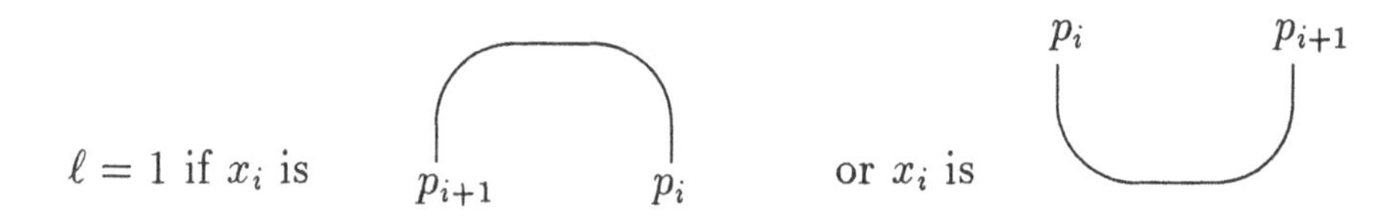

and

$\ell = -1$ if x_i is ... or x_i is

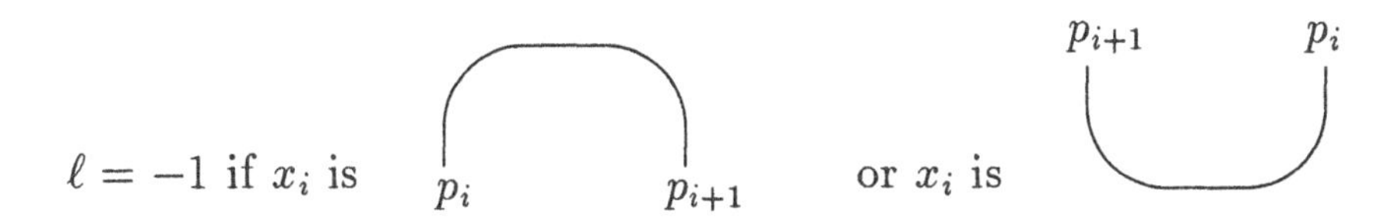

3) If x_i is the first line of a crossing $(x_i, x_j) \in \mathcal{C}$, set

$$\mathsf{h}_i = \sum 1 \otimes \cdots \otimes s^{\ell}(R^{(1)}) \otimes \cdots \otimes R^{(2)} \otimes \cdots \otimes 1$$

(1's except in the ith and j^{th} positions) when $i < j$, where

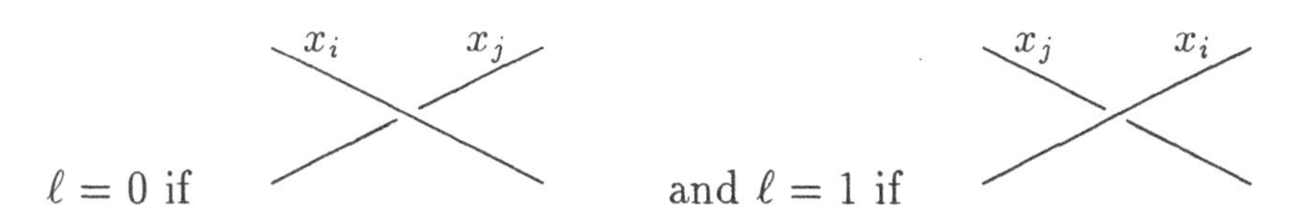

and

$$\mathsf{h}_i = \sum 1 \otimes \cdots \otimes R^{(2)} \otimes \cdots \otimes s^{\ell}(R^{(1)}) \otimes \cdots \otimes 1$$

(1's except in the ith and j^{th} positions) when $i > j$, where

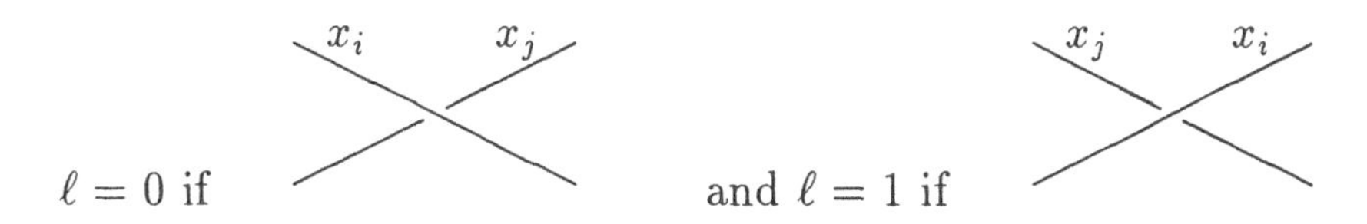

Now set $\mathsf{h} = \mathsf{h}_1 \cdots \mathsf{h}_n$. Starting with h, apply the "straightening" rule

$$\cdots \otimes h \otimes \mathsf{s}^{\ell} \otimes \cdots \longrightarrow \cdots \otimes \mathsf{s}^{\ell} \otimes s^{\ell}(h) \otimes \cdots$$

where $h \in H$ and s is the antipode of H, to obtain an element of $\mathcal{K}^{(|\mathcal{A}|)} \otimes \mathcal{H}^{(|\mathcal{C}|)}$. Applying the multiplication map $\mathcal{H}^{(n)} \longrightarrow \mathcal{H}$ to the "straightened" tensor sum gives us an element of the form $\mathsf{s}^{-2d}w \in \mathcal{H}' = k[\mathsf{s}^2, \mathsf{s}^{-2}]H$. The element $G^d w$ we seek is the image of $\mathsf{s}^{-2d}w$ under the map $\mathcal{H}' \to H$ determined by $\mathsf{s}^{2\ell}h \mapsto G^{-\ell}h$.

Observe that $s^2(w) = w$ by (11). Since $s^2(h) = GhG^{-1}$ for all $h \in H$ by part b) of Proposition 2, it follows that $Gw = wG$. If H_R is commutative, then $G^d w = G^e v^n$ for some integers e, n.

Observe that "words" w which arise from knot diagrams of the form below, and are evaluated as indicated, span a subalgebra of H.

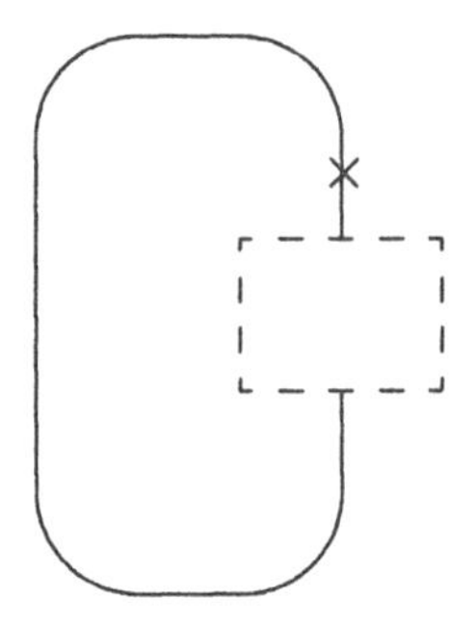

The box with the broken border contains the portion of the diagram not showing.

Let λ be a non-zero right integral for H^*. The values $\lambda(v^n)$ and $\lambda(v^{-n})$ for $n \geq 1$ arise in connection with the following diagrams:

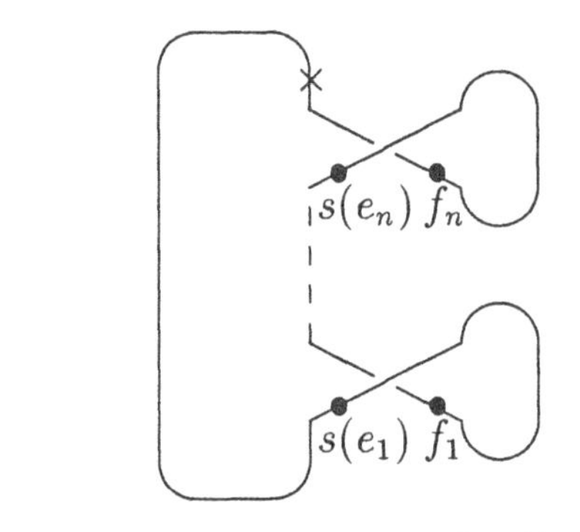

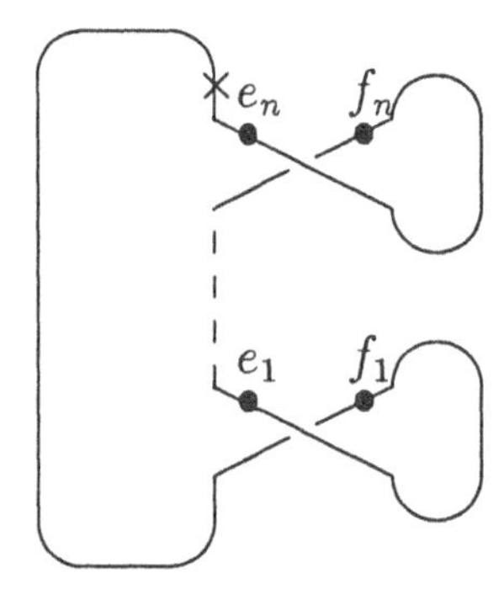

$d = n - 1, \qquad w = (s^{-1}(u))^n$

$\mathrm{Tr}_{\mu_G}(K) = \mu_G(G^d w) = \lambda(v^n)$

$d = n - 1, \qquad w = u^{-n}$

$\mathrm{Tr}_{\mu_G}(K) = \mu_G(G^d w) = \lambda(v^{-n}).$

For these calculations note that $u = Gv$ means that $u^{-n} = G^{-n}v^{-n}$ and $(s^{-1}(u))^n = G^{-n}v^n$. The values for $\lambda(v^n)$ and $\lambda(v^{-n})$ are computed for $\mathcal{U}_q(sl_2)'$ in a special case in [9]. See [13] also.

The reader may wonder why the diagrams above do not represent the unknot. *Ambient* isotopy allows for each of the four figures

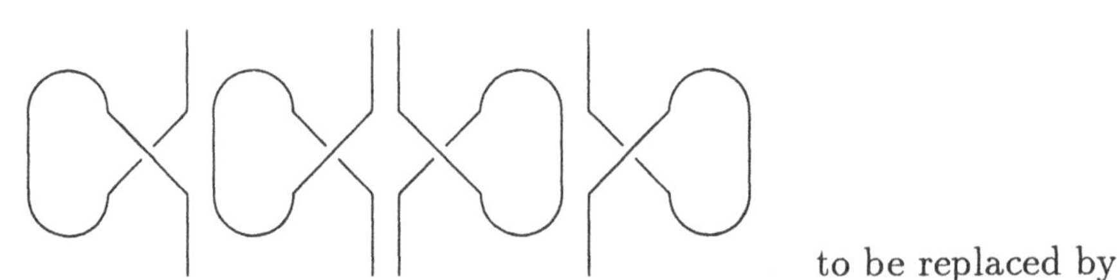

whereas regular isotopy does not. If the segments with a curl were pieces of rope, then pulling the ends to straighten them would introduce a twist.

Suppose that $N > 1$ and k has a primitive N^{th} root of unity ω. Recall that $(k[\mathcal{G}], R_1, \Theta(\omega^{-1}))$ is a ribbon Hopf algebra, where $\mathcal{G} = (a)$ is the cyclic group of order N. Here $G = 1$ so $\mu_G = \lambda$. Set $v = \Theta(\omega^{-1})$. We may assume that λ is defined by $\lambda(a^i) = \delta_{0,i}$. Therefore $\mu_G(v^n) = \sum_{\ell=0}^{N-1} \frac{1}{N}\omega^{-n\ell^2}$ for all $n \in Z$. Let $\chi : k[\mathcal{G}] \longrightarrow k$ be the algebra homomorphism determined by $\chi(a) = \omega^\ell$. Suppose that the characteristic of k is not 2. Then $\chi_s(v^n) = \frac{1}{2}(\omega^{n\ell^2} + \omega^{-n\ell^2})$ for all $n \in Z$.

4 The Algebra of Knot Invariants

In this section we explore the connections between the algebra of knot invariants and separation properties of sets of invariants. The ideas we discuss are very simple and quite general, but do provide a framework for considering what is meant by "new" invariant.

Let $\mathcal{K}$ be a non-empty set, R an equivalence relation on $\mathcal{K}$ and let A be any ring with unity. We let $\mathcal{A} = \mathrm{Inv}_R(\mathcal{K}, A)$ be the set of all functions $f : \mathcal{K} \longrightarrow A$ such that $f(K) = f(K')$ whenever $K, K' \in \mathcal{K}$ belong to the same equivalence class. We are primarily interested in the case when $\mathcal{K}$ is the set of all knots, the relation R is either ambient or regular isotopy and $A = C[X, X^{-1}]$, where C is the field of complex numbers. Knot invariants which are of considerable interest belong to $\mathcal{A}$ in this case.

Observe that $\mathcal{A}$ is a ring with unity under pointwise operations. If A is a k-algebra then $\mathcal{A}$ is as well. Suppose that A is a k-algebra, where k is a field, $\mathcal{K}$ is the set of knots and R is regular isotopy. Then the set of regular isotopy invariants Tr_χ defined for finite-dimensional ribbon Hopf algebras over k in the previous section spans a subalgebra of $\mathcal{A}$ by (18).

Let $\mathcal{F}$ denote the set of functions $\sigma : A \longrightarrow A$. For $\sigma \in \mathcal{F}$ and $f \in \mathcal{A}$, observe that the composite $\sigma \circ f \in \mathcal{A}$. Let S be a subset of $\mathcal{A}$. We say that S is *σ-closed* if $\sigma \circ f \in S$ for all $\sigma \in \mathcal{F}$ and $f \in S$. Notice that $\mathcal{A}$ is σ-closed and that the intersection of σ-closed subsets of $\mathcal{A}$ is closed. Therefore S is contained in a unique minimal σ-closed subset of $\mathcal{A}$, which we denote by $\overline{S}$. We denote the subring of $\mathcal{A}$ which S generates by (S). The collection of σ-closed subrings of $\mathcal{A}$ contains $\mathcal{A}$ and is closed under intersections. Therefore there is a unique minimal σ-closed subring of $\mathcal{A}$ which contains S, which we denote by $(S)_\sigma$. In constructive terms $(S)_\sigma$ can be described as the union of the chain

$$(S) \subseteq \overline{(S)} \subseteq (\overline{(S)}) \subseteq \overline{(\overline{(S)})} \subseteq \dots .$$

If $S = \{f_1, \dots, f_n\}$ we will frequently write $(f_1, \dots, f_n)_\sigma$ for $(S)_\sigma$.

Lemma 4 *Suppose that $S \subseteq \mathcal{A} = \mathrm{Inv}_R(\mathcal{K}, A)$, where $\mathcal{K}$ is a non-empty set, R is an equivalence relation on $\mathcal{K}$ and A is a ring with unity. Let $K, K' \in \mathcal{K}$. If $f(K) = f(K')$ for all $f \in S$, then $f(K) = f(K')$ for all $f \in (S)_\sigma$.*

The proof follows with the observation that $\mathcal{B} = \{f \in \mathcal{A} \,|\, f(K) = f(K')\}$ is a σ-closed subring of $\mathcal{A}$ which contains S.

In terms of knot invariants, if knots K and K' are not distinguished by all $f \in S$ then they are not distinguished by all $f \in (S)_\sigma$.

Regarding $a \in A$ as the function on $\mathcal{K}$ defined by $a(K) = a$ for $K \in \mathcal{K}$, we can consider A as a subring of $\mathcal{A}$. We define a relation on $\mathcal{A}$ by $f \geq g$ if and only if $\sigma \circ f = g$ for some $\sigma \in \mathcal{F}$. It is easy to see that

$$f \geq a \quad \text{for all } f \in \mathcal{A} \text{ and } a \in A, \tag{19}$$

$$f \geq f \tag{20}$$

and

$$\text{if} \quad f \geq g \quad \text{and} \quad g \geq h, \quad \text{then} \quad f \geq h. \tag{21}$$

Thus the relation is a preorder. Since $a \geq b$ for all $a, b \in A$, this preorder is not a partial order. Suppose that $S \subseteq \mathcal{A}, f \in (S)_\sigma$ and $g \in \mathcal{A}$. Then by definition $g \in (S)_\sigma$ whenever $f \geq g$.

In terms of distinguishing equivalence classes, $f \geq g$ means that f is at least as good as g.

Lemma 5 *Suppose that $f, g \in \mathcal{A} = \mathrm{Inv}_R(\mathcal{K}, A)$, where $\mathcal{K}$ is a non-empty set, R is an equivalence relation on $\mathcal{K}$ and A is a ring with unity. Then:*

a) *$f \geq g$ if and only the following condition is satisfied: whenever $K, K' \in \mathcal{K}$ and $f(K) = f(K')$, $g(K) = g(K')$.*

b) *$f \geq g$ and $g \geq f$ if and only the following condition is satisfied: for all $K, K' \in \mathcal{K}$, $f(K) = f(K')$ if and only if $g(K) = g(K')$.*

PROOF: Let $f, g \in \mathcal{A}$. We need only show that the condition of part a) implies $f \geq g$. The proof is straightforward, but we include it to illuminate the connection between the preorder and partitions of $\mathcal{K}$.

Assume that the condition holds. For $h \in \mathcal{A}$ let $\sim_h$ denote the equivalence relation on $\mathcal{K}$ defined by $K \sim_h K'$ if and only if $h(K) = h(K')$, let $[K]_h$ denote the equivalence class containing $K \in \mathcal{K}$, let $\mathcal{K}_h$ denote the set of equivalence classes of $\sim_h$ and let $\mathcal{K}_h \simeq \operatorname{Im} h$ be the identification given by $[K]_h \mapsto h(K)$. By assumption the map $\mathcal{K}_f \longrightarrow \mathcal{K}_g$ $([K]_f \mapsto [K]_g)$ is well-defined. Extend the composite

$$\operatorname{Im} f \simeq \mathcal{K}_f \longrightarrow \mathcal{K}_g \simeq \operatorname{Im} g$$

to $\sigma \in \mathcal{F}$. Then $g = \sigma \circ f$. This completes the proof.

Part a) of Lemma 5 has several notable consequences. Suppose that $f \in \mathcal{A}$. It is easy to see that $\mathcal{B} = \{g \in \mathcal{A} \,|\, f \geq g\}$ is a σ-closed subring of $\mathcal{A}$. Therefore

$$(f)_\sigma = \{g \in \mathcal{A} \,|\, f \geq g\},$$

which we refer to as a cyclic σ-closed subring of $\mathcal{A}$. If $\tau : A \longrightarrow A$ is one-one, then $\tau \circ f \geq f$. Consequently $(f)_\sigma = (\tau \circ f)_\sigma$. If $\mathcal{A}$ is a k-algebra, then a σ-closed subring of $\mathcal{A}$ is a k-subalgebra as well.

Let $S \subseteq \mathcal{A}$ be a subset and $f \in \mathcal{A}$. Whether or not $f \in (S)_\sigma$ can be decided by a comparison of f with the elements of S, when $(S)_\sigma$ is cyclic.

Proposition 5 *Suppose that $S \subseteq \mathcal{A} = \operatorname{Inv}_R(\mathcal{K}, A)$, where $\mathcal{K}$ is a non-empty set, R is an equivalence relation on $\mathcal{K}$ and A is a ring with unity. Suppose that $f \in \mathcal{A}$ and that $(S)_\sigma$ is cyclic. Then the following are equivalent:*

a) $f \in (S)_\sigma$.

b) *Whenever $K, K' \in \mathcal{K}$ and $g(K) = g(K')$ for all $g \in S$, $f(K) = f(K')$.*

PROOF: Let $K, K' \in \mathcal{K}$. We have noted that $\mathcal{B} = \{g \in \mathcal{A} \,|\, g(K) = g(K')\}$ is a σ-closed subring of $\mathcal{A}$. Thus part a) implies part b) in any event.

Suppose that the condition of part b) holds. By assumption $(S)_\sigma = (h)_\sigma$ for some $h \in \mathcal{A}$. We need only show that $h \geq f$. Suppose that $h(K) = h(K')$. Then for each $g \in S$ it follows that $g(K) = g(K')$ since $h \geq g$. Thus by assumption $f(K) = f(K')$. Therefore $h \geq f$ by part a) of Lemma 5.

The previous proposition fails to be true without the assumption that $(S)_\sigma$ is cyclic. We give a very simple example reminiscent of the Vassiliev invariants. Suppose that the equivalence classes of R are singletons and that there is an infinite strictly increasing chain of subsets

$$\mathcal{K}_1 \subset \mathcal{K}_2 \subset \mathcal{K}_3 \subset \ldots \subset \mathcal{K}$$

whose union is $\mathcal{K}$. If $\mathcal{B}_i$ is the set of all functions in $\mathcal{A}$ which are constant on $\mathcal{K} \backslash \mathcal{K}_i$ then

$$\mathcal{B}_1 \subset \mathcal{B}_2 \subset \mathcal{B}_3 \subset \ldots \subset \mathcal{A}$$

is a strictly increasing chain of σ-closed subrings of $\mathcal{A}$ whose union is therefore a σ-closed subring $\mathcal{B}$ of $\mathcal{A}$. The condition of part b) of Proposition 5 holds for $S = \mathcal{B}$ and all $f \in \mathcal{A}$, but $\mathcal{B} \neq \mathcal{A}$.

Observe that elements of $\mathcal{B}$ separate points of $\mathcal{K}$, even though $\mathcal{B}$ is not dense in the sense that $\overline{\mathcal{B}} = \mathcal{A}$.

When A is a finite ring $(f)_\sigma$ is always a finite set with at most $|A|^{|A|}$ elements. Thus finitely generated σ-closed subrings of $\mathcal{A}$ may not be cyclic. However, under fairly general circumstances finite generation does imply cyclic.

Lemma 6 *Suppose that $f_1, \ldots, f_n \in \mathcal{A} = \mathrm{Inv}_R(\mathcal{K}, A)$, where $\mathcal{K}$ is a non-empty set, R is an equivalence relation on $\mathcal{K}$ and A is a ring with unity. If there are infinite subsets $T, T' \subseteq A$ such that $|\mathrm{Im}\, f_i| \leq |T|, |T'|$ for all $1 \leq i \leq n$ and the multiplication map $T \times T' \longrightarrow A$ is one-one, then*

$$(f_1, \ldots, f_n)_\sigma = (f)_\sigma$$

for some $f \in \mathcal{A}$ such that $|\mathrm{Im}\, f| \leq |T|, |T'|$.

PROOF: By induction on n we may assume that $n = 2$. We can choose $\sigma_1, \sigma_2 \in \mathcal{F}$ which are one-one such that $\sigma_1 \circ f_1$ and $\sigma_2 \circ f_2$ have their image in T and T' respectively. Since $(f_1, f_2)_\sigma = (\sigma_1 \circ f_1, \sigma_2 \circ f_2)_\sigma$, we may assume that $\mathrm{Im}\, f_1 \subseteq T$ and $\mathrm{Im}\, f_2 \subseteq T'$. Let $f \in \mathcal{A}$ be defined by $f(K) = f_1(K) f_2(K)$ for $K \in \mathcal{K}$. Then $f \in (f_1, f_2)_\sigma$. By hypothesis and part a) of Lemma 5 we have that $f \geq f_1, f_2$. Therefore $f_1, f_2 \in (f)_\sigma$. Since $|T \times T'| = max\{|T|, |T'|\}$, the proof is complete.

Let $N = \{0, 1, 2, \ldots\}$ be the set of natural numbers, which can always be thought of as a multiplicative subset of A when the characteristic of A

is 0. There are many countably infinite subsets T, T' of N which satisfy the hypothesis of the previous proposition. For example, let $1 < p, q \in N$ be two relatively prime integers, let T be the set of powers of p and T' be the set of powers of q. Since the set of equivalence classes of knots under either ambient or regular isotopy is countable, we have the following application of Lemma 6 and Proposition 5:

Proposition 6 *Suppose that $\mathcal{A} = \mathrm{Inv}_R(\mathcal{K}, A)$, where $\mathcal{K}$ is the set of knots, R is the equivalence relation ambient or regular isotopy on $\mathcal{K}$, and A is a ring with unity of characteristic 0. Let $f, f_1, \ldots, f_n \in \mathcal{A}$. Then:*

a) *$(f_1, \ldots, f_n)_\sigma$ is cyclic.*

b) *$f \notin (f_1, \ldots, f_n)_\sigma$ if and only if there are $K, K' \in \mathcal{K}$ such that $f_i(K) = f_i(K')$ for $1 \leq i \leq n$ but $f(K) \neq f(K')$.*

Part b) can be interpreted as saying that what it means for f to be fundamentally different from $f_1, \ldots, f_n$ in an algebraic sense is what it means in a topological sense.

Suppose that the field k has characteristic 0, $\mathcal{K}$ is the set of all knots and R is regular isotopy. Let (H, R, v) be a finite-dimensional ribbon Hopf algebra over k. Then the set S of all the regular isotopy invariants Tr_χ arising from (H, R, v) is a finite-dimensional subspace of $\mathcal{A}$. By Proposition 6 we have that $(S)_\sigma$ is cyclic. Thus there is a single invariant $f \in (S)_\sigma$ which has the same knot distinguishing capabilities as the family of invariants Tr_χ.

5 Unimodular Ribbon Quasitriangular Envelopes for a Class of Finite-Dimensional Pointed Hopf Algebras

In this section we construct and discuss properties of certain finite-dimensional pointed Hopf algebras which we denote by $H_{(N,\nu,\omega)}$. They constitute a very extensive and interesting class. Included are the basic examples described and studied by Taft in [29]. We construct a quasitriangular envelope $U_{(N,\nu,\omega)}$ for $H_{(N,\nu,\omega)}$ which is also a unimodular ribbon Hopf algebra. Among these envelopes is $\mathcal{U}_q(sl_2)'$ when q is a root of unity. The Hopf algebra $U_{(N,\nu,\omega)}$ is a quotient of the double $D(H_{(N,\nu,\omega)})$ in a very simple way.

We use q-polynomials and the Diamond Lemma to establish the existence $H_{(N,\nu,\omega)}$ and $U_{(N,\nu,\omega)}$ and to derive some of their basic properties. In the process we reformulate some of the basic arguments of [29]. It should be noted that Taft rethought his examples in light of q-polynomials in [28]. The n^2-dimensional example of [29] is $T_{n,\omega} = H_{(n,1,\omega)}$. The Hopf algebra $T_{2,\omega}$ is due to Sweedler, and $T_{n,\omega}$ for $n > 2$ is a generalization of Sweedler's example due to Taft.

We give a rather detailed discussion of $U_{(N,\nu,\omega)}$ in to relation to Hennings' invariant. We show that his invariant is defined for $U_{(N,\nu,\omega)}$ with some ribbon element except when $N = 2M$ and M is even and ν is odd.

5.1 Construction and Properties of $H_{(N,\nu,\omega)}$

We begin by constructing a 4-parameter family of pointed Hopf algebras over the field k. Let n, N and ν be positive integers such that n divides N and $1 \leq \nu < N$. Suppose that $q \in k$ is a primitive n^{th} of unity and let r the order of q^ν. We define a Hopf algebra $H = H_{n,q,N,\nu}$ over k as follows. As an algebra H is generated by a and x which satisfy the relations

$$a^N = 1, \qquad x^r = 0 \qquad \text{and} \qquad xa = qax.$$

The coalgebra structure of H is determined by

$$\Delta(a) = a \otimes a \qquad \text{and} \qquad \Delta(x) = x \otimes a^\nu + 1 \otimes x.$$

The antipode of H is the algebra map $s : H \longrightarrow H^{op}$ determined by

$$s(a) = a^{-1} \qquad \text{and} \qquad s(x) = -xa^{-\nu} = -q^{-\nu}a^{-\nu}x.$$

Consequently $s^2(a) = a$, $s^2(x) = q^{-\nu}x$ and thus $s^2(h) = a^\nu h a^{-\nu}$ for all $h \in H$. In particular s^2 is an inner automorphism.

Note that when $r = 1$ the Hopf algebra H is the group algebra of the cyclic group of order N.

We will go through the fundamental exercise of constructing H from basic principles, providing in the process a model for a simple and direct construction of $U_{(N,\nu,\omega)}$ – in particular of $\mathcal{U}_q(sl_2)$' – in Section 5.2.

Let $(C, \Delta_C, \epsilon_C)$ be a coalgebra over k. The universal mapping property of the tensor algebra $(T(C), \imath)$ of C over k accounts for a unique bialgebra structure $(T(C), \Delta_{T(C)}, \epsilon_{T(C)})$ on $T(C)$ such that $\imath : C \longrightarrow T(C)$ is a coalgebra map. We may assume that $\imath$ is the inclusion.

Let I be an ideal of $T(C)$ and $\pi : T(C) \longrightarrow T(C)/I \equiv B$ be the projection. Assume that $(\pi \otimes \pi) \circ \Delta_{T(C)}$ and $\epsilon_{T(C)}$ vanish on generators of I. Then $(\pi \otimes \pi) \circ \Delta_{T(C)}$ and $\epsilon_{T(C)}$ lift to unique algebra maps $\Delta : B \longrightarrow B \otimes B$ and $\epsilon : B \longrightarrow k$ such that $(\pi \otimes \pi) \circ \Delta_{T(C)} = \Delta \circ \pi$ and $\epsilon_{T(C)} = \epsilon \circ \pi$. Since π is onto, the algebra B is a bialgebra with underlying coalgebra structure (B, Δ, ϵ). Thus π is also a bialgebra map.

Observe that B is a Hopf algebra if and only if the linear map $\pi|_C : C \longrightarrow B$ has an inverse s in the convolution algebra $\mathrm{Hom}_k(C, B)$ whose unique extension to an algebra map $S : T(C) \longrightarrow B^{op}$ vanishes on generators of I. If $\pi|_C$ has such an inverse, then the algebra map $s : B \longrightarrow B^{op}$ determined by $s \circ \pi = S$ is an antipode for B. The necessity of the condition is clear. For sufficiency, consider two bialgebras H and H' and algebra maps $f : H \longrightarrow H'$ and $g : H \longrightarrow H'^{op}$. The calculations $f{*}g(ab) = \sum f(a_{(1)})(f{*}g(b))g(a_{(2)})$ and $g{*}f(ab) = \sum g(b_{(1)})(g{*}f(a))f(b_{(2)})$ for $a, b \in H$ show that f and g are inverses in the convolution algebra $\mathrm{Hom}_k(H, H')$ if and only if $f * g \equiv \epsilon 1 \equiv g * f$ on a set of algebra generators for H.

To construct $H_{n,q,N,\nu}$ we let C be the 4-dimensional coalgebra over k with basis $\{A, B, D, X\}$ whose structure is determined by

$$\Delta(A) = A \otimes A, \quad \Delta(B) = B \otimes B, \quad \Delta(D) = D \otimes D \quad \text{and} \quad \Delta(X) = X \otimes B + D \otimes X,$$

and let I be the ideal of $T(C)$ generated by

$$A^N - 1, \quad B - A^\nu, \quad D - 1, \quad XA - qAX \quad \text{and} \quad X^r.$$

In light of the preceding discussion, the only detail in the construction of $H_{n,q,N,\nu}$ which may require a little effort to work out is $(\pi \otimes \pi) \circ \Delta_{T(C)}(X^r) = 0$, or equivalently $(x \otimes a^\nu + 1 \otimes x)^r = 0$, where $x = \pi(X)$ and $a = \pi(A)$. Let $\mathcal{A} = x \otimes a^\nu$ and $\mathcal{X} = 1 \otimes x$. Then $\mathcal{X}\mathcal{A} = q^\nu \mathcal{A}\mathcal{X}$. By part b) of Proposition 1 we compute

$$(x \otimes a^\nu + 1 \otimes x)^r = (x \otimes a^\nu)^r + (1 \otimes x)^r = x^r \otimes a^{\nu r} + 1 \otimes x^r = 0.$$

Proposition 7 *Suppose that $n > 1$ and the field k has a primitive n^{th} root of unity q. Let $H = H_{n,q,N,\nu}$ be the Hopf algebra described above. Then:*

a) *H is a pointed coalgebra and $G(H) = (a)$.*

b) *The set $\beta = \{a^\ell x^m \,|\, 0 \leq \ell < N, 0 \leq m < r\}$ is linear basis for H. Therefore $\dim H = Nr$.*

c) $\Delta(a^\ell x^m) = \sum_{i=0}^{m} \binom{m}{i}_{q^\nu} a^\ell x^{m-i} \otimes a^{\ell+(m-i)\nu} x^i$ *for all* $0 \leq \ell < N$ *and* $0 \leq m < r$.

d) *A non-zero left integral for* H *is*

$$\Lambda_\ell = (1 + a + a^2 + \ldots + a^{N-1})x^{r-1}$$

and a non-zero right integral for H *is*

$$\Lambda_r = (1 + q^{r-1}a + q^{2(r-1)}a^2 + \ldots + q^{(N-1)(r-1)}a^{N-1})x^{r-1}.$$

e) *The distinguished grouplike element* α *of* H^* *is determined by* $\alpha(a) = q^{r-1}$ *and* $\alpha(x) = 0$. *Thus* H *is not unimodular when* $r > 1$.

PROOF: Part a) follows by Lemma 1. Part d) follows from part b) and the observation that $\Lambda_r = x^{r-1}(1 + a + a^2 + \ldots + a^{N-1})$ is a right integral for H. Part e) follows from part d). To show part c), we write $\Delta(x) = \mathcal{A} + \mathcal{X}$, where $\mathcal{A} = x \otimes a^\nu$ and $\mathcal{X} = 1 \otimes x$, and apply part a) of Proposition 1 to $\Delta(x^m) = (\mathcal{A} + \mathcal{X})^m$.

To show part b) we use the Diamond Lemma on $T(C) = k\{A, B, D, X\}$, where $A < B < D < X$ and the substitution rules are:

$$A^N \leftarrow 1, \quad B \leftarrow A^\nu, \quad D \leftarrow 1, \quad XA \leftarrow qAX \quad \text{and} \quad X^r \leftarrow 0.$$

The only ambiguities to consider are the overlap ambiguities:

$$(A^\ell A^{N-\ell})A^\ell = A^\ell(A^{N-\ell}A^\ell), \quad (X^\ell X^{r-\ell})X^\ell = X^\ell(X^{r-\ell}X^\ell),$$

$$(XA)A^{N-1} = X(AA^{N-1}) \qquad \text{and} \qquad (X^{r-1}X)A = X^{r-1}(XA).$$

These are easily resolved.

Proposition 8 *Suppose that* $n > 1$ *and the field* k *has a primitive* n^{th} *root of unity* q. *Let* $H = H_{n,q,N,\nu}$ *and assume that* $r > 1$. *Then* $H \simeq H^*$ *as Hopf algebras if and only if* k *has a primitive* N^{th} *root of unity* ω *such that* $q = \omega^\nu$.

PROOF: Assume that $\pi : H \longrightarrow H^*$ is a Hopf algebra isomorphism, let $A = \pi(a)$, $\omega = A(a)$ and $X = \pi(x)$. We first show that ω is a primitive N^{th} root of unity. Let $\ell \geq 0$. Since $a \in G(H)$ it follows that $A^\ell \in G(H^*) = \text{Alg}_k(H, k)$. Since x is nilpotent $A^\ell(x) = 0$, and since a is a grouplike element $A^\ell(a) = \omega^\ell$. Now $A^\ell \neq \epsilon$ for $1 \leq \ell < N$ and $A^N = \epsilon$. Therefore ω is a primitive N^{th} root of unity.

We next show that $q = \omega^\ell$. Applying both sides of the equation $XA = qAX$ to $h \in G(H)$ we deduce that $X(h) = qX(h)$ since $A(h) \neq 0$. Since $q \neq 1$ it follows that $X(h) = 0$ for all $h \in G(H)$. Suppose that $X(hx) = 0$ for all $h \in G(H)$. Then using part c) of Proposition 7 it follows by induction on j that $X^j(hx^i) = 0$ for all $h \in G(H)$ whenever $1 \leq j < r$ and $0 \leq i \leq j$. But then $X^{r-1} = 0$, a contradiction. Therefore $X(hx) \neq 0$ for some $h \in G(H)$. Now applying both sides of $XA = qAX$ to hx we conclude that $A(ha^\nu) = qA(h)$, and hence $\omega^\nu = q$.

Conversely, suppose that $\omega \in k$ is a primitive N^{th} root of unity such that $q = \omega^\nu$. Let $A : H \longrightarrow k$ be the algebra homomorphism determined by

$$A(a) = \omega \qquad \text{and} \qquad A(x) = 0,$$

and define a functional $X : H \longrightarrow k$ by

$$X(a^\ell x^m) = \delta_{1,m} \qquad \text{for all} \qquad 0 \leq \ell < N, \ 0 \leq m < r.$$

Since A is an algebra homomorphism it follows that $\Delta(A) = A \otimes A$. Establishing $X(uv) = X(u)A^\nu(v) + \epsilon(u)X(v)$ for $u, v \in \mathfrak{B}$ described in part b) of Proposition 7 will show that $\Delta(X) = X \otimes A^\nu + \epsilon \otimes X$. There are basically two cases to work out: $u = a^i$, $v = a^j x$ and $u = a^i x$, $v = a^j$. Using part c) of Proposition 7 one can derive the formulas

$$A^\ell X^m(a^u x^v) = \delta_{m,v}(m)_{q^\nu}!\omega^{\ell u} \tag{22}$$

and $XA(a^u x^v) = \delta_{1,v}\omega^{u+\nu}$ for all $\ell, u \in Z$ and $0 \leq m \leq r$ and $0 \leq v < r$, the former by induction on m. We conclude from the formulas that $A^N = \epsilon, X^r = 0$ and $XA = qAX$.

To complete the proof we need only show that the set $\{A^\ell X^m \,|\, 0 \leq \ell < N,\ 0 \leq m < r\}$ is linearly independent. Suppose that $\sum_{\ell=0}^{N-1}\sum_{m=0}^{r-1} \alpha_{\ell,m} A^\ell X^m = 0$, where $\alpha_{\ell,m} \in k$. Evaluating both sides of this equation on $a^u x^v$ we have that $\sum_{\ell=0}^{N-1} \alpha_{\ell,v}\omega^{\ell u} = 0$ by (22). Thus for all $0 \leq v < r$ it follows that

$(\alpha_{0,v}, \ldots, \alpha_{N-1,v})\mathcal{A} = 0$, where $\mathcal{A} = (a_{ij})_{0 \le i,j < N}$ is the $N \times N$ matrix defined by $a_{ij} = \omega^{ij}$. But $\mathcal{A}$ is invertable since ω is a primitive N^{th} root of unity. Therefore $\alpha_{\ell,v} = 0$ for all $0 \le \ell < N$ and $0 \le v < r$.

From this point on we are interested in the case when $H = H_{n,q,N,\nu}$ is isomorphic to H^*. Suppose that N, ν are positive integers satisfying $1 < N, 1 \le \nu < N$ and N does not divide ν^2. Suppose further that $\omega \in k$ is a primitive N^{th} root of unity. Then we set $H_{(N,\nu,\omega)} = H_{n,q,N,\nu}$, where $q = \omega^\nu$. Observe that the order of $q^\nu = \omega^{\nu^2}$ is $r = N/(N, \nu^2)$. Thus

$$\operatorname{Dim} H_{(N,\nu,\omega)} = \frac{N^2}{(N, \nu^2)}.$$

The condition $r > 1$ is equivalent to the condition N does not divide ν^2.

By virtue of Propositions 7 and 8

$$\lambda_\ell = (\epsilon + A + A^2 + \ldots + A^{N-1})X^{r-1}$$

is a non-zero left integral for $H^*_{(N,\nu,\omega)}$ and

$$\lambda_r = (\epsilon + q^{r-1}A + q^{2(r-1)}A^2 + \ldots + q^{(N-1)(r-1)}A^{N-1})X^{r-1}$$

is a non-zero right integral for $H^*_{(N,\nu,\omega)}$. These have much simpler descriptions in terms of the dual basis of ß for $H^*_{(N,\nu,\omega)}$. We shall write $\{\overline{v_1}, \ldots, \overline{v_n}\}$ for the dual basis of a basis $\{v_1, \ldots, v_n\}$ of a finite-dimensional vector space.

Proposition 9 *Suppose that $N > 1$ and the field k has a primitive N^{th} root of unity ω. Let $H_{(N,\nu,\omega)}$, ß, λ_ℓ and λ_r be defined as above. Then:*

a) $\lambda_\ell = ((r-1)_{q^\nu}!N)\overline{x^{r-1}}$ *and* $\lambda_r = ((r-1)_{q^\nu}!N)\overline{a^{-\nu(r-1)}x^{r-1}}$.

b) *The distinguished grouplike element of H is $g = a^{-\nu(r-1)}$.*

PROOF: Let $b \in H$ and $p \in H^*$. From our discussion of (2) it follows that $qp = q(b)p$ for all $q \in H^*$ if and only $\sum h_{(1)}p(h_{(2)}) = p(h)b$ for all $h \in H$, and from our discussion of (3) it follows that $pq = q(b)p$ for all $q \in H^*$ if and only if $\sum p(h_{(1)})h_{(2)} = p(h)b$ for all $h \in H$. Thus using part c) of Proposition 7 we see that $\overline{x^{r-1}}$ and $\overline{a^{-\nu(r-1)}x^{r-1}}$ are left and right integrals respectively for $H^*_{(N,\nu,\omega)}$, and that $g = a^{-\nu(r-1)}$. Now left or right integrals are unique up to scalar multiple. The calculation $\lambda_\ell(x^{r-1}) = (r-1)_{q^\nu}!N = \lambda_r(a^{-\nu(r-1)}x^{r-1})$ completes the proof.

By [8, Proposition 7] the double $(D(T_{n,\omega}), \mathcal{R})$ has a ribbon element if and only if n is odd. Hennings observed earlier that n odd implies $(D(T_{n,\omega}), \mathcal{R})$ has a ribbon element [5]. There are many examples of self-dual pointed Hopf algebras $H = H_{(N,\nu,\omega)}$ such that $(D(H), \mathcal{R})$ has a ribbon element.

Theorem 3 *Suppose that $N > 1$ and the field k has a primitive N^{th} root of unity ω and let $H = H_{(N,\nu,\omega)}$. Then $(D(H), \mathcal{R})$ has a ribbon element except when N is even and ν is odd.*

PROOF: Let g and α be the distinguished grouplike elements of H and H^* respectively. Then $g = a^{-\nu(r-1)}$ by part b) of Proposition 9, and it follows that $\alpha(a) = q^{r-1} = \omega^{\nu(r-1)}$ by part e) of Proposition 7. Let s be the antipode of H. Then $s^2(a) = a$ and $s^2(x) = q^{-\nu}x$. By Theorem 2 we have that $(D(H), \mathcal{R})$ has a ribbon element if and only if there are $\ell \in G(H)$ and $\beta \in G(H^*)$ such that $\ell^2 = g$, $\beta^2 = \alpha$ and $s^2(h) = \ell(\beta \rightharpoonup h \leftharpoonup \beta^{-1})\ell^{-1}$ for all $h \in H$. Both sides of this last equation describe algebra endomorphisms of H which agree on the grouplike elements of H. Thus the last equation holds generally if and only if it holds for $h = x$.

Let $\ell \in G(H)$ and $\beta \in G(H^*)$. Now $G(H) = (a)$ and $G(H^*)$ are cyclic of order N. Therefore $\ell = a^p$ and $\beta(a) = \omega^q$ for some $p, q \in Z$. The equation $s^2(x) = \ell(\beta \rightharpoonup x \leftharpoonup \beta^{-1})\ell^{-1}$ is equivalent to $\omega^{-\nu^2} = \omega^{\nu(q-p)}$. Therefore $(D(H), \mathcal{R})$ has a ribbon element if and only if there are solutions p, q to the equations

$$\omega^{2p} = \omega^{-\nu(r-1)}, \qquad \omega^{2q} = \omega^{\nu(r-1)} \qquad \text{and} \qquad \omega^{-\nu^2} = \omega^{\nu(q-p)}.$$

Suppose that $\omega^{2q} = \omega^{\nu(r-1)}$, and set $p = -q$. Since $1 = q^{\nu r} = \omega^{\nu^2 r}$, it follows that $\omega^{\nu(q-p)} = \omega^{2q\nu} = \omega^{\nu^2(r-1)} = \omega^{-\nu^2}$. Therefore $(D(H), \mathcal{R})$ has a ribbon element if and only if there is a solution q to the equation $\omega^{2q} = \omega^{\nu(r-1)}$. If ν is even, or if ν and N are odd, this is the case. If ν is odd and N is even, then r, which is the order of $q^\nu = \omega^{\nu^2}$, is even. Therefore $\nu(r-1)$ is odd. There is no solution in this case. This completes the proof.

Gelaki showed in [4] that $H = T_{n,\omega}$ does not admit a quasitriangular structure when $n > 2$ by determining the Hopf algebra maps $F : H^* \longrightarrow H^{cop}$. To do this, he showed that $H^* \simeq H$ and determined all Hopf algebra maps $f : H \longrightarrow H^{cop}$. We will generalize his result, following the lines of his argument and describe the Hopf algebra endomorphisms of $H = H_{(N,\nu,\omega)}$ and Hopf algebra maps $f : H^* \longrightarrow H^{cop}$ in the process.

For a Hopf algebra H over k, we let $\mathrm{Aut}_{Hopf}(H)$ denote the group of Hopf algebra automorphisms of H and $\mathrm{End}_{Hopf}(H)$ denote the semigroup of Hopf algebra endomorphisms of H under composition. We will let Z_n^* denote the multiplicative group of units of the ring Z_n.

Theorem 4 *Suppose that N, $\check{N} > 1$ and the field k has primitive N^{th} and $\check{N}^{th}$ roots of unity ω and $\check{\omega}$ respectively. Let $H = H_{(N,\nu,\omega)}$ and $\check{H} = H_{(\check{N},\check{\nu},\check{\omega})}$ be defined as above. Then:*

a) *Suppose $h \in G(H)$ and $z \in H$ satisfy $\Delta(z) = z \otimes h + 1 \otimes z$. Then $z = \alpha(h-1)$ for some $\alpha \in k$, or $h = a^\nu$ and $z = \alpha(a^\nu - 1) + \beta x$ for some $\alpha, \beta \in k$.*

b) *Let $f : H \longrightarrow \check{H}$ be a map of Hopf algebras. Then $f(x) = \beta_f \check{x}$ for some $\beta_f \in k$. If $\beta_f \neq 0$, then $f(a^\nu) = \check{a}^{\check{\nu}}$.*

c) *$H \simeq \check{H}$ as Hopf algebras if and only if the following hold:*

   i) *$N = \check{N}$.*

   ii) *$\check{\omega} = \omega^t$ where $1 \leq t < N$ is relatively prime to N.*

   iii) *There is a solution s relatively prime to N to the congruences $\nu s \equiv \check{\nu} \pmod N$ and $ts^2 \equiv 1 \pmod n$, where n is the order of $q = \omega^\nu$.*

d) *Let $U^* = \{s \in Z_N^* \mid s \equiv 1 \pmod n\}$. Then there is an isomorphism of groups $U^* \times k^* \simeq \mathrm{Aut}_{Hopf}(H)$ given by $(s, \beta) \mapsto f$, where $f(a) = a^s$ and $f(x) = \beta x$.*

e) *$H_{(N,\nu,\omega)} \simeq H_{(N,\nu,\check{\omega})}$ if and only if $q = \check{q}$.*

f) *Suppose $f : H \longrightarrow H^{cop}$ is a map of Hopf algebras, and assume that $f(x) \neq 0$. Then $f(a) = a^s$ where s is an odd integer and $N = 2\nu$. Furthermore $f(a^\nu) = a^{-\nu}$, $f(x) = \beta a^{-\nu} x$ for some $\beta \in k$ and ν is odd.*

g) *$H \simeq H^{cop}$ if and only if $N = 2\nu$ and ν is odd.*

PROOF: Part a) follows in a straightforward manner after expressing z in terms of ß and using part c) of Proposition 7 to compare the expansions of $\Delta(z)$ and $z \otimes h + 1 \otimes z$. To show part b), let $f : H \longrightarrow \check{H}$ be a Hopf algebra map. We may assume that $\check{z} = f(x) \neq 0$. Since $f(a) \in G(\check{H})$, which is a commutative group, the relation $f(x)f(a) = qf(a)f(x)$ shows that $\check{z}$ can not commute with the grouplike element $f(a)$ of $\check{H}$. Therefore $\check{z} \notin k[G(\check{H})]$. Since $\Delta(\check{z}) = \check{z} \otimes f(a^\nu) + 1 \otimes \check{z}$, by part a) we conclude that $f(a^\nu) = \check{a}^{\check{\nu}}$ and $f(x) = \check{z} = \alpha(\check{a}^{\check{\nu}} - 1) + \beta\check{x}$ for some $\alpha, \beta \in k$. Reconsidering the relation $f(x)f(a) = qf(a)f(x)$, we see that $\alpha = 0$ since $f(a)$ is a grouplike element of $\check{H}$, $q \neq 1$ and $\check{r} > 1$.

To show part c), we first assume the conditions. Then $a^{\nu s} = a^{\check{\nu}}$ and $\check{q}^s = \omega^{t\check{\nu}s} = \omega^{t\nu s^2} = \omega^\nu = q$. In particular $\check{r} = r$, and therefore $\operatorname{Dim} H = \operatorname{Dim} \check{H}$. Now it is easy to see that the linear map $f : H \longrightarrow \check{H}$ defined by $f(a^\ell x^m) = \check{a}^{s\ell}\check{x}^m$ is in fact an algebra map. Since s and N are relatively prime, $\check{a}^s$ generates $(\check{a})$. Therefore f is an algebra isomorphism. To see that f is a coalgebra map is a matter of noting that $\Delta(\check{x}) = \check{x} \otimes \check{a}^{\check{\nu}} + 1 \otimes \check{x} = \check{x} \otimes \check{a}^{s\nu} + 1 \otimes \check{x}$.

To show that the conditions are necessary, we note that any Hopf algebra isomorphism $f : H \longrightarrow \check{H}$ must restrict to an isomorphism of $k[G(H)]$ and $k[G(\check{H})]$, and satisfy $f(x) = \beta\check{x}$ for some $0 \neq \beta \in k$ and $f(a^\nu) = \check{a}^{\check{\nu}}$ by part a). Thus $N = \check{N}$ and $f(a) = \check{a}^s$ for some s relatively prime to N. Since $f(a^\nu) = \check{a}^{\check{\nu}}$, we have $a^{\nu s} = \check{a}^{\check{\nu}}$, and thus $\omega^{\nu s} = \omega^{\check{\nu}}$. Now $N = \check{N}$ also implies that $(\omega) = (\check{\omega})$. Therefore $\check{\omega} = \omega^t$ for some $1 \leq t < N$ which is relatively prime to N. Applying f to $xa = qax$ we derive $\check{q}^s\check{a}^s\check{x} = \check{x}\check{a}^s = q\check{a}^s\check{x}$. Therefore $\check{q}^s = q$. But then $\check{q}^s = \check{\omega}^{\check{\nu}s} = \omega^{t\check{\nu}s} = \omega^{t\nu s^2}$. As $q = \omega^\nu$ we have the equation $\omega^{ts^2\nu} = \omega^\nu$. At this point is is easy to complete the proof of part c), and to establish parts d) and e).

We now show part f). Assume that $f : H \longrightarrow H^{cop}$ is a map of Hopf algebras, and suppose that $f(x) \neq 0$. Let s be the antipode of H, and consider the composite $H \xrightarrow{f} H^{cop} \xrightarrow{s} H^{op} = H_{(N,\nu,\omega)}$. Then $z = s(f(x))$ satisfies the hypothesis of part b). Therefore $z = \beta' x$ for some $0 \neq \beta' \in k$ and $s(f(a)) = a^\nu$. It now follows that $f(x) = \beta a^{-\nu}x$ for some $0 \neq \beta \in k$ and $f(a^\nu) = a^{-\nu}$. Now $f(a) = a^s$ for some $s \in Z$, and thus $a^{s\nu} = a^{-\nu}$. Applying f to $xa = qax$ we see that $q = q^s$, and hence $a^\nu = a^{s\nu}$. Therefore $a^\nu = a^{-\nu}$, and consequently $a^{2\nu} = 1$. This means that N divides 2ν. Since N does not divide ν^2, it follows that ν is odd and N is even. Hence $N = 2\nu$, and consequently $q^2 = 1$. Since $q = q^s$ and $q \neq 1$, it now follows that s is odd. This completes the proof of part f). Since part g) is an immediate consequence of part f), the proof of the proposition is complete.

Let $H = H_{(N,\nu,\omega)}$ and $G = G(H)$. Then the Hopf algebra map $\Pi : H \longrightarrow k[G]$ determined by $\Pi(a) = a$ and $\Pi(x) = 0$ is a projection onto the group algebra $k[G]$. Therefore $H_{(N,\nu,\omega)}$ has the structure of a product $B \times k[G]$ which is described in [21].

If $f \in \mathrm{End}_{Hopf}(H)$ and $f(x) = 0$, then $f = \mathsf{f} \circ \Pi$ for some $\mathsf{f} \in \mathrm{End}_{Hopf}(k[G])$. Note that the bialgebra endomorphisms f of H which satisfy $f(x) = 0$ form a sub-semigroup of $\mathrm{End}_{Hopf}(H)$ which can be identified with the semigroup $\mathrm{End}_{Group}(G)$.

The bialgebra endomorphisms f of H which satisfy $f(x) \neq 0$ form a sub-semigroup $\mathrm{End}_{Hopf}(H)'$ of $\mathrm{End}_{Hopf}(H)$ as well. Let $U = \{s \in Z_N \,|\, s \equiv 1\,(mod\ n)\}$, where n is the order of $q = \omega^\nu$. Then by the proof of parts b) and c) of the previous theorem we see that there is an isomorphism of semigroups $U \times k^* \simeq \mathrm{End}_{Hopf}(H)'$ given by $(s, \beta) \mapsto f$, where $f(a) = a^s$ and $f(x) = \beta x$.

Let $f : H \longrightarrow H^{cop}$ be a Hopf algebra map, and assume that $N \neq 2\nu$ or that ν is even. Then $f(x) = 0$ by part f) of the previous theorem. Thus $\mathrm{Im}\, f \subseteq k[G]$ and f factors through Π as above. In particular $f \in \mathrm{End}_{Hopf}(H)$.

We conclude this section by determining which of the Hopf algebras $H_{(N,\nu,\omega)}$ are quasitriangular.

Corollary 3 *Suppose that $N > 1$ and the field k has a primitive N^{th} root of unity ω and let $H = H_{(N,\nu,\omega)}$. Then*

a) *H admits a quasitriangular structure if and only if $N = 2\nu$ and ν is odd.*

Suppose that $N = 2\nu$ and ν is odd.

b) *The quasitriangular structures which H admits are described by $(H, R_{s,\beta})$ where*

$$R_{s,\beta} = \frac{1}{N}\Big(\sum_{i,\ell=0}^{N-1} \omega^{-i\ell} a^i \otimes a^{s\ell}\Big) + \frac{\beta}{N}\Big(\sum_{i,\ell=0}^{N-1} \omega^{-i\ell} a^i x \otimes a^{s\ell+\nu} x\Big),$$

$1 \leq s < N$ is odd and $\beta \in k$.

c) $(H, R_{s,\beta})$ *is triangular if and only if* $s = \nu$.

d) $(H, R_{s,\beta}, v)$ *is a ribbon Hopf algebra, where* $v = a^{\nu}\Theta(\omega^{-s})$.

PROOF: We will sketch an argument, leaving most of the details to the reader. Recall that the quasitriangular structures which H admits can be thought of as the bialgebra maps $F : H^* \longrightarrow H^{cop}$ which satisfy (9); that is

$$\sum(p_{(1)} \rightharpoonup h)F(p_{(2)}) = \sum F(p_{(1)})(h \leftharpoonup p_{(2)})$$

for all $p \in H^*$ and $h \in H$.

Suppose that $F : H^* \longrightarrow H^{cop}$ is a bialgebra map and set $G = (a)$. By Proposition 8 there is an isomorphism of Hopf algebras $H \simeq H^*$ which sends a to A and x to X as defined above. Thus we can write F as a composite of Hopf algebra maps $H^* \simeq H \xrightarrow{f} H^{cop}$.

Assume further that (9) is satisfied by F. The reader can check that this condition holds generally if and only if it holds for subcoalgebras C of H and P of H^* which generate H and H^* respectively as algebras. Clearly (9) holds whenever $p = \epsilon$ or $h = 1$. Thus (9) holds for $H = H_{(N,\nu,\omega)}$ if and only if it holds when $p = A, A^{\nu}$ or X and $h = a, a^{\nu}$ or x; that is

$$(A \rightharpoonup h)f(a) = f(a)(h \leftharpoonup A),$$

$$(A^{\nu} \rightharpoonup h)f(a^{\nu}) = f(a^{\nu})(h \leftharpoonup A^{\nu})$$

and

$$(X \rightharpoonup h)f(a^{\nu}) + hf(x) = f(x)(h \leftharpoonup A^{\nu}) + (h \leftharpoonup X)$$

hold for $h = a, a^{\nu}$ and x. These conditions translate to

$$qxf(a) = f(a)x, \quad qf(x)a = af(x), \quad f(a^{\nu}) = a^{\nu} \quad \text{and} \quad xf(x) = f(x)x.$$

Now $f(a) = a^s$ for some integer s. Thus the first and third equations yield $q^{s+1} = 1$ and $q^s = q$. Therefore $q^2 = 1$, so s is odd and N divides 2ν. Consequently ν is odd and $N = 2\nu$. Note that $f(x) = \beta a^{-\nu}x$ for some $\beta \in k$ by part f) of the previous theorem.

Suppose that $N = 2\nu$, where ν is odd, and $1 \leq s < N$ is an odd integer. Then the order of $q = \omega^{\nu}$ is $r = 2$. Let $\beta \in k$. Then there is a unique bialgebra map $f : H \longrightarrow H^{cop}$ determined by $f(a) = a^s$ and $f(x) = \beta a^{-\nu}x = \beta a^{\nu}x$.

Let $F : H^* \longrightarrow H^{cop}$ be the composite $H^* \simeq H \xrightarrow{f} H^{cop}$. We will determine the $R \in H \otimes H$ such that $F = f_R$. Let $G = (a)$. We have noted (6) that the elements $e_\ell = \sum_{i=0}^{N-1} \frac{1}{N}\omega^{-i\ell}a^i$ form an orthonormal set of idempotents in the group algebra $k[G]$. By (22) it follows that $A^\ell X^m(\frac{1}{(j)_{q^\nu}!}e_i x^j) = \delta_{\ell,i}\delta_{m,j}$ for all $0 \leq \ell, i < N$ and $0 \leq m, j < r$. Thus for any linear map $T : H^* \longrightarrow H$ we have $T = f_R$, where $R = \sum_{\ell=0}^{N-1}\sum_{m=0}^{r-1} \frac{1}{(m)_{q^\nu}!}e_\ell x^m \otimes T(A^\ell X^m)$. Therefore for $R_{s,\beta}$ satisfying $F = f_{R_{s,\beta}}$, we compute

$$R_{s,\beta} = \sum_{\ell=0}^{N-1}\sum_{m=0}^{r-1} \frac{1}{(m)_{q^\nu}!} e_\ell x^m \otimes a^{s\ell}(\beta a^{-\nu}x)^m.$$

This concludes the proof of parts a) and b).

Consider the quasitriangular Hopf algebra $(H, R_{s,\beta})$. Since $x^2 = 0$ it follows by the formula above that $u_{s,\beta} = \sum s(R^{(2)}_{s,\beta})R^{(1)}_{s,\beta} = \Theta(\omega^{-s})$. It is easy to see that $\hbar = 1$. Since $s^2(h) = a^{-\nu}ha^\nu$ for all $h \in H$, it follows that $v = a^\nu\Theta(\omega^{-s})$ is a ribbon element for $(H, R_{s,\beta})$. Since $\Theta(\omega^{-s})$ is a grouplike element if and only if $\omega^{-2s} = 1$, we have established parts c) and d). This completes the proof.

For Sweedler's example $T_{2,\nu} = H_{(2,1,\omega)}$ observe that $R_{1,\beta}$ is the R_β of Section 2.1.

5.2 Construction and Properties of $U_{(N,\nu,\omega)}$

Let N and ν be positive integers such that $1 \leq \nu < N$ and N does not divide ν^2. Suppose that $\omega \in k$ is a primitive N^{th} of unity. Let $q = \omega^\nu$ and r be the order of $q^\nu = \omega^{\nu^2}$. We will define a Hopf algebra $U_{(N,\nu,\omega)}$ as follows. As an algebra $U = U_{(N,\nu,\omega)}$ is generated by a, x and y which satisfy the relations

$$a^N = 1, \qquad x^r = 0, \qquad y^r = 0,$$

$$xa = qax, \qquad ya = q^{-1}ay \qquad \text{and}$$

$$yx - q^{-\nu}xy = a^{2\nu} - 1.$$

The coalgebra structure of $U_{(N,\nu,\omega)}$ is determined by

$$\Delta(a) = a \otimes a, \quad \Delta(x) = x \otimes a^\nu + 1 \otimes x \quad \text{and} \quad \Delta(y) = y \otimes a^\nu + 1 \otimes y.$$

The antipode of U is the algebra map $s : U \longrightarrow U^{op}$ determined by

$$s(a) = a^{-1}, \qquad s(x) = -q^{-\nu}a^{-\nu}x \qquad \text{and} \quad s(y) = -q^{\nu}a^{-\nu}y.$$

Note that $s^2(a) = a, s^2(x) = q^{-\nu}x, s^2(y) = q^{\nu}y$ and $s^2(h) = a^{\nu}ha^{-\nu}$ for all $h \in U$. In particular s^2 is an inner automorphism.

The Hopf algebra $U_{(N,\nu,\omega)}$ is constructed in the same manner as is $H_{(N,\nu,\omega)}$. We start with the 5-dimensional coalgebra C over k with basis $\{A, B, D, X, Y\}$ whose structure is determined by

$$\Delta(A) = A \otimes A, \quad \Delta(B) = B \otimes B, \quad \Delta(D) = D \otimes D,$$

$$\Delta(X) = X \otimes B + D \otimes X \quad \text{and} \quad \Delta(Y) = Y \otimes B + D \otimes Y,$$

and we let I be the ideal of $T(C)$ generated by

$$A^N - 1, \quad B - A^{\nu}, \quad D - 1, \quad X^r, \quad Y^r,$$

$$XA - qAX, \quad YA - q^{-1}AY \quad \text{and} \quad YX - q^{-\nu}XY - A^{2\nu} + 1.$$

The details are straightforward and left to the reader.

We will show that $\text{ß} = \{a^{\ell}x^m y^p \,|\, 0 \leq \ell < N,\ 0 \leq m, p < r\}$ is a linear basis for $U_{(N,\nu,\omega)}$ by applying the Diamond Lemma to $T(C) = k\{A, B, D, X, Y\}$ with $A < B < D < X < Y$ and the substitution rules

$$A^N \leftarrow 1, \quad B \leftarrow A^{\nu}, \quad D \leftarrow 1, \quad X^r \leftarrow 0, \quad Y^r \leftarrow 0,$$

$$XA \leftarrow qAX, \quad YA \leftarrow q^{-1}AY \quad \text{and}$$
$$YX \leftarrow q^{-\nu}XY + A^{2\nu} - 1.$$

There are ten ambiguities to resolve, all of them overlap:

$$(A^{\ell}A^{N-\ell})A^{\ell} = A^{\ell}(A^{N-\ell}A^{\ell}),$$

$$(X^{\ell}X^{r-\ell})X^{\ell} = X^{\ell}(X^{r-\ell}X^{\ell}), \qquad (Y^{\ell}Y^{r-\ell})Y^{\ell} = Y^{\ell}(Y^{r-\ell}Y^{\ell}),$$

$$(XA)A^{N-1} = X(AA^{N-1}), \qquad (YA)A^{N-1} = Y(AA^{N-1}),$$

$$(X^{r-1}X)A = X^{r-1}(XA), \quad (Y^{r-1}Y)A = Y^{r-1}(YA),$$

$$(YX)A = Y(XA),$$

$$(YX)X^{r-1} = Y(XX^{r-1}) \qquad \text{and} \qquad (Y^{r-1}Y)X = Y^{r-1}(YX).$$

All but the last two are trivial to resolve. By induction on ℓ the monomial YX^ℓ reduces to

$$q^{-\nu\ell}X^\ell Y + (1+q^\nu+\ldots+q^{\nu(\ell-1)})A^{2\nu}X^{\ell-1} - (1+q^{-\nu}+\ldots+q^{-\nu(\ell-1)})X^{\ell-1}$$

and the monomial $Y^\ell X$ reduces to

$$q^{-\nu\ell}XY^\ell + q^{-\nu(\ell-1)}(1+q^{-\nu}+\ldots+q^{-\nu(\ell-1)})A^{2\nu}Y^{\ell-1} - (1+q^{-\nu}+\ldots+q^{-\nu(\ell-1)})Y^{\ell-1}$$

for $1 \leq \ell < r$. These reductions give immediate resolutions of the last two ambiguities.

Proposition 10 *Suppose that $N > 1$ and the field k has a primitive N^{th} root of unity ω. Let $U = U_{(N,\nu,\omega)}$. Then:*

a) *U is a pointed coalgebra and $G(U) = (a)$.*

b) *The set $\beta = \{a^\ell x^m y^p \,|\, 0 \leq \ell < N,\ 0 \leq m,p < r\}$ is linear basis for U. Thus* $\mathrm{Dim}\, U_{(N,\nu,\omega)} = \frac{N^3}{(N,\nu^2)^2}$.

c)

$$\Delta(a^\ell x^m y^p) = \sum_{i=0}^{m}\sum_{j=0}^{p} \binom{m}{i}_{q^\nu} \binom{p}{j}_{q^{-\nu}} q^{\nu i(p-j)} a^\ell x^{m-i} y^{p-j} \otimes a^{\ell+\nu(m-i)+\nu(p-j)} x^i y^j$$

for all $0 \leq \ell < N$ and $0 \leq m,p < r$.

d) *$\Lambda = (1+a+a^2+\ldots+a^{N-1})x^{r-1}y^{r-1}$ is a non-zero two-sided integral for U. Thus U is unimodular.*

e) *$H_{(N,\nu,\omega)}$ is the sub-Hopf algebra of $U_{(N,\nu,\omega)}$ generated by a and x.*

PROOF: Part a) follows by Lemma 1. We have just sketched a proof for part b). Part e) follows from part b). Part c) follows from part e) and part c) of Proposition 7.

To show part d), we use the fact that the space of left integrals of a finite-dimensional Hopf algebra H over k is a one-dimensional ideal of H. Let $G = (a)$. Then by part b) we have that U is a free left $k[G]$-module with basis $\{x^m y^p \,|\, 0 \leq m,p < r\}$. Suppose that Λ is a non-zero left integral for

U, and write $\Lambda = \sum_{m=m_0}^{r-1} \sum_{p=p_0}^{r-1} c_{m,p} x^m y^p$, where $c_{m,p} \in k[G]$ and $c_{m_0,p_0} \neq 0$. Let α be the distinguished grouplike element of U^*. Then ϵ and α vanish on x and y since x and y are nilpotent. Therefore $x\Lambda = \epsilon(x)\Lambda = 0 = \Lambda\alpha(y) = \Lambda y$ which means that $m_0 = r - 1 = r_0$. Hence $\Lambda = cx^{r-1}y^{r-1}$ for some $c \in k[G]$. Now $a\Lambda = \epsilon(a)\Lambda = \Lambda$ implies that c is a left integral for $k[G]$. We may assume that $c = 1 + a + a^2 + \ldots + a^{N-1}$. Since $\Lambda a = \Lambda$ it follows that $\alpha(a) = 1$. Thus $\alpha = \epsilon$, and U is unimodular. This concludes the proof.

Imitating the proof of Proposition 9, we can easily express left and right integrals of $U^*_{(N,\nu,\omega)}$ in terms of the dual basis of ß and determine the distinguished grouplike element of $U_{(N,\nu,\omega)}$.

Proposition 11 *Suppose that $N > 1$ and the field k has a primitive N^{th} root of unity ω. Let ß be the basis for $U = U_{(N,\nu,\omega)}$ described above. Then:*

a) $\lambda_\ell = \overline{x^{r-1}y^{r-1}}$ *and* $\lambda_r = \overline{a^{-2\nu(r-1)}x^{r-1}y^{r-1}}$ *are non-zero left and right integrals respectively for U^*.*

b) *The distinguished grouplike element of U is $g = a^{-2\nu(r-1)}$.*

c) $G(U^*) \simeq Z_{(N,2\nu)}$.

PROOF: We need only show part c). Let m be a positive integer and consider the multiplicative group of mth roots of unity G_m of k. Since $G(U^*)$ consists of algebra maps, it follows from the relations defining the algebra structure of U that there is a group isomorphism $G(U^*) \simeq G_{2\nu} \cap G_N = G_{(N,2\nu)}$ given by $\eta \mapsto \eta(a)$. Since G_N is cyclic of order N, it follows that $G_{2\nu} \cap G_N$ has order $(N, 2\nu)$.

Note that $G(U^*) = \{\epsilon\}$ may very well happen. This is the case if and only if N is odd and ν and N are relatively prime by the previous proposition.

We now turn our attention to the connection between $D(H_{(N,\nu,\omega)})$ and $U_{(N,\nu,\omega)}$. Set $H = H_{(N,\nu,\omega)}$. Recall that $D(H) = H^{*cop} \otimes H$ as a coalgebra, and that $\imath_H : H \longrightarrow D(H)$ and $\imath_{H^*} : H^{*cop} \longrightarrow D(H)$ defined by $\imath_H(h) = \epsilon \otimes h$ and $\imath_{H^*}(p) = p \otimes 1$ for $h \in H$ and $p \in H^*$ are one-one Hopf algebra maps. Let $s, S = s^*$ and $\mathbf{s}$ be the antipodes of H, H^* and $D(H)$ respectively. Let $Y = S^{-1}(X) = -A^{-\nu}X$ and define

$$\mathbf{a} = \imath_H(a), \quad \mathbf{x} = \imath_H(x), \quad \mathbf{A} = \imath_{H^*}(A) \quad \text{and} \quad \mathbf{y} = \imath_{H^*}(Y).$$

Then the relations

$$\mathbf{a}^N = 1, \qquad \mathbf{x}^r = 0, \qquad \mathbf{x}\mathbf{a} = q\mathbf{a}\mathbf{x},$$

$$\mathbf{A}^N = 1, \qquad \mathbf{y}^r = 0 \qquad \text{and} \qquad \mathbf{A}^{-1}\mathbf{y} = q\mathbf{y}\mathbf{A}^{-1}$$

hold in $D(H)$ since they hold in H or H^*. Using (4), (5) and (22) we derive

$$\mathbf{A}\mathbf{a} = \mathbf{a}\mathbf{A} \qquad \text{and} \qquad \mathbf{y}\mathbf{x} = q^{-\nu}\mathbf{x}\mathbf{y} + \mathbf{A}^{-\nu}\mathbf{a}^{\nu} - 1,$$

$$\mathbf{x}\mathbf{A} = q^{-1}\mathbf{A}\mathbf{x} \qquad \text{and} \qquad \mathbf{a}\mathbf{y} = q\mathbf{y}\mathbf{a}.$$

These ten relations give a complete description of the algebra structure of $D(H)$. Therefore there is a surjective algebra map

$$\pi = \pi_{(N,\nu,\omega)} : D(H_{(N,\nu,\omega)}) \longrightarrow U_{(N,\nu,\omega)}$$

determined by

$$\pi(\mathbf{a}) = a, \quad \pi(\mathbf{x}) = x, \quad \pi(\mathbf{A}) = a^{-1} \quad \text{and} \quad \pi(\mathbf{y}) = y.$$

It is easy to see that π is a coalgebra map. Therefore π is a map of Hopf algebras. By considering dimensions it follows that $\ker \pi = (\mathbf{A}^{-1} - \mathbf{a})$. Hence

$$D(H_{(N,\nu,\omega)})/(\mathbf{A}^{-1} - \mathbf{a}) \simeq U_{(N,\nu,\omega)}$$

as Hopf algebras. Since π is onto, $(U_{(N,\nu,\omega)}, R)$ is quasitriangular, where $R = (\pi \otimes \pi)(\mathcal{R})$.

Now let H be any finite-dimensional Hopf algebra over k. Let $\{h_1, \ldots, h_n\}$ be a linear basis for H and $\{\overline{h_1}, \ldots, \overline{h_n}\}$ be the dual basis for H^*. Recall that $\mathcal{R} = \sum_{i=1}^n (\epsilon \otimes h_i) \otimes (\overline{h_i} \otimes 1)$ and $\mathsf{u} = \sum \mathsf{s}(\mathcal{R}^{(2)})\mathcal{R}^{(1)} = \sum_{i=1}^n S^{-1}(\overline{h_i}) \otimes h_i$.

Now let $H = H_{(N,\nu,\omega)}$ again, let $\text{ß} = \{a^{\ell}x^m \mid 0 \leq \ell < N, 0 \leq m < r\}, G = (a)$ and $\mathcal{G} = (A)$. For $\ell \in Z$ set

$$E_\ell = \sum_{i=0}^{N-1} \frac{\omega^{-i\ell}}{N} A^i \quad \text{and} \quad e_\ell = \sum_{i=0}^{N-1} \frac{\omega^{-i\ell}}{N} a^i.$$

Then $\{E_0, \ldots, E_{N-1}\}$ and $\{e_0, \ldots, e_{N-1}\}$ are orthonormal sets of idempotents for the group algebras $k[\mathcal{G}]$ and $k[G]$ respectively by (6). Note that

$$y e_\ell = e_{\ell+\nu} y \tag{23}$$

for all $\ell \in Z$. The elements in the dual basis of ß for $H^*_{(N,\nu,\omega)}$ are described by

$$\overline{a^\ell x^m} = \frac{1}{(m)_{q^\nu}!} E_\ell X^m \qquad \text{for all} \qquad 0 \le \ell < N, 0 \le m < r$$

by (22). Therefore

$$\mathcal{R} = \sum_{\ell=0}^{N-1} \sum_{m=0}^{r-1} \frac{1}{(m)_{q^\nu}!} (\epsilon \otimes a^\ell x^m) \otimes (E_\ell X^m \otimes 1) \tag{24}$$

and

$$\mathsf{u} = \sum_{m=0}^{r-1} \sum_{\ell=0}^{N-1} \frac{1}{(m)_{q^\nu}!} (Y^m E_{-\ell} \otimes a^\ell x^m). \tag{25}$$

Recall that $\Theta : k \longrightarrow k[G]$ is defined by $\Theta(\beta) = \sum_{\ell=0}^{N-1} \beta^{\ell^2} e_\ell$ for all $\beta \in k$.

Theorem 5 *Suppose that $N > 1$ and the field k has a primitive N^{th} root of unity ω. Let $(U_{(N,\nu,\omega)}, R)$ be defined as above, and set $u = \sum s(R^{(2)})R^{(1)}$, where s be the antipode of U. Then:*

a) $R = \sum_{m=0}^{r-1} \sum_{\ell=0}^{N-1} \frac{1}{(m)_{q^\nu}!} a^\ell x^m \otimes s^{-1}(y^m e_\ell)$.

b) $u = \sum_{m=0}^{r-1} \frac{1}{(m)_{q^\nu}!} y^m \Theta(\omega) x^m$.

c) $u^{-1} = \sum_{m=0}^{r-1} \frac{(-1)^m q^{m(m-3)/2}}{(m)_{q^\nu!}} \Theta(\omega^{-1}) y^m x^m$.

d) *$(U_{(N,\nu,\omega)}, R, v)$ is a ribbon Hopf algebra, where $v = a^{\nu(r-1)}u$.*

Proof: Parts a) and b) follow by (24) and (25) respectively since $\pi_{(N,\nu,\omega)}$ is a Hopf algebra map. To show part d) we first recall that $U = U_{(N,\nu,\omega)}$ is unimodular. By part b) of Proposition 11 the distinguished grouplike element of U is $g = a^{-2\nu(r-1)}$. Thus equation (16) is simply $\hbar = g^{2\nu(r-1)}$. By Proposition 2 it follows that v is a ribbon element of (U, R) if and only if $v = \ell u$, where $\ell \in G(U)$ satisfies $\ell^2 = a^{2\nu(r-1)}$ and $s^2(h) = \ell^{-1}h\ell$ for all $h \in U$. Now $\ell = a^{\nu(r-1)}$ satisfies these conditions.

We now show part c). Since $u^{-1} = \sum R^{(2)} s^2(R^{(1)})$ by (12), we have by part a) and (11) that $u^{-1} = \sum_{m=0}^{r-1} \sum_{\ell=0}^{N-1} \frac{1}{(m)_{q^\nu}!} y^m e_\ell s^3(a^\ell x^m)$. Now $s(x) = -xa^{-\nu}$ and $a^{-\nu}x = q^\nu x a^{-\nu}$. Therefore $s^3(x) = -q^{-\nu} x a^{-\nu}$ and $(xa^{-\nu})^m =$

$q^{\nu(m-1)m/2}x^m a^{-\nu m}$ for $m \geq 0$. Since $s^3(a^\ell x^m) = s^3(x^m)a^{-\ell}$, we compute $s^3(a^\ell x^m) = (-1)^m q^{\nu m(m-3)/2} x^m a^{-\nu m-\ell}$ for $m \geq 0$. Thus we find

$$\begin{aligned} y^m e_\ell s^3(a^\ell x^m) &= (-1)^m q^{\nu m(m-3)/2} e_{\ell+\nu m} y^m x^m a^{-\nu m-\ell} \\ &= (-1)^m q^{\nu m(m-3)/2} a^{-(\ell+\nu m)} e_{(\ell+\nu m)} y^m x^m \end{aligned}$$

using (23). Thus the proof of part c) follows by (7). This completes the proof of the theorem.

Observe that $(U_{(N,\nu,\omega)}, R)$ is never triangular since u is never a grouplike element by part a) of Proposition 10. We have noted that $H = H_{(N,\nu,\omega)}$ can be identified with the subalgebra of $U = U_{(N,\nu,\omega)}$ generated by a and x. Let $\imath : H \longrightarrow U$ be the inclusion. Then $(U, R, \imath)$ is a quasitriangular envelope for H.

Observe that $(U_{\mathsf{q}}(sl_2)', R) = (U_{(n,2,\omega)}, R)$, where $\mathsf{q} = \omega^{-2} = q^{-1}$. For if we set $e = ya^{-1}/(q^2-1)$, $f = xa^{-1}$ and $k = a^{-1}$, then $t = \omega^{-1}$ is a primitive n^{th} root of unity, $\mathsf{q} = t^2$, r is the order of $t^4 = \mathsf{q}^2$, the algebra structure of $U_{(n,2,\omega)}$ is defined by the relations

$$k^n = 1, \quad ke = \mathsf{q}ek, \quad kf = \mathsf{q}^{-1}fk, \quad e^r = 0, \quad f^r = 0 \quad \text{and}$$

$$ef - fe = \frac{k^2 - k^{-2}}{\mathsf{q} - \mathsf{q}^{-1}},$$

and the coalgebra structure of $U_{(n,2,\omega)}$ is determined by

$$\Delta(k) = k \otimes k, \quad \Delta(e) = e \otimes k^{-1} + k \otimes e \quad \text{and} \quad \Delta(f) = f \otimes k^{-1} + k \otimes f.$$

The fact that the quasitriangular structure is the usual one is proven in [14, Section 7].

By expanding the formulas for u and u^{-1} described in Theorem 5 we obtain by (23):

$$u = \sum_{m=0}^{r-1} \frac{1}{(m)_{q^\nu}! N} \left(\sum_{i,\ell=0}^{N-1} \omega^{\ell^2 - i(\ell+\nu m)} a^i \right) y^m x^m \tag{26}$$

and

$$u^{-1} = \sum_{m=0}^{r-1} \frac{(-1)^m q^{\nu(m-3)m/2}}{(m)_{q^\nu}! N} \left(\sum_{i,\ell=0}^{N-1} \omega^{-\ell^2 - i\ell} a^i \right) y^m x^m. \tag{27}$$

Let $\lambda_\ell = \overline{x^{r-1}y^{r-1}}$ and $\lambda_r = \overline{a^{-2\nu(r-1)}x^{r-1}y^{r-1}}$ be the left and right integrals respectively for $U^*_{(N,\nu,\omega)}$ described in Proposition 11. We will find the values of λ_r on v and v^{-1}. First we show that

$$\lambda_r(a^\ell y^m x^p) = q^{-\nu}\delta_{\ell,-2\nu(r-1)}\delta_{m,r-1}\delta_{p,r-1} \tag{28}$$

for all $0 \le \ell < N$ and $0 \le m,p < r$, where we regard ℓ and $-2\nu(r-1)$ as elements of Z_N. Since both λ_r and $\lambda_\ell \circ s$ are non-zero right integrals for $U^*_{(N,\nu,\omega)}$, they are scalar multiples of each other. Let Λ be the integral for $U_{(N,\nu,\omega)}$ defined in Proposition 10. Since $U_{(N,\nu,\omega)}$ is unimodular it follows that $s(\Lambda) = \Lambda$ by part a) of [22, Proposition 3]. Thus $\lambda_\ell(s(\Lambda)) = \lambda_\ell(\Lambda) = 1 = \lambda_r(\Lambda)$ which means $\lambda_\ell \circ s = \lambda_r$. Therefore $\lambda_r(a^\ell y^m x^p) = 0$ unless $m = p = r-1$. Since $s(x) = -q^{-\nu}a^{-\nu}x$ and $s(y) = -ya^{-\nu}$, we find that

$$\begin{aligned} s(y^{r-1}x^{r-1}) = s(x)^{r-1}s(y)^{r-1} &= q^{-\nu}a^{-\nu(r-1)}x^{r-1}y^{r-1}a^{-\nu(r-1)} \\ &= q^{-\nu}a^{-2\nu(r-1)}x^{r-1}y^{r-1}. \end{aligned}$$

Thus (28) is established.

By using (28), (26) and (27) we see that

$$\lambda_r(a^j u) = \frac{q^{-\nu}}{(r-1)_{q^\nu}!N}(\sum_{\ell=0}^{N-1}\omega^{\ell^2+(j+2\nu(r-1))(\ell+\nu(r-1))})$$

and

$$\lambda_r(a^j u^{-1}) = (-1)^{r-1}q^{-\nu}(\frac{q^{\nu(r-1)(r-4)/2}}{(r-1)_{q^\nu}!N})(\sum_{\ell=0}^{N-1}\omega^{-\ell^2+(j+2\nu(r-1))\ell})$$

for all $0 \le j < N$. Using the identity $\ell^2 + (j+2\nu(r-1))(\ell+\nu(r-1)) = (\ell+\nu(r-1))(\ell+\nu(r-1)+j) + \nu^2(r-1)^2$ we conclude that

$$\lambda_r(a^j u) = \frac{1}{(r-1)_{q^\nu}!N}(\sum_{\ell=0}^{N-1}\omega^{\ell(\ell+j)}) \tag{29}$$

and using the identity $-\ell^2 + (j+2\nu(r-1))\ell = -(\ell-\nu(r-1))(\ell-\nu(r-1)-j) + \nu^2(r-1)^2 + \nu(r-1)j$ we conclude that

$$\lambda_r(a^j u^{-1}) = (-1)^{r-1}q^{(r-1)j}\frac{q^{\nu(r-1)(r-4)/2}}{(r-1)_{q^\nu}!N}(\sum_{\ell=0}^{N-1}\omega^{-\ell(\ell+j)}) \tag{30}$$

for all $j \in Z$. Hennings' invariant is defined on a very large class of Hopf algebras of type $U_{(N,\nu,\omega)}$.

Proposition 12 *Suppose that $N > 1$ and the field k contains a primitive N^{th} root of unity ω. Let $(U_{(N,\nu,\omega)}, R)$ be the quasitriangular Hopf algebra defined above and set $U = U_{(N,\nu,\omega)}$. Suppose that λ a non-zero right integral for U^* and $u = \sum s(R^{(2)})R^{(1)}$. Then:*

a) *$v = a^{\nu(r-1)}$ is a ribbon element for (U,R), and $\lambda(v) \neq 0 \neq \lambda(v^{-1})$ except when $N = 2M$ and either M is even and ν is odd or M is odd and ν is even.*

b) *If $N = 2M$ where M is even and ν is odd, then $\lambda(v) = 0 = \lambda(v^{-1})$ for all ribbon elements v for (U,R).*

c) *Suppose that $N = 2M$ where M is odd and ν is even. Then $v = a^{M+\nu(r-1)}u$ is a ribbon element for (U,R) which satisfies $\lambda(v) \neq 0 \neq \lambda(v^{-1})$.*

PROOF: We first note that v is a ribbon element for (U,R) if and only if $v = a^j u$, where $\omega^{2j} = \omega^{2\nu(r-1)}$ and $\omega^{\nu j} = \omega^{-\nu^2}$. These conditions are the equivalents of $\ell^2 = \hbar = a^{2\nu(r-1)}$ and $s^2(h) = \ell^{-1}h\ell$ for all $h \in U$, where $\ell = a^j$. Let $v = a^j u$ be a ribbon element for (U,R). We may assume that $\lambda = \lambda_r$. By (29) it follows that $\lambda(v) \neq 0$ if and only if $\sum_{\ell=0}^{n-1} \omega^{\ell(\ell+j)} \neq 0$, and likewise by (30) it follows that $\lambda(v^{-1}) \neq 0$ if and only if $\sum_{\ell=0}^{N-1} \omega^{-\ell(\ell+j)} \neq 0$.

We use Lemma 2 to draw several conclusions at this point. First of all, if N is odd then $\lambda(v) \neq 0 \neq \lambda(v^{-1})$. Suppose that $N = 2M$ is even. If M is even, then $\lambda(v) \neq 0 \neq \lambda(v^{-1})$ if and only if j is even. If M is odd, then $\lambda(v) \neq 0 \neq \lambda(v^{-1})$ if and only if j is odd.

When N is even, observe that ν is odd if and only if $\nu(r-1)$ is odd. Part a) now follows. To see part b), note that $\omega^{j\nu} = \omega^{-\nu^2}$ means $N = 2M$ divides $(j+\nu)\nu$. Thus if ν is odd then j must be odd also. Part c) is clear.

Let $U = U_{(N,\nu,\omega)}$. Factorizability of (U,R) reduces to factorizability of $(k[G(U)], R_{N-1})$.

Theorem 6 *Suppose that $N > 1$ and the field k contains a primitive N^{th} root of unity ω. Then $(U_{(N,\nu,\omega)}, R)$ is factorizable if and only if N is odd.*

PROOF: Let $U = U_{(N,\nu,\omega)}$ and $K = k[G]$, where $G = G(U)$. We will show that factorizability of (U,R) is equivalent to several conditions, the last of which is $\Theta(\omega^{-1}) \leftharpoonup K^* = K$. It therefore follows that (U, R) is factorizable if and only if $(k[G], R_{N-1})$ is factorizable.

By Corollary 1 we have that (U, R) is factorizable if and only if $u^{-1} \leftharpoonup U^* = U$. Let Λ be a non-zero integral for U. Then $\Lambda \leftharpoonup U^* = U$. Thus to show that (U, R) is factorizable it is necessary and sufficient to find a $p \in U^*$ such that $\Lambda = u^{-1} \leftharpoonup p$. Now $\Lambda = s(\Lambda) \in Ky^{r-1}x^{r-1}$ by part d) of Proposition 10. Thus we are lead to consider the free K-module $M = K \oplus Kyx \oplus Ky^2x^2 \oplus \cdots \oplus Ky^{r-1}x^{r-1}$. Since the monomials of the form $a^\ell y^m x^p$, where $0 \leq \ell < N$ and $0 \leq m, p < r$, form a linear basis for U, we can define functionals Y and X on U by $Y(a^\ell y^m x^p) = \delta_{m,1}\delta_{p,0}$ and $X(a^\ell y^m x^p) = \delta_{m,0}\delta_{p,1}$. Let $A \in G(U^*)$ be the algebra homomorphism determined by $A(a) = \omega$. Then it is not hard to show that

$$A^\ell Y^m X^p(a^u y^i x^j) = \delta_{m,i}\delta_{p,j}\omega^{\ell u}(m)_{q^{-\nu}}!(p)_{q^\nu}!$$

for all $0 \leq \ell, u < N$ and $0 \leq m, p, i, j < r$. The last argument used in the proof of Proposition 8 shows that $\{A^\ell Y^m X^p \,|\, 0 \leq \ell < N,\, 0 \leq m, p < r\}$ is a linear basis for U^*. For $0 \leq p \leq m < r$ set

$$\beta_{m,p} = \frac{(m)_{q^{-\nu}}!(p)_{q^\nu}!}{(m-p)_{q^{-\nu}}!(m-p)_{q^\nu}!} q^{-\nu(m-p)}.$$

Then using part c) of Proposition 7 it follows for all $c \in K$ that

$$cy^m x^m \leftharpoonup A^\ell Y^p X^p = \beta_{m,p} a^{2\nu p}(c \leftharpoonup A^\ell) y^{m-p} x^{m-p}$$

for $p \leq m$ and has the value 0 otherwise. Now set

$$m_p = \sum_{j=0}^{r-1-p} \frac{(-1)^{j+p} q^{-j\nu} q^{\nu((j+p)(j+p-3)/2)}(j+p)_{q^{-\nu}}!}{(j)_{q^{-\nu}}!(j)_{q^\nu}!} y^j x^j$$

for $0 \leq p < r$. Observe that $\{m_0, \ldots, m_{r-1}\}$ is a basis for M as a free K-module. By part c) of Theorem 5 we find that

$$u^{-1} \leftharpoonup KY^p X^p = a^{2\nu p}(\Theta(\omega^{-1}) \leftharpoonup K^*) m_p$$

for $0 \leq p < r$.

Now suppose that $u^{-1} \leftharpoonup U^* = U$. Then $\mathrm{Dim}\,(u^{-1} \leftharpoonup KY^p X^p) = N$. Therefore $\Theta(\omega^{-1}) \leftharpoonup K^* = K$. Now assume this condition holds. Then

$u^{-1} \leftharpoonup KY^pX^p = Km_p$ for all $0 \leq p < r$. Therefore $\Lambda \in M \subseteq u^{-1} \leftharpoonup U^*$. We have shown that (U, R) is factorizable if and only if $\Theta(\omega^{-1}) \leftharpoonup K^* = K$.

This last equation is equivalent to $\text{Rank}\,\Delta(\Theta(\omega^{-1})) = N$. We will give two proofs that this is the case if and only if N is odd. Since $\Theta(\omega^{-1}) = \sum_{i,\ell=0}^{N-1} \frac{1}{N}\omega^{-\ell(\ell+i)}a^i$, it follows that $\text{Rank}\,\Delta(\Theta(\omega^{-1})) = N$ if and only if

$$\sum_{\ell=0}^{N-1} \omega^{\ell(\ell+i)} \neq 0$$

for all $0 \leq i < N$. This is the case if and only if N is odd by Lemma 2. For a second proof, observe that the rank condition is equivalent to $(k[G], R_{N-1})$ is factorizable, or $(k[G], R_1)$ is factorizable, replacing ω by ω^{-1}. We showed at the end of Section 2.3 that this is the case if and only if N is odd.

By the theorem $(U_{\mathsf{q}}(sl_2)', R) = (U_{(n,2,\omega)}, R)$ is factorizable if and only if n is odd. By Proposition 12 it follows that the 3-manifold invariant described above is defined for $(U_{\mathsf{q}}(sl_2)', R, v)$, where v is a suitably chosen ribbon element.

References

[1] George Bergman, The diamond lemma for ring theory, *Adv. Math.* **29** (1978), 178–218.

[2] V. G. Drinfel'd, On almost cocommutative Hopf algebras, *Leningrad Math J.* (translation) **1** (1990), 321–342.

[3] V. G. Drinfel'd, Quantum Groups, *Procedings of the International Congress of Mathematicians,* Berkeley, California, USA (1987), 798–820.

[4] Shlomo Gelaki, Master's thesis, Ben Gurion Univ. of the Negev, Israel.

[5] M. A. Hennings, Invariants of links and 3-manifolds obtained from Hopf algebras, *preprint.*

[6] Louis H. Kauffman, Gauss codes, quantum groups and ribbon Hopf algebras, *Reviews in Math. Physics* (to appear).

[7] Louis H. Kauffman, "Knots and Physics", World Scientific, New Jersey, 1991.

[8] Louis H. Kauffman and David E. Radford, A necessary and sufficient condition for a finite-dimensional Drinfel'd double to be a ribbon Hopf algebra, *J. of Algebra* **159** (1993), 98–114.

[9] Louis H. Kauffman and David E. Radford, Hopf algebras and 3-manifold invariants, *research announcement* (paper to appear).

[10] S. Majid, Physics for algebraists: non-commutative and non-cocommutative Hopf algebras by a bicrossproduct construction, *J. of Algebra* **129** (1990), 1-91.

[11] S. Majid, Quasitriangular Hopf algebras and Yang–Baxter equations, *Int. J. Mod. Physics A* **5** (1990), 1-91.

[12] Susan Montgomery, Some remarks on filtrations of Hopf algebras, *preprint*.

[13] Tomotada Ohtsuki, Invariants of 3-manifolds derived from framed universal invariants of framed links, *preprint*, Tokyo University.

[14] D. E. Radford, Generalized double crossproducts associated with the quantized enveloping algebras, *preprint* (1991).

[15] David E. Radford, Irreducible Representations of $\mathcal{U}_q(g)$ arising from $\mathrm{Mod}^{\bullet}_{C^{1/2}}$, *Conf. Proc.* Quantum deformations of algebras and their representations, Bar Ilan Univ. and Weizmann Inst. Sci., Israel 1992.

[16] David E. Radford, Minimal quasitriangular Hopf algebras, *J. of Algebra* **157** (1993), 281–315.

[17] David E. Radford, On the antipode of a quasitriangular Hopf algebra, *J. of Algebra* **151** (1992), 1–11.

[18] David E. Radford, On the coradical of a finite-dimensional Hopf algebra, *Proc. Amer. Math. Soc.* **53** (1975), 9–15.

[19] David E. Radford, On the quasitriangular structures of a semisimple Hopf algebra, *J. of Algebra* **141** (1991), 354–358.

[20] David E. Radford, The order of the antipode of a finite-dimensional Hopf algebra is finite, *Amer. J. of Math.* **98** (1976), 333-355.

[21] David E. Radford, The structure of a Hopf algebra with a projection, *J. of Algebra* **92** (1985), 322–347.

[22] David E. Radford, The trace function and Hopf algebras, *J. of Algebra* (to appear).

[23] N. Yu. Reshetikhin, Invariants of links and quantized universal enveloping algebras, *LOMI preprint* (1988).

[24] N. Yu. Reshetikhin and M. A. Semenov-Tian-Shansky, Quantum R-matrices and factorization problems, *J. Geom. Physics* **5** (1988), 533-550.

[25] N. Yu. Reshetikhin and V. G. Turaev, Invariants of 3-manifolds via link polynomials and quantum groups, *Invent. Math.* **103** (1991), 547–597.

[26] N. Yu. Reshetikhin and V. G. Turaev, Ribbon graphs and their invariants derived from quantum groups, *Comm. Math. Physics* **127** (1990), 1–26.

[27] M. E. Sweedler, "Hopf Algebras," Mathematics Lecture Notes Series, Benjamin, New York, 1969.

[28] E. J. Taft, Noncocommutative sequences of divided powers, in *Conf. Proc.* Lie algebras and related topics, Lecture Notes in Math. Series, Springer Verlag, 1982.

[29] E. J. Taft, The order of the antipode of a finite-dimensional Hopf algebra, *Proc. Nat. Acad. Science USA* **68** (1971), 2631–2633.

Hopf Galois Extensions, Crossed Products, and Clifford Theory

H.-J. SCHNEIDER University of Munich, Munich, Germany

0 INTRODUCTION

The purpose of this expository article is to survey some of the recent developments in the area of Hopf Galois extensions and crossed products of Hopf algebras generalizing Clifford theory for groups [Cl], [CR], Lie algebras in characteristic zero [Bl] and finite group schemes [Vo1]. Moreover, proofs are given of some new or not easily accessible results.

Faithfully flat Hopf Galois extensions can be viewed as noncommutative torsors [D G] or principal fibre bundles where the role of the group is played by a Hopf algebra. A remarkable part of Clifford theory and also of the theory of torsors and of affine quotients of affine group schemes does in fact follow from theorems on arbitrary Hopf algebras or Hopf Galois extensions proved by completely different "noncommutative" methods.

In section 1 of this article, Hopf Galois extensions and crossed products are defined, some examples are given and some recent basic results are described generalizing theorems on affine quotients of affine group schemes.

Section 2 explains the important idea of Hopf modules. Galois descent theory introduced by A. Weil is seen to be a special case of the formalism of Hopf modules describing descent data.

In 1937 Clifford [Cl] investigated the relation between representations of a group G and of a normal subgroup N over a field k. His approach was then generalized by many authors (see [CR, §11 C, § 19 C]). As it turned out, the Hopf algebra structure of the quotient $k[G/N]$ is a crucial tool in this theory.

The extension of the group algebras $R = k[N] \subset A = k[G]$ is H-Galois where the Hopf algebra H is the quotient algebra $k[G/N]$. The induced module $M \otimes_R A$, M a right R-module, is a typical example of a Hopf module for any Hopf Galois extension.

In the group algebra case, there is a left and right R-linear retraction of the inclusion map $R \subset A$. Hence M is always a direct summand in $M \otimes_R A$ as an R-module. However, in the general case of Hopf algebras H, this is false. This difficulty already appears in the pioneering work of Voigt [Vo1] on induced representations of finite group schemes, where A is a finite-dimensional cocommutative Hopf algebra (for example the restricted universal enveloping algebra of a finite-dimensional p-Lie algebra) and $R \subset A$ is a normal Hopf subalgebra.

In section 3, the endomorphism rings of Hopf modules are studied. As in classical Clifford theory for groups, many problems are reduced to the study of the algebraic structure of crossed products which is usually a difficult task. Results on the crossed product structure of endomorphism rings of induced representations for groups [CR], strongly G-graded algebras [Da] and finite group schemes [Vo1] are generalized to the case of Hopf Galois extensions. In this section, 3.2, 3.11 and 3.12-14 are new.

Section 4 concerns the Krull-Schmidt decomposition of induced modules in the context of Hopf Galois extensions.

In 1959 Green [G] proved the following indecomposability theorem for group algebras $k[N] \subset k[G]$, G finite, G/N a p-group where $p > 0$ is the characteristic of the algebraically closed field k: If M is a finite-dimensional indecomposable R- module, then the induced module $M \otimes_R A$ is indecomposable over A. A version of Green's theorem is given for stable modules and Hopf Galois extensions. The result in 4.5 is new. And in the non-stable case a counterexample is described in detail.

Part of Clifford theory for groups is an irreducibility criterion showing how to produce an irreducible representation of the group starting from an irreducible representation of a normal subgroup. In 1969 Blattner [Bl] proved a Lie algebra analog of the irreducibility criterion for groups considering induced representations $M \otimes_R A$ where $A = U(\mathfrak{g})$ is the universal enveloping of a Lie algebra, $R = U(\mathfrak{n})$, $\mathfrak{n}$ a Lie ideal in $\mathfrak{g}$, and the characteristic of the field k is zero.

As for groups, the extension $U(\mathfrak{n}) \subset U(\mathfrak{g})$ is H-Galois where the Hopf algebra H is the quotient algebra $U(\mathfrak{g}/\mathfrak{n})$. In section 5, a generalization of these irreducibility criteria is described for H-Galois extensions. In particular, the notion of the stabilizer of representations of groups or Lie algebras in characteristic zero [Bl] is defined for Hopf Galois extensions.

One advantage of looking at Hopf Galois extensions is to obtain applications in various special cases, thereby obtaining a better understanding of each of them.

For example, the largest inner subcoalgebra of a Hopf algebra acting on an algebra can be seen to be the stabilizer of a certain representation. Thus the theorems on the existence of the largest inner Hopf subalgebra [Ma1], [Ma2] follow from more general results on the stabilizer of modules. Similarly, the Hopf Galois version of the irreducibility criterion for modules yields results on the simplicity of the crossed product algebra $R \#_\sigma H$ when the (weak) action of H on R is "outer" [BeM2].

In the special case of skew group algebras, the simplicity of the algebra $R \# k[G]$ when R is simple and G is outer was shown by Jacobson 1943 [Ja] for division rings R and by Azumaya 1946 [Az] in general.

The last section contains a report on joint work with S. Montgomery, which is still in progress. For arbitrary H-crossed products, two new "actions" are introduced. They allow to establish some basic results on H-crossed products such as the extension to the Martindale quotient ring.

I would like to thank D. Voigt and also my former students K.-P. Greipel, M. Josek and K. Hoffmann for many helpful and fruitful discussions concerning the material of this article, and my coauthor S. Montgomery for stimulation.

NOTATIONS (cf. [Sw 2], [A]). Algebras and coalgebras are always defined over the fixed base field k, and $\otimes = \otimes_k$ Hom $=$ Hom$_k$.

If T is a ring, $\mathcal{M}_T$ resp. ${}_T\mathcal{M}$ denotes the category of right resp. left T-modules.

If C is a coalgebra with comultiplication $\Delta : C \longrightarrow C \otimes C$ and counit $\varepsilon : C \longrightarrow k$, the usual convention $\Delta(c) = \sum c_1 \otimes c_2$ is used, and C^+ denotes the kernel of ε. Similarly, if X is a right C-comodule with comodule structure map $\Delta_X : X \longrightarrow X \otimes C$, $\Delta_X(x) = \sum x_0 \otimes x_1$. The left C^*- module structure on X given by $p \cdot x = \sum x_0\, p(x_1)$, $x \in X$, $p \in C^* = \mathrm{Hom}(C,k)$, is called the *adjoint module structure.*

For an algebra A, Hom(C,A) is an algebra with *convolution product* given by $(f*g)(c) = \sum f(c_1)g(c_2)$ for all $c \in C$ and f, $g \in \mathrm{Hom}(C,A)$, and the $*$-inverse of a linear map f is usually denoted by f^{-1}.

Throughout this paper, H denotes a Hopf algebra with comultiplication Δ, counit ε and antipode S.

A *right H-comodule algebra* A is an algebra A together with the structure Δ_A of a right H-comodule on A such that $\Delta_A : A \longrightarrow A \otimes H$ is an algebra map, i.e., $\Delta_A(xy) = \sum x_0 y_0 \otimes x_1 y_1$ for all x, $y \in A$, and $\Delta_A(1) = 1 \otimes 1$. Dually, an algebra A is a *left H-module algebra* if A is a left H-module such that $h \cdot (xy) = \sum (h_1 \cdot x)(h_2 \cdot y)$ and $h \cdot 1 = \varepsilon(h)1$ for all $h \in H$ and x, $y \in A$. If H is finite-dimensional, the two notions are equivalent; the *adjoint coaction* of an H-module algebra is again given by $p \cdot a = \sum a_0 p(a_1)$ for all $p \in H^*$, $a \in A$, where $p \cdot a$ denotes the action of p on a.

1 HOPF GALOIS EXTENSIONS AND CROSSED PRODUCTS

Let A be a right H-comodule algebra, and $R := A^{coH} := \{a \in A \mid \sum a_0 \otimes a_1 = a \otimes 1\}$ the subalgebra of all H-coinvariant elements. Then $R \subset A$ is called an H-*extension*.

1.1 DEFINITION [KT]. The H-extension $R \subset A$ is called a *right H-Galois extension* or A is called *H-Galois* if the canonical map $can : A \otimes_R A \longrightarrow A \otimes H$, $can(x \otimes y) = \sum x y_0 \otimes y_1$, is bijective.

This definition was first introduced by Kreimer and Takeuchi [KT] in 1981 for finite-dimensional Hopf algebras H extending previous definitions by Chase and Sweedler [CS] and Chase, Harrison and Rosenberg [CHR].

1.2 EXAMPLES. (1) Let $H = k^G$ be the dual Hopf algebra of the group algebra of a finite group G. Then an extension of commutative fields $R \subset A$ over k is H-Galois if and only if $R \subset A$ is a finite Galois extension with Galois group G.

(2) When A and H both are commutative, the definition can be translated into the language of algebraic geometry [DG], [Wa]. Let $X := \mathrm{Spec}(A)$ resp. $G := \mathrm{Spec}(H)$ be the affine scheme resp. group scheme represented by A resp. H. The comodule algebra structure $\Delta_A : A \longrightarrow A \otimes H$ is dual to an action $\mu := \mathrm{Spec}(\Delta_A) : X \times G \longrightarrow X$ of the group scheme G on X. Define $Y := \mathrm{Spec}(R)$, $R = A^{coH}$, and $\pi := \mathrm{Spec}(i) : X \longrightarrow Y$ where i is the inclusion map $R \subset A$. The canonical map $can : A \otimes_R A \longrightarrow A \otimes H$ represents the map $X \times G \longrightarrow X \times_Y X$, given on rational points by $(x,g) \longmapsto (x, xg)$, where $xg := \mu(x,g)$. Then X is a *G-torsor over Y* or a *principal fibre bundle over Y with group G* if and only if $R \subset A$ is a faithfully flat H-Galois extension.

(3) Let G be a group and $H = k[G]$ the group algebra. An H-extension $R \subset A$ is H-Galois if and only if A is a strongly graded G-algebra with $A_1 = R$ [Da].

(4) Let A be a Hopf algebra and R a Hopf subalgebra which is *normal* in A, i.e. for all $x \in A$ and $y \in R$ the elements $\sum x_1 y S(x_2)$ and $\sum S(x_1) y x_2$ are contained in R. Then $H := A/R^+A$ is a quotient Hopf algebra of A, and $A \longrightarrow A \otimes H$, $x \longmapsto \sum x_1 \otimes \overline{x}_2$, is a right H-comodule algebra. Assume A is left or right faithfully flat over R. Then $R \subset A$ is an H-Galois extension [T2]. In particular, if N is a normal subgroup of the group G resp. $\mathfrak{n}$ is an ideal of the Lie algebra $\mathfrak{g}$, then $k[N] \subset k[G]$ resp. $U(\mathfrak{n}) \subset U(\mathfrak{g})$ is a faithfully flat $k[G/N]$- resp. $U(\mathfrak{g}/\mathfrak{n})$-Galois extension.

Recall the notion of integrals for an algebra A with an augmentation ε, i.e. an algebra map $\varepsilon : A \longrightarrow k$. An element $a \in A$ is called a *left integral* in A if $x\,a = \varepsilon(x)a$ for all $x \in A$. Right integrals are defined in a symmetric fashion. In a fundamental paper in 1969, Larson and Sweedler [LS] proved that the subspace of all left resp. right integrals of any finite-dimensional Hopf algebra H is one-dimensional. Moreover, they showed that there are left integrals Λ in H and λ in the dual Hopf algebra H^* such that $\lambda(\Lambda) = 1$.

1.3 THEOREM [KT]. *Let H be finite-dimensional and $R \subset A$ a right H-extension. Assume the canonical map $A \otimes_R A \longrightarrow A \otimes H$ is surjective. Then $R \subset A$ is H- Galois and A is finitely generated and projective as a left and right R-module.*

PROOF. The following direct proof is a special case of [S3, 3.1]. By [LS] there are left integrals Λ in H and λ in H^* such that $\lambda(\Lambda) = 1$. Then $t : A \longrightarrow R$, $t(a) := \lambda \cdot a$, is a well defined left and right R-linear map. Here, the action of λ on a in A is the adjoint action given by $\lambda \cdot a = \sum a_0 \lambda(a_1)$. Since *can* is surjective and the antipode of H is bijective, also the map $can' : A \otimes_R A \longrightarrow A \otimes H$, $x \otimes y \longmapsto \sum x_0 y \otimes x_1$, is surjective [KT].

Now choose finitely many elements r_i, l_i in A such that $can'(\sum r_i \otimes l_i) = 1 \otimes \Lambda$. Then for all $x \in A$

$$\sum t(x r_i)\, l_i = \sum x_0 r_{i0}\, \lambda(x_1 r_{i1})\, l_i = \sum x_0\, \lambda(x_1 \Lambda) = \sum x_0\, \varepsilon(x_1) = x.$$

In particular, A is finitely generated and projective over R as a left module. This part of the proof is shown as in [OS1], [U1].

To prove that *can* is injective, assume $\sum x_j \otimes y_j \in A \otimes_R A$ such that $\sum x_j y_{j0} \otimes y_{j1} = 0$. Then for all j, $y_j = \sum t(y_j r_i)\, l_i$ by the above equation with y_j instead of x. Hence, in $A \otimes_R A$

$$\sum x_j \otimes y_j = \sum x_j \otimes t(y_j r_i)\, l_i = \sum x_j t(y_j r_i) \otimes l_i = \sum x_j y_{j0}\, r_{i0}\, \lambda(y_{j1} r_{i1}) \otimes l_i$$

by the definition of t. But the last expression is zero since $\sum x_j y_{j0} \otimes y_{j1} = 0$.

Similarly one shows that A is also right projective over R. □

1.4 REMARK. Let H be finite-dimensional and Λ a non-zero left or right integral in H. Then Λ is a left H^*-basis of H where the left H^*-action is adjoint to Δ (cf. [LS]). Let $R \subset A$ be a right H-extension and assume that $1 \otimes \Lambda$ is contained in the image of *can* or *can'*. Then $R \subset A$ is H-Galois.

PROOF. The image of *can* resp. *can'* is a left H^*-submodule of $A \otimes H$, where $p(a \otimes h) := a \otimes p \cdot h$, $p \cdot h = \sum h_1 p(h_2)$, and a left resp. right A-submodule, where $x(a \otimes h) := xa \otimes h$ resp. $(a \otimes h)x := ax \otimes h$ for all $p \in H^*$, $h \in H$ and x, $a \in A$. Hence *can* resp. *can'* is surjective, since H is generated by Λ as a left H^*-module. By 1.3, the extension is H-Galois. □

1.5 COROLLARY [KT]. *Assume in 1.3 that R is commutative. Then A is faithfully flat as a left and right R-module.*

In the commutative case of example 1.2 (2), the *action* $X \times G \longrightarrow X$ is called *free* if the map $X \times G \longrightarrow X \times_Y X$ given by $(x,g) \longmapsto (x,xg)$ is a closed imbedding, i.e. if the canonical map $A \otimes_R A \longrightarrow A \otimes H$ is surjective. By 1.3 and 1.5, X is a G-torsor over Y, if the action is free and the group scheme G is finite. Thus 1.3 is a noncommutative generalization of the main theorem on quotients of affine schemes by finite group schemes when the action is free [DG, III, § 2, 6.1].

Assume H is a finite-dimensional cocommutative Hopf algebra and $H' \subset H$ a Hopf subalgebra. Then $H^* \longrightarrow (H')^*$ is a surjective map of commutative Hopf algebras and $\mathrm{Spec}((H')^*) \subset \mathrm{Spec}(H)$ is an inclusion of finite affine group schemes. By applying the quotient theory of finite group schemes (i.e. 1.3), it was shown in [OS1] that H is free as a left and right H'-module. In 1989 Nichols and Zoeller obtained the following fundamental noncommutative generalization.

1.6 THEOREM [NZ]. *Let H be a finite-dimensional Hopf algebra and $H' \subset H$ a Hopf subalgebra. Then H is free as a left and right H'-module.*

Depending on 1.6 and results on Hopf algebras with cocommutative coradical in [S5] (see [OS2] in the cocommutative case), the method of proof of 1.3 can be generalized to the *non-normal* case. If $I \subset H$ is a coideal and a right ideal, then the canonical epimorphism $p : H \longrightarrow \overline{H} := H/I$, $p(h) = \overline{h}$, the residue class of h, is a right H-linear map of coalgebras. For any right H-comodule algebra A, define the subalgebra of $\overline{H}$-coinvariants by $A' := \{ a \in A \mid (1 \otimes p)\Delta_A(a) = a \otimes \overline{1} \}$. The extension $A' \subset A$ is called $\overline{H}$-*Galois* if the canonical map $A \otimes_{A'} A \longrightarrow A \otimes \overline{H}$, $x \otimes y \longmapsto \sum x\, y_0 \otimes \overline{y}_1$, is bijective.

As an example, let $H = k[G]$ be a group algebra, $H' = k[G']$ the group algebra of a subgroup and $I = H'^{+}H$. Then $H' \subset H$ is $\overline{H}$-Galois.

1.7 THEOREM [S3]. *Let H' be a Hopf subalgebra of H, $p : H \longrightarrow \overline{H} := H/H'^{+}H$ the canonical map, $R \subset A$ a right H-Galois extension and define A' as before. If H is finite-dimensional or the coradical of H is cocommutative and $\overline{H}$ is finite-dimensional, then*

(1) *$A' \subset A$ is $\overline{H}$-Galois and $A' = \Delta_A^{-1}(A \otimes H')$.*

(2) *There are finitely many elements $r_i, l_i \in A$, an automorphism β of A' and a left A'-linear and right β-semilinear map $t : A \longrightarrow A'$ such that for all $x \in A$*
$$x = \sum t(x\, r_i)\, l_i = \sum r_i\, (\beta^{-1} t)(l_i\, x).$$

(3) *If A is faithfully flat as a left or right R-module, then A is faithfully flat as a left and right A'-module.*

By 1.7 (2), $A' \subset A$ is a β-*Frobenius extension* in the sense of Nakayama and Tsuzuku [NT]. This implies the

1.8 COROLLARY (*Frobenius reciprocity*). *In the situation of 1.7, let X be a right A-module and Y a right A'-module. Let $Y_{\beta^{-1}}$ be the twisted A'-module Y, where $a' \in A'$ is operating on $y \in Y$ as $y\beta^{-1}(a')$. Then*
$$\Phi : \mathrm{Hom}_{A'}(X,Y) \longrightarrow \mathrm{Hom}_A(X, Y_{\beta^{-1}} \otimes_{A'} A),\ \Phi(f)(x) := \sum f(x\, r_i) \otimes l_i,$$
is bijective.

Moreover, in 1.7, the elements r_i, l_i and t, β can be constructed as follows. Let λ be a non zero left integral of the dual algebra of $\overline{H}$. Choose $\Lambda \in H$ such that $\Lambda\, \lambda = \varepsilon$. Here $(\overline{H})^*$ is a left H-module by $(h\, \varphi)(\overline{g}) = \varphi(\overline{g}\, h)$ for all $g, h \in H$ and $\varphi \in (\overline{H})^*$. Then $h'\lambda = \chi(h')\,\lambda$ for all $h' \in H'$, where $\chi : H' \longrightarrow k$ is an algebra homomorphism. Now choose r_i, l_i such that $\sum r_{i0}\, l_i \otimes r_{i1} = 1 \otimes \Lambda$ and define

$$t(a) := \sum a_0\, \lambda(\overline{a}_1),\ \beta(a') := \sum a'_0\, \chi(a'_1)$$

for all $a \in A$, $a' \in A'$.

In general, the automorphism β in 1.7 arises in an unavoidable way (cf. [OS1]).

The next theorem is a noncommutative generalization of results of [CPS] and [Ob] on affine quotients in the case of free actions and arbitrary (non finite) affine groups.

1.9 THEOREM [S1]. *Let H be a Hopf algebra with bijective antipode and $R \subset A$ a right H-extension. Then the following are equivalent:*

(1) (a) *The canonical map $A \otimes_R A \longrightarrow A \otimes H$ is surjective.*
    (b) *A is an injective right H-comodule.*

(2) (a) *$R \subset A$ is a right H-Galois extension.*
    (b) *A is faithfully flat as a left or right R-module.*

A Hopf Galois extension $R \subset A$ will be called *faithfully flat* if A is a faithfully flat left and right R-module. When the antipode is bijective, by 1.9 (2), the Galois extension is faithfully flat if it is faithfully flat on one side. In general, Hopf Galois extensions need not be faithfully flat (for examples with finite-dimensional H see [KT], [DT2]). But in practice, they often are cleft (hence faithfully flat) in the sense of

1.10 DEFINITION [Sw1]. A right H-extension $R \subset A$ is H-*cleft* if there is a right H-colinear map $\gamma : H \longrightarrow A$ which is invertible in the convolution algebra $\mathrm{Hom}(H, A)$.

Cleft extensions are characterized in the class of all Galois extensions in

1.11 THEOREM [DT] (*cf.* [BCM]). *Let $R \subset A$ be a right H-extension. Then the following are equivalent.*

(1) *A is H-cleft.*

(2) (a) *A is H-Galois.*
    (b) *$A \cong R \otimes H$ as left R-modules and right H-comodules.*

In (2), the module and comodule structures on $R \otimes H$ are defined by $r(x \otimes h) := rx \otimes h$, $x \otimes h \longmapsto x \otimes \Delta(h)$ for all $r, x \in R$ and $h \in H$. The H-extension $R \subset A$ has the *normal basis property* if (2) (b) holds. If A is H-cleft with map γ as in 1.10, then the inverse of the canonical map $A \otimes_R A \longrightarrow A \otimes H$ is given in 1.11 by mapping $a \otimes h$ onto $\sum a\, \gamma^{-1}(h_1) \otimes \gamma(h_2)$. Thus the choice of the non canonical map γ yields a description of can^{-1}.

Let A be H-cleft. Then A can be identified with an "H-crossed product" [DT], cf. [BCM]. Choose an H-colinear and invertible map $\gamma : H \longrightarrow A$ such that $\gamma(1) = 1$.. Then for all $g, h \in H$ and $r \in R$

$$h \cdot r := \sum \gamma(h_1)\, r\, \gamma^{-1}(h_2) \text{ and } \sigma(g,h) := \sum \gamma(g_1)\, \gamma(h_1)\, \gamma^{-1}(g_2 h_2)$$

are H-coinvariant elements, and $\sigma : H \otimes H \longrightarrow R$, $g \otimes h \longmapsto \sigma(g,h)$, is $*$-invertible.

Now define an H-comodule algebra $R \#_\sigma H$ as follows. The underlying vector space is $R \otimes H$ with elements written as $r \# h = r \otimes h$ and with comultiplication $R \otimes \Delta$ and multiplication

$$(r \# g)(s \# h) := \sum r(g_1 \cdot s)\sigma(g_2, h_1) \# g_3 h_2.$$

Then $R \subset R \#_\sigma H$, $r \mapsto r \# 1$, is H-cleft, hence an H-Galois extension, and

$$R \#_\sigma H \cong A,\ r \# h \mapsto r\gamma(h),$$

is an isomorphism of algebras and H-comodules leaving R fixed.

In general, given an algebra R, a "weak action" $\cdot : H \otimes R \longrightarrow R$ and an invertible "cocycle" $\sigma : H \otimes H \longrightarrow R$ satisfying certain axioms (see [DT], [BCM]), the *H-crossed product* $R \#_\sigma H$ is defined as above.

1.12 THEOREM. (1) [BM] *Let $R \#_\sigma H$ be an H-crossed product. Then $\gamma : H \longrightarrow R \#_\sigma H$, $\gamma(h) := 1 \# h$, is right H-colinear and invertible. Hence $R \#_\sigma H$ is H-cleft.*
(2) [DT1] *Conversely, cleft H-Galois extensions are isomorphic to H-crossed products as described before.*

Thus H-crossed products and H-cleft extensions are equivalent notions. Instead of working directly in the complicated algebra structure of $R \#_\sigma H$, in many cases it suffices to use the Galois property together with faithful flatness of the crossed product over R. Clearly, R $\#_\sigma$ H is free as a left R-module.

2 HOPF MODULES DESCRIBING DESCENT DATA

Let $R \subset A$ be a right H-extension and M a right R-module. The key to the induced A-module $M \otimes_R A$ is its right H-comodule structure

$$M \otimes \Delta_A : M \otimes_R A \longrightarrow M \otimes_R A \otimes H,\ m \otimes a \mapsto \sum m \otimes a_0 \otimes a_1.$$

The A-module and H-comodule structures on $M \otimes_R A$ are compatible in the following sense.

2.1 DEFINITION. A *right (A,H)-Hopf module* [T], [D1] is a right A-module X which is also a right H-comodule such that the structure map $\Delta_X : X \longrightarrow X \otimes H$ is A-linear where A is operating diagonally on $X \otimes H$, i.e. $\Delta_X(xa) = \sum x_0 a_0 \otimes x_1 a_1$ for all $x \in X$, $a \in A$. Let $\mathcal{M}_A^H$ be the category of right (A,H)-Hopf modules with A-linear and H-colinear maps as morphisms.

If $I \subset H$ is a coideal and a right ideal and $\overline{H} := H/I$, then *right $(A,\overline{H})$-Hopf modules* are defined in the same way, and $\mathcal{M}_A^{\overline{H}}$ denotes the category of all right $(A,\overline{H})$-Hopf modules.

Clearly, $M \otimes_R A$ is a right (A,H)-Hopf module. Let A' be the algebra of $\overline{H}$-coinvariants as in the first section. Then $M \otimes_{A'} A$ is a right $(A,\overline{H})$-Hopf module with comodule structure map $m \otimes a \longmapsto \sum m \otimes a_0 \otimes \overline{a}_1$ for all $m \in M$, $a \in A$.

As an example, let $R = k \subset K = A$ be a finite Galois extension of fields with Galois group G. Then $k \subset K$ is a right k^G-Galois extension where the adjoint left $k[G]$-action on K is given by $(g,x) \longmapsto g(x)$, $g \in G = \mathrm{Aut}(K/k)$, $x \in K$. A vector space X over K is a right (K,k^G)-Hopf module if and only if for all $g \in G$ there is a k-linear automorphism $g\cdot : X \longrightarrow X$ such that $g\cdot(xa) = (g\cdot x)g(a)$ and $g\cdot(h\cdot x) = (gh)\cdot x$ for all $x \in X$, $a \in K$ and g, $h \in G$. Hence, in this case Hopf modules describe descent data in Galois descent theory introduced by A. Weil [KO, II, § 5]. Galois descent for field extensions is a very special case of the following descent theorem for $\overline{H}$-Galois extensions.

2.2 THEOREM [S1] (cf. [DT2]). *Let I be a coideal and a right ideal of H and $\overline{H} := H/I$. Let A be a right H-comodule algebra and $A' := A^{co\overline{H}}$. Then the following are equivalent:*

(1) $\mathcal{M}_{A'} \longrightarrow \mathcal{M}_A^{\overline{H}}$, $M \longmapsto M \otimes_{A'} A$, *is an equivalence.*

(2) (a) $A' \subset A$ *is an* $\overline{H}$-*Galois extension.*

(b) A *is left faithfully flat over* A'.

The proof of (1) ⇒ (2) is very easy, and (2) ⇒ (1) follows from the consideration of two commutative diagrams of exact sequences.

Theorem 2.2 shows that faithfully flat Hopf Galois extensions are a distinguished class of Hopf Galois extensions. The theorem was known before in many special cases under various names. It includes inseparable descent [KO, II, § 6]. In the case $I = 0$ and A, H commutative it is called the generalized Taylor lemma in [Vo1, 5.2]. In the case $I = 0$ and H a group algebra of a group G, Hopf modules are G-graded modules and 2.2 was shown by Dade [Da, 2.8]. Theorem 2.2 also contains the imprimitivity theorem (for induced modules) of Koppinen and Neuvonen [KN] where $A' = H'$ is a Hopf subalgebra of $A = H$ and $\overline{H}$ is finite-dimensional. In this case, Hopf modules are described by certain $(\overline{H})^*$-module structures called systems of imprimitivity based on H' (see [S1, 3.9 (4)] and [U2]). For restricted Lie algebras, the result of Koppinen and Neuvonen is analogous to, and was inspired by, Blattner's imprimitivity theorem for Lie algebras in characteristic zero [Bl]. Finally, in case $A = H = \overline{H}$ and $A' = k$, 2.2 is the fundamental theorem on (H,H)-Hopf modules in [Sw2].

In short, the descent theorem 2.2 is a basic formality showing the usefulness of Hopf modules. The idea of Hopf modules sometimes allows to solve "non-linear" problems in the "linear" abelian category of Hopf modules (as in the proof of 1.8).

The next result gives examples where (a) and (b) in 2.2 (2) are satisfied.

2.3 THEOREM [S3], [D3]. *Let $R \subset A$ be a right H-Galois extension such that A is left faithfully flat over R. Let I be a coideal and a right ideal of H, $\overline{H} := H/I$ and $A' := A^{co\overline{H}}$. If H is left faithfully coflat over $\overline{H}$, then $A' \subset A$ is a right $\overline{H}$-Galois extension and A is left faithfully flat over A'.*

3 ENDOMORPHISM RINGS OF HOPF MODULES

In this section, let H be a Hopf algebra with bijective antipode and $R \subset A$ a right H-Galois extension.

For right (A,H)-Hopf modules X the following endomorphism rings are of interest:

$$\mathrm{End}_A^H(X) \subset \mathrm{End}_A(X) \subset \mathrm{End}_R(X).$$

Here, $\mathrm{End}_A^H(X)$ is the ring of all right A-linear and H-colinear endomorphisms of X.

If $X = M \otimes_R A$ is an induced module such that the canonical map $M \longrightarrow (M \otimes_R A)^{coH}$, $m \longmapsto m \otimes 1$, is an isomorphism (by 2.2, this holds for any R-module M if A is faithfully flat over R), then $\mathrm{End}_A^H(X) \cong \mathrm{End}_R(M)$. Decompositions of $\mathrm{End}_A(X)$ resp. $\mathrm{End}_R(X)$ correspond to decompositions of the induced module $M \otimes_R A$ over A resp. R.

3.1 DEFINITION. (1) For all $h \in H$ choose finitely many elements $r_i(h)$, $l_i(h)$ such that $can(\sum r_i(h) \otimes l_i(h)) = 1 \otimes h$, where $can : A \otimes_R A \longrightarrow A \otimes H$ is the canonical isomorphism.

(2) Let X, Y be right A-modules. The diagonal actions of H on $\mathrm{Hom}_R(X,Y)$ are defined by

$$(f \cdot h)(x) := \sum f(x\, r_i(h))\, l_i(h) \text{ and } h \cdot f := f \cdot S^{-1}(h)$$

for all $f \in \mathrm{Hom}_R(X,Y)$, $x \in X$ and $h \in H$.

If $A = H$ with comodule structure Δ, then the right diagonal action is given by $(f \cdot h)(x) = \sum f(x\, S(h_1))\, h_2$. For group-like elements h this is the usual diagonal action in representation theory of groups where $(f \cdot h)(x) = f(x h^{-1})\, h$.

In general, (2) defines right and left H-module structures on $\mathrm{Hom}_R(X,Y)$. If $X = Y$, then $\mathrm{End}_R(X)$ is a left H-module algebra by (2) with $\mathrm{End}_A(X)$ as the algebra of H-invariants (cf. [S2, 3.5]).

3.2 THEOREM. *Assume H is finite-dimensional. Let X be a right (A,H)-Hopf module. Then*

$$\Phi : \mathrm{End}_A(X) \# H^* \longrightarrow \mathrm{End}_R(X),\ \Phi(f \# p) := f(p \cdot x),$$

is an isomorphism of right H^*-comodule algebras. Here, the right H^*-coaction on $\mathrm{End}_R(X)$ *is adjoint to the left H-action in* 3.1, *and the left H^*-module algebra structure on* $\mathrm{End}_A(X)$ *(defining the smash product) is given by $(p \cdot f)(x) := \sum p_1 \cdot f(S(p_2) \cdot x)$ for all $p \in H^*$, $f \in$* $\mathrm{End}_A(X)$ *and $x \in X$, where the H^*-action on X is adjoint to the given H-comodule structure of X.*

PROOF. Define $\gamma : H^* \longrightarrow \mathrm{End}_R(X)$ by $\gamma(p)(x) := p \cdot x$. Then γ is an algebra homomorphism since X is a left H^*-module. The following computation shows that γ is right H^*-colinear, or equivalently left H-linear, where the adjoint left action of H on H^* is given by $h \cdot p := \sum p_1\, p_2(h)$: For all $h \in H$, $p \in H^*$ and $x \in X$

$$\gamma(S(h)\cdot p)(x) = \sum (p_1\, p_2(S(h)))\cdot x = \sum x_0\, p_1(x_1)\, p_2(S(h)) = \sum x_0\, p(x_1\, S(h)), \text{ and}$$

$$\begin{aligned}(S(h)\cdot\gamma(p))(x) &= \sum \gamma(p)(x\, r_i(h))\, l_i(h) = \sum x_0\, r_i(h)_0 p(x_1\, r_i(h)_1)\, l_i(h)\\ &= \sum x_0\, r_i(h_2)\, l_i(h_2)\, p(x_1\, S(h_1)) \quad \text{by [S2, 3.4, (2)(e)]}\\ &= \sum x_0\, p(x_1\, S(h)) \quad \text{by [S2, 3.4, (2)(c)].}\end{aligned}$$

Hence, by 1.12, Φ is an isomorphism of H^*-comodule algebras where the left action of H^* on $\mathrm{End}_A(X) = \mathrm{End}_R(X)^{coH^*}$ is given by $(p\cdot f)(x) = \sum (\gamma(p_1) f \gamma(S(p_2)))\ (x) = \sum p_1\cdot f(S(p_2)\cdot x)$ (the cocycle is trivial since γ is an algebra map). □

If $X = A$, then Φ is the well known isomorphism $A \# H^* \cong \mathrm{End}_R(A_R)$ [KT, (1.7)(3)].

The comodule structure of $\mathrm{End}_A(X)$, adjoint to the left H^*-action in 3.2, admits a generalization to infinite-dimensional Hopf algebras. If X is a right (A,H)-Hopf module, define $\delta : \mathrm{End}_A(X) \longrightarrow \mathrm{Hom}_A(X, X \otimes H)$ by $\delta(f)(x) := \sum f(x_0)_0 \otimes f(x_0)_1\, S(x_1)$. Note that $\delta(f)$ is A-linear where $X \otimes H$ is a right A-module by $(x \otimes h)a := xa \otimes h$.

3.3 DEFINITION [U3], cf. [S2]. Let X be a right (A,H)-Hopf module. Then $E := \mathrm{END}_A(X)$ is the set of all $f \in \mathrm{End}_A(X)$ satisfying the following condition: There is an element $\sum f_0 \otimes f_1$ in $\mathrm{End}_A(X) \otimes H$, denoted by $\Delta_E(f)$, such that $\delta(f)(x) = \sum f_0(x) \otimes f_1$ for all $x \in X$, or equivalently, $\delta(f) \in \mathrm{End}_A(X) \otimes H \subset \mathrm{Hom}_A(X, X \otimes H)$.

3.4 LEMMA [U3], cf. [S2]. *Let X be a right (A,H)-Hopf module. Then $E = \mathrm{END}_A(X)$ is a right H-comodule algebra with coaction Δ_E and $\mathrm{END}_A(X)^{coH} = \mathrm{End}_A^H(X)$.*

The coaction looks more natural in the following form

$$\Delta_X(f(x)) = \sum f_0(x_0) \otimes f_1\, x_1 \text{ for all } x \in X$$

saying that X is a Hopf module in ${}_E\mathcal{M}^H$. Note that $\mathrm{END}_A(X) = \mathrm{End}_A(X)$ if H is finite-dimensional or X is A-finitely generated. When H is finite-dimensional one can check that the H-coaction on $\mathrm{END}_A(X)$ is adjoint to the left action of H^* in 3.2. The coaction in 3.3 was introduced in [S2] in the case of induced modules X and assuming $\mathrm{End}_A(X) = \mathrm{END}_A(X)$. The results in [S2] can easily be generalized replacing $\mathrm{End}_A(X)$ by $\mathrm{END}_A(X)$.

If A is left faithfully flat over R and M is a right R-module, then the natural map $\iota_M : M \longrightarrow M \otimes_R A$, $m \longmapsto m \otimes 1$, is injective. This map splits in the case of group algebras H but not in general. The next theorem explains this phenomenon since group algebras are cosemisimple and all their comodules are injective.

3.5 THEOREM [S2]. *Let M be a right R-module such that the natural map $M \longrightarrow (M \otimes_R A)^{coH}$ is an isomorphism. Then the following are equivalent:*

(1) *$\iota_M : M \longrightarrow M \otimes_R A$ is an R-split monomorphism.*

(2) *$\mathrm{END}_A(M \otimes_R A)$ is an injective H-comodule.*

PROOF (Sketch). (1) $\Rightarrow$ (2). If $g : M \otimes_R A \longrightarrow M$ is an R-linear map such that $g\iota_M = id$, define $\gamma : H \longrightarrow \mathrm{END}_A(M \otimes_R A)$ by $\gamma(h)(m \otimes 1) := \sum g(m \otimes r_i(h)) \otimes l_i(h)$. Then γ is a well defined right H-colinear map and $\gamma(1) = id$. This proves (2) by [D2].

(2) $\Rightarrow$ (1). By [D2], there is a right H-colinear map $\gamma : H \longrightarrow \mathrm{END}_A(M \otimes_R A)$ such that $\gamma(1) = id$. Then a right R-linear map $g : M \otimes_R A \longrightarrow M$ satisfying $g\iota_M = id$ is defined by $\sum \gamma(S^{-1}(a_1))(m \otimes a_0) = g(m \otimes a) \otimes 1$ in $M \otimes_R A$ for all $a \in A$, $m \in M$. □

In case $M = R$, 3.5 is [D2, (2.4)].

3.6 EXAMPLE. Let k be algebraically closed of characteristic $p > 0$. Let A be the algebra generated by x, y, z satisfying the relations $xy - yx = z$, $xz = zx$, $yz = zy$, $x^p = 0$, $y^p = 0$ and $z^p = z$. Then A is a Hopf algebra where x, y and z are primitive. This is the restricted enveloping algebra of the 3-dimensional nilpotent restricted Lie algebra $kx + ky + kz$, and the elements $x^i y^j z^l$, $0 \le i, j, l \le p - 1$, form a k-basis of A.

Consider the Hopf subalgebras $U := k[y] \subset R := k[y,z] \subset A$. Then R is a commutative normal Hopf subalgebra and A/R^+A is isomorphic to the p-dimensional Hopf algebra $H := k[x]$ generated by x with relation $x^p = 0$ such that x is primitive. Thus $R \subset A$ is a faithfully flat H-Galois extension.

The R-module $R/U^+R \cong k[z]$ is semisimple. Since $k[z] \cong k^p$, the decomposition of R/U^+R can be described as follows. Let $\zeta_1, \cdots, \zeta_{p-1}$ be the $(p - 1)$-th roots of unity in k and let $\zeta_0 := 0$. For all i let M_i be the 1-dimensional R-module with k-basis m_i and $m_i y = 0$, $m_i z = \zeta_i m_i$. Then $R/U^+R \cong \oplus_i M_i$. In this situation, the natural map

$$\iota_i : M_i \longrightarrow M_i \otimes_R A \text{ is } R\text{-split if and only if } i = 0.$$

PROOF. Suppose there is an R-linear map $g : M_i \otimes_R A \longrightarrow M_i$ such that $g\iota_i = id$. Then $m_i z = g(m_i \otimes z) = g(m_i \otimes (xy - yx)) = g(m_i \otimes x)y - g(m_i y \otimes x) = 0$. Hence, $i = 0$. Conversely, $id \otimes \varepsilon$ is a splitting map in case $i = 0$. □

3.7 DEFINITION [S2]. A right R-module M is *H-stable* if there is a right R-linear and H-colinear isomorphism $M \otimes_R A \cong M \otimes H$, where the module and comodule structures on $M \otimes H$ are defined by $(m \otimes h)r = mr \otimes h$ and $id \otimes \Delta$.

Note that R is H-stable if and only if $R \subset A$ satisfies the normal basis property. If $H = k[G]$ is a group algebra, then H-stable modules are called *G-invariant* in [Da].

3.8 REMARK. (1) Let G be a group, N a normal subgroup of G and $R = k[N] \subset A = k[G]$ the $k[G/N]$-Galois extension with coaction given by $g \longmapsto g \otimes \overline{g}$, $g \in G$. Then a right R-module M is $k[G/N]$-stable if and only if M is G-stable in the classical sense, i.e. for all $g \in G$, $M_g \cong M$ as R-modules, where $M_g = M$ with twisted N-action $m \cdot g' := mgg'g^{-1}$ for all $g' \in N$, $m \in M$.

(2) If H, A and M are finite-dimensional, then it follows from the theorem of Krull-Schmidt that M is H-stable if and only if $M \otimes_R A$ is R-isomorphic to some M^n, $n \ge 1$ [S2, 3.2 (4)].

3.9 THEOREM [S2]. *Let M be a right R-module such that the natural map $M \longrightarrow (M \otimes_R A)^{coH}$ is an isomorphism. Then the following are equivalent:*

(1) *M is H-stable.*

(2) $\mathrm{END}_A(M \otimes_R A) \cong \mathrm{End}_R(M) \#_\sigma H$ *is an H-crossed product.*

PROOF (Sketch). This was shown in [S2, 3.6] (for left modules) assuming $\mathrm{END}_A(M \otimes_R A) = \mathrm{End}_A(M \otimes_R A)$. The (left version of the) same proof works here. To prove (1) ⇒ (2), let $\Phi : M \otimes_R A \cong M \otimes H$ be an R-linear and H-colinear isomorphism and define $q : M \otimes_R A \longrightarrow M$ by $q := (id \otimes \varepsilon)\Phi$. Then $\gamma : H \longrightarrow \mathrm{END}_A(M \otimes_R A)$, $\gamma(h)(m \otimes 1) := \sum q(m \otimes r_i(h)) \otimes l_i(h)$ is a right H-colinear map with $*$-inverse given by $\gamma^{-1}(h)(m \otimes 1) := \Phi^{-1}(m \otimes S(h))$. Hence $\mathrm{END}_A(M \otimes_R A)$ is H-cleft and an H-crossed product by 1.11. □

Theorem 3.9 generalizes [Da, 5.14], where H is a group algebra. Voigt [Vo1, 12.6] proved the implication (1) ⇒ (2) of 3.9 in the situation where A is a finite-dimensional cocommutative Hopf algebra over an algebraically closed field, $R \subset A$ is a normal Hopf subalgebra and $H = A/R^+A$ using completely different but equivalent definitions of stable modules and crossed products.

If H is irreducible, then M is H-stable if and only if $M \longrightarrow M \otimes_R A$ is R-split. More generally, from 3.8 and 3.9 one obtains the

3.10 COROLLARY [S2]. *Let M be a right R-module. If A is faithfully flat over R and H is pointed, then the following are equivalent:*

(1) *M is H-stable.*

(2) (a) *$\iota_M : M \longrightarrow M \otimes_R A$ is an R-split monomorphism.*

(b) *For all group-like elements $g \in H$, $M \cong M \otimes_R A_g$ as right R-modules, where $A_g := \Delta_A^{-1}(A \otimes g)$.*

3.11 COROLLARY. *Assume A is faithfully flat over R and H is finite-dimensional. Let $p : H \longrightarrow \overline{H}$ be a surjective Hopf algebra map. Then A is a right $\overline{H}$-comodule algebra via p, and $A' := A^{co\overline{H}} \subset A$ is a faithfully flat $\overline{H}$-Galois extension. Let M be a right R-module. If M is H-stable, then $M \otimes_R A'$ is $\overline{H}$-stable.*

PROOF. By 1.6 (and duality), H is H-cofree. Hence, by 2.3, $A' \subset A$ is a faithfully flat H-Galois extension. Since M is H-stable, $\mathrm{End}_A(M \otimes_R A)$ is an H-crossed product by 3.9. Hence $\mathrm{End}_A(M \otimes_R A) \cong \mathrm{End}_A(M \otimes_R A' \otimes_{A'} A)$ is cleft over $\overline{H}$ by the transitivity result in [S3, 2.2] depending on the freeness theorem 1.6 of Nichols and Zoeller. Thus $M \otimes_R A'$ is $\overline{H}$-stable by 3.9. □

The natural question as to when $\mathrm{END}_A(M \otimes_R A)$ is H-Galois was answered recently by van Oystaeyen and Zhang [vOZ] in the case where H is finite-dimensional and unimodular and A is faithfully flat over R. To do so, they first study a different question again for finite-dimensional unimodular H: When is $\mathrm{End}_R(X)$ H^*-Galois, X any right A-module (not a Hopf module as in 3.2)?

Recall that $\mathrm{End}_R(X)$ is a left H-module algebra by 3.1, hence a right H^*-comodule algebra by adjunction.

Next a direct proof of their result will be given working for arbitrary finite-dimensional Hopf algebras.

3.12 THEOREM [vOZ]. *Let H be finite-dimensional and X a right A-module. Then the following are equivalent:*

(1) $\mathrm{End}_A(X) \subset \mathrm{End}_R(X)$ *is a right H^*-Galois extension.*

(2) $X \otimes_R A$ *is A-isomorphic to a direct summand of some X^n, $n \geq 1$.*

PROOF. Fix a natural number n. As in the proof of 1.3, let λ and Λ be left integrals in H^* and H such that $\lambda(\Lambda) = 1$, and choose elements r_i, l_i, $1 \leq i \leq m$ for some m, such that $\sum r_{i0} l_i \otimes r_{i1} = 1 \otimes \Lambda$ in $A \otimes H$. Hence, in the notation introduced in 3.1, $\sum r_i \otimes l_i = \sum r_i(S^{-1}(\Lambda)) \otimes l_i(S^{-1}(\Lambda))$. Let $can_X : X \otimes_R A \longrightarrow X \otimes H$, $x \otimes a \longmapsto \sum x a_0 \otimes a_1$, be the isomorphism $X \otimes_A can$.

By 1.7, any pair of right A-linear maps $\varphi : X \otimes_R A \longrightarrow X^n$, $\psi : X^n \longrightarrow X \otimes_R A$ is given by R-linear maps $f_j, g_j \in \mathrm{End}_R(X)$, $1 \leq j \leq n$, such that $\varphi(x \otimes a) = (f_j(x)a)$ for all $x \in X$ and $a \in A$, and $\psi((x_j)) = \sum_{i,j} g_j(x_j r_i) \otimes l_i$ for all $x_1, \cdots, x_n \in X$.

Then the following equivalences hold:

$$\psi\,\varphi = id$$

$\Longleftrightarrow$ For all $x \in X$, in $X \otimes_R A$: $x \otimes 1 = \psi((f_j(x))) = \sum_{i,j} g_j(f_j(x) r_i) \otimes l_i$

$\Longleftrightarrow$ For all $x \in X$, in $X \otimes H$:

$$\begin{aligned} x \otimes 1 &= \textstyle\sum g_j(f_j(x) r_i)\, l_{i0} \otimes S(l_{i1}) \text{ by applying } (id \otimes S)\, can_X \\ &= \textstyle\sum g_j(f_j(x) r_i(S^{-1}(\Lambda_2)))\, l_i(S^{-1}(\Lambda_2)) \otimes \Lambda_1 \quad \text{by [S2, 3.4 (2)(d)]} \\ &= \textstyle\sum (\Lambda_2 \cdot g_j)(f_j(x)) \otimes \Lambda_1 \\ &= \textstyle\sum (g_{j0}\, g_{j1}(\Lambda_2))(f_j(x)) \otimes \Lambda_1 \\ &= \textstyle\sum g_{j0}(f_j(x)) \otimes g_{j1}(\Lambda_2)\, \Lambda_1 \end{aligned}$$

$\Longleftrightarrow$ In $\mathrm{End}_R(X) \otimes H^*$: $id \otimes \lambda = \sum g_{j0}\, f_j \otimes g_{j1}$,

where the last equivalence follows from the fact that Λ is a basis of the left H^*-module H (with module structure $p \cdot h = \sum h_1\, p(h_2)$, $p \in H^*$, $h \in H$, adjoint to Δ), and $\lambda \cdot \Lambda = \sum \Lambda_1\, \lambda(\Lambda_2) = 1$, λ being a left integral in H^* such that $\lambda(\Lambda) = 1$.

Hence, $X \otimes_R A$ is an A-direct summand in some X^n if and only if $1 \otimes \lambda$ lies in the image of $can' : F \otimes_E F \longrightarrow F \otimes H^*$, $g \otimes f \longmapsto \sum g_0 f \otimes g_1$, where $F := \mathrm{End}_R(X)$ and $E := \mathrm{End}_A(X)$. But the latter condition means that $E \subset F$ is H^*-Galois by 1.4. □

In the next corollary, H is finite-dimensional, and any right (A,H)-Hopf module X will be considered as a right $A \# H^*$-module (the left action of H^* being adjoint to the H-coaction) by $x(a \# p) := S(p) \cdot (xa)$, $x \in X$, $a \in A$, $p \in H^*$ (cf. [CFM] where S^{-1} is used instead of S). This defines an equivalence of categories between $\mathcal{M}_A^H$ and $\mathcal{M}_{A \# H^*}$.

Recall that the endomorphism ring $\text{End}_A(X)$ is a right H-comodule algebra by 3.3 (adjoint to the left H^*-action in 3.2).

3.13 COROLLARY [vOZ]. *Let H be finite-dimensional and let X be a right (A,H)-Hopf module. Then the following are equivalent:*

(1) $\text{End}_A^H(X) \subset \text{End}_A(X)$ *is a right H-Galois extension.*

(2) $X \otimes_A (A \# H^*)$ *is $(A \# H^*)$-isomorphic to a direct summand of some X^n, $n \geq 1$.*

PROOF. Apply 3.12 to X and the H^*-Galois extension $A \subset A \# H^*$. Note that the resulting right H-comodule structure on $\text{End}_A(X)$ coincides with the structure in (1) defined in 3.3. □

3.14 COROLLARY [vOZ]. *Assume H is finite-dimensional and A is faithfully flat over R. Let M be a right R-module. Then the following are equivalent:*

(1) $\text{End}_R(M) \subset End_A(M \otimes_R A)$ *is a right H-Galois extension.*

(2) $M \otimes_R A$ *is R-isomorphic to a direct summand of some M^n, $n \geq 1$.*

PROOF. By 2.2, $\mathcal{M}_A^H \cong \mathcal{M}_{A \# H^*} \longrightarrow \mathcal{M}_R$, $X \longmapsto X^{coH} = X^{H^*}$, is an equivalence. Now apply 3.13 to $X := M \otimes_R A$ and observe that $X^{coH} \cong M$ and the H-coinvariant elements of $X \otimes_A (A \# H^*) \cong X \# H^*$ are R-isomorphic to X. □

For group algebras H, 3.13 and 3.14 are shown in [Da, 4.6, 4.7].

If H, A and M are finite dimensional and if M is an indecomposable R-module, then by 3.8 (2), condition (2) in 3.14 is equivalent to M being H-stable.

4 DECOMPOSITION OF INDUCED MODULES

In this section, assume H is finite-dimensional and $R \subset A$ is a right H-Galois extension.

The character group of H is $\text{Alg}(H,k) = G(H^*)$, the group of all group-like elements of H^*. The adjoint left action of H^* on A (given by $p \cdot a = \sum a_0 p(a_1)$, $p \in H^*$, $a \in A$) defines a group homomorphism from the character group of H to the algebra automorphisms of A

$$G(H^*) \longrightarrow \text{Aut}(A),\ \varphi \longmapsto (a \longmapsto \varphi \cdot a).$$

For any right A-module X and $\varphi \in G(H^*)$ let X_φ be the twisted A-module, where $a \in A$ acts on $x \in X_\varphi = X$ as $x\varphi(a)$.

For any ring T let $S(T)$ be the set of all T-isomorphism classes $[X]$ of the simple right T-modules X. If $S(T)$ is finite, let $s(T)$ denote the number of elements in $S(T)$.

4.1 DEFINITION. The natural right action of $G(H^*)$ on S(A) is given by

$[V]\varphi := [V_\varphi]$, V any simple right A-module, φ any character of H.

4.2 LEMMA. *Let X,Y be simple right A-modules. Then the following are equivalent:*

(1) *$[X]$ and $[Y]$ are both contained in the same orbit under the natural action of $G(H^*)$.*

(2) *There is a one-dimensional left H-submodule in $\mathrm{Hom}_R(X,Y)$ (the H-module structure is defined in 3.1).*

PROOF. Let f be any non-zero element in $\mathrm{Hom}_R(X,Y)$. Then $H\cdot f$ is one-dimensional if and only if there is a character $\varphi \in G(H^*)$ such that $h\cdot f = \varphi(h)f$ for all $h \in H$. The latter condition means that f defines an A-linear map $X \longrightarrow Y_\varphi$ (cf. [S2, 3.5] for the case of left modules) hence an isomorphism $X \cong Y_\varphi$. □

In view of 4.2, it is interesting to study the one-dimensional H-submodules of $\mathrm{Hom}_R(X,Y)$, where X, Y are simple A-modules. This will be done in the next theorem due to M. Josek [Jo] generalizing [S2, 6.1].

Recall that a ring T is left perfect if and only if $T/rad(T)$ is semisimple and any non zero right T-module contains a simple submodule, or if and only if the set of all finitely generated submodules of any right T-module satisfies the descending chain condition [Re].

4.3 THEOREM [Jo]. *Assume A is faithfully flat over R, the ring R is left perfect and H^* is pointed (or equivalently, any simple H-module is one-dimensional). Then*

$$s(A) \leq s(R)\, s(H), \text{ and } s(A) \text{ divides } s(H), \text{ if } s(R) = 1.$$

PROOF. Since all simple H-modules are one-dimensional, any non-zero H-module contains a one-dimensional submodule. Hence, by 4.2, if X, Y are simple right A- modules, then $[X]$ and $[Y]$ are equivalent under the action of $G(H^*)$ if and only if $\mathrm{Hom}_R(X,Y)$ is non-zero.

Let r be the number of orbits of $S(A)$ under the action of $G(H^*)$. Then $s(A) \leq r\, s(H)$, and $s(A)$ divides $s(H)$, if $r = 1$. It remains to be shown that $r \leq s(R)$. This follows from the remark (to be proved below) that any simple right R-module is an R-epimorphic image of some simple right A-module:

Let X, Y be simple non-equivalent right A-modules. Choose simple R-submodules $U \subset X$, $V \subset Y$, and simple right A-modules X', Y' such that U resp. V is an epimorphic image of X' resp. Y'. Now assume $U \cong V$ as R-modules. Then there is a non-zero map $X' \longrightarrow U \cong V \subset Y$ in $\mathrm{Hom}_R(X',Y)$, hence $X' \sim Y$ under the action of $G(H^*)$. But since also $X' \sim X$ and $Y' \sim Y$, one gets the contradiction $X \sim Y$.

One possibility to prove the above remark is to use Frobenius reciprocity. If M is a simple right R-module, then $M \otimes_R A$ contains a simple right A-submodule X (since also A is left perfect). By 1.8, $\mathrm{Hom}_R(X,M) \cong \mathrm{Hom}_A(X,M \otimes_R A) \neq 0$. □

4.4 REMARK. For any group-like element $g \in G(H)$ and any simple right R-module M, the right R-module $M \otimes_R A_g$, $A_g = \{ a \in A \mid \Delta_A(a) = a \otimes g \}$, is again simple and $M \mapsto M \otimes_R A_g$ defines an action of the group $G(H)$ on $S(R)$ [S2, 1.5].

Now assume H is pointed. If X is a simple right A-module and U, V are simple right R-modules such that U and V are both R-isomorphic to submodules of X then there is a $g \in G(H)$ such that $U \cong V \otimes_R A_g$ (cf. [S2, 2.2]).

Hence, if H and H^* are both pointed and R is left perfect, the proof of 4.3 shows that there is a bijection between $S(A)/G(H^*)$ and $S(R)/G(H)$ (cf. [Jo]).

A similar result was shown by Chin for prime ideals instead of simple modules: If H and H^* are both pointed and R is any left H-module algebra, then there is a bijection between Spec$(R \# H)/G(H^*)$ and Spec(R)/G(H) [Ch1, 2.3].

If T is a ring and M is a right T-module of finite length, then $\mathrm{End}_T(M)$ is semiprimary, hence left and right perfect, and $s_T(M) := s(\mathrm{End}_T(M))$ is the number of isomorphism classes of indecomposable right T-modules occuring in a Krull-Schmidt decomposition of M.

In the third section, the endomorphism rings of induced modules are described as crossed products in 3.2, 3.9. Hence, from 4.3 one obtains information on the decomposition of induced modules.

4.5 COROLLARY. *Let X be a right (A,H)-Hopf module and assume X is an R-module of finite length.*

(1) *If H^* is pointed, then $s_A(X) \leq s_R(X)\, s(H)$.*
(2) *If H is pointed, then $s_R(X) \leq s_A(X)\, s(H^*)$.*
(3) *If H and H^* are both local algebras, then $s_R(X) = s_A(X)$.*

PROOF. Let $E := \mathrm{End}_A(X) \subset F := \mathrm{End}_R(X)$. By 3.2, $E \# H^* \cong F$. Hence (2) follows from 4.3 applied to the H^*-Galois extension $E \subset F$.

By duality [BM], one obtains (1) in the same way. Since $E \subset F$ is a free H^*-Galois extension of rank $n = [H : k]$, $F \# (H^*)^* \cong M_n(E)$ by [KT, (1.7)] or 3.2. Since $s_A(X) = s(E) = s(M_n(E))$, (1) again follows from 4.3.

Finally, (3) is an immediate consequence of (1) and (2), since $s(H) = 1 = s(H^*)$ when H and H^* are both local. □

4.6 COROLLARY [S2]. *Assume H is local, and A is faithfully flat over R. If M is an indecomposable H-stable right R-module of finite length, then $s_A(M \otimes_R A) = 1$, hence $M \otimes_R A$ is A-isomorphic to some Q^n, $n \geq 1$, Q an indecomposable right A-module.*

PROOF. This follows from 4.5 (1) with $X := M \otimes_R A$, since $s_R(M) = s_R(M \otimes_R A)$ for any H-stable module M, and $s_R(M) = 1 = s(H)$ by assumption. Alternatively, apply 4.3 to the H-extension $\mathrm{End}_R(M) \subset \mathrm{End}_A(M \otimes_R A)$ which is an H-crossed product by 3.9. □

Consider the special case of 4.6 where $R = k[N] \subset A = k[G]$, N being a normal subgroup of the finite group G and $\mathrm{char}(k) = p > 0$. Then $H = k[G/N]$ is local if and only if G/N is a p-group. If M is a finite-dimensional right $k[N]$-module which is indecomposable and G-stable, then, by 4.6, there is an indecomposable right $k[G]$-

module Q such that $M \otimes_{k[N]} k[G] \cong Q^n$ for some $n \geq 1$ as $k[G]$-modules. This is Green's indecomposability theorem for stable modules over arbitrary fields.

If k is algebraically closed, then by Green's theorem, $n = 1$. However, in the general situation of 4.6, $n > 1$ is possible also over algebraically closed fields as was shown by Voigt [Vo1, 12.9] for restricted Lie algebras.

Green's theorem for group algebras and arbitrary representations follows easily from the stable case. But the generalization of 4.6 to non-stable modules does not hold as was shown by M. Josek in his thesis [Jo] by a class of interesting examples. The easiest case of his examples will be described below. Its proof is simplified by the use of 4.5 (3).

4.7 LEMMA. *Let R be a right H-comodule algebra and $A = R \# H^*$, where the action of H^* on R is adjoint to the given coaction of H. Let M be a right R-module. Then there is an R-isomorphism*

$$M \otimes_R A \cong M \otimes H,$$

where R acts diagonally on the right hand side via the comodule algebra structure of R, $(m \otimes h)r = \sum m r_0 \otimes h r_1$ for all $m \in M$, $h \in H$, $r \in R$.

PROOF. The isomorphism $M \otimes_R (R \# H^*) \cong M \otimes H^*$, $m \otimes r \# p \longmapsto mr \otimes p$, is right R-linear, where R acts on $M \otimes H^*$ via Δ_R and H^* is a right H-module by $(ph)(x) := p(hx)$, for all $p \in H^*$ and $h, x \in H$. Hence the lemma follows from the right H-isomorphism $H^* \cong H$ [LS]. □

4.8 EXAMPLE [Jo, 8.7]. Let $\operatorname{char}(k) = p > 0$ and $H := k[t]$, the p-dimensional Hopf algebra with $t^p = 0$ and t primitive.

(1) Let $k[T]$ be the algebra of polynomials in one variable T. Then the canonical map $\pi : k[T] \longrightarrow k[t]$, $\pi(T) := t$, is a surjective Hopf algebra map, where T is primitive. Hence, $k[T]$ is a right $k[t]$-comodule algebra with structure map $(id \otimes \pi)\Delta : k[T] \longrightarrow k[T] \otimes k[t]$.

For all $n \geq 1$, let W_n be the indecomposable $k[T]$-module $W_n := k[T]/(T^n)$. Then

$$W_{p+1} \otimes k[t] \cong W_{2p} \oplus (W_p)^{p-1}$$

is the Krull-Schmidt decomposition of the $k[T]$-module $W_{p+1} \otimes k[t]$, where $k[T]$ is acting diagonally via $(id \otimes \pi)\Delta$ as in 4.7.

PROOF. For all $i \geq 0$, let w_i be the residue class of T^i in W_{p+1}. By definition of the action, for all $w \in W_{p+1}$ and $P \in k[T]$

$$\sum (w\, S(P_1) \otimes 1)\, P_2 = w \otimes \pi(P) \text{ in } W_{p+1} \otimes k[t].$$

In particular, taking $w = w_i$ and $P = T^j$, $i, j \geq 0$,

$$w_i \otimes t^j = \sum_{l=0}^{j} \binom{j}{l} (-1)^l \, (w_{i+l} \otimes 1)\, T^{j-l}.$$

Therefore, the elements $(w_i \otimes 1)T^j$, $0 \le i \le p$, $0 \le j < p$, form a system of $(p+1)p$ generators of the $(p+1)p$-dimensional vector space $W_{p+1} \otimes k[t]$, hence a basis. The decomposition is now obvious. Note that $(w_0 \otimes 1)T^p = w_p \otimes 1$ and $(w_i \otimes 1)T^p = 0$ for all $i \ge 1$. Then the k-linear span of $(w_0 \otimes 1)T^j$, $(w_p \otimes 1)T^j$, $0 \le j < p$, is a $k[T]$-subspace isomorphic to W_{2p}, and for all $1 \le i \le p-1$, the span of $(w_i \otimes 1)T^j$, $0 \le j < p$, is isomorphic to W_p. □

(2) Let R be the commutative algebra generated by x, y with relations $x^p = 0$ and $y^p = 0$. Then

$$\Delta_R : R \longrightarrow R \otimes k[t],\ \Delta_R(x) := x \otimes 1,\ \Delta_R(y) := x \otimes t + y \otimes 1,$$

defines a $k[t]$-comodule algebra structure on R. As in 4.7, let $A := R \# H^*$ be the corresponding smash product. For all $n \ge 1$, let M_n be the $2n$-dimensional indecomposable right R-module with k-basis $u_1, \cdots, u_n, v_1, \cdots, v_n$ and R-action given by $u_i x = v_i$, $u_i y = v_{i+1}$, $v_i x = 0 = v_i y$, $1 \le i,\ j \le n$, $v_{n+1} := 0$. (M_n is one of the indecomposable Kronecker modules over the factor ring $k[X,Y]/(X^2, XY, Y^2)$ of $R = k[X,Y]/(X^p, Y^p)$). Then

$$M_{p+1} \otimes k[t] \cong M_{2p} \oplus (M_p)^{p-1}$$

as right R-modules, where R is acting diagonally on $W_{p+1} \otimes k[t]$ via Δ_R.

PROOF. For any right $k[T]$-module U, define $F(U) := U \times U$ as a vectorspace with right R-module structure given by $(u, v)x := (0, u)$, $(u, v)y := (0, uT)$. Then F is a functor from $k[T]$-modules to R-modules, where $F(f) := (f, f)$ on homomorphisms f. It is easy to see that F is additive, $F(W_n) \cong M_n$ and $F(W_n \otimes k[t]) \cong M_n \otimes k[t]$ for all n. Hence the decomposition in (2) follows from (1). □

(3) Let $R \subset R \# H^*$ be the H^*-Galois extension (smash product) in (2), and $M := M_{p+1}$. Then H and H^* are both local algebras, M is indecomposable and

$$s_A(M \otimes_R A) = 2.$$

PROOF. By (2) and 4.7, $s_R(M \otimes_R A) = 2$. Since $H \cong H^*$ is local, $s_A(M \otimes_R A) = s_R(M \otimes_R A)$ by 4.5 (3). □

Moreover, in this example R is a Hopf algebra isomorphic to the dual of $k[T]/(T^q)$, $q := p^2$, where the residue class of T is primitive, and A is the semidirect product of cocommutative Hopf algebras.

5 THE STABILIZER OF REPRESENTATIONS AND AN IRREDUCIBILITY CRITERION

In this section, let H be a Hopf algebra with bijective antipode and $R \subset A$ a right faithfully flat H-Galois extension.

For any subcoalgebra $C \subset H$ let $A(C) := \{\ a \in A \mid \Delta_A(a) \in A \otimes C\ \}$ be the largest right C-subcomodule of A. Note that $A(C)$ is a left and right R-submodule of A.

5.1 DEFINITION [S1]. Let M be a right R-module and $C \subset H$ a subcoalgebra. Then C *stabilizes* M if there is a right R-linear and C-colinear isomorphism

$$M \otimes_R A(C) \cong M \otimes C.$$

Here, the module structures are defined by $(m \otimes a)r := m \otimes ar$, $(m \otimes c)r := mr \otimes c$ for all $m \in M$, $a \in A(C)$, $c \in C$, $r \in R$ and the comodule structures are given by $id \otimes \Delta_A$ and $id \otimes \Delta$.

For example, if $C = k \cdot 1$, then $A(C) = R$ and C stabilizes M. The module M is called *purely unstable* if $k \cdot 1$ is the only subcoalgebra stabilizing M.

5.2 THEOREM [S2], [S4]. *Let M be a right R-module and $Z := \mathrm{End}_R(M)$ its endomorphism ring. Then*

(1) *There is a largest subcoalgebra H_{st} of H stabilizing H.*

(2) *If* (a) *H is pointed*

or (b) *the coradical of H is cocommutative and $Z/rad(Z)$ is artinian,*

then H_{st} is a Hopf subalgebra of H.

5.3 DEFINITION [S2]. In 5.2, H_{st} is called the *stabilizer* of M (with respect to the H- *Galois extension* $R \subset A$).

Note that $Z/rad(Z)$ is artinian when M is a module of finite length. If (a) and (b) in (2) are not satisfied, there are examples where H_{st} is not a Hopf subalgebra [S4].

5.3 REMARK. (1) Let $H = k[G]$ be a group algebra. Then $A = \oplus_g A_g$, $A_1 = R$, is strongly graded. It follows easily from the definition that $H_{st} = k[G_{st}]$, where G_{st} is the subgroup of all elements $g \in G$ such that $M \otimes_R A_g \cong M$ as R-modules. Note that $A_g = A(kg)$ by definition.

(2) Let G be a group and N a normal subgroup of G. Then $k[N] \subset k[G]$ is a right $k[G/N]$-Galois extension as in 3.8 (1). In this case, $(G/N)_{st} = G_{st}/N$, where G_{st} is the subgroup of all elements $g \in G$ such that $M_g \cong M$ (see 3.8 (1) for the definition of the twisted module M_g).

(3) Assume $\mathrm{char}(k) = 0$. Let $\mathfrak{g}$ be a Lie algebra and $\mathfrak{n}$ an ideal of $\mathfrak{g}$. Then $U(\mathfrak{n}) \subset U(\mathfrak{g})$ is a right $H = U(\mathfrak{g}/\mathfrak{n})$-Galois extension, and the H-stabilizer of an $U(\mathfrak{n})$-module is $U(\mathfrak{g}_{st}/\mathfrak{n})$, where $\mathfrak{g}_{st}$ is the stabilizer defined by Blattner [Bl], [Di, 5.3.1].

(4) Let A be a Hopf algebra, R a normal Hopf subalgebra and $H := A/R^+A$ the quotient Hopf algebra. Let M be a right R-module and assume (a) or (b) in 5.2 (2). In this situation there are two Hopf Galois extensions:

(i) $R \subset R \# A$, the smash product where A is operating on R by the adjoint action, $ad(a)(r) := \sum a_1 r S(a_2)$.

(ii) $R \subset A$, the right H-Galois extension of 1.2 (4).

If A_{st} is the A-stabilizer of M in the sense of (i), and H_{st} is the H-stabilizer of M in the sense of (ii), then $A_{st}/R^+A_{st} \subset H_{st}$. If A is cocommutative, then $A_{st}/R^+A_{st} = H_{st}$.

PROOF. First note that R is contained in A_{st}, since $M \otimes_R (R \# R) \longrightarrow M \otimes R$, $m \otimes r \# s \longmapsto \sum mrs_1 \otimes s_2$, is bijective.

Since the coradical of H is cocommutative, A is faithfully flat over any Hopf subalgebra A' [T1, 3.2]. Hence, by the left version of [T2, Th. 1], the natural map $H' := A'/R^+A' \longrightarrow A/R^+A$ is injective and $A' \cong A \,\square_H\, H'$. When A is cocommutative, any Hopf subalgebra of H is of the form A'/R^+A', A' some Hopf subalgebra of A.
Therefore it suffices to show for all Hopf subalgebras A' of A:

$$A' \subset A_{st} \text{ if and only if } A'/R^+A' \subset H_{st}.$$

Let $R \subset A'$ be a Hopf subalgebra and $\pi : A' \longrightarrow A'/R^+A' =: H'$ the canonical epimorphism.

Recall the definition of the cotensorproduct $V \,\square_C\, W$ of a right C-comodule and a left C-comodule over any coalgebra C: $V \,\square_C\, W$ is the kernel of $\Delta_V \otimes id - id \otimes \Delta_W$ in $V \otimes W$.

Then

$$M \otimes_R (R \# A') \longrightarrow M \otimes_R (A' \,\square_{H'}\, A'),\ m \otimes r \# a \longmapsto \sum m \otimes ra_1 \otimes a_2,$$

where A' is an H'-comodule via π, is an isomorphism of right A'-comodules with comodule structures $id \otimes id \otimes \Delta$. This map is also right R-linear where the module structure on $M \otimes_R (A' \,\square_{H'}\, A')$ is given by multiplication on the second factor and $M \otimes_R (R \# A')$ is a right R-module by restriction of the $R \# A'$-module structure, i.e. $(m \otimes r \# a)s = \sum m \otimes r\, ad(a_1)(s) \# a_2$ for all $m \in M$, $r, s \in R$ and $a \in A'$.

Furthermore, the natural map $M \otimes_R (A' \,\square_{H'}\, A') \longrightarrow (M \otimes_R A') \,\square_{H'}\, A'$ is bijective since A' is left coflat over H. Thus there is an isomorphism

$$\Phi : M \otimes_R (R \# A') \cong (M \otimes_R A') \,\square_{H'}\, A'$$

of right R-modules and A'-comodules.

If H' stabilizes M in the sense of (ii), then $M \otimes_R A' \cong M \otimes H'$, and $(M \otimes_R A') \,\square_{H'}\, A' \cong M \otimes H' \,\square_{H'}\, A' \cong M \otimes A'$. Hence, by the isomorphism Φ, A' stabilizes M in the sense of (i). The converse follows from (a version of) [S2, 4.2] applied to the surjective coalgebra map $\pi : A' \longrightarrow H'$. □

(5) In the situation of (4), when A and M are finite-dimensional, A is cocommutative and the field k is algebraically closed, the existence of the stabilizer as a Hopf subalgebra was shown by Voigt [Vo1]. He used a completely different description in the context of group schemes. Then Greipel [Gr] proved 5.2 for smash products and pointed Hopf algebras being mainly interested in the case (4) (i).

5.4 EXAMPLE. In example 4.8 (3), R, A and H are finite- dimensional cocommutative Hopf algebras, and $A = R \# H^*$ is a smash product. The R-module M is indecomposable and there are two non-isomorphic summands in the R-Krull-Schmidt decomposition of $M \otimes_R A$. Hence M cannot be H^*-stable. Since the dimension of H^* is a prime p, M is purely unstable. The number of R-resp. A- isomorphism classes of summands in the Krull-Schmidt decomposition is 1 for M and 2 for the induced module $M \otimes_R A$.

In the case of finite group algebras and finite-dimensional indecomposable modules $M \otimes_R A(H_{st})$ resp. $M \otimes_R A$ have $A(H_{st})$- resp. A-Krull-Schmidt decompositions of the same type [CR, (19.20)]. However, as the above example shows, even in the case of finite-dimensional cocommutative Hopf algebras, the stabilizer does not control the type of the Krull-Schmidt decomposition.

But the stabilizer does control the Jordan-Hölder series also for pointed cocommutative Hopf algebras. This follows from the criterion for irreducibility of induced representations generalizing Blattner's result [Bl] for Lie algebras in charcteristic 0.

5.5 THEOREM [S2], [S4]. *Let M be a simple right R-module with H-stabilizer H_{st} and $S := A(H_{st})$. Assume that M is centrally simple, i.e. k is the center of* $\mathrm{End}_R(M)$, *and H is pointed and equal to the sum of its irreducible components.*

(1) *If M is purely unstable, i.e. $S = R$, then $M \otimes_R A$ is a simple right A-module and* $\mathrm{End}_R(M) \cong End_A(M \otimes_R A)$.

(2) *If H is cocommutative and Q is a simple S-module which is isomorphic to a direct sum of copies of M as an R-module, then $Q \otimes_S A$ is a simple right A-module and* $\mathrm{End}_S(Q) \cong \mathrm{End}_A(Q \otimes_S A)$.

Moreover, in the proof of 5.5 (2), it is shown that any R-submodule of $M \otimes_R A$ isomorphic to M is contained in $M \otimes_R S$. Hence $\mathrm{End}_A(M \otimes_R A) \cong \mathrm{End}_S(M \otimes_R S)$, and 5.5 and 3.9 imply

5.6 COROLLARY [S2]. *Let H be pointed and cocommutative and M a centrally simple right R- module. Then* $\mathrm{End}_A(M \otimes_R A) \cong \mathrm{End}_R(M) \#_\sigma H_{st}$ *is an H_{st}-crossed product.*

5.7 EXAMPLE. In example 3.6, the Hopf algebra H is finite-dimensional, cocommutative and irreducible, hence pointed. It was shown that the canonical map $M_i \longrightarrow M_i \otimes_R A$ is not R-split for the one-dimensional simple R-modules M_i, $1 \leq i \leq p-1$. Hence the M_i's are not stable. Since the dimension of H is a prime p, for all $1 \leq i \leq p-1$, M_i is purely unstable. By 5.6, $M_i \otimes_R A$ is a simple A-module. But as an R-module, $M_i \otimes_R A$ is not semisimple since M_i is not a direct summand.

Thus Clifford's theorem on the semisimplicity of the restriction of simple representations fails in the general case. Instead, for pointed Hopf algebras there is a weaker version of Clifford's theorem describing the quotients of a Jordan-Hölder series [S2, 2.2].

5.8 REMARK. (1) In 5.5, the assumption on the center of $\mathrm{End}_R(M)$ is not really necessary. Let M be a simple right R-module and let K be the center of its endomorphism ring. Then K is a field and $R \otimes K \subset A \otimes K$ is an $H \otimes K$-Galois extension by field extension. M resp. $M \otimes_R A$ are modules over $R \otimes K$ resp. $A \otimes K$ in the natural way. Then M is also simple as an $R \otimes K$-module and the center of $\mathrm{End}_{R \otimes K}(M) = \mathrm{End}_R(M)$ is K. Note also that $M \otimes_{R \otimes K} (A \otimes K) \cong M \otimes_R A$ as modules over $A \otimes K$. Thus 5.5 can be applied after field extension.

(2) As in classical Clifford theory, in the situation of 5.5 there are two steps to construct simple A-modules. In the first step, compute H_{st} and S. In the second step, find a simple S-module Q which is M-primary, i.e. R-isomorphic to a direct sum of copies of M (for example an S-submodule of $M \otimes_R S$). Then, by 5.5 (2), $Q \otimes_S A$ is simple over A. The second step is also called stable Clifford theory.

Let $E := \mathrm{End}_S(M \otimes_R S)$ and $E' := \mathrm{End}_R(M)$. Since E is an H_{st}-crossed product, the functor ${}_E\mathcal{M}^{H_{st}} \longrightarrow {}_{E'}\mathcal{M}$, $X \longmapsto X^{coH_{st}}$, is an equivalence (cf. 2.2). In particular, the map $E \otimes_{E'} M \longrightarrow M \otimes_R S$, $f \otimes m \longmapsto f(m \otimes 1)$, is an isomorphism since it induces an isomorphism on the coinvariant elements. Hence for any right E-module V, $V \otimes_{E'} M \cong V \otimes_E (M \otimes_R S)$ is M-primary. Then the functor $V \longmapsto V \otimes_E (M \otimes_R S)$ from $\mathcal{M}_E$ to the category of all right S-modules which are M-primary is an equivalence (cf. [Da, 7.4]).

Thus M-primary simple right S-modules Q are given by simple right modules over the H_{st}-crossed product $E \cong \mathrm{End}_R(M) \#_\sigma H_{st}$.

(3) 5.5 was shown by Voigt [Vo1] in the case where A is a finite-dimensional cocommutative Hopf algebra over an algebraically closed field, R is a normal Hopf subalgebra and M is finite-dimensional.Then 5.5 (2) was proved by Greipel [Gr] in the situation of 5.3 (4) (i), where R is a normal Hopf subalgebra of the cocommutative and pointed Hopf algebra A.

Using the irreducibility criterion for infinitesimal group schemes, Voigt [Vo2] obtained the following characterization of solvable infinitesimal group schemes extending results of Schue [Sch] and Strade [St] on restricted Lie algebras.

A finite-dimensional Hopf algebra H is called *solvable* if there exists a sequence of Hopf subalgebras $k = H_0 \subset H_1 \subset \cdots \subset H_n = H$, $n \geq 0$, such that each H_i is normal in H_{i+1} with abelian quotient $H_{i+1}/H_i^+ H_{i+1}$.

H is called *monomial* if for any simple right H- module there is a Hopf subalgebra H' of H and a one-dimensional right H'-module U such that $U \otimes_{H'} H \cong M$ as right H-modules.

5.9 THEOREM [Vo1], [Vo2]. *Let k be algebraically closed, $char(k) = p > 0$, and H a finite-dimensional cocommutative and irreducible Hopf algebra.*

(1) *If H is monomial, then H is solvable.*

(2) *Assume $p > 2$ and H is solvable. Then H is monomial.*

For the remainder of this section, assume $A = R \#_\sigma H$ is an H-crossed product with weak action $\cdot : H \otimes R \longrightarrow R$ and invertible cocycle σ, and R is a subalgebra of a given algebra Q.

5.10 DEFINITION. A subcoalgebra C in H is *Q-inner on R* if there is an invertible map $u : C \longrightarrow Q$ such that $c \cdot r = \sum u(c_1)\, r\, u^{-1}(c_2)$ for all $c \in C$, $r \in R$.

It turns out that C is Q-inner if and only if C is contained in the stabilizer of a certain representation. To this end consider the (Q,R)-bimodule Q as a right $R \otimes Q^{op}$-module by $x(r \otimes q^{op}) := qxr$ for all $x, q \in Q$, $r \in R$, and R as a subalgebra of $R \otimes Q^{op}$ by mapping $r \in R$ onto $r \otimes 1$. Then $R \otimes Q^{op} \subset (R \otimes Q^{op}) \#_\sigma H$ is an H-crossed product with the same cocycle σ and weak action $h \cdot (r \otimes q^{op}) := h \cdot r \otimes q^{op}$.

5.11 LEMMA [S4]. *Let C be a subcoalgebra of H. Then the following are equivalent:*

(1) *C is Q-inner on R.*

(2) *C is contained in the H-stabilizer of Q as a right $R \otimes Q^{op}$-module with respect to the H-Galois extension $R \otimes Q^{op} \subset (R \otimes Q^{op}) \#_\sigma H$.*

Hence, theorem 5.2 can be formulated in the special situation of 5.11 (2) as follows. Note that $\mathrm{End}({}_Q Q_R)$ is the centralizer $C_Q(R)$ of R in Q.

5.12 COROLLARY. *Let Z be the centralizer of R in Q.*

(1) *There is a largest Q-inner subcoalgebra H_{inn}.*

(2) *If* (a) *H is pointed*

or (b) *the coradical of H is cocommutataive and $Z/rad(Z)$ is artinian,*

then H_{inn} is a Hopf subalgebra of H.

5.12 was first proved by Masuoka [M1], [M2] assuming in 5.12 (2) that $R = Q$ and $A = R \# H$ is a smash product.

In [S4], an example is given where H_{inn} is not a Hopf subalgebra.

It is also interesting to apply the irreducibility criterion 5.5 in this setting. In 5.5, the center of $\mathrm{End}_R(M)$ was assumed to be k. Since $\mathrm{End}({}_Q Q_R) \cong C_Q(R)$, this assumption is too restrictive. However, as was noted in 5.8 (1), the assumption on the center of $\mathrm{End}_R(M)$ is not really necessary. Instead extend the base field. Thus the next definition seems to be very natural.

5.13 DEFINITION [S4]. Let K be a subfield of the center of Q.

(1) *A K*-subcoalgebra C of $H \otimes K$ is *(Q,K)-inner* if there is a K-linear invertible map $u : C \longrightarrow Q$ such that $c \cdot r = \sum u(c_1)\, r\, u^{-1}(c_2)$ for all $c \in C$, $r \in R$. Here, $c \cdot r = \sum (c_i \cdot r)\alpha_i$, $c = \sum c_i \otimes \alpha_i$, where c_i resp. α_i are elements in H resp. K.

(2) H is *(Q,K)-outer on R* if K is the only (Q,K)-inner subcoalgebra of $H \otimes K$.

5.14 THEOREM [S4]. *Let K be a subfield of the center of Q.*

(1) *A K-subcoalgebra C of $H \otimes K$ is (Q,K)-inner if and only if C is contained in the $H \otimes K$-stabilizer of Q as a right $R \otimes Q^{op} \otimes K$-module with respect to the $H \otimes K$-Galois extension $R \otimes Q^{op} \otimes K \subset (R \otimes Q^{op} \otimes K) \#_{K,\sigma} (H \otimes K)$ obtained from $(R \otimes Q^{op}) \#_\sigma H$ by field extension.*

(2) *Assume H is pointed. Then H is (Q,K)-outer on R if and only if*

(a) *1 is the only group-like element g of H such that there exists a unit u in Q satisfying $g \cdot r = u\, r\, u^{-1}$ for all $r \in R$.*

(b) If $x_1, \cdots, x_n, n \geq 1$, *are k-linearly independent primitive elements in H and $\alpha_1, \cdots, \alpha_n$ are in K such that there is an element $q \in Q$ satisfying $\sum \alpha_i (x_i \cdot r) = q\, r - r\, q$ for all $r \in R$, then $\alpha_i = 0$ for all i.*

Conditions (a) and (b) in 5.14 (2) appeared in the work of Kharchenko, see [BeM1], [BeM2]. From the point of view of stabilizers they simply mean that a certain representation is purely unstable.

5.5 (1) now has the following two corollaries [S4].

5.15 COROLLARY [BeM2]. *Let R be a simple algebra with center K. Assume $H \otimes K$ is pointed and equal to the sum of its irreducible components. If H is (R,K)-outer on R, then $R \#_\sigma H$ is a simple algebra.*

5.15 was first obtained by Jacobson [Ja] and Azumaya [Az] for skew group algebras, i. e. in the case where H is a group algebra and the cocycle σ is trivial.

The simplicity result in 5.15 does not hold for arbitrary pointed Hopf algebras, as was shown by example 3.5 in [Mo1]. Thus, also 5.5 (1) does not hold if H is just pointed.

5.16 COROLLARY [BeM2]. *Let R be a prime left Goldie algebra, Q the classical ring of left fractions of R and K the center of Q. Assume $H \otimes K$ is pointed and equal to the sum of its irreducible components. If H is (Q,K)-outer on R, then $I \cap R \neq 0$ for all non-zero ideals I of $R \#_\sigma H$.*

In [BeM2], 5.16 is shown in the more general case where R is any prime algebra and Q is the symmetric Martindale quotient of R. The Goldie condition in 5.16 guarantees that Q is a simple (Q,R)-bimodule and 5.5 applies directly. 5.15 and 5.16 are proved in [BeM2] in a completely different and more computational way using conditions (a) and (b) in 5.14.

6 BASIC THEORY OF CROSSED PRODUCTS

In this section, assume the antipode of H is bijective and let A be a right H-comodule algebra, $R := A^{coH}$ the algebra of coinvariants and $\gamma : H \longrightarrow A$ a fixed invertible and right H-colinear map such that $\gamma(1) = 1$.

As explained in the first section, A can be identified with an H-crossed product $R \#_\sigma H$, where the weak action and the invertible cocycle σ are defined by

$$h \cdot r := \sum \gamma(h_1)\, r\, \gamma^{-1}(h_2),\ \sigma(g,h) := \sum \gamma(g_1)\, \gamma(h_1)\, \gamma^{-1}(g_2 h_2),\ g, h \in H,\ r \in R.$$

Then $\gamma(h) = 1 \# h$ for all $h \in H$.

There are two more "actions" which are important for the structure of $R \#_\sigma H$.

6.1 DEFINITION [MS]. For all $h \in H$ and $r \in R$, define

$$h \wedge r := \sum \gamma(h_1)\, r\, \gamma(S(h_2)),$$
$$h \vee r := \sum \gamma^{-1}(h_2)\, r\, \gamma^{-1}(S(h_1)).$$

When γ is an algebra map, i.e. $A \cong R \# H$ is a smash product, then $h \wedge r = h \cdot r$ and $h \vee r = S(h) \cdot r$. But in the general case, $\vee$ and $\wedge$ cannot be expressed by the weak action. The identities in the next lemma extend well known identities for smash products first noted in [Co2, Lemma 1], [CF, Lemma 1].

6.2 LEMMA [MS]. *For all $r, s \in R$, $h \wedge r$ and $h \vee r$ are contained in R and*

(1) $h \wedge r = \sum (h_1 \cdot r)(h_2 \wedge 1)$,
$h \vee r = \sum (h_2 \vee 1)(S(h_1) \cdot r)$.

(2) $r\, \gamma(S(h)) = \sum \gamma^{-1}(h_1)(h_2 \wedge r)$,
$\gamma^{-1}(h)\, r = \sum (h_2 \vee r)\, \gamma(S(h_1))$.

(3) $\sum h_1 \wedge (r\,(h_2 \vee s)) = (h \cdot r)\, s$,
$\sum h_1 \vee ((h_2 \wedge r)\, s) = r\,(S(h) \cdot s)$.

6.3 REMARK. (1) By 6.2 (1), the new "actions" differ from the weak action by the factors $h \wedge 1$ and $h \vee 1$. In terms of σ, these are given by

$$h \wedge 1 = \sum \sigma(h_1, S(h_2)),\ h \vee 1 = \sum \sigma^{-1}(S(h_1), h_2).$$

(2) By the isomorphism $R \#_\sigma H \cong A$, $r \# h \longmapsto r\,\gamma(h)$, any element in A can be represented in the form $\sum r_i\, \gamma(h_i)$. Since A^{op} is H^{op}-cleft by the mapping $H^{op} \longrightarrow A^{op}$, $h^{op} \longmapsto \gamma^{-1}(S^{-1}(h))^{op}$, elements in A also have a γ^{-1}-representation $\sum \gamma^{-1}(S^{-1}(h_i))\, r_i$. The identities in 6.2 (2) allow to compute the γ-representation from the γ^{-1}-representation and conversely.

Using the explicit description of $h \wedge 1$ and $h \vee 1$ in (1), the first identity in 6.2 (2) can be written as

$$r \# S(h) = \sum (\sigma^{-1}(S(h_2), h_3) \# S(h_1))\,(h_4 \cdot r)\,\sigma(h_5, S(h_6)).$$

In a sense, this formula is a commutation rule for elements in R and in $1 \# H$.

(3) By 6.2 (3), $\sum h_1 \wedge (h_2 \vee r) = \varepsilon(h)\, r = \sum h_1 \vee (h_2 \wedge r)$. Hence the mapping $\wedge : H \longrightarrow \mathrm{End}(R)$ is the inverse of $\vee$ under convolution.

A subset I of R is called H-*stable* if $h \cdot r \in I$ for all $h \in H$ and $r \in I$. The next corollary follows from 6.1. It was known before only under additional assumptions [Co2], [Ch2],

[Mo1].

6.4 COROLLARY [MS]. *Let I be an H-stable ideal in R. Then $AI = IA$, and the left and right annihilators of I are also H-stable.*

In the study of crossed products of group algebras, one of the "absolutely necessary ingredients" [Pa, p. 83] is the extension of the crossed product to the Martindale quotient ring.

For Hopf algebras, it is more difficult to extend the crossed product to the Martindale quotient ring. This was first done in [Co1], [Co2] for smash products and in [Ch2], [Mo2] for crossed products under additional assumptions on the action. Using $\wedge$ and $\vee$, the action can now be extended in full generality.

Recall the definition of the right Martindale quotient [Pa]. Let $\mathcal{F}(R)$ be the set of all ideals of R having left and right annihilator zero. Then Q^r is the direct limit of the system $\mathrm{Hom}_R(I_R, R_R)$, $I \in \mathcal{F}(R)$. Explicitly, elements of Q^r are equivalence classes $[f]$ of right R-linear maps $f : I \longrightarrow R$, $I \in \mathcal{F}(R)$. Here $[f] = [f_1]$, $f_1 : I_1 \longrightarrow R$, if f and f_1 agree on $I \cap I_1$. If f resp. g represent elements in Q^r defined on I resp. J, then $[f] + [g]$ is represented by the map $I \cap J \longrightarrow R$, $x \longmapsto f(x) + g(x)$, and $[f] \cdot [g]$ is represented by $JI \longrightarrow R$, $x \longmapsto f(g(x))$. Then Q^r is a ring, and R is a subring of Q^r by mapping $r \in R$ onto the class of left multiplication on R by r.

6.5 DEFINITION [Mo2]. Let $\mathcal{F}$ be a multiplicatively closed subset of $\mathcal{F}(R)$, i.e $R \in \mathcal{F}$ and for all $I, J \in \mathcal{F}$ also $IJ \in \mathcal{F}$. The weak action of H on R is *$\mathcal{F}$-continuous* if for all $h \in H$ and $I \in \mathcal{F}$ there is some $J \in \mathcal{F}$ such that $h \cdot J \subset I$.

In particular, the weak action is $\mathcal{F}_H$-continuous, where $\mathcal{F}_H$ is the set of all H-stable ideals in $\mathcal{F}(R)$. If H is pointed, then the weak action is even $\mathcal{F}(R)$-continuous [Mo2].

If $\mathcal{F}$ is multplicatively closed, let $Q^r(\mathcal{F})$ denote the subring of Q^r whose elements are represented by mappings defined on ideals in $\mathcal{F}$.

6.6 THEOREM [MS]. *Let $\mathcal{F} \subset \mathcal{F}(R)$ be multiplicatively closed. Assume the weak action of H on R is $\mathcal{F}$-continuous. Then there is a unique H-crossed product action on $Q^r(\mathcal{F})$ extending the weak action on R with cocycle $\sigma : H \otimes H \longrightarrow R \subset Q^r(\mathcal{F})$.*

For all $h \in H$ and $[f] \in Q^r(\mathcal{F})$, $f : I \longrightarrow R$, $I \in \mathcal{F}$, $h \cdot [f]$ is represented by $h \cdot f : J \longrightarrow R$, where J is an ideal in $\mathcal{F}$ such that $h_2 \vee J \subset I$ for all h_2 in $\Delta(h) = \sum h_1 \otimes h_2$, and

$$(h \cdot f)(r) := \sum h_1 \wedge f(h_2 \vee r) \text{ for all } r \in J.$$

Explicitly, the extended action is given by the rather involved formula

$$(h \cdot f)(r) = \sum h_1 \cdot f(\sigma^{-1}(S(h_5), h_6)\ (S(h_4) \cdot r))\ \sigma(h_2, S(h_3)).$$

Similarly, the extended action of the left Martindale quotient is given by

$$(h \cdot f)(r) := \sum S^{-1}(h_2) \vee f(S^{-1}(h_1) \wedge r).$$

As in the classical case, let $Q(\mathcal{F})$ be the symmetric Martindale quotient consisting of all $q \in Q^r(\mathcal{F})$ such that $Iq \subset R$ for some $I \in \mathcal{F}$. The next result shows that $Q^r(\mathcal{F})$-inner actions are in fact $Q(\mathcal{F})$-inner. Thus the symmetric ring of quotients is large enough.

6.7 THEOREM [MS]. *Assume the situation of 6.6.*

(1) *$Q(\mathcal{F})$ is an H-stable subalgebra of $Q^r(\mathcal{F})$. Hence the crossed product extends uniquely to $Q(\mathcal{F}) \#_\sigma H$.*

(2) *If $C \subset H$ is a subcoalgebra such that $S(C) = C$ and $u : C \longrightarrow Q^r(\mathcal{F})$ is an invertible map such that $c \cdot r = \sum u(c_1)\, r\, u^{-1}(c_2)$ for all $c \in C$ and $r \in R$, then for all c in C, $u(c)$ and $u^{-1}(c)$ are contained in $Q(\mathcal{F})$.*

6.7 extends results in [Co2] for smash products.

6.8 REMARK. (1) Let A be any right H-comodule algebra and assume A is left flat over $R := A^{coH}$. For any right (A,H)-Hopf modules X,Y the H-comodule $\mathrm{HOM}_A(X,Y)$ is defined as in 3.3 in the case $X = Y$. Let $\mathcal{F}_H$ be the set of all $I \in \mathcal{F}(R)$ such that IA is an ideal in A. Let $Q_H^{r,f}(A)$ be the direct limit of the system

$$\mathrm{HOM}_A(I \otimes_R A, A),\ I \in \mathcal{F}_H.$$

Then $Q_H^{r,f}(A)$ is a right H-comodule algebra and $Q^r(\mathcal{F}_H)$ can be identified with the subalgebra of all H-coinvariant elements. Thus the coaction of A can always be extended to the "finite" Martindale quotient ring $Q_H^{r,f}(A)$ of A.

(2) Now assume in (1) that there is an invertible and H-colinear map $\gamma : H \longrightarrow A$. Then the extended map $\gamma : H \longrightarrow A \subset Q_H^{r,f}(A)$ is invertible and H-colinear. Hence $Q_H^{r,f}(A) \cong Q^r(\mathcal{F}_H) \#_\sigma H$ is an H-crossed product. The induced action on $Q^r(\mathcal{F}_H)$ can be seen to be exactly the one defined above.

In the case of H-module algebras, Matczuk [Mat] introduced H-centrally closed algebras generalizing results on centrally closed algebras of Martindale [M], Erickson, Martindale and Osborn [EMO] and Krempa [Kr].

Using the new "actions" $\wedge$ and $\vee$, most of the results in [Mat] can be extended to arbitrary crossed product actions.

As an application of the crossed product theory in this section, a partial answer to the following open problem can be given:

Assume R is semiprime. Is $R \#_\sigma H$ semiprime when H is semisimple?

By the theorem of Fisher and Montgomery [FM], the answer is positive for group algebras. The answer is also positive if R is H-prime, i.e. the product of non-zero H-stable ideals of R is non-zero, and H is Q-inner on R, where Q is the symmetric Martindale quotient with respect to $\mathcal{F}_H$.

6.9 THEOREM [MS]. *Assume R is H-prime, H is finite-dimensional and semisimple and H is Q-inner on R. Then $R \#_\sigma H$ is semiprime.*

This extends results of [BCM], [BM] and [Ch2].

REFERENCES

[A] E. Abe, *Hopf Algebras*, Cambridge Univ. Press, Cambridge, 1980.

[Az] G. Azumaya, *New formulation of the theory of simple rings*, Proc. Japan Acad. 22 (1946), 325-345.

[BeM1] J. Bergen and S. Montgomery, *Smash products and outer derivations*, Israel J. Math. 53 (1986), 321-345.

[BeM2] ———, *Ideals and quotients in crossed products of Hopf algebras*, J. Algebra, to appear.

[Bl] R. J. Blattner, *Induced and produced representations of Lie algebras*, Trans. Amer. Math. Soc. 144 (1969), 457-474.

[BCM] R. J. Blattner, M. Cohen and S. Montgomery, *Crossed products and inner actions of Hopf algebras*, Trans. Amer. Math. Soc. 298 (1986), 671-711.

[BM] R. J. Blattner and S. Montgomery, *Crossed products and Galois extensions of Hopf algebras*, Pacific J. Math. 137 (1989), 37-54.

[Ch1] W. Chin, *Spectra of smash products*, Israel J. Math. 72 (1990), 84-98.

[Ch2] ———, *Crossed products and generalized inner actions of Hopf algebras*, Pacific J. Math. 150 (1991), 241-259.

[CS] S. U. Chase and M. E. Sweedler, *Hopf Algebras and Galois Theory*, Lecture Notes in Math. vol. 97, Springer, Berlin, 1969.

[CHR] S. U. Chase, D. K. Harrison and A. Rosenberg, *Galois theory and Galois cohomology of commutative rings*, Mem. Amer. Math. Soc. 52 (1965), 15-33.

[Cl] A. H. Clifford, *Representations induced in an invariant subgroup*, Annals of Math. 38 (1937), 533-550.

[CPS] E. Cline, B. Parshall and L. Scott, *Induced modules and affine quotients*, Math. Ann. 230 (1977), 1-14.

[Co1] M. Cohen, *Hopf algebras acting on semiprime algebras*, Contemporary Math. 43 (1985), 49-61.

[Co2] ———, *Smash products, inner actions and quotient rings*, Pacific J. Math. 125 (1986), 45-65.

[CF] M. Cohen and D. Fishman, *Hopf algebra actions*, J. Algebra 100 (1986), 363-379

[CFM] M. Cohen, D. Fishman and S. Montgomery, *Hopf Galois extensions, smash products, and Morita equivalence*, J. Algebra 133 (1990), 351-372.

[CR] C. W. Curtis and I. Reiner, *Methods of Representation Theory*, Vol. I, Wiley, New York, 1981.

[Da] E. C. Dade, *Group-graded rings and modules*, Math. Z. 174 (1980), 241-262.

[DG] M. Demazure and P. Gabriel, *Groupes Algébriques*, Masson, Paris, 1970.

[Di] J. Dixmier, *Algèbres enveloppantes*, Cahiers Scientifiques XXXVII, Gauthier-Villars, Paris, 1974.

[D1] Y. Doi, *On the structure of relative Hopf modules*, Comm. Algebra 11 (1983), 243-255.

[D2] ——, *Algebras with total integrals*, Comm. Algebra 13 (1985), 2137 - 2159.

[D3] ——, *Unifying Hopf modules*, J. Algebra 153 (1992), 373 - 385.

[DT1] Y. Doi and M. Takeuchi, *Cleft comodule algebras for a bialgebra*, Comm. Algebra 14 (1986), 801 - 818.

[DT2] ——, *Hopf - Galois extensions of algebras, the Miyashita - Ulbrich action, and Azumaya algebras*, J. Algebra 121 (1989), 488 - 516.

[EMO] T. S. Erickson, W. S. Martindale and J. M. Osborn, *Prime nonassociative algebras*, Pacific J. Math. 60 (1975), 49 - 63.

[FM] J. Fisher and S. Montgomery, *Semiprime skew group rings*, J. Algebra 52 (1978), 241 - 247.

[G] J. A. Green, *On the indecomposable representations of a finite group*, Math. Z. 70 (1959), 430 - 445.

[Gr] K. P. Greipel, *Induzierte Darstellungen punktierter Hopfalgebren*, thesis, Universität München, 1984.

[Ja] N. Jacobson, *Theory of Rings*, Amer. Math. Soc., Providence, 1943.

[Jo] M. Josek, *Zerlegung induzierter Darstellungen für Hopf - Galois - Erweiterungen*, thesis, Algebra Berichte Nr. 65, Fischer, München, 1992.

[KO] M.-A. Knus and M. Ojanguren, *Théorie de la Descente et Algèbres d'Azumaya*, Lecture Notes in Math. vol. 389, Springer, Berlin, 1974.

[KN] M. Koppinen and T. Neuvonen, *An imprimitivity theorem for Hopf algebras*, Math. Scand. 41 (1977), 193 - 198.

[KT] H. F. Kreimer and M. Takeuchi, *Hopf algebras and Galois extensions of an algebra*, Indiana Math. J. 30 (1981), 675 - 692.

[Kr] J. Krempa, *On semisimplicity of tensor products*, Ring Theory, Proceedings of the 1978 Antwerp Conference, 105 - 122.

[LS] R. G. Larson and M. E. Sweedler, *An associative orthogonal bilinear form for Hopf algebras*, Amer. J. Math. 91 (1969), 75 - 94.

[M] W. S. Martindale, III, *Prime rings satisfying a generalized polynomial identity* J. Algebra 12 (1969), 576 - 584.

[Ma1] A. Masuoka, *Existence of a unique maximal subcoalgebra whose action is inner*, Israel J. Math. 72 (1990), 149 - 157.

[Ma2] ——, *On Hopf algebras with cocommutative coradicals*, J. Algebra 144 (1991), 451 - 466.

[Mat] J. Matczuk, *Centrally closed Hopf module algebras*, Comm. Algebra 19 (1991), 1909 - 1918.

[Mo1] S. Montgomery, *Hopf Galois extensions*, Contemporary Math. vol. 124 (1992), 129 - 140.

[Mo2] ——, *Biinvertible actions of Hopf algebras*, Israel J. Math., to appear.

[MS] S. Montgomery and H.-J. Schneider, work in progress.

[NZ] W. D. Nichols and M. B. Zoeller, *A Hopf algebra freeness theorem*, Amer. J. Math. 111 (1989), 381 - 385.

[Ob] U. Oberst, *Affine Quotientenschemata nach affinen algebraischen Gruppen und induzierte Darstellungen*, J. Algebra 44 (1977), 503-538.

[OS1] U. Oberst and H.-J. Schneider, *Über Untergruppen endlicher algebraischer Gruppen*, manuscripta math. 8 (1973), 217-241.

[OS2] ——, *Untergruppen formeller Gruppen von endlichem Index*, J. Algebra 31, (1974), 10-44.

[vOZ] F. van Oystaeyen and Y. Zhang, *H-module endomorphism rings*, preprint 1992.

[Pa] D. S. Passman, *Infinite Crossed Products*, Academic Press, San Diego, 1989.

[Re] G. Renault, *Algèbre non commutative*, Gauthier-Villars, Paris, 1975.

[S1] H.-J. Schneider, *Principal homogeneous spaces for arbitrary Hopf algebras*, Israel J. Math. 72 (1990), 167-195.

[S2] ——, *Representation theory of Hopf Galois extensions*, Israel J. Math. 72 (1990), 196-231.

[S3] ——, *Normal basis and transitivity of crossed products for Hopf algebras*, J. Algebra 152 (1992), 289-312.

[S4] ——, *On inner actions of Hopf algebras and stabilizers of representations*, J. Algebra, to appear.

[S5] ——, work in progress.

[Sch] J. Schue, *Representations of solvable Lie p-algebras*, J. Algebra 38 (1976), 253-267.

[St] H. Strade, *Darstellungen auflösbarer Lie-p-Algebren*, Math. Ann. 232 (1978), 15-32.

[Sw1] M. E. Sweedler, *Cohomology of algebras over Hopf algebras*, Trans. Amer. Math. Soc. 133 (1968), 205-239.

[Sw2] ——, *Hopf Algebras*, Benjamin, New York, 1969.

[Ta1] M. Takeuchi, *A correspondence between Hopf ideals and sub-Hopf algebras*, manuscripta math. 7 (1972), 251-270.

[Ta2] ——, *Relative Hopf modules — equivalences and freeness criteria*, J. Algebra 60 (1979), 452-471.

[U1] K.-H. Ulbrich, *Galoiserweiterungen von nicht-kommutativen Ringen*, Comm. Algebra 10 (1982), 655-672.

[U2] ——, *On modules induced or coinduced from Hopf subalgebras*, Math. Scand. 67 (1990), 177-182.

[U3] ——, *Smash products and comodules of linear maps*, Tsukuba J. Math. 14 (1990), 371-378.

[Vo1] D. Voigt, *Induzierte Darstellungen in der Theorie der endlichen, algebraischen Gruppen*, Lecture Notes in Math. vol. 592, Springer, Berlin, 1977.

[Vo2] ——, *Groupes algébriques infinitésimaux résolubles et leurs représentations linéaires irréductibles*, Séminaire Paul Dubreil, Lecture Notes in Math. vol. 740, Springer, Berlin, 1979, 48-68.

[Wa] W. C. Waterhouse, *Introduction to Affine Group schemes*, Springer, Berlin, 1979.

Algebraic Aspects of Linearly Recursive Sequences

EARL J. TAFT Rutgers University, New Brunswick, New Jersey

ABSTRACT. We consider linearly recursive sequences $\{a_n\}$ with coordinates a_n in a field, where either $n \geq 0$ or $n \in Z$, the integers. This is an expository article stressing various structures of these spaces of sequences as bialgebra, Hopf algebra or Lie bialgebra. Besides theoretical considerations, we also consider algorithmic procedures for computing the diagonal of such a sequence, and for determining whether or not such a sequences is invertible under several possible products.

1. INTRODUCTION.

Let F be a field. A *linearly recursive sequence* $\{a_n\}_{n\geq 0}$ with coordinates $a_n \in F$ is one which satisfies a linearly recursive relation with constant coefficients, i.e., there is a monic polynomial $h(x) = x^r - h_1x^{r-1} - \cdots - h_r$ of degree $r \geq 1$ in $F[x]$ such that $a_n = h_1a_{n-1} + h_2a_{n-2} + \cdots + h_ra_{n-r}$ for all $n \geq r$. We note that we do not demand that $h_r \neq 0$, as do some authors. If $h_r \neq 0$, then we may view $a_0, \ldots, a_{r-1}$ as initial data, with subsequent coordinates a_n for $n \geq r$ determined by the initial data and the recursive relation. If $h_r = 0$, we may still view $a_0, \ldots, a_{r-1}$ as initial data, but there will be some initial terms which do not affect the rest of the sequence. For example, a *geometric sequence* is of the form $(ar^n)_{n\geq 0}, = (a, ar, ar^2, ar^3, \ldots)$ and satisfies the recursive relation $x - r$. The sequence $(\pi, 1, 2, 4, 8, 16, 32, 64, \ldots)$ over the real numbers $\mathbb{R}$ satisfies $x^2 - 2x$, so while the initial data is given by $a_0 = \pi$ and $a_1 = 1$, only a_1 really affects the remainder of the sequence.

Linearly recursive sequences are classical objects in mathematics, and have been long and extensively studied from many points of view. It would be impossible to attempt such a survey here, but we do recommend the article [P] as general reading about these sequences.

Perhaps the most classical point of view of linearly recursive sequences is through their generating functions $\sum_{i=0}^{\infty} a_iy^i$ in the power-series algebra $F[[y]]$. In this form, they are simply the rational fractions $\frac{p(y)}{q(y)}$, where $q(0) \neq 0$, where $q(y)$ is inverted as a power-series. For example, the geometric series $(ar^n)_{n\geq 0}$ for $a \neq 0$ and $r \neq 0$ is $\frac{a}{1-ry}$. The sequence $(n)_{n\geq 0}$ of non-negative integers satisfies $x^2 - 2x + 1 = (x-1)^2$ and its generating function is $\frac{1}{(y-1)^2}$. A favorite type of linearly recursive sequence is the Fibonacci sequence $(1, 1, 2, 3, 5, 8, 13, 21, \ldots)$ with recursive relation $x^2 - x - 1$

and (in this example), initial date $a_0 = a_1 = 1$. The generating function here is $\frac{1}{1-y-y^2}$. (There is a reversal of coefficients from the recursive relation.) Our remark that we allow $h_r = 0$ in the recursive relation means that $p(y)$ in the generating function $\frac{p(y)}{q(y)}$ may have positive degree. For example $(0, 1, 2, 4, 8, \dots)$ has generating function $\frac{y}{1-2y}$. An important example of this are the unit sequences u_n in which the n-th coordinate is 1 and all other coordinates are zero. u_n satisfies x^{n+1} and has generating function y^n. Thus we are allowing all finite sequences (i.e., those eventually zero), whose generating functions are polynomials in y.

In practice, a given sequence is usually recognized as recursive (we sometimes drop the adverb "linearly" from now on) by displaying a recursive relation. It is sometimes useful to know the *minimal* recursive relation $h(x) = x^r - h_1 x^{n-1} - \cdots - hr$, i.e., with $r \geq 1$ minimal. If a sequences satisfies $h(x)$, then it satisfies any polynomial multiple of $h(x)$, i.e., the set of all recursive relations satisfied by a given recursive sequence is an ideal in $F[x]$. Choosing r minimal thus is the same as choosing a (monic) generator of this ideal.

Since recursive sequences correspond to rational series, i.e., $\frac{p(y)}{q(y)}$ with $q(0) \neq 0$, it is evident that they form an F-algebra, i.e., a subalgebra of $F[[y]]$. The fact that the sum of two recursive sequences is recursive can be explained in terms of their recursive relations. If (a_n) and (b_n) satisfy $h(x)$ and $k(x)$ respectively, then $(a_n + b_n)$ satisfies the least common multiple of $h(x)$ and $k(x)$. As a subalgebra of $F[[y]]$, the product is the ordinary power series one. However, this product does not in general support a compatible coproduct (see Section 2). Hence we will consider instead several other products on the F-vector space of linearly recursive sequences. The easier one to define is the Hadamard (or coordinate-wise) product $(a_n)(b_n) = (c_n)$, where $c_n = a_n b_n$. If (a_n) satisfies $h(x)$, and (b_n) satisfies $k(x)$, let $\alpha_1, \dots, \alpha_t$ be the distinct roots of $h(x)$ in the algebraic closure $\bar{F}$ of F, with multiplicities $r_1, \dots, r_t$ respectively, and $\beta_1, \dots, \beta_u$ the distinct roots of $k(x)$ with respective multiplicities $s_1, \dots, s_u$. Then $(a_n b_n)$ will satisfy $\prod_{i,j=1}^{t,u} (x - \alpha_i \beta_j)^{r_i + s_j - 1}$, which has coefficients in F. Under the Hadamard product, the recursive sequences form a bialgebra.

Another product we will consider is the divided-power (or Hurwitz) product $(a_n)(b_n) = (d_n)$, where $d_n = \sum_{i=0}^{n} \binom{n}{i} a_i b_{n-i}$. The reason for the name is that if we write the generating function of (a_n) as $\sum_{n=0}^{\infty} a_n Z^{(n)}$, then the product is given by $Z^{(n)} Z^{(m)} = \binom{n+m}{n} Z^{(n+m)}$. At characteristic zero, we can think of $Z^{(n)}$ as $\frac{y^n}{n!}$. Our recursive sequences form a Hopf algebra under the divided-power product. Using the notation of the previous paragraph for the roots of the recursive relations satisfied by (a_n) and (b_n), the divided-power product of (a_n) and (b_n) will satisfy the recursive relation $\prod_{i,j=1}^{t,u} (x - (\alpha_i + \beta_j))^{r_i + s_j - 1}$, which has coefficients in F.

The coalgebra structure we will use on the space of recursive sequences will result from realizing this space as an appropriate dual (the so-called "continuous" dual) of the algebra $F[x]$ (see sections 4 and 5). We will also discuss our recursive sequences as a Lie coalgebra (see section 12) where they will appear as an appropriate dual to a certain Lie algebra (see section 14). They will furthermore support various Lie bialgebra structures (see sections 12 and 16). It turns out that there is a relation between the associative and the Lie points of view (section 15).

So far we have considered sequences (a_n) with $n \geq 0$. We shall also have occasion to consider (a_n) with $n \in Z$. Here the recursive relation $h(x) = x^r - h_1 x^{r-1} - \cdots - h_r$ means that $a_n = h_1 a_{n-1} + \cdots + h_r a_{n-r}$ for all $n \in Z$. Thus here we must have $h_r \neq 0$. Here the coalgebra structure on these recursive sequences will be dual to the algebra structure of the Laurent polynomial algebra $F[x, x^{-1}]$, and there will be a Hopf algebra structure on these sequences where the product is the Hadamard product. In particular, this space does not include the finite sequences (almost everywhere zero).

While we will not give formal proofs, we will try to give references which will be useful to the general reader. Another good reference for linearly recursive sequences is [C-M-P].

2. ALGEBRAS, COALGEBRAS, BIALGEBRAS AND HOPF ALGEBRAS.

By *F-algebra*, we shall always mean an associative F-algebra A with unit element, i.e., an associative multiplication $m : A \otimes A \to A$ (all tensor products are over F) and a unit element, thought of as a map $\mu : F \to A$. Our basic examples will be $F[x]$ and $F[x, x^{-1}]$.

An F-coalgebra C will have a comultiplication (or diagonal) $\Delta : C \to C \otimes C$ which is coassociative, i.e., $(I \otimes \Delta)\Delta = (\Delta \otimes I)\Delta$ as maps from C to $C \otimes C \otimes C$, and a counit $\varepsilon : C \to F$ satisfying $c = \sum \varepsilon(c_i)c_i' = \sum \varepsilon(c_i')c_i$ for all $c \in C$, where $\Delta c = \sum_i c_i \otimes c_i'$. An element c in C is called *group-like* if $c \neq 0$ and $\Delta c = c \otimes c$. Necessarily $\epsilon(c) = 1$. A basic example is FS for any non-empty set S, which has the elements of S as a basis, $\Delta s = s \otimes s$ and $\epsilon(s) = 1$ for $s \in S$, i.e., the elements of S are group-like. $G(C)$ will denote the set of group-like elements of a coalgebra C. FS is called the *group-like coalgebra* on S.

A *bialgebra* B is simultaneously an algebra and a coalgebra, with connecting axioms that $\Delta : B \to B \otimes B$ and $\epsilon : B \to F$ should be algebra homomorphisms. A basic example is a monoid algebra FM of a monoid M. The algebra structure is determined by the multiplication in M, and the coalgebra structure is that of the group-like coalgebra on the set M. The polynomial algebra $F[x]$ is of this nature, where $\Delta(x^n) = x^n \otimes x^n$ and $\epsilon(x) = 1$ for all $n \geq 0$.

A *Hopf algebra* H is a bialgebra which has an *antipode* S, i.e., a map $S : H \to H$ satisfying $\sum S(h_i)h_i' = \epsilon(h)1 = \sum h_i S(h_i')$ for all $h \in H$, where $\Delta h = \sum h_i \otimes h_i'$. See [A] or [S] as general references for coalgebras, bialgebras and Hopf algebras. A basic example is the group algebra FG, where $S(g) = g^{-1}$ for g in G. The algebra $F[x, x^{-1}]$ of Laurent polynomials is of this nature, where $\Delta(x^n) = x^n \otimes x^n$ and $\epsilon(x^n) = 1$ for all $n \in Z$. For us, another important examples will be $F[x]$, where $\Delta x^n = \sum_{i=0}^{n} \binom{n}{i} x^i \otimes x^{n-i}$ and $\epsilon(x^n) = \delta_{0n}$ (1 if $n = 0$, 0 otherwise). Here $\Delta x = 1 \otimes x + x \otimes 1$ and $\epsilon(x) = 0$, i.e., x is *primitive* (this notion makes sense in any bialgebra). The antipode S has is given by $S(x^n) = (-x)^n$ for $n \geq 0$. In any bialgebra, the primitive elements form a Lie algebra under $[p, q] = pq - qp$. Another basic example of a Hopf algebra is the universal enveloping algebra $U(L)$ of a Lie algebra L, where the elements of L are primitive in $U(L)$, and $S(x) = -x$ for x in L, i.e., S is the principal involution of $U(L)$.

We will mention examples involving divided-power algebras, coalgebras, bialgebras and Hopf algebras as they arise naturally in our discussion of duality and linearly recursive sequences.

We note finally that an algebra is *commutative* if $ab = ba$ for $a, b \in A$, i.e, $m(\text{twist}) = m$, where twist: $A \otimes A \to A \otimes A$ sends $a \otimes b \mapsto b \otimes a$. A coalgebra is *cocommutative* if (twist) $\Delta = \Delta$, i.e., $\sum c_i \otimes c_i' = \sum c_i' \otimes c_i$ for all c in C, $\Delta c = \sum c_i \otimes c_i'$. Our examples involving $F[x]$ and $F[x, x^{-1}]$ are both commutative and cocommutative. FM, FG and $U(L)$ are cocommutative, but not necessarily commutative (unless M, G or L is).

General references to the ideas in this section 2 are [A] and [S]

3. DUALITY: FROM COALGEBRAS TO ALGEBRAS.

Let (C, Δ, ϵ) be a coalgebra, (A, m, μ) an algebra. The space $\text{Hom}_F(C, A)$ is an algebra under the *convolution product* $*$: $f * g = m(f \otimes g)\Delta$, i.e., $(f * g)(c) = \sum f(c_i)g(c_i')$ for $\Delta c = \sum c_i \otimes c_i'$. The unit element is $\mu\epsilon$.

In particular, if $A = F$, then the coalgebra C has a dual algebra $C^* = \text{Hom}\ (C, F)$ with convolution product and unit element ϵ. We describe two example of dual algebras.

First consider the bialgebra $F[x]$ with x^n group-like for $n \geq 0$. We consider only the coalgebra structure of $F[x]$. An element f in $F[x]^*$ is uniquely determined by its action on the x^n for $n \geq 0$, so we may identify f with the sequence (f_n), where $f_n = f(x^n)$. Since x^n is group-like for $n \geq 0$, the convolution product is the Hadamard product on the space of all sequences, i.e., $(f_n)(g_n) = (h_n)$, where $h_n = f_n g_n$. The unit element is the constant (geometric) sequence (1).

For our second example, consider the coalgebra part of the Hopf algebra structure on $F[x]$ with x primitive, i.e., $\Delta x^n = \sum_{i=0}^{n} \binom{n}{i} x^i \otimes x^{n-i}$ for $n \geq 0$. Then $(f * g)(x^n) = \sum \binom{n}{i} f(x^i) g(x^{n-i})$, i.e., $(f * g)_n = \sum_{i=0}^{n} \binom{n}{i} f_i g_{n-i}$, the divided-power product of $(f_n) and (g_n)$. Thus if we let $\{Z^{(n)}\}_{n \geq 0}$ denote the dual basis in $k[x]^*$ to $\{x^n\}_{n \geq 0}$, then we can identify $f \in k[x]^*$ either as a sequence (f_n) or as a formal divided-power series $\sum_{n=0}^{\infty} f_n Z^{(n)}$, with product described by $Z^{(n)} Z^{(m)} = \binom{n+m}{n} Z^{(n+m)}$. The unit element is $Z^{(0)}$, i.e., the sequence $(1, 0, 0, 0, \dots)$. We call this dual algebra the algebra of divided-power series (in one variable).

In general, we note that if C is cocommutative, then C^* is commutative.

4. DUALITY: FROM ALGEBRAS TO COALGEBRAS.

For a coalgebra C, with $\Delta : C \to C \otimes C$, we have the transpose map $\Delta^* : (C \otimes C)^* \to C^*$. If we combine this with the canonical embedding of $C^* \otimes C^* \mapsto (C \otimes C)^*$, the composite is precisely the convolution product on C^*.

Let A be an algebra. Dualizing $m : A \otimes A \to A$, we get $m^* : A^* \to (A \otimes A)^*$, but in general $m^*(A^*)$ is not contained in $A^* \otimes A^*$ (considered as canonically embedded in $(A \otimes A)^*$). Thus we look for a perhaps smaller subspace A^0 such that $m^*(A^0) \subset A^0 \otimes A^0$, and maximal with respect to this property. One obvious way would be to define a subspace V of A^* to be *good* if $m^*(V) \subset V \otimes V$, and then define A^0 as the sum of all such good subspaces of A^*. This approach turns out to be useful for many categories of algebras (not necessarily associative), and we will also do this later for Lie algebras (section 13). However, for associative algebras, A^0 can be constructed element-wise, i.e., $A^0 = \{f \in A^* | m^*(f) \in A^* \otimes A^*\}$. This description of A^0 makes it palatable that A^0 is a coalgebra whose comultiplication is the restriction of m^* to A^0. The counit of A^0 is evaluation at the unit element of A. This description of A^0 is proved in Proposition 6.0.3 of [S], where several other descriptions of A^0 are also discussed.

One of the descriptions of A^0 involves the ideal structure of A, thus $A^0 = \{f \in A^* | f(J) = 0$ for some cofinite ideal J of $A\}$. The cofinite condition is that A/J be finite-dimensional. This is the starting point for the discussion of A^0 in [Sw]. It will be our method of introducing the coalgebra structure on linearly recursive sequences.

The other description of A^0 relates to the idea of a representative function on a monoid M. A function $f : M \to F$ is called representative if the set of right and left translates of f by elements in M spans a finite-dimensional F-space. Then A^0 consists of those functions in A^* which are representative with respect to the multiplicative monoid A, i.e., the linear representative functions on A. This viewpoint is the principal one of Hochschild (e.g., [H]). It will be useful to us in getting an algorithm for diagonalizing a linearly recursive sequence. The basic reason for this is that the finite-dimensional span of an f in A^0 turns out to be the subcoalgebra generated by f. In fact, it turns out that any coalgebra C is a coalgebra of representative functions. The usual embedding of C into its double-dual C^{**}, is actually an embedding of C in C^{*0}, and one can use this to prove the fundamental theorem of coalgebras, namely that every element in C is contained in a finite-dimensional subcoalgebra of C.

To repeat in concrete notation, if $f \in A^0$, then $\Delta f = \sum g_i \otimes h_i$ for $g_i, h_i \in A^0$ means that $f(ab) = \sum g_i(a)h_i(b)$ for all a, b in A. Also $\epsilon_{A^0}(f) = f(1)$. We remark that if A is commutative, then A^0 is cocommutative.

5. LINEARLY RECURSIVE SEQUENCES AS A COALGEBRA.

Consider the algebra $A = F[x]$. Then $A^0 = \{f \in A^* | f(J)\} = 0$ for some cofinite ideal J of $A\}$. But such a J is an arbitrary non-zero ideal of $F[X]$, and has a (unique) monic generator $h(x) = x^r - h_1 x^{n-1} - \cdots - h_r$. To say that $f(J) = 0$ thus means that f is linearly recursive with recursive relation $h(x)$. Thus $A^0 = F[x]^0$ is identified as those sequences $f = (f_n)$ in A^*, where $f_n = f(x^n)$, which are linearly recursive.

As an example, let $Z^{(n)}$ be the unit functional which has the value 1 on x^n, and 0 on x^i for $i \neq n$. Then we claim $\Delta Z^{(n)} = \sum_{i=0}^n Z^{(i)} \otimes Z^{(n-i)}$ for each $n \geq 0$. To see this, note that $\Delta Z^{(n)}(x^j \otimes x^k) = Z^{(n)}(x^{j+k}) = \delta_{n,j+k}$, whereas $\sum_{i=0}^n Z^{(i)}(x^j)Z^{(n-i)}(x^k) = \begin{cases} 0 & \text{if } j+k \neq n \\ 1 & \text{if } j+k = n \end{cases} = \delta_{n,j+k}$. Thus $F[x]^0$ contains the finite sequences, i.e., the F-span of the $\{Z^{(n)} | n \geq 0\}$ with $\Delta Z^{(n)} = \sum_{i=0}^n Z^{(i)} \otimes Z^{(n-i)}$ and $\epsilon(Z^{(n)}) = \delta_{0,n}$. This F-span is called the *divided-power coalgebra* $F[Z^{(0)}, Z^{(1)}, Z^{(2)}, \ldots]$. If F has characteristic 0, this divided-power coalgebra is isomorphic to the coalgebra $F[y]$ with y primitive (a Hopf algebra) where $Z^{(n)} \leftrightarrow \frac{y^n}{n!}$. For $\Delta\left(\frac{y^n}{n!}\right) = \sum_{i=0}^n \frac{\binom{n}{i} y^i \otimes y^{n-i}}{n!} = \sum_{i=0}^n \frac{y^i}{i!} \otimes \frac{y^{n-i}}{(n-i)!}$, and $\epsilon\left(\frac{y^n}{n!}\right) = \delta_{o,n}$.

Next, consider the algebra $A = F[x, x^{-1}]$ of Laurent polynomials. We identify f in A^* with the sequence $(f_n)_{n \in Z}$ where $f_n = f(x^n)$. Every non-zero ideal of $F[x, x^{-1}]$ is generated by a monic polynomial $h(x) = x^r - h_1 x^{r-1} - \cdots - h_r$, where $h_r \neq 0$ (since x is invertible). Thus $f = (f_n)_{n \in Z}$ is linearly recursive with recursive relation $h(x)$ with $h(0) \neq 0$. Since the $Z^{(n)}$ for $n \in Z$ are not in $F[x, x^{-1}]^0$ for $n \in Z$, the simplest example we can give now are geometric sequences $(ar^n)_{n \in Z}$ for $a \neq 0$, $r \neq 0$ in F. This satisfies the recursive relation $x - r$. If $a = 1$, we call the sequence $(r^n) = e(r)$, since at characteristic zero, $\sum_{n=-\infty}^{\infty} r^n Z^{(n)} = \sum_{n=-\infty}^{\infty} \frac{r^n y^n}{n!} = \sum_{n=-\infty}^{\infty} \frac{(ry)^n}{n!}$. Note that $\Delta e(r) = e(r) \otimes e(r)$, since applying both sides to $x^i \otimes x^j$ yields $r^{i+j} = r^i r^j$, and $e(r) \neq 0$. So $e(r)$ is a group-like element of $F[x, x^{-1}]^0$. The geometric sequence $e(r)$ is often called an exponential sequence. The same calculation for $F[x]^0$ shows that the sequence $e(r) = (r^n)_{n \geq 0}$ for $r \in F$ (here r may be 0) is a group-like element in $F[x]^0$. It is common to use the same notation $e(r)$ for $r \neq 0$ in both $F[x]^0$ and $F[x, x^{-1}]^0$, since if F has characteristic 0, in $F[x]^0$ we have $e(r) = (r^n)_{n \geq 0}$ with generating function $\sum_{n=0}^{\infty} r^n Z^{(n)} = \sum_{n=0}^{\infty} \frac{r^n y^n}{n!} = e^{ry}$.

Finally, we remark that the algebra embedding of $F[x]$ into $F[x, x^{-1}]$ yields a coalgebra embedding β of $F[x, x^{-1}]^0$ into $F[x]^0$, which is simply restriction to $F[x]$. The image of this coalgebra map are those linearly recursive sequences with minimal recursive relation with non-zero constant term.

There is an interesting coalgebra map α from $F[x]^0$ to $F[x, x^{-1}]^0$, called the back-solving map. If $f \in F[x]^0$, write its minimal recursive relation as $h(x) = x^k q(x)$ where $q(0) \neq 0$. Then $\alpha(f)$ is obtained by suppressing $f_0, f_1, \ldots, f_{k-1}$, and back-solving to the left from the k-th coordinate f_k indefinitely to the left, using the recursive relation $q(x)$. This can be done since $q(0) \neq 0$. For example, the sequence $(\pi, \sqrt{2}, 1, 1, 2, 3, 5, 8, 13, \ldots)$ which satisfies $x^4 - x^3 - x^2 = x^2(x^2 - x - 1)$, back-solves by deleting π and $\sqrt{2}$ to the sequence $(\ldots, 2, -1, 1, 0, 1, 1, 2, 3, 5, 8, 13, \ldots)$, where the 0 is the first coordinate ($n = 1$), satisfying $x^2 - x - 1$ in $k[x, x^{-1}]^0$. The kernel of α consists of the finite sequences.

As a general reference for identifying $k[x]^0$, see [P-T]. The back-solving map will be used in Section 10 to determine the linearly recursive sequences which are invertible under the Hadamard product (see [L-T]).

6. LINEARLY RECURSIVE SEQUENCES AS BIALGEBRA OR HOPF ALGEBRA.

We recall (see Section2) that $F[x]$ is a bialgebra with x group-like, and is a Hopf algebra with x primitive. Also $F[x, x^{-1}]$ is a Hopf algebra with x group-like. Considering only the algebra structures of $F[x]$ and $F[x, x^{-1}]$, we have identified the coalgebras $F[x]^0$ and $F[x, x^{-1}]^0$ as linearly recursive sequences. These coalgebras are subspaces of the convolution algebras $F[x]^*$ and $F[x, x^{-1}]^*$. Since we have discussed two coalgebras structures on $F[x]$, $F[x]^*$ has two convolution products, namely the Hadamard product and the divided-power product. The product on $F[x, x^{-1}]^*$ is the Hadamard product. The following proposition thus indicates some bialgebra or Hopf algebra structures for linearly recursive sequences.

Proposition:. *(a) Let B be a bialgebra. then B^0 is a subalgebra of B^* under the convolution product. B^0 is a bialgebra with comultiplication dual to the multiplication in B and with the convolution product.*

(b) Let H be a Hopf algebra. Then H^0 is a Hopf algebra.

H^0 is a bialgebra as described in (a). The antipode of H^0 is the restriction of S^*, the transpose of the antipode S of H, to H^0.

This proposition is easy to prove, using the description of B^0 as linear representative functions on B. We note here its implications for linearly recursive sequences.

If $B = F[x]$ with x group-like, then B^0 is the bialgebra of linearly recursive sequences $(a_n)_{n\geq 0}$ under the Hadamard product. Similarly, if $H = F[x, x^{-1}]^0$ with x group-like, then H^0 is the Hopf algebra of linearly recursive sequences $(a_n)_{n\in Z}$ under the Hadamard product. Finally, if $H = F[x]$ with x primitive, then H^0 is the Hopf algebra of linearly-recursive sequences under the divided-power product. This is a subalgebra of the algebra of divided-power series, and contains as subcoalgebra the divided-power coalgebra $F[Z^{(0)}, Z^{(1)}, Z^{(2)}, \ldots]$. The antipode of H^0 sends $Z^{(n)}$ to $(-1)^n Z^{(n)}$ for $n \geq 0$.

7. GLOBAL STRUCTURE OF LINEARLY RECURSIVE SEQUENCES.

We consider the Hopf algebra $H = F[x]$ with x primitive. We wish to construct a certain basis for H^0, the Hopf algebra of linearly recursive sequences with divided-power product. As before, we identify f in $H^0 \subset H^*$ with $\sum_{n=0}^{\infty} f_n Z^{(n)}$, where $f_n = f(x^n)$. $\{Z^{(n)} | n \geq 0\}$ is the dual basis to $\{x^n | n \geq 0\}$. The $Z^{(n)}$ are in H^0, and multiply and comultiply as divided-powers. In any Hopf algebra, the group-like elements form a multiplicative group. In Section 5, we noted that the geometric sequences $e(r) = (r^n)_{n\geq 0}$, $r \in F$, are group-like. It is easy to see these are the only group-like elements in H^0. The group-operation is $e(r)e(s) = e(r+s)$, i.e., the group-like elements $G(H^0)$ of H^0 are isomorphic to the additive group of F. We also note that H^0 contains the divided-power polynomial Hopf algebra $F[Z^{(0)}, Z^{(1)}, Z^{(2)}, \ldots]$, which corresponds precisely to the finite sequences. This is the so-called irreducible component of 1 in H^0 (see [S]). This is related to the fact, easy to check, that the primitive elements of H^0 are the scalar multiples of $Z^{(1)}$.

Now assume that F is algebraically closed. If follows from the above discussion, and Theorem 8.1.5 of [S], that $H^0 = F[G(H^0)] \otimes F[Z^{(0)}, Z^{(1)}, Z^{(2)}, \ldots]$, i.e., the tensor product of the group-algebra of the group of geometric sequences and the divided-power polynomial Hopf algebra. This means that the recursive sequence space has a basis $\{e(r)Z^{(n)} | r \in F, n \geq 0\}$, where the product is the divided-power one. This is an integral basis, in the sense that all the Hopf structure (multiplication, unit element, comultiplication, count, antipode) for these basis elements involve only integer coefficients. For example, $e(r)Z^{(n)} \cdot e(s)Z^{(m)} = \binom{n+m}{n} e(r+s)Z^{(n+m)}$, $\Delta(e(r)Z^{(n)}) = \sum_{i=0}^{n} e(r)Z^{(i)} \otimes e(r)Z^{(n-i)}$, $\epsilon(e(r)Z^{(n)}) = \delta_{0,n}$ and $S(e(r)Z^{(n)}) = (-1)^n e(-r)Z^{(n)}$.

This basis is not so well-known (perhaps because it involves divided-power ideas) as another more classical one which we now discuss. Here we take F as $\mathbb{C}$, the complex numbers. We now consider the bialgebra $B = F[x]$ with x group-like. Then B^0 is the bialgebra of linearly recursive sequences with Hadamard product. Since Theorem 8.1.5 of [S] requires a Hopf algebra, we extend $B = F[x]$ to the Hopf algebra $H = F[x, x^{-1}]$, x group-like. Then recall (Section 5) that H^0 is all linear recursive sequences $(a_n)_{n\in Z}$, where the minimal recursion relation has non-zero constant term. The group-like elements of H^0 are the geometric sequences $e(r) = (r^n)_{n\in Z}$ for $r \neq 0$ in F, with group structure $e(r)e(s) = e(rs)$, i.e., isomorphic to the multiplicative group of F. The primitive elements of H^0 are the scalar multiples of $P = (n)_{n\in Z}$, i.e., the sequence of integers, and the irreducible component of 1 is the polynomial algebra $F[P]$. This means that H^0 has a basis $\{e(r)P^i | r \in F^*, i \geq 0\}$, where the product is the Hadamard product. Thus $e(r)P^i$ is the sequence $(r^n n^i)_{n\in Z}$. This is an integral basis of $F[x, x^{-1}]^0$, and is classically well-known.

To return to $F[x]^0$, note the maps β: $F[x,x^{-1}]^0 \to F[x]^0$ and α: $F[x]^0 \to F[x,x^{-1}]^0$ described in Section 5. These are bialgebra maps, $\alpha\beta = I$, so β is injective with image the recursive sequences with minimal recursive relation having non-zero constant term, and α is surjective with kernel consisting of the finite sequences. Thus $F[x]^0$, the linearly recursive sequences $(a_n)_{n\geq 0}$ have a basis $\{(r^n n^i)_{n\geq 0} | r \in F,\ r \neq 0,\ i \geq 0\} \cup \{u_j | j \geq 0\}$, the latter being a basis for the finite sequences, i.e., $u_j(x^t) = \delta_{jt}$ for $t \geq 0$. We use u_j rather than $Z^{(j)}$ since the product is the Hadamard product, not the divided-power product, i.e., $u_j u_k = \delta_{jk} u_j$. The coalgebra structure of the $\{u_j | j \geq 0\}$ is still that of the divided-power coalgebra, so this is also an integral basis of $F[x]^0$. Surprisingly, this basis is useful in the discussion later about a Lie coalgebra structure for linearly recursive sequences (Section 14).

We also note that $F[x^n, x^{-1}]^0$ has a basis $\{\alpha(e(r)Z^{(n)}) | r \in F^*,\ n \geq 0\}$. Here $e(r)Z^{(n)}$ is their divided-power product. While this follows from what we have already done, we prefer to discuss it in Section 15 in the context of Lie coalgebras.

8. LOCAL STRUCTURE OF LINEARLY RECURSIVE SEQUENCES.

We wish to describe how to diagonalize a given linearly recursive sequence $f = (f_n)$ in $F[x]^0$. We noted in Section 4 that f generates a finite-dimensional subcoalgebra C_f of $F[x]^0$, and that C_f is the finite-dimensional span of the translates of f by elements in $F[x]$. Let $h(x) = x^r - h_1 x^{r-1} - \cdots - h_r$ be the minimal recursive relation of f. Let $Sf = xf$, i.e., $(Sf)(p(x)) = f(xp(x))$, $S^2 f = x^2 f$, etc. S is a shift to the left, i.e., $S(f_0, f_1, f_2, \dots) = (f_1, f_2, f_3, \dots)$. Then $f, Sf, \dots, S^{r-1}f$ are a basis for C_f.

Here is an algorithm for computing Δf in $F[x]^0 \otimes F[x]^0$. See [P-T] for a proof, and a more general algorithm for computing $\Delta(S^t f)$. We form the r by r Hankel matrix $H(f)$ of f whose rows consist of the first r coordinates of $f, Sf, \dots, S^{r-1}f$ respectively. $H(f)$ is a symmetric invertible matrix. Let $H(f)^{-1} = (s_{ij})$, $0 \leq i,j \leq r-1$. Then $\Delta f = \sum_{i,j=0}^{r-1} s_{ij}(S^i f) \otimes (S^j f)$.

For example, let $F = (1,1,2,3,5,8,13,\dots)$ be the Fibonacci sequence with initial data $F_0 = F_1 = 1$. F satisfies $x^2 - x - 1$, and $SF = (1,2,3,5,8,13,21,\dots)$. $H(F)$ is the 2 by 2 matrix $\begin{bmatrix} 1 & 1 \\ 1 & 2 \end{bmatrix}$ with inverse $\begin{bmatrix} 2 & -1 \\ -1 & 1 \end{bmatrix}$. Thus $\Delta F = 2(F \otimes F) - F \otimes SF - SF \otimes F + SF \otimes SF$. Another example is $\sin Z = (0,1,0,-1,0,1,0,-1,0,\dots)$ satisfying $x^2 + 1$. $S(\sin Z) = \cos Z = (1,0,-1,0,1,0,-1,0,\dots)$. $H(\sin Z) = \begin{bmatrix} 0 & 1 \\ 1 & 0 \end{bmatrix} = (H(\sin Z))^{-1}$. So $\Delta(\sin Z) = \sin Z \otimes \cos Z + \cos Z \otimes \sin Z$ (the coalgebra version of the formula for the sine of a sum of two angles). Similarly, $\Delta(\cos Z) = \cos Z \otimes \cos Z - \sin Z \otimes \sin Z$. The 2-dimensional coalgebra with basis $\sin Z$, $\cos Z$ is called the *trigonometric coalgebra.* As a final example, we note that our algorithm does not always provide an integral coalgebra basis formula. For example let $f = (0,1,4,9,16,\dots) = (n^2)_{n\geq 0}$. f satisfies $(x-1)^3$ and $H(f) = \begin{bmatrix} 0 & 1 & 4 \\ 1 & 4 & 9 \\ 4 & 9 & 16 \end{bmatrix}$ has inverse with upper left entry $\frac{31}{28}$ (say $F = \mathbb{R}$ here). On the other hand, C_f does have an integral basis (as coalgebra) consisting of $e(1), e(1)Z^{(1)}$ and $e(1)Z^{(2)}$ (divided power product).

Finally we note that our algorithm can sometimes provide interesting identities on the coordinates of a linearly recursive sequence, simply by applying it to $x^i \otimes x^j$

for $i, j \geq 0$. These identities are often of a numerical or combinatorial nature. For example, let $F = (1, 1, 2, 3, 5, 8, 13, \dots)$ be the Fibonacci sequence diagonalized in the previous paragraph. Then our formula $\Delta F = 2(F \otimes F) - F \otimes SF - SF \otimes F + SF \otimes SF$, when applied to $x^i \otimes x^j$, yields $F_{i+j} = 2F_iF_j - F_iF_{j+1} - F_{i+1}F_j + F_{i+1}F_{j+1}$ for all $i, j \geq 0$, where F_n is the n-th Fibonacci number for $n \geq 0$, $F_0 = F_1 = 1$.

9. LINEARLY RECURSIVE SEQUENCES INVERTIBLE UNDER THE DIVIDED-POWER PRODUCT.

We have seen that linearly recursive sequences $(a_n)_{n \geq 0} = \sum_{n=0}^{\infty} a_n Z^{(n)}$ form a Hopf algebra under the divided-power product which is the continuous dual H^0 of the Hopf algebra $H = F[x]$ with x primitive. The determination of the invertible elements of H^0 is fairly easy. The answer depends on the characteristic of F. We sketch the argument here. Details can be found in [T2].

First it is easy to see that we can assume F is algebraically closed. Each group-like element $e(r) = \sum_{n \geq 0} r^n Z^{(n)}$ for $r \in F$ is invertible, with $e(r)^{-1} = e(-r)$. Thus if a is any non-zero element of F, then $ae(r)$ is also invertible. We claim that these are the only invertible elements when F has characteristic zero. To see this, recall that in Section 7, we showed that H^0 has a basis $\{e(r)Z^{(n)} | r \in F,\ n \geq 0\}$. When F has characteristic zero, replace $Z^{(n)}$ by $\frac{y^n}{n!}$. Then a linearly recursive sequence is a polynomial in y with coefficients in the group-algebra $k[G]$, where $G = \{e(r) | r \in F\}$ is isomorphic to the additive group of F. Such a polynomial is invertible if and only if its constant term is invertible and all other coefficients are nilpotent. But G is a torsion-free abelian group. Standard theorems in group-algebras then give that $k[G]$ is an integral domain, and its invertible elements are precisely of the form ag, $a \neq 0$ in F, g in G. This is our desired conclusion.

Now let F be of positive characteristic p. Let $f = \sum_{n \geq 0} f_n Z^{(n)}$ be linearly recursive. If f is invertible under the divided-power product, then clearly $f_0 \neq 0$. We claim that this is also sufficient for f to be invertible. First note that a standard p-multinomial calculation shows that $f^p = \sum_{n \geq 0} f_n^p (Z^{(n)})^p$. But for $n > 0$, $(Z^{(n)})^p = \frac{(pn)!}{(n!)^p} Z^{(pn)}$, and one sees that this integer coefficient is divisible by p. Thus $f^p = f_0^p$, so that $f^{-1} = (f_0^{-1})^p f^{p-1}$ is linearly recursive.

In summary, we have proved that at characteristic zero, a linearly recursive sequence $f = (f_n)_{n \geq 0}$ has a linearly recursive inverse under the divided-power product if and only if f is a non-zero geometric sequence (ar^n) for a and r in F, $a, r \neq 0$. At positive characteristic, f has a linearly recursive inverse under the divided-power product if and only if $f_0 \neq 0$.

We note that these criteria are algorithmically effective. Thus suppose we are given a sequence $f = (f_n)_{n \geq 0}$, and suppose we know a recursive relation $h(x)$ satisfied by f. We first check f_0, which is part of the initial data. If $f_0 = 0$, then f is not invertible. If $f_0 \neq 0$, then f is invertible if F has positive characteristic. If $f_0 \neq 0$ and F has characteristic 0, examination of f_0 and f_1 will predict the ratio r in F if f were a geometric sequence. Then one checks whether or not $x - r$ divides $h(x)$, to determine whether or not f is invertible. Thus our criteria are effective in a fairly trivial manner.

10. LINEARLY RECURSIVE SEQUENCES INVERTIBLE UNDER THE HADAMARD PRODUCT.

The determination of those linearly recursive sequences which are invertible under the Hadamard product is more intricate than under the divided-power product. We will indicate the answer here, as well as give an effective algorithm to determine Hadamard invertibility of a given linearly recursive sequence. Complete details can be found in [L-T]. The result had been obtained by analytic methods when F has characteristic zero in [B] and in general in [R]. However our discussion uses the Hopf algebra ideas developed in this article.

We first indicate the result. Clearly a necessary condition is that no coordinate be 0. A geometric sequence $(ar^n)_{n\geq 0}$ with $a, r \in F^*$ satisfies $x - a$, and its Hadamard inverse $(a^{-1}r^{-n})_{n\geq 0}$ satisfies $x - a^{-1}$. More generally, consider an interlacing of a finite number of such never-zero geometric sequences. If $(a_1 r_1^n), \ldots, (a_k r_k^n)$ are such sequences for $n \geq 0$, their interlacing is the sequence $(a_1, \ldots, a_k, a_1 r_1, \ldots, a_k r_k, a_1 r_1^2, \ldots, a_k r_k^2, \ldots)$. This interlacing satisfies $(x^k - r_1)(x^k - r_2)\ldots(x^k - r_k)$, so is linearly recursive. Its Hadamard inverse is also such an interlacing. The result is that a linearly recursive sequence is Hadamard invertible if and only if after a finite number of non-zero coordinates, it becomes an interlacing of a finite number of never-zero geometric sequences.

Recall that linearly recursive sequences under the Hadamard product are the continuous bialgebra dual $F[x]^0$ to $F[x]$ with x group-like. We first transfer the problem to $F[x, x^{-1}]^0$, the continuous Hopf algebra dual to $F[x, x^{-1}]$ (Sections 5 and 6). To do this, we use the bialgebra maps β: $F[x, x^{-1}]^0 \to F[x]^0$ (restriction to $F[x]$) and α: $F[x]^0 \to F[x, x^{-1}]^0$ (the backsolving map) introduced in Section 5. Recall that the kernel of α consists of the finite sequences. Let $f = (f_n)_{n\geq 0}$ be linearly recursive. Write its recursive relation $h(x) = x^k q(x)$ where $q(0) \neq 0$. Then it follows that f is Hadamard invertible if and only if $f_0, \ldots, f_{k-1}$ are all non-zero and $\alpha(f)$ is Hadamard invertible in $F[x, x^{-1}]^0$ (see Proposition 2.1 of [L-T]). The transferred problem is to show that a linearly recursive sequence $f = (f_n)_{n\in Z}$ in $F[x, x^{-1}]^0$ is Hadamard invertible in $F[x, x^{-1}]^0$ if and only if f is an interlacing of finitely many never-zero geometric sequences.

We sketch the proof of this for F of characteristic zero. A complete proof, including the case of positive characteristic, can be found in Section 3 of [L-T]. It is easy to see that we can assume that F is algebraically closed. We have noted in Section 7 that $F[x, x^{-1}]^0$ has a basis of sequences $\{e(r)P^i | r \in F^*, i \geq 0\}$, where $P = (n)_{n\in Z}$ is the basic primitive sequence of integers, and $G = \{e(r) | r \in F^*\}$ is the group of group-like elements isomorphic to F^*. In Section 7 we developed this basis for $F = \mathbb{C}$ to stress its classical nature, but of course F can be any algebraically field of characteristic zero. Thus an f in $F[x, x^{-1}]^0$ is a polynomial in P with coefficients in the group-algebra $F[G]$.

If $f \in F[x, x^{-1}]^0$ is an interlacing of never-zero geometric sequences, then it is Hadamard invertible. Now let $f \in F[x, x^{-1}]^0$ be Hadamard invertible. Let H be the subgroup of F^* generated by all the roots of the minimal recursive relations of f and f^{-1} (recall that these minimal recursive relations have non-zero constant terms). Since H is a finitely-generated abelian group, $H = K \times L$, K a finite abelian group (in fact K is cyclic), L a finitely-generated free abelian group. Then f and f^{-1} are polynomials in P with coefficients in $F[H] = F[K] \otimes F[L]$. Since $F[H]$ is an integral domain, f and f^{-1} are in $F[K] \otimes F[L]$. $F[K]$ is a finite-dimensional separable algebra. We identify a basis of orthogonal idempotents of $F[K]$. Let $m = |K|$, ω a primitive m-th root of unity in F. Then $K = \{e(\omega^j) | 0 \leq j \leq m-1\}$. We define characters χ_i of K, $0 \leq i \leq m-1$, by $\chi_i(e(\omega^j)) = \omega^{ij}$ for $0 \leq j \leq m-1$. Let $z_i = \frac{1}{m}\sum_{j=0}^{m-1} \omega^{-ij} e(\omega^j)$. Then $z_0, z_1, \ldots, z_{m-1}$ is a basis of orthogonal idempotents for $F[K]$. As a linearly recursive sequence in $F[x, x^{-1}]^0$, z_i has 1 in all coordinates whose index is congruent to i modulo m, and 0 elsewhere. Thus $F[K]$ are those periodic sequences of period m. Now the invertible elements of $F[H]$ are all of the form ah where $a \neq 0$ in F and $h \in H$. But $F[H] = F[U] \otimes F[L] = \sum_{i=0}^{m-1} \oplus F[L] z_i$. Thus $f = \sum_{i=0}^{m-1} a_i e(r_i) z_i$, where $a_i \in F^*$, $r_i \in L$. This says that f is the interlacing of the m geometric sequences $(a_i r_i^n)_{n \in Z}$.

Finally, we discuss an algorithm to show that our procedure is effective. Thus let $f = (f_n)_{n \geq 0}$ be linearly recursive with recursive relation $h(x) = x^k q(x)$, $q(0) \neq 0$. We first check if $f_0, \ldots, f_{k-1}$ are each non-zero. If so, we back-solve f to $\alpha(f)$ in $F[x, x^{-1}]^0$. Let m be the order of the torsion subgroup of the subgroup of F^* generated by the roots of $q(x)$. Then $\alpha(f)$ is Hadamard invertible if and only if it is the interlacing of m never-zero geometric sequences (see Theorem 4.2 of [L-T]). So if we have sufficient knowledge of the roots of $q(x)$, we can predict how many geometric sequences need to be interlaced if $\alpha(f)$ is to be Hadamard invertible. But then starting at the zero-th coordinate of $\alpha(f)$, examination of the first $2m$ coordinates will predict the ratios $r_1, \ldots, r_m$ of these geometric sequences, and thus that if f were Hadamard invertible, it would satisfy the relation $(x^m - a_1) \ldots (x^m - a_m)$. Finally we check whether or not $q(x)$ divides this polynomial.

We gave two examples of this algorithm. First let $f = (1, 1, 1, 2, 1, 1, 4, 1, 1, \ldots)$ with initial data $f_0 = f_1 = f_2 = f_4 = f_5 = 1$ and $f_3 = 2$, with relation $h(x) = x^6 - 3x^3 + 2 = (x^3 - 2)(x^3 - 1)$. Since $\sqrt[3]{2}$ has infinite order, $m = 3$ which predicts the 3 ratios 2, 1 and 1. Thus the predicted relation is $(x^3 - 2)(x^3 - 1)^2$. Since $h(x)$ divides this polynomial, $\alpha(f)$, and thus f, is Hadamard invertible. Note that $h(x)$ may be a proper divisor of the predicted relation. Our second example is $f = (-1, 2, 8, 20, 44, \ldots)$ with $f_0 = -1$, $f_1 = 2$ and relation $x^2 - 3x + 2 = (x-1)(x-2)$. Here $m = 1$, so if f and $\alpha(f)$ were Hadamard invertible, they would already be geometric sequences. Since they clearly are not, f is not Hadamard invertible, i.e., $(-1, \frac{1}{2}, \frac{1}{8} \frac{1}{20}, \frac{1}{44}, \ldots)$ is not linearly recursive. As a final example, consider the Fibonacci sequence $(1, 1, 2, 3, 5, 8, 13, \ldots)$ satisfying $x^2 - x - 1$. Note that $\alpha(f) = (\ldots, 2, -1, 1, 0, 1, 1, 2, 3, 5, 8, 13, \ldots)$ has a zero in the (-1)-st coordinate. Thus $\alpha(f)$, and hence f, is not Hadamard invertible. This can also be obtained by our algorithm.

11. LINEARLY RECURSIVE SEQUENCES IN SEVERAL VARIABLES.

Although this article deals with $F[x]^0$, i.e., linearly recursive sequences in one variable, we indicate here some supplemental remarks to the Hadamard invertibility question of Section 10 for linearly recursive sequences in several variables. A discussion of this problem can be found in $[T3]$. A general discussion of these sequences as a bialgebra can be found in $[C - P]$.

Let A be the algebra $F[x_1, \ldots, x_n]$ of polynomials in n commuting variables. Then $A = F[x_1] \otimes F[x_2] \otimes \cdots \otimes F[x_n]$ has $A^0 = F[x_1]^0 \otimes F[x_2]^0 \otimes \cdots \otimes F[x_n]^0$. A will be a Hopf algebra if $x_1, \ldots, x_n$ are primitive, with $A^*$the algebra of divided-power series in n variables. A will be a bialgebra if $x_1, \ldots, x_n$ are group-like. If we view each f in A^* as a multisequence, or tableau $(f_{i_1 i_2 \ldots i_n})$ for all $i_j \geq 0$, where $f_{i_1 i_2 \ldots i_n} = f(x_1^{i_1} x_2^{i_2} \ldots x_n^{i_n})$, then the product in A^* is now the Hadamard (or pointwise) product. In either case, $A^0 = F[x_1]^0 \otimes \cdots \otimes F[x_n]^0$ as a bialgebra. We extend the Hadamard invertible result of Section 10 for $n = 1$ to the case $n > 1$.

Let $f = (f_{i_1 i_2 \ldots i_n})$, $f_{i_1 \ldots i_n} = f(x_1^{i_1} \ldots x_n^{i_n})$. A "row" of f is a sequence $(f_{i_1 \ldots i_{k-1} j i_{k+1} \ldots i_n})_{j \geq 0}$ for a fixed $i_1, \ldots i_{k-1}, i_{k+1}, \ldots, i_n \geq 0$, which we say is parallel to the x_k-axis. We use the description of A^0 as $\{f \text{ in } A^* | f(J) = 0 \text{ for some}$ cofinite ideal of $A\}$. For such an f and J, each x_i for $1 \leq i \leq n$ has all its powers spanning a finite-dimensional subspace of A/J. Thus each row of f parallel to the x_i-th row satisfies a minimal (monic) relation $h_i(x)$ in $F[x]$. Thus J contains the cofinite elementary ideal Γ generated by $h_1(x_1), h_2(x_2), \ldots, h_n(x_n)$, We use Γ to backsolve f.

Thus let H be the Hopf algebra $F[x_1, x_1^{-1}, \ldots, x_n, x_n^{-1}]$ with $x_1, \ldots x_n$ group-like. Then elements f of H^* are tableaux $(f_{i1 \ldots i_n})$ for $i_1, \ldots, i_n \in Z, f_{i_1, \ldots, i_n} = f(x_1^{i_1} \ldots x_n^{i_n})$. $H = F[x_1, x_1^{-1}] \otimes \cdots \otimes F[x_n, x_n^{-1}]$, so $H^0 = F[x_1, x_1^{-1}]^0 \otimes \cdots \otimes F[x_n, x_n^{-1}]^0$. Each row of f in H^0 parallel to the x_i-th axis is a doubly-infinite sequence satisfying a relation $q_i(x_i)$ with $q_i(0) \neq 0$. As before, we have maps β: $F[x_1, x_1^{-1}, \ldots, x_n, x_n^{-1}]^0 \to F[x_1, \ldots, x_n]$ given by restriction, and α: $F[x_1, \ldots, x_n]^0 \to F[x_1, x_1^{-1}, \ldots, x_n, x_n^{-1}]^0$, the backsolving map. If $f \in F[x_1, \ldots, x_n]^0$, $\alpha(f)$ is obtained by writing each relation $h_i(x) = x^{k_i} q_i(x)$ with $q_i(0) \neq 0$. Then we use $q_i(x)$ to backsolve f along the x_i-th row of f starting from the k_i-th coordinate. The result is that f is Hadamard invertible if and only if all $f_{i_1 \ldots i_n} \neq 0$ and each row of f is eventually an interlacing of never-zero geometric sequences. In this case, all rows of f parallel to a given axis satisfy a single recursive relation, and are themselves Hadamard invertible. This transfers the problem to showing that each row of $\alpha(f)$ is an interlacing of never-zero geometric sequences, and for this we can use the procedures of Section 10.

Our procedure for $n > 1$ is also effective, i.e., there is a finite algorithm to determine if a given f in $F[x_i, \ldots, x_n]^0$ is Hadamard invertible. It is ultimately based on the algorithm for the case $n = 1$ explained in Section 10. Let $f \in F[x_1, \ldots, x_n]^0$, $f = (f_{i_1 i_2 \ldots i_n})$ for $i_j \geq 0$, $j = 1, 2, \ldots, n$. Each row parallel to the x_i-th axis satisfies $h_i(x) = x^{k_i} q_i(x)$ with $q_i(0) \neq 0$ for $1 \leq i \leq n$. Starting at the position $k_1 k_2 \ldots k_n$, consider the finite hypercube which includes all initial data beyond this point, i.e., in each row parallel to the x_i-th axis, include positions k_i to degree $h_i(x) - 1$. For each of the (finite number of) rows in and in front of this hypercube, backsolve to obtain the corresponding row in $\alpha(f)$. If one of these rows is not an interlacing of non-zero geometric sequences, then $\alpha(f)$, and thus f, is not Hadamard invertible. If each of these back-solved rows is an interlacing of non-zero geometric sequences, then $\alpha(f)$ is Hadamard invertible, because outside of these rows, the recursive relation will guarantee the interlacing of geometric sequences in the other rows. Then we have only to check that the entries deleted in the back-solving were not zero, and that f is Hadamard invertible, since the reciprocal sequence f^{-1} will satisfy $x_i^{k_i}$ times an appropriate interlacing relation for each i,

$1 \leq i \leq n$. For example, consider f as the tableau

	1	x_1	x_1^2	x_1^3	x_1^4	x_1^5	x_1^6	$\ldots$
1	$*$	$*$	$*$	$\ldots$	$\ldots$	$\ldots$	$\ldots$	$\ldots$
x_2	$*$	$*$	$*$	$\ldots$	$\ldots$	$\ldots$	$\ldots$	$\ldots$
x_2^2	$*$	1	5	2	10	4	20	$\ldots$
x_2^3	$*$	7	9	14	18	28	36	$\ldots$
x_2^4	$\vdots$	3	15	6	30	12	60	$\ldots$
x_2^5	$\vdots$	21	27	42	54	84	108	$\ldots$
$\vdots$	$\vdots$	$\vdots$	$\vdots$	$\vdots$	$\vdots$	$\vdots$	$\vdots$	

for $n = 2$. The horizontal rows satisfy $x^3 - 2x = x(x^2 - 2)$ and the vertical rows satisfy $x^2(x^2 - 3)$. Starting with the hypercube $\begin{bmatrix} 1 & 5 \\ 7 & 9 \end{bmatrix}$ we back-solve its 2 horizontal rows and its 2 vertical rows getting each such back-solved row as an interlacing of never-zero geometric sequences. Then we check the eight positions labeled $*$ in front of the hypercube. If these 8 entries are all non-zero, then the linearly recursive sequence represented by this tableau will be Hadamard invertible.

We note that our definition here of linearly recursive sequence in n variables does not encompass some tableaux in $F[x_1, \ldots, x_n]^*$ that one might like to include. We illustrate this remark for the case $n = 2$. The simplest such tableau is perhaps the one with 1's on the main diagonal and 0's elsewhere, i.e., $f_{ij} = \delta_{ij}$. The i-th horizontal row satisfies x^{i+1}, and so does the i-th vertical row, for each $i \geq 0$. But these relations depend on i. Another example not included is the tableau whose i-th horizontal row for $i \geq 0$ is the sequence of $(i+1)$-st powers of the positive integers, so that it satisfies $(x-1)^{i+2}$. The j-th vertical row for $j \geq 0$ is a geometric sequence with initial coordinate $j+1$ and ratio $j+1$, so that it satisfies $x - (j+1)$. Again, these two sets of relations depend on i and on j, so the tableau does not represent an element of $F[x_1, x_2]^0$.

12. LIE COALGEBRAS AND LIE BIALGEBRAS.

A *Lie algebra* L over F has a skew-symmetric (bracket) product $[\,,\,]$ satisfying the Jacobi identity $[[a,b],c] + [[b,c],a] + [[c,a]b] = 0$ for a, b, c in L. If A is an algebra, then A^- is the Lie algebra which is A as a vector space and bracket product $[a,b] = ab - ba$. Dually, a *Lie coalgebra* M has a (cobracket) comultiplication δ: $M \to M \wedge M$, the skew-symmetric rank 2 tensors in $M \otimes M$, which satisfies the co-Jacobi identity $(1 + \sigma + \sigma^2)(1 \otimes \delta)\delta = 0$, where $\sigma = (123)$ in S_3 acts in the usual way on $M \otimes M \otimes M$. If C is a coalgebra, then C^- is the Lie coalgebra with is C as a vector space and cobracket $\delta(c) = \sum c_i \otimes c_i' - \sum c_i' \otimes c_i$, where $\Delta c = \sum c_i \otimes c_i'$ in $C \otimes C$, i.e., $\delta = (1 - tw)\Delta$, where tw is the twist map of $C \otimes C$ to $C \otimes C$ (Section 2). Whereas coalgebras are locally-finite, i.e., every element lies in a finite-dimensional subcoalgebra, this is not alway true for Lie coalgebras. Thus we define $Loc\, M$ to be the sum of all finite-dimensional Lie subcoalgebras of M. We will see examples where $Loc\, M \neq M$.

A *Lie bialgebra* L is a Lie algebra L which also has the structure of a Lie coalgebra (L, δ) such that δ is a one-cocycle in $Z'(L, L \wedge L)$, where L acts on (the right) of

$L \wedge L$ by the adjoint action $[a \wedge b, c] = [a, c] \wedge b + a \wedge [b, c]$. The cocycle condition is that $\delta([a, b]) = [\delta a, b] - [\delta b, a]$. If δ is the coboundary $\delta = \delta_r$ determined by r in $L \wedge L$, i.e., $\delta(a) = [r, a]$ for a in L, we say that L is a *coboundary Lie bialgebra.*

Let L be a Lie algebra. Some important Lie bialgebra structures on L arise from solutions r in $L \wedge L$ of the *classical Yang-Baxter equation*

$$\text{(CYBE)} \qquad [r_{12}, r_{13}] + [r_{12}, r_{23}] + [r_{13}, r_{23}] = 0,$$

where, if $r = \sum a_i \otimes b_i$, $r_{12} = \sum a_i \otimes b_i \otimes 1$, $r_{13} = \sum a_i \otimes 1 \otimes b_i$ and $r_{23} = \sum 1 \otimes a_i \otimes b_i$. The left-hand side of (CYBE) is in $L \otimes L \otimes L$, for example $[r_{12}, r_{13}] = \sum_{i,j} [a_i a_j] + b_i \otimes b_j$. The importance of (CYBE) is that if r is a solution, then $\delta = \delta_r$ satisfies the co-Jacobi identity, and thus $(L, [\,,\,], \delta_r)$ has the structure of a *triangular* coboundary Lie bialgebra (see [T4] for a complete proof). More generally, a necessary and sufficient condition for $r \in L \wedge L$ is that the left-hand side of (CYBE) be in the kernel of the adjoint representation of L on $L \otimes L \otimes L$.

Basic ideas about coalgebras can be found in [M1]. Lie bialgebras are mentioned in [D].

13. DUALITY BETWEEN LIE ALGEBRAS AND LIE COALGEBRAS.

The passage from Lie coalgebras to Lie algebras via convolution is similar to the associative case. Thus if M is a Lie coalgebra with cobracket $\delta m = \sum m_i \otimes m_i'$ for m in M, then M^* is a Lie algebra under the convolution product $(fg)(m) = \sum f(m_i) g(m_i')$ for f, g in M^*.

The passage from Lie algebras to Lie coalgebra involves the idea of a good subspace described in Section 4. Thus, if L is a Lie algebra, with bracket $[\,,\,]$, a subspace V of L^* is *good* if $[\,,\,]^*(V) \subseteq V \otimes V$, where we identify $L^* \otimes L^*$ in $(L \otimes L)^*$ in the usual way. Then L^0 is the sum of all the good subspace of L^* and L^0 is a Lie coalgebra under the restriction of $[\,,\,]^*$ to L^0.

Recall that from an algebra A, A^0 has an alternate description as $\{f \text{ in } A^* | f(J) = 0 \text{ for some cofinite ideal } J \text{ of } A\}$. For a Lie algebra L, it turns out (see [M1]) that $\{f \in L^* | f(J) = 0 \text{ for some cofinite ideal } J \text{ of } L\}$ is equal to $\mathrm{Loc}(L^0)$, the locally finite part of L^0 (Section 12). In the next section 14, we will identify the space of linearly recursive sequences as L^0 for L the Lie algebra of derivations of $F[x]$.

Finally if L is a Lie bialgebra, then L^0 is a Lie subalgebra of L^* under convolution product, and, in fact, L^0 is itself a Lie bialgebra. This is proved in [T4]. In Section 16, we will indicate various Lie algebra structures on the Lie coalgebra of linearly recursive sequences, under each of which it has the structure of a Lie bialgebra.

14. THE LIE COALGEBRA OF LINEARLY RECURSIVE SEQUENCES.

Let W_1 be the Witt algebra, i.e., the Lie algebra of derivations of the algebra $F[x]$. W_1 has a basis $\{e_n\}_{n \geq -1}$, where $e_n = x^{n+1} \frac{d}{dx}$. The bracket product is given by $[e_n, e_m] = (m - n) e_{n+m}$. If F has characteristic zero, then W_1 is a well-known infinite-dimensional simple Lie algebra, from which it follows that $\mathrm{Loc}(W_1^0) = 0$ (Section 13). Since we are going to identify W_1^0 as the Lie coalgebra of linearly recursive sequences, we have an example where Loc M is much smaller that M.

Since W_1 has the basis $\{e_n\}_{n \geq -1}$, we can identify the space W_1^* with sequences $f = (f_n)_{n \geq -1}$, where $f_n = f(e_n)$. The shift in indexing to start with -1 instead of 0 as before causes no special problem.

From now on, F has characteristic not equal to two. The identification of W_1^0 with the space of linearly recursive sequences can be found in [N1]. Rather than outline the technical argument here, we give a partial classical explanation when F is the complex numbers. Recall that the space of linearly recursive sequences has a basis of $\{(r^n n^i)|r \in F^*, i \geq 0\} \cup \{u_n|n \geq -1\}$, where $\{u_n\}_{n\geq -1}$ is the dual basis to the $\{e_n\}_{n\geq -1}$, i.e., $\{u_n\}$ span the finite sequences. The indexing of a sequence now starts with $n = -1$. We first note that the $\{u_n\}_{n\geq -1}$ span a good subspace of W_1^*. We use here τ for cobracket, since later we shall use δ for W_1 itself. Then the cobracket formula for the finite sequences is $\tau(u_n) = \sum_{j=-1}^{n+1}(n-2j)u_j \otimes u_{n-j}$. This is proved by applying both sides to $e_j \otimes e_k$ for $j,k \geq -1$ (recall that τ is a restriction of $[\,,\,]^*$). The left-hand side gives $u_n((k-j)e_{j+k}) = 0$ unless $j+k=n$, in which case we get $k-j = n-2j$. The right-hand side clearly yields the same result.

Next we note that $\{(r^n n^i)|r \in F^*, i \geq 0\}$ also span a good subspace of W_1^*, which shows that every linearly recursive sequence is in W_1^0 (see also [M2]). The cobracket formula here is that $\tau(r^n n^i) = \sum_{j=0}^{n+1}(\binom{i}{j} - \binom{i}{j-1})(r^n n^j) \otimes (r^n n^{i+1-j})$, indicating the combinatorial nature of the cobracket. For example, for a geometric sequence $e(r) = (r^n)$ for $r \neq 0$, $\tau(r^n) = (r^n)\otimes(nr^n) - (nr^n)\otimes(r^n)$, since evaluating at $e_j \otimes e_k$, it means that $(k-j)r^{j+k} = r^j(kr^k) - (jr^j)r^k$. For the sequence $(n)_{n\geq -1}$, the cobracket formula says $\tau(n) = (1)\otimes(n^2) - (n^2)\otimes(1)$, since $(k-j)(j+k) = k^2 - j^2$. For $(n^2)_{n\geq -1}$, we have $\tau(n^2) = (1)\otimes(n^3) + (n)\otimes(n^2) - (n^2)\otimes(n) - (n^3)\otimes(1)$, which says that $(k-j)(j+k)^2 = k^3 + jk^2 - j^2k - j^3$. To see a coefficient different from ± 1, note $\tau(n^3) = (1)\otimes(n^4) + 2(n)\otimes(n^3) - 2(n^3)\otimes(n) - (n^4)\otimes(1)$, since $(k-j)(j+k)^3 = k^4 + 2jk^3 - 2j^3k - j^4$. We leave the proof of the general formula for $\tau(r^n n^i)_{n\geq -1}$ for $r \in F^*$, $i \geq 0$ to the reader. It was also observed in [M2] that W_1^0 contains all linearly recursive complex sequences. To see that every f in W_1^0 is a linearly recursive sequence, we refer the reader to [N1].

The space $(f_n)_{n\in Z}$ of linearly recursive sequences with minimal recursive relation with non-zero constant term can similarly be regarded as W^0, where W is the Lie algebra of derivations of $F[x, x^{-1}]$. The same cobracket formula for $\tau(r^n n^i)$ proves that every linearly recursive sequence is in W^0. Recall that the dual basis elements u_n for $n \in Z$ are not linearly recursive as doubly-infinite sequences. Note also that for W_1^0, the formula $\tau(u_n) = \sum_{j+k=n}(k-j)u_j \otimes u_k$ where $j,k \geq -1$ makes no sense here for $j,k \in Z$.

In the next section 15, we discuss an alternative identification of W_1^0 and W^0 as linearly recursive sequences, which also enables one to compute the cobracket of such a sequence, once one has computed the (coassociative) comultiplication of the sequence. The latter was already discussed in Section 8.

15. RELATION BETWEEN LIE DUALITY AND ASSOCIATIVE DUALITY.

We indicate here a way of relating our coalgebra of linearly recursive sequences as the continuous dual of the algebra $F[x]$ with its structure as the Lie coalgebra W_1^0 dual to the Lie algebra W_1. This identification will also yield a cobracket formula $\pi(f)$ for f in W_1^0 in terms of the diagonalization Δf of f viewed in $F[x]^0$. Details can be found in [N2].

Let A be an algebra between $F[x]$ and $F[x, x^{-1}]$. This will enable us to also treat the case of doubly-infinite linearly recursive sequences as $F[x,x^{-1}]^0$ versus W^0. We assume that F has characteristic not equal to two. We can extend the derivation $\frac{d}{dx}$ of $F[x]$ to a derivation D of A. Let A_D denote the Lie algebra structure on A given by the bracket product $[\,,\,]_D$, where $[a,b]_D = aD(b) - D(a)b$. Thus we can form the dual Lie coalgebra $(A_D)^0$.

Considering A as an algebra with multiplication m, the derivation condition on D is that $Dm = m(1 \otimes D + D \otimes 1)$. If f is in A^0, then m^*f is in $A^0 \otimes A^0$, and $(m^*D^*)(f) = (1 \otimes D^* + D^* \otimes 1)m^*(f)$, so that $D^*(A^0) \subset A^0$. Let D^0: $A^0 \to A^0$ be the restriction of D^* to A^0. D^0 is a *coderivation* of the coalgebra A^0, i.e., $\Delta D^0 = (1 \otimes D^0 + D^0 \otimes 1)\Delta$. We can now give A^0 the structure of a Lie coalgebra by defining a cobracket τ by $\tau = (1 \otimes D^0 - D^0 \otimes 1)\Delta$, where Δ is the comultiplication on A^0. We denote this Lie coalgebra by $(A^0)_{D^0}$.

For $A = F[x]$, we have realized A^0 as the coalgebra of linearly recursive sequences $f = (f_n)_{n\geq 0}$ with $f_n = f(x^n)$ for $n \geq 0$. We have also noted that W_1^0 is the Lie coalgebra of these same linearly recursive sequences $f = (f_n)_{n\geq 0}$, where we reindex so that $f_n = f(e_{n-1})$ for $n \geq 0$, i.e., $n - 1 \geq -1$. The identification of f in W_1^* as $(f_n)_{n\geq 0}$ with $f_n = f(e_{n-1})$ is precisely the one obtained by identifying W_1 with $F[x]_D$ as a Lie algebra. To see this, just note that $[e_n, e_m] = (m-n)e_{m+n}$ corresponds in $F[x]_D$ to $[x^{n+1}, x^{m+1}]_D = x^{n+1}(m+1)x^m - (n+1)x^n x^{m+1} = (m-n)x^{n+m+1}$. Thus the Lie coalgebra W_1^0 discussed in Section 14 can be viewed as $(F[x]_D)^0$. Similarly, the Lie coalgebra W^0 of doubly-infinite linearly recursive sequences can be identified with $(F[x,x^{-1}]_D)^0$.

We now return to an algebra A between $F[x]$ and $F[x,x^{-1}]$. The Lie structure $[a,b]_D = aD(b) - D(a)b$ on A_D means that $[\,,\,]_D^* = (1 \otimes D - D \otimes 1)^*m^*$, so that for $f \in A^0$, $[\,,\,]_D^*(f) = \tau(f) \in A^0 \otimes A^0$, so that $(A^0)_{D^0}$ is a Lie subcoalgebra of $(A_D)^0$. In Corollary 3 of [N2] it is proved that, in fact, $(A^0)_{D^0} = (A_D)^0$ as Lie coalgebras. Note that we can identify A_D with the Lie algebra Der A of derivations of A by identifying a in A_D with aD, the derivation of A given by $(aD)(b) = a(D(b))$. Thus we have that $(\text{Der } A)^0 = (A_D)^0 = (A^0)_{D^0}$ as Lie coalgebras.

We now apply this to $A = F[x]$. Thus $W_1^0 = (\text{Der } A)^0 = (A_D)^0 = (A^0)_{D^0}$ is the Lie coalgebra of linearly recursive sequences. Our definition of $(A^0)_{D^0}$ indicates how to compute the cobracket $\tau(f)$ for f in W_1^0, provided one knows the diagonalization Δf of f in A^0. Thus $\tau(f) = (1 \otimes D^0 - D^0 \otimes 1)\Delta$, where D^0 is the restriction to A^0 of D^*. Recall that $D = \frac{d}{dx}$. So for $f = (f_n)_{n\geq 0}$ in A^*, where $f_n = f(x^n)$, $(D^*f)(x^n) = f(nx^{n-1}) = nf_{n-1}$. Thus $D^*f = (nf_{n-1})_{n\geq 0}$ (we interpret f_{-1} as 0). In particular, if $f \in A^0$ is linearly recursive, then $D^0 f = (nf_{n-1})_{n\geq 0}$.

In Section 8 we gave an algorithm for computing Δf for f in A^0. Thus we also have an algorithm for computing the cobracket $\tau(f)$. For example, consider the dual bases $\{Z^{(n)}\}_{n\geq 0}$ of $\{x^n\}_{n\geq 0}$, i.e., $Z^{(n)}(x^m) = \delta_{n,m}$. We saw that the $\{Z^{(n)}\}_{n\geq 0}$ span a divided-power coalgebra, which consists of all finite sequences. Thus $\Delta Z^{(n)} = \sum_{\substack{i+j=n \\ i,j\geq=0}} Z^{(i)} \otimes Z^{(j)}$. Note that $D^0(Z^{(n)}) = (n+1)Z^{(n+1)}$. Our algorithm then shows that $\tau(Z^{(n)}) = \sum_{\substack{i+j=n+1 \\ i,j\geq 0}} (j-i)Z^{(i)} \otimes Z^{(j)}$ for each $n \geq 0$. A

Lie coalgebra with a basis $\{Z^{(n)}\}_{n\geq 0}$ with $\tau(Z^{(n)}) = \sum_{\substack{i+j=n+1\\ i,j\neq 0}} (j-i)Z^{(i)} \otimes Z^{(j)}$ is called a *divided-power Lie coalgebra.* If $r \in F$, then $\{e(r)Z^{(n)}\}_{n\geq 0}$ are also the basis of a divided-power Lie coalgebra (the product $e(r)Z^{(n)}$ is the divided-power product), as well as the basis of a divided-power coalgebra. If F is algebraically closed, then Theorem 7 of [N2] shows that the Lie coalgebra W_1^0 is the direct sum over all r in F of divided-power Lie coalgebras U_r where U_r has a basis $\{u_{r,n}\}_{n\geq 0}$, where $u_{r,n}(x^s) = \binom{s}{n}a^{s-n}$. But $(e(r)Z^{(n)})(x^s) = (e(r)\otimes Z^{(n)})\sum_{i=0}^{s}\binom{s}{i}x^i \otimes x^{s-i} = \binom{s}{n}r^{s-n}$, i.e., U_r has a basis $\{e(r)Z^{(n)}|n \geq 0\}$ as above. Thus we have recovered our basis $\{e(r)Z^{(n)}|r \in F,\, n \geq 0\}$ for the space of linearly recursive sequences described in Section 7.

We illustrate the cobracket algorithm for the Fibonacci sequence $F = (1,1,2,3,5,8,13,21,\dots)$. Then $SF = (1,2,3,5,8,13,\dots)$ and $\Delta F = 2(F\otimes F) - F\otimes SF - SF\otimes F + SF\otimes SF$ (see Section 8). Also $D^0F = (0,1,2,6,12,25,48,91,\dots)$ and $D^0(SF) = (0,1,4,9,20,40,78,\dots)$. Thus $\tau(F) = 2(F\otimes D^0F - D^0F\otimes F) - (F\otimes D^0(SF) - D^0(SF)\otimes F) - (D^0F\otimes SF - SF\otimes D^0F) + (SF\otimes D^0(SF) - D^0(SF)\otimes SF)$.

For $A = F[x,x^{-1}]$, a similar treatment of doubly-infinite linearly recursive sequences is possible, using $W^0 = (\text{Der } A)^0 = (A_D)^0 = (A^0)_{D^0}$ as Lie coalgebras. Here the dual basis elements $\{Z^{(n)}|n \in Z\}$ are not in W^0, but $e(r)Z^{(n)}$ is in W^0 for $r \in F^*$, $n \geq 0$, since it satisfies the relation $(x-r)^{n+1}$. Another way to see this is to apply the backsolving map α: $F[x]^0 \to F[x,x^{-1}]^0$ to $e(r)Z^{(n)}$, getting $e(r)Z^{(n)}$ in $F[x,x^{-1}]^0$. The kernel of α is the finite sequences (which amounts to allowing $r=0$), so α is injective on the $e(r)Z^{(n)}$ for $r \neq 0$ in F, $n \geq 0$. Since α is surjective, the $\{e(r)Z^{(n)}|r \in F^*,\, n \geq 0\}$ form a basis for $F[x,x^{-1}]^0$. For this basis, the same diagonalization and cobracket formulas hold as in $F[x]^0$ and W_1^0, i.e., for a fixed $r \in F^*$, $\{e(r)Z^{(n)}|n \geq 0\}$ is the basis of a divided-power coalgebra and a divided-power Lie coalgebra. Thus $F[x,x^{-1}]^0$ and W_1^0 are the direct sums of these divided-power coalgebras and divided-power Lie coalgebras respectively.

16. LIE BIALGEBRA STRUCTURES OF LINEARLY RECURSIVE SEQUENCES.

Recall from Section 12 that a triangular coboundary Lie bialgebra L is a Lie bialgebra whose cobracket $\delta = \delta_r$ is a coboundary in $B^1(L, L\wedge L)$ where r in $L \wedge L$ is a solution of the classical Yang-Baxter equation. We now indicate such a structure for the Witt algebras W_1 and W. Details can be found in [T4].

W_1 has a basis $\{e_n\}_{n\geq -1}$ where $e_n = x^{n+1}\frac{d}{dx}$ and $[e_n, e_m] = (m-n)e_{n+m}$ for $n, m \geq -1$. For each $i \geq -1$, the element $r_i = e_0 \wedge e_i$ in $W_1 \wedge W_1$ is a solution of the classical Yang-Baxter equation. See also [M4]. Thus setting $\delta_i = \delta_{r_i}$, we get a triangular coboundary Lie bialgebra structure $W_1^{(i)}$ on W_1. If F has characteristic zero, these structures are mutually distinct, i.e., if $i \neq j$, $i,j \geq -1$, then $W_1^{(i)}$ and $W_1^{(j)}$ are non-isomorphic as Lie coalgebras. To see this, we note that $\delta_i = \delta_{r_i}$ for $r_i = e_0 \wedge e_i$ gives the formula that $\delta_i(e_n) = n(e_n \wedge e_i) + (n-i)(e_0 \wedge e_{n+i})$. In particular, $\delta_i(e_i) = 0$. From this we determine the local structure Loc $W_1^{(i)}$. If $i = -1$, Loc $W_1^{(-1)} = W_1^{(-1)}$ since for each $n \geq -1$, $e_{-1}, e_0, \dots, e_n$ span a finite-dimensional Lie subcoalgebra. If $i = 0$, $\delta_0 = 0$. If $i = 1$, Loc $W_1^{(1)}$ is 3-dimensional with basis e_{-1}, e_0, e_1. For $i \geq 2$, Loc $W_1^{(i)}$ is 2-dimensional with basis e_0, e_i. Thus we can assume that $i \geq 2$, $j \geq 2$ and $i \neq j$. Any Lie coalgebra isomorphism T: $W_1^{(i)} \to W_1^{(j)}$ would restrict to an isomorphism of Loc $W_1^{(i)}$ to Loc $W_1^{(j)}$. One sees that $T(e_i) = \frac{j}{i} e_j$ and $T(e_0) = ae_0 + be_j$ for a, b in F, $a \neq 0$. Then a calculational argument yields $T(e_{-1}) \in$ Loc $W_1^{(j)}$, a contradiction. See [T4] for details.

Returning to F of any characteristic not two, recall that we identified W_1^0 as the Lie coalgebra of linearly recursive sequences $f = (f_n)_{n \geq -1}$, where $f_n = f(e_n)$. Thus for each $i \geq -1$, $(W_1^{(i)})^0$ gives a Lie bialgebra structure on the space of linearly recursive sequences. The Lie coalgebra structure is independent of i, since it depends only on the Lie algebra structure of W_1. This was discussed in Sections 14 and 15. The Lie algebra product is the convolution product dual to the cobracket δ_i of $W_1^{(i)}$. For $i = 0$, $\delta_0 = 0$ so $W_1^{(0)}$ is an abelian Lie algebra. For $i \neq 0$, we indicate the Lie product for the dual basis $\{u_n\}_{n \geq -1}$ to $\{e_n\}_{n \geq -1}$, i.e., for the finite sequences. $[u_0, u_n] = (n - 2i)u_{n-i}$ for $n \neq 0$, $[u_n, u_i] = nu_n$ for $n \neq 0, i$ and all other products are zero. If $f = (f_n)_{n \geq -1} = \sum_{n \geq -1} f_n u_n$ and $g = (g_n)_{n \geq -1} = \sum_{n \geq -1} g_n u_n$, then we can compute any coordinate of $[f, g]$ by the above formulas, but if f and g are linearly recursive sequences, we do not know any algorithm for computing $[f, g]$ as a linearly recursive sequence, i.e., the initial data and a recursive relation for $[f, g]$ (recall that in Section 15 we gave an algorithm for computing the cobracket $\tau(f)$ of a linearly recursive sequence f). However, from the coordinate formula for the bracket product in $(W_1^{(i)})^*$, one can see that $(W_1^{(i)})^*$ is a 3-step solvable Lie algebra, and of course so $(W_1^{(i)})^0$, its Lie subalgebra of linearly recursive sequences.

For W, with basis $\{e_n\}_{n \in Z}$, for each $i \in Z$, the element $e_0 \wedge e_i$ satisfies the classical Yang-Baxter equation, so we have a triangular coboundary Lie bialgebra $W^{(i)}$. The formula for the cobracket is similar to the one in $W_1^{(i)}$. Also, $(W^{(i)})^*$ is a 3-step solvable Lie algebra for $i \neq 0$, and so is its Lie subalgebra $(W^{(i)})^0$ of doubly-infinite linearly recursive sequences. Again we have no algorithm to describe $[f, g]$ as a doubly-infinite linearly recursive sequence when f and g are such sequences.

In [T4], we also discuss the Virasoro algebra V which is a central extension $V = W \oplus Fc$ of W, where c is central and $[e_n, e_m] = (m - n)e_{n+m} + \frac{1}{12}(m^3 - m)\delta_{n,-m}c$. It turns out that V^0 and W^0 are isomorphic Lie coalgebras in the sense that if $f \in V^0$, then $f(c) = 0$, and $(f_n = f(e_n))_{n \in Z}$ is a doubly-infinite linearly recursive sequence. Using $e_0 \wedge e_i$ for $i \in Z$ to give a Lie bialgebra structure $V^{(i)}$ on V, $(V^{(i)})^0$ and $(W^{(i)})^0$ are isomorphic Lie bialgebras, so 3-step solvable Lie algebras for $i \neq 0$. For each $i \in z$, the element $c \wedge e_i$ of $V \wedge V$ satisfies the classical Yang-Baxter equation, yielding a Lie bialgebra structure on V with cobracket δ_i' satisfying $\delta_i'(e_n) = (n - i)(c \wedge e_{i+n})$ and $\delta_i'(c) = 0$. The dual Lie algebra product is described coordinatewise by $[c^*, u_n] = (n - 2i)u_{n-i}$, and all other products involving the dual basis $\{u_n^*\}_{n \in Z}$ and c^* are zero. This is a metabelian (2-step solvable) Lie algebra (even for $i = 0$).

References

[A] E. Abe, *Hopf Algebras*, Cambridge Tracts in Mathematics 74, Cambridge University Press, Cambridge, 1977.

[B] B. Benzaghou, *Algèbres de Hadamard*, Bull. Soc. Math. France **98** (1970), 209–252.

[C-M-P] L. Cerlienco, M. Mignotte et F. Piras, *Suites récurrentes linéaires*, L'Enseignement Mathématique **33** (1987), 67–108.

[C-P] L. Cerlienco and F. Piras, *On the continuous dual of a polynomial bialgebra*, Comm. Algebra **19** (1991), 2707–2727.

[D] V.G. Drinfeld, *Quantum groups*, Proceedings Int. Congress Math., Berkeley, 1986.

[H] G.Hochschild, *Introduction to Affine Algebraic Groups*, Holden-Day, San Francisco, 1971.

[L-T] R.G. Larson and E.J. Taft, *The algebraic structure of linearly recursive sequences under Hadamard product*, Israel J. Math. **72** (1990), 118–132.

[M1] W. Michaelis, *Lie coalgebras*, Advances in Math. **38** (1980), 1-54.

[M2] ———, *On the dual Lie coalgebra of the Witt algebra*, Proceedings of the XVII International Colloquium on Group Theoretical Methods in Physics (Y. Saint-Aubin and L. Vinet, eds.), World Scientific, Singapore, 1989, pp. 435–439.

[M3] ———, *An example of a non-zero Lie coalgebra M for which $Loc(M) = 0$*, J. Pure Appl. Algebra **68** (1990), 341–348.

[M4] ———, *A class of infinite-dimensional Lie bialgebras containing the Virasoro algebra*, Advances in Math. (to appear).

[N1] W. Nichols, *The structure of the dual Lie coalgebra of the Witt algebra*, J. Pure Appl. Algebra **68** (1990), 359-364.

[N2] ———, *On Lie and associated duals*, J. Pure Appl. Algebra **87** (1993), 313–320.

[P-T] B. Peterson and E.J. Taft, *The Hopf algebra of linearly recursive sequences*, Aequationes Math. **20** (1980), 1–17.

[P] A. van der Poorten, *Some facts that should be better known, especially about rational functions*, Number Theory and Applications (R.A. Mollin, ed.), Kluwer Acad. Publ., Dordrecht, 1989.

[R] C. Reutenauer, *Sur les éléments inversibles de l'algèbre de Hadamard des séries rationelles*, Bull. Soc. Math. France **110** (1982), 225–232.

[S] M. Sweedler, *Hopf Algebras*, Benjamin, New York, 1969..

[T1] E.J. Taft, *Reflexivity of algebras and coalgebras*, Amer. Journal Math. **94** (1972), 1111-1130.

[T2] ———, *Hurwitz invertibility of linearly recursive sequences*, Congressus Numerantium **73** (1990), 37–40.

[T3] ———, *Hadamard invertibility of linearly recursive sequences in several variables*, Discrete Math. (to appear).

[T4] ———, *Witt and Virasoro algebras as Lie bialgebras*, J. Pure Appl. Algebra **87** (1993), 301–312.

Relations of Representations of Quantum Groups and Finite Groups

MITSUHIRO TAKEUCHI University of Tsukuba, Tsukuba, Ibaraki, Japan

This expository paper is based the author's talk at DePaul University on the relationship between

(I) Unipotent representations of $\mathrm{GL}_n(q)$, where q is a power of a prime p, and
(II) Polynomial representations (of degree n) of quantum GL_d, where d is any integer greater than or equal to n.

The quantization occurring in (II) refers to the number q in (I). In both representations, the same base field whose characteristic is not p is used.

The (classical) Schur algebra $S_K(d, n)$ is associated with polynomial representations of GL_d of degree n. If E is a d-dimensional K-vector space, the symmetric group $W = S_n$ acts naturally on $E^{\otimes n}$ on the right. Schur's reciprocity law tells that the Schur algebra $S_K(d, n)$ is isomorphic to the endomorphism algebra of the right KW module $E^{\otimes n}$. Here, the base field K is arbitrary. This gives a direct relationship between degree n polynomial representations of GL_d (for various d) and representations of the group algebra KW. We refer to Green [7] for this classical theory.

Dipper–James [3] have introduced the q-Schur algebra and shown unipotent representations of $\mathrm{GL}_n(q)$ can be described with it. Dipper–Donkin [2] have further given a connection with polynomial representations of quantum GL_d.

Throughout this paper, we fix the base field K, a nonzero element q in K, and the number n. We put $W = S_n$.

We begin by describing quantum GL_d and introduce the q-Schur algebra and the Hecke algebra. The q-analog of Schur's reciprocity law suggests that we generalize the concept of q-Schur algebra. We show the representation theory of the q-Schur algebra. We show the representation theory of the q-Schur algebra as developed in [4,5] applies to our generalized q-Schur algebras. Finally, we talk about relations with unipotent representations of $\mathrm{GL}_n(q)$ in case q is a power of a prime.

1 QUANTUM GL_d

Various versions of quantum GL_d have appeared in the literature. The two-parameter quantization I give in [10] is most convenient to use. This includes the standard one (with the same parameters) and the Dipper–Donkin one (with 1 as one of the parameters) as special cases.

Take two elements α, β in K such that $\alpha\beta = q$. Define the K-algebra $A_{K,\alpha,\beta}(d)$ by generators $x_{11}, \ldots, x_{dd}$ and the following relations:

$$x_{ik}x_{ij} = \alpha x_{ij}x_{ok} \qquad \text{if } j < k,$$

$$x_{jk}x_{ik} = \beta x_{ik}x_{jk} \qquad \text{if } i < j,$$

$$x_{jk}x_{il} = \beta\alpha^{-1}x_{il}x_{jk}, \quad x_{jl}x_{ik} - x_{ik}x_{jl} = (\beta - \alpha^{-1})x_{il}x_{jk}$$

$$\text{if } i < j, k < l.$$

This is a polynomial algebra in $x_{11}, \ldots, x_{dd}$. That is, if we give an arbitrary total ordering on these generators, then the set of all monomials (relative to the given ordering) forms a basis. This algebra has a bialgebra structure such that we have

$$\Delta(x_{ik}) = \sum_{j=1}^{d} x_{ij} \otimes x_{jk}, \quad \epsilon(x_{ij}) = \delta_{ij}.$$

There is a grouplike element g called the *quantum determinant*. It is defined by

$$g = \sum_{\sigma} (-\alpha)^{-l(\sigma)} x_{1,\sigma(1)} \ldots x_{d,\sigma(d)} = \sum_{\sigma} (-\beta)^{-l(\sigma)} x_{\sigma(1),1} \ldots x_{\sigma(d),d}$$

the sum over all σ in S_d. The quantum determinant g satisfies

$$x_{ij}g = (\beta\alpha^{-1})^{i-j} g x_{ij}$$

so that the localization $A_{K,\alpha,\beta}(d)[g^{-1}]$ is defined by Ore's method. If we extend the bialgebra structure to this localization by letting g^{-1} be a grouplike element, then this becomes a *Hopf algebra*. The two-parameter quantization of GL_d relative to (α, β) is defined to be the quantum group associated with the Hopf algebra $A_{K,\alpha,\beta}(d)[g^{-1}]$.

By a *polynomial representation* of the quantum GL_d, we mean a right comodule for $A_{K,\alpha,\beta}(d)$. As an algebra, it is graded:

$$A_{K,\alpha,\beta}(d) = \oplus_{n=0}^{\infty} A_{k,\alpha,\beta}(d, n)$$

where $A_{k,\alpha,\beta}(d, n)$ denotes the nth component, which is seen to be a subcoalgebra of finite dimension. Right comodules for $A_{K,\alpha,\beta}(d, n)$ are called polynomial representations of degree n.

The coalgebra structures of $A_{K,\alpha,\beta}(d)$ and $A_{K,\alpha,\beta}(d,n)$ depend only on the product $q=\alpha\beta$. This means if we choose another pair of parameters α', β' such that $q=\alpha'\beta'$, then we have coalgebra isomorphisms

$$A_{K,\alpha,\beta}(d) \simeq A_{K,\alpha',\beta'}(d), \quad A_{K,\alpha,\beta}(d,n) \simeq A_{K,\alpha',\beta'}(d,n).$$

This property, called *hyperbolic invariance*, is due to DuParshall–Wang [6]. This shows we can well-define the dual algebra

$$S_{K,q}(d,n) = A_{K,\alpha,\beta}(d,n)^*$$

called the *q-Schur algebra*. The category of all left $S_{K,q}(d,n)$ modules, $\underline{\underline{\mathrm{mod}}}S_{K,q}(d,n)$, is identified with the category of all right $A_{K,\alpha,\beta}(d,n)$ comodules, i.e., polynomial representations of degree n of the quantum GL_d.

2 THE HECKE ALGEBRA, A q-ANALOGUE OF KW

The *Hecke algebra* $\mathcal{H}=\mathcal{H}_{K,q}(W)$ is defined by generators $T_1,\ldots,T_{n-1}$ and the following relations:

(i) $(T_i-q)(T_i+1)=0$,
(ii) $T_iT_{i+1}T_i = T_{i+1}T_iT_{i+1}$,
(iii) $T_iT_j = T_jT_i \quad$ if $|i-j|>1$.

A transposition of the form $s_a=(a,a+1)$ is called *basic*. We can write any permutation π in W as a product of basic transpositions: $\pi = s_{i_1}\ldots s_{i_l}$ with $l=l(\pi)$. For such a reduced expression, the product $T_\pi = T_{i_1}\ldots T_{i_l}$ gives a well-defined element in $\mathcal{H}$, and the set of T_π, π in W, forms a basis for $\mathcal{H}$. When $q=1$, the Hecke algebra $\mathcal{H}$ reduces to the group algebra KW. If q is a power of a prime number, then $\mathcal{H}_{K,q}(W)$ coincides with the *Iwahori–Hecke algebra* $H_K(G,B)$ where $G=\mathrm{GL}_n(q)$ and B is the upper Borel subgroup.

3 THE RECIPROCITY LAW, OR THE DOUBLE CENTRALIZER PROPERTY

Let E be a d-dimensional K-vector space with basis $e_1,\ldots,e_d$. The K-space $E^{\otimes n}$ has a basis

$$e_i = e_{i_1}\otimes\ldots\otimes e_{i_n}, \quad i=(i_1,\ldots,i_n) \text{ in } I(d,n)$$

where $I(d,n)$ denotes the set of all sequences i with $1\le i_1,\ldots,i_n\le d$. This space has the right $A_{K,\alpha,\beta}(d,n)$ comodule structure:

$$e_j \mapsto \sum_{i \text{ in } I(d,n)} e_i\otimes x_{ij}$$

where

$$x_{ij} = x_{i_1j_1}\ldots x_{i_nj_n}.$$

Hence it has a left $S_{K,q}(d,n)$ module structure.

On the other hand, there is a right $\mathcal{H}$ module structure on $E^{\otimes n}$ described as follows: Let i be a sequence in $I(d,n)$ and let $s = (a, a+1)$ be a basic transposition. We define

$$e_i T_S = \begin{cases} q e_{is} & \text{if } i_a = i_{a+1}, \\ \beta e_{is} & \text{if } i_a < i_{a+1}, \\ (q-1)e_i + \alpha e_{is} & \text{if } i_a > i_{a+1}. \end{cases}$$

One checks that this makes $E^{\otimes n}$ into a right $\mathcal{H}$ module by using the defining relations for $\mathcal{H}$. One sees the module structure also depends only on the product $q = \alpha\beta$.

The actions of $S_{K,q}(d,n)$ and $\mathcal{H}$ on $E^{\otimes n}$ commute with each other. The canonical algebra map

$$S_{K,q}(d,n) \longrightarrow \mathrm{End}_{\mathcal{H}}(E^{\otimes n})$$

is an isomorphism. This is a q-analog of Schur's reciprocity law [7,p.28]. See [5,6] for the proof. This characterizes the q-Schur algebra as the endomorphism algebra of some $\mathcal{H}$ module. It follows immediately that $S_{K,q}(d,n)$ is semisimple if $\mathcal{H}$ is semisimple. This occurs when $[n][n-1]\cdots[1] \neq 0$, where $[i] = 1 + q + q^2 + \cdots + q^{i-1}$.

4 COMPOSITIONS AND PARTITIONS

A composition of n means a sequence $\lambda = (\lambda_1, \lambda_2, \ldots)$ of integers ≥ 0 whose sum is n. The set of all compositions of n will be denoted by $\mathcal{C}(n)$. Let $\Lambda(d,n)$ denote the subset of all compositions λ of n such that $\lambda_a = 0$ if $a > d$.

A *partition* of n means a composition λ of n such that $\lambda_1 \geq \lambda_2 \geq \lambda_3 \geq \cdots$. Let $\mathcal{P}(n)$ be the set of all partitions of n. This is the union of $\Lambda^+(d,n) = \Lambda(d,n) \cap \mathcal{P}(n)$, $d \geq 1$. It also coincides with $\Lambda^+(n,n)$. For two compositions λ, μ of n, we write $\lambda \sim \mu$ if there is a permutation σ such that $\lambda_{\sigma(i)} = \mu_i$ for all i. Obviously, every equivalence class of $\sim$ contains a unique partition.

The *dual* partition λ' of a composition λ is defined by

$$\lambda'_a = \text{ the number of } b \text{ with } \lambda_b \geq a.$$

For two compositions λ, μ of n, we have $\lambda \sim \mu$ if and only if $\lambda' = \mu'$.

The set $\mathcal{P}(n)$ has a ordering, called the *dominance ordering*, which plays a role here. Let λ, μ by two partitions of n. We write $\lambda \trianglelefteq \mu$ if

$$\lambda_1 + \cdots + \lambda_a \leq \mu_1 + \cdots + \mu_a \qquad \text{for all } a \geq 1.$$

We have $\lambda \trianglelefteq \mu$ if and only if $\mu' \trianglelefteq \lambda'$.

Let λ be a composition of n. The interval of integers $\{1, 2, \ldots, n\}$ divides into a union of subintervals

$$\{1, 2, \ldots, \lambda_1\}, \ \{\lambda_1 + 1, \lambda_2 + 2, \ldots, \lambda_1 + \lambda_2\}, \ \ldots$$

The subgroup of those permutations that leave these subintervals invariant is denoted by Y_λ and called the *Young subgroup* associated with λ. Let x_λ (resp. y_λ) by the sum of all T_π (resp. $(-q)^{-l(\pi)}T_\pi$) for π in Y_λ. These are q-analogues of the *Young symmetrizer* (resp.

anti-symmetrizer). A permutation d is *distinguished* relative to λ if d^{-1} is increasing on each subinterval above. Let $\mathcal{D}_\lambda$ denote the set of all permutations distinguished relative to λ.

The right ideals $x_\lambda\mathcal{H}$ and $y_\lambda\mathcal{H}$ will play important roles. They have K-bases $x_\lambda T_f$ and $y_\lambda T_d$ with d in $\mathcal{D}_\lambda$. For two compositions λ, μ of n, we have $\lambda \sim \mu$ if and only if $x_\lambda\mathcal{H} \simeq x_\mu\mathcal{H}$ or $y_\lambda\mathcal{H} \simeq y_\mu\mathcal{H}$ as right $\mathcal{H}$ modules.

5 GENERALIZATION OF THE q-SCHUR ALGEBRA

Let Λ be a finite set of composition admitting some redundancy. Strictly speaking it means a pair consisting of a finite set Λ with a map of it into $\mathcal{C}(n)$. Let M_Λ (resp. $M^\#_\Lambda$) be the direct sum of right $\mathcal{H}$ modules $x_\lambda\mathcal{H}$ (resp. $y_\lambda\mathcal{H}$) for λ in Λ. These right $\mathcal{H}$ modules have isomorphic endomorphism algebra. We put

$$S_\Lambda = \mathrm{End}_\mathcal{H}(M_\Lambda) \simeq \mathrm{End}_\mathcal{H}(M^\#_\Lambda).$$

One of the most important examples of S_Λ is provided by the q-analog of Schur's reciprocity law (Section 3). Take $\Lambda(d,n)$ as Λ. The right $\mathcal{H}$ module $E^{\otimes n}$ (ibid.) is isomorphic to M_λ. Hence the q-analog of Schur's reciprocity law implies

$$S_q(d,n) = S_{\Lambda(d,n)}.$$

In this sense, our algebras S_Λ are generalizations of the q-Schur algebra. The original q-Schur algebra [3] corresponds to the case $\Lambda = \mathcal{P}(n)$.

The representation theory of algebras S_Λ parallels [5] (see [11]). The main results will be reviewed in the following.

For each λ in Λ, let $\xi_\lambda : M_\Lambda \to x_\lambda\mathcal{H}$ be the projection onto the λ component. We get orthogonal idempotents ξ_λ (λ in Λ) in S_Λ whose sum is 1. If V is a left S_Λ module, it is the direct sum of K subspaces $V^\lambda = \xi_\lambda V$, the λ-*weight space*. If two compositions λ, μ in Λ are equivalent under $\sim$, there are f, g in S_Λ such that $\xi_\lambda = fg$ and $\xi_\mu = gf$. This implies $\dim_K V^\lambda = \dim_K V^\mu$. In case $\Lambda = \lambda(d,n)$, this means the *formal character*

$$\phi_V = \sum_{\lambda \text{ in } \Lambda(d,n)} \dim_K V^\lambda X_1^{\lambda_1} \cdots X_d^{\lambda_d}$$

is a symmetric function.

The algebra S_Λ is the direct sum of subspaces $\xi_\lambda S_\Lambda \xi_\mu$, λ, μ in Λ and we may identify

$$\xi_\lambda S_\Lambda \xi_\mu = \mathrm{Hom}_\mathcal{H}(x_\mu\mathcal{H}, x_\lambda\mathcal{H})$$

which has a canonical K basis (independent of q) whose basis elements are in bijective correspondence with the *row-standard μ-tableaux of type λ* [4].

6 WEYL MODULES

Weyl modules play a crucial role in the representation theory of S_Λ. If λ is a composition of n, the K space $x_\lambda\mathcal{H}y_\lambda$, is one dimensional with basis $z_\lambda = x_\lambda T_{\pi_\lambda} y_{\lambda'}$, where the permutation

π_λ is defined in a manner indicated by the following example:

$$\pi_\lambda : \begin{matrix} 1 & 5 & \\ 2 & & \\ 3 & 6 & 8 \\ 4 & 7 & \end{matrix} \longrightarrow \begin{matrix} 1 & 2 & \\ 3 & & \\ 4 & 5 & 6 \\ 7 & 8 & \end{matrix} \qquad \text{where } \lambda = (2, 1, 3, 2).$$

If λ is in Λ, we may think of z_λ as an element in M_Λ. The submodule $S_\Lambda z_\lambda = W_\lambda$ of $M_\Lambda y_{\lambda'}$ is called the *Weyl module* corresponding to λ. If two weights λ, μ in Λ are equivalent under $\sim$, then $W_\lambda = W_\mu$. Let Λ^+ be the set of partitions α of n such that $\alpha \sim \lambda$ for some λ in Λ. It follows that we can well-define W_α to be W_λ.

Let λ be a composition in Λ and μ a partition in Λ^+. The λ-weight space $(W_\mu)^\lambda = \xi_\lambda S_\Lambda \xi_\mu z_\mu$ has a canonical K basis whose basis elements are in bijective correspondence with the row-standard and strictly column-standard μ-tableaux of type λ (*semi-standard basis theorem* [5]). This implies $\dim_K (W_\mu)^\lambda$ does not depend on K or q. In case $\Lambda = \Lambda(d, n)$, it follows that the Weyl module W_μ has the *Schur function* $\underline{\underline{S}}_\mu$ [7] as its formal character.

Weyl modules are *highest weight modules* in the sense that $(W_\mu)^\mu$ is one-dimensional and $(W_\mu)^\lambda \neq 0$ implies $\lambda \trianglelefteq$ for λ, μ in Λ^+.

7 IRREDUCIBLE (OR SIMPLE) S_Λ MODULES

If $[n]! \neq 0$, the Weyl module W_μ is irreducible for all μ in Λ^+. In general, there is a unique maximal submodule $W_\mu^{\max}$ of W_μ, and in fact, $W_\mu^{\max} = W_\mu \cap W_\mu^\perp$ relative to some symmetric no degenerate invariant inner product on $M_\Lambda y_{\mu'}$. The quotient modules $F_\mu = W_\mu / W_\mu^{\max}$ are *absolutely irreducible self-dual* and mutually nonisomorphic for μ in Λ^+.

In order for the modules F_μ to exhaust all irreducible S_Λ modules, we require some assumption on Λ^+.

(#) Let λ, μ be two partitions of n with $\lambda \trianglerighteq \mu$. If μ is in Λ^+, then λ is also in Λ^+.

For instance, $\Lambda = \Lambda(d, n)$ satisfies this property. If the set Λ^+ satisfies (#), then the set of modules F_μ, μ in Λ^+, makes a complete set of representatives for the isomorphism classes of all irreducible S_Λ modules. In this case, we denote by $d_{\lambda\mu}$ the multiplicity of F_μ as a composition factor of W_λ for λ, μ in Λ^+. The *decomposition matrix* $(d_{\lambda\mu})$ is lower triangular in the sense that $d_{\lambda\lambda} = 1$ and $d_{\lambda\mu} \neq 0$ implies $\lambda \trianglerighteq \mu$. (Condition (#) means Λ^+ is an ideal in the set of all dominant weights (cf. [12]).

8 UNIPOTENT REPRESENTATIONS

We are now going to apply the above theory of (generalized) q-Schur algebras to *unipotent representations* of the finite general linear group $G = \mathrm{GL}_n(q)$. In the following, we assume q is a power of a prime p, and the base field K contains a primitive pth root of 1. This implies $\mathrm{char}(K) \neq p$. Let B be the upper Borel subgroup of G and consider the left ideal $M = KG \cdot [B]$ of the group algebra KG generated by $[B]$, the sum of all elements in B. Let

$$\psi : KG \longrightarrow \mathrm{End}_K(M)$$

denote the corresponding representation. A representation of G over K is called *unipotent* if it vanishes on $\mathrm{Ker}(\psi)$. It is the same as representations of the algebra $\psi(KG)$.

Unipotent representations are equivalent to the representations of some (generalized) q-Schur algebra S_Λ. More precisely, there is an idempotent e in KG such that the functor

$$\underline{\underline{\mathrm{mod}}}\psi(KG) \longrightarrow \underline{\underline{\mathrm{mod}}}\psi(eKGe), \qquad V \mapsto eV$$

is a *category equivalence*, and there is a finite set Λ of compositions of n (admitting some redundancy) such that $\Lambda^+ = \mathcal{P}(n)$ and that

$$\psi(eKGe) \simeq S_\Lambda.$$

This is the main result of [11] (see also [3]). We show how to realize the idempotent e and the labelling Λ below.

Let U^- be the subgroup of G consisting of all lower unitriangular matrices. Let $\chi_1 = 1, \chi_2, \ldots, \chi_q$ be the distinct homomorphisms of the additive group $(\mathbb{F}_q, +)$ into $K^\times$. We consider the set of all sequences of integers

$$c = (c_1, \ldots, c_{n-1}), \qquad 1 \le c_a \le q \text{ for } 1 \le a < n.$$

We associate a homomorphism χ_c and an idempotent e_c in KG with such a sequence c as follows:

$$\chi_c : U^- \longrightarrow K^\times, \quad (g_{ij}) \mapsto \prod_{a=1}^{n-1} \chi_{c_a}(g_{a+1,a}),$$

$$e_c = |U^-|^{-1} \sum_{g \text{ in } U^-} \chi_c(g) g$$

The idempotents e_c are orthogonal to one another and we have

$$\sum_c e_c = |(U^-, U^-)|^{-1} \sum_{g \text{ in } (U^-, U^-)}^{g}$$

with the commutator subgroup (U^-, U^-) of U^-. Let e be this idempotent. Then we have the above category equivalence. This the algebra $\psi(KG)$ is *Morita equivalent* to $\psi(eKGe)$.

By definition, the endomorphism algebra $\mathrm{End}_{KG\diamond}(M)$ is identified with the Iwahori–Hecke algebra $H_K(G, B)$ and it is isomorphic to our Hecke algebra $\mathcal{H} = \mathcal{H}_{K,q}(W)$ (Section 2). We will let $\mathcal{H}$ act on M on the right. Then the action of T_π, π in W, is given by $[B]T_\pi = [B\pi B]$, the sum of all elements in the double coset $B\pi B$.

If c is a sequence as above, there is a composition λ^c of n such that

$$e_c M \simeq y_{\lambda^c}\mathcal{H} \qquad \text{as right } \mathcal{H} \text{ modules.}$$

Let $\{\alpha_1, \alpha_2, \ldots, \alpha_{h-1}\}$ with $1 \le \alpha_1 < \alpha_2 < \cdots < \alpha_{h-1} < n$, be the set of $1 < a < n$ such that $c_a = 1$. We have

$$\lambda^2 = (\alpha_1, \alpha_2 - \alpha_1, \ldots, \alpha_{h-1} - \alpha_{h-2}, n - \alpha_{h-1}, 0, 0, \ldots).$$

Let Λ be the collection of compositions λ^c indexed by all sequence $c = (c_1, \ldots, c_{n-1})$ with $1 \le c_a \le q$. Then $\Lambda^+ = \mathcal{P}(n)$ and we have

$$eM \simeq \oplus_c y_{\lambda^c}\mathcal{H} = M_\Lambda^\# \qquad \text{as right } \mathcal{H} \text{ modules.}$$

This $\mathcal{H}$ module satisfies the following important property:

Unipotent version of Schur's reciprocity law [11,3.1]

The canonical map $\psi : eKGe \to \mathrm{End}_{\mathcal{H}}(eM)$ is *surjective*. It follows that

$$\psi(eKGe) \simeq \mathrm{End}_{\mathcal{H}}(M_{\Lambda}^{\#}) \simeq S_{\Lambda}$$

hence the algebra $\psi(KG)$ is Morita equivalent to the q-Schur algebra S_{Λ}. More precisely, there is a category equivalence

$$\underline{\underline{\mathrm{mod}}}\psi(KG) \longrightarrow \underline{\underline{\mathrm{mod}}}S_{\Lambda}, \quad V \mapsto eV.$$

In [9,11.11] some $\psi(KG)$ module S_{λ} (an analog of the *Specht module*) and its irreducible quotient D_{λ} are defined with each partition λ of n. One sees the above equivalence sends S_{λ} (resp. D_{λ}) to the Weyl module $W_{\lambda'}$ (resp. its quotient $F_{\lambda'}$) corresponding to the dual partition λ'. The set D_{λ} for all partitions λ of n exhausts all irreducible unipotent representations as a result of Section 7, and the decomposition number $d_{\lambda\mu}$ (ibid.) can be interpreted as the multiplicity of $D_{\mu'}$ as a composition factor of $S_{\lambda'}$ (cf. [3,4.9]).

REFERENCES

1. R. Dipper, Polynomial representations of finite general linear groups in non-describing characteristic, Progress in Math., Vol. 95 (1991), 343–370.
2. R. Dipper and S. Donkin, Quantum GL_n, preprint.
3. R. Dipper and G. James, The q-Schur algebra, Proc. London Math. Soc. (3) 59 (1989), 23–50.
4. ———, Representations of Hecke algebras of general linear groups, Proc. London Math. Soc. (3) 52 (1986), 20–52.
5. ———, q-Tensor space and q-Weyl modules, Trans. AMS, Vol. 327 (1991), 251–282.
6. J. Du, B. Parshall, and J. P. Wang, Two-parameter quantum linear groups and the hyperbolic invariance of q-Schur algebras, Preprint.
7. J. A. Green, Polynomial representations of GL_n, L. N. in Math. 830, Springer 1980.
8. G. James, The decomposition matrices of $GL_n(q)$ for $n \leq 10$, Proc. London Math. Soc. (3) 60 (1990), 225–265.
9. ———, Representations of general linear groups, London Math. Soc. L. N. 94, Cambridge 1984.
10. M. Takeuchi, A two-parameter quantization of GL_n, Proc. Japan Acad., 66, Ser. A (1990), 112–114.
11. ———, Dipper–James theory from a new point of view, preprint.
12. Cline, Parshall, and Scott, Integral and graded quasi-hereditary algebras I, J. Algebra 131 (1990), 126–160.